AF905478

Donald L. Smith · Chantal Hamel (Eds.) *Crop Yield*

Physiology and Processes

Springer-Verlag Berlin Heidelberg GmbH

Donald L. Smith · Chantal Hamel (Eds.)

Crop Yield
Physiology and Processes

With 103 Figures

Springer

Professor Donald L. Smith
Plant Science Department

Professor Chantal Hamel
Natural Resourch Department

Mc Gill University
Macdonald Campus
21, 111 Lakeshore Road
Ste. Anne de Bellevue
QC, Canada H9X 3V9

ISBN 978-3-642-63644-8 ISBN 978-3-642-58554-8 (eBook)
DOI 10.1007/978-3-642-58554-8

Library of Congress Cataloging-in-Publication Data
Crop yield : physiology and processes / Donald L. Smith, Chantal Hamel (eds.).
p. cm.
Includes bibliographical references and index.

1. Crops - Physiology. 2. Crop yields. I. Smith, Donald L., 1953- . II. Hamel, Chantal, 1956- .
SB112.5.C76 1998
631.5'58 - dc21

This work is subject to copyright. AII rights reserved, whether the whole or part of the material is concerned, specifically the rights of translation, reprinting, reuse of iIlustrations, recitation, broadcasting, reproduction on microfilm or in any other way, and storage in data banks. Duplication of this publication or parts thereof is permitted only under the provisions of the German Copyright Law of September 9,1965, in its current version, and permis sion for use must always be obtained from Springer-Verlag Berlin Heidelberg GmbH. Violations are liable for prosecution under the German Copyright Law.

Springer-Verlag Berlin Heidelberg 1999
Originally published by Springer-Verlag Berlin Heidelberg New York in 1999
Softcover reprint of the hardcover 1st edition 1999

The use of general descriptive names, registered names, trademarks, etc. in this publication does not imply, even in the absence of a specific statement, that such names are exempt from the relevant protective laws and regulations and therefore free for general use.

Product liability: The publisher cannot guarantee the accuracy of any information about dosage and application thereof contained in this book. In every individual case the user must check such information by consulting the relevant literature.

Typesetting: Best-set Typesetter Ltd., Hong Kong
Cover Design: Design & Production GmbH, Heidelberg
SPIN 10536493 31/3137 - 5 4 3 2 1 0 - Printed on acid-free paper

Preface

This book has been prepared for those seeking a better understanding of the functioning of crop plants, particularly the processes that lead to the generation of products valued by human beings. The contributors, who are among the world's foremost experts on the important crops upon which humanity depends for food or fibre, address the relevant processes for their specific crop.

Currently, the world population is continuing to increase. It is projected to plateau around the middle of the next century, and while there is considerable controversy regarding the population level when this plateau is achieved, most estimates are in the area of 10 000 000 000. At present, there are about 800 000 000 people in the world who do not have secure access to food.

Over the last 50 years various aspects of agricultural research have been combined to increase the output of world crops approximately 2.5-fold. Given the need to feed the increasing population, and to provide better access, it is predicted that during the next 50 years the agricultural research community must repeat this achievement.

Molecular genetics provides us with a wealth of information regarding a wide range of organisms. Currently, genome sequencing of new organisms is being completed at a rate of about one organism every 2 weeks. Recently, sequencing of the major cereal crops, wheat, rice and maize, has been undertaken. It seems certain that before long all of the world's major crops will be sequenced. The next great challenge will be to understand how all the proteins coded for by the genes in the genome combine to make up the physiology of each crop species. This begins at the cellular level, and moves to the whole plant and ecophysiology level to have its greatest impact. We are in the golden age of biology, and we will soon be able to acquire all the information at any level of biological organization. This will be at the genome level initially, progressing soon to the proteome and physiology levels. This is, and will continue to be, a time of great challenge and great opportunity for those of us working in agricultural research. Thus, physiology will play a key role in the coming decades.

Contents

Sugarcane
S.E. Lingle

Sugar Beet
T.H. Thomas

Physiological Control of Growth and Yield in Wheat: Analysis and Synthesis

R.K.M. Hay[1]

1 Introduction

The 1970s was a decade of revolution for crop physiology and, in particular, for the physiology of temperate cereals. Up to that time, it had proven difficult to correlate annual crop yield with those factors, notably radiation supply, which were known from controlled experiments to influence photosynthetic performance (Evans et al. 1975). Furthermore, it appeared that the tools of crop growth analysis developed in the field by Watson at Rothamsted had exhausted their usefulness: the product of leaf area index and net assimilation rate rarely gave sensible crop growth rates, particularly at high leaf area indices (Watson et al. 1963; Russell et al. 1989). At the same time, interpretation of the results of laboratory work on the environmental control of photosynthesis, and on the roles of sources and sinks in determining grain growth, posed formidable intellectual challenges, principally because of the plasticity of the cereal plant. Competing teams were able to demonstrate sink or source limitation of photosynthesis and yield almost at will by varying plant material and environmental conditions (Evans 1993).

The revolution came about primarily as a result of the advent of cheap and robust solarimeters (e.g., Szeicz et al. 1964) which, for the first time, made it possible to make continuous long-term measurements of the interception of photosynthetically active radiation (PAR) by crop canopies in the field. As predicted by pioneering work in the 1960s (Monteith 1977), measurements of this kind in stands of temperate cereals showed that not only was biomass yield at harvest linearly related to the cumulative sum of PAR *intercepted* by the canopy, but also that the slope of the relationship (the radiation-use efficiency, ε, in g DM/MJ PAR) did not vary markedly among unstressed wheat and barley crops (e.g., in the Midlands of England: Gallagher and Biscoe 1978a). Furthermore, for each increment in the growth of a cereal crop up to anthesis, the rate of dry matter accumulation was linearly related to the amount of PAR intercepted; this was unexpected in view of the known dependence of the rate of photosynthesis on leaf age and environmental factors, notably irradiance, nitrogen and water supply.

[1] Scottish Agricultural Science Agency, East Craigs, Edinburgh EH12 8NJ, UK

Crop Yield, Physiology and Processes
D. L. Smith, C. Hamel (Eds.)
© Springer-Verlag Berlin Heidelberg 1999

Observations of this kind were important for a number of reasons. First, they provided a powerful confirmation that good crop physiology could (and should, if at all possible) be done in the field, given appropriate instrumentation. Clear-cut and decisive relationships had been established from field data. Secondly, the idea that, for temperate cereal crops, ε was a *relatively stable* crop parameter (with ontogeny, and between crops), as were individual grain weight (Gallagher and Biscoe 1978b) and harvest index (Hay 1995; Moot et al. 1996), tended to divert some attention away from studies of crop photosynthesis; more interest was focussed on the time course of interception of PAR, and on the factors determining the grain population density of the crop. Consequently, in the 1980s and early 1990s, considerable resources (Garfield 1990) were devoted to *quantitative* studies, and subsequent modelling, of cereal organ development (initiation, expansion and senescence of leaves, tillers, spikelets and florets) (Gallagher 1979; Kirby and Appleyard 1984; Hay and Kirby 1991; Ritchie and NeSmith 1991). The driving force for much of the simulation work was the need to predict the effects of climatic change on crop yield (e.g., Butterfield and Morison 1992; Goudriaan 1996). Only in the late 1980s did the balance move back towards understanding the influence of the environment on crop photosynthesis and ε (e.g., Kiniry et al. 1989).

Since the crop physiology of wheat yield has been the subject of a number of comprehensive reviews in the past decade (e.g., Lupton 1987; Hay and Walker 1989; Hay and Kirby 1991; Ritchie 1991; Slafer and Rawson 1994), this chapter will concentrate upon the major concepts which have arisen out of the revolution in crop physiology: quantitative analysis of organ development (and the synthesis of the results into simulation models), and the factors determining radiation-use efficiency of wheat crops, with an emphasis on temperate wheat crops under productive conditions. One objective is to provide a progress report on the hypothesis that environmental and genetic factors exercise their control over the yield of wheat crops largely through their effects on the interception of PAR, rather than radiation-use efficiency.

2 Components of Yield

The grain yield (Y) of a wheat crop can be expressed in terms of the following components (Monteith 1977; Hay and Walker 1989):

$$Y = Q \times F \times \varepsilon \times H, \quad (1)$$

where Q is the total quantity of incident solar radiation received over the life of the crop; F is the fraction of Q intercepted by the canopy; ε is the radiation use efficiency of the canopy; and H is the harvest index (correctly expressed in terms of the biomass of the stand *including below-ground parts* although, in practice, involving only the above-ground parts).

Of these components, Q is determined by factors outside the control of the agriculturalist, and H, and its rate of increase after anthesis, have high heritabilities in the absence of severe stress (i.e., in most productive environments) (Hay 1995; Moot et al. 1996; Wheeler et al. 1996). Consequently, genetic and environmental factors (including those influencing yield via the soil and root system) normally affect yield through two shoot parameters: F and ε. Since a great deal is now known about the control of canopy development (Sect. 4), and ε tends to remain constant through the life of a crop (Sect. 6), Eq. (1) can be used to predict crop yield, for example on a running basis, as well as to simulate and interpret yield in retrospect.

Grain yield can also be expressed in terms of a more traditional set of components:

$$Y = \text{ear population density} \times \text{number of grains per ear} \times \text{mean grain weight}, \quad (2)$$

which, since mean grain weight is a relatively stable varietal characteristic (Hay and Walker 1989), can usefully be simplified to:

$$Y = \text{grain population density} \times \text{mean grain weight}. \quad (3)$$

Equations (2) and (3) have been used widely to explain, in retrospect, how genetic and environmental factors have affected harvested grain yield by acting at different stages in crop development (generation and death of tillers, spikelets and florets; grain filling; Hay and Walker 1989). However, with increased understanding of the quantitative aspects of reproductive development, Eq. (3) could also prove useful in the modelling and prediction of grain yield and its components. As we shall see in Section 7, this has proved to be more difficult than the modelling of radiation interception.

3 Wheat Development

The development of the plants of a wheat crop involves two closely coordinated processes giving rise to the sources (leaves) and sinks (fertile spikelets/florets) for photosynthate (Hay and Kirby 1991). Both types of organ are initiated at the apex of each fertile stem, but the total number of primordia initiated, and the partitioning of these primordia between vegetative and reproductive organs, depend upon genetic and environmental factors, principally temperature and photoperiod. Once initiated, the survival of spikelets/florets appears to depend upon a wide range of plant and environmental factors, including nitrogen status, carbohydrate supply, water relations and temperature; in contrast, all the leaf primordia on a fertile wheat shoot survive to give mature leaves, and the senescence of each leaf does not normally begin until at least one phyllochron after full expansion. The relationships between these two coordinated processes, on mainstems and subsidiary tillers, can be illustrated by the model winter wheat plant in Fig. 1; an alternative and ingenious approach to illustrating the coordination of apical development is presented

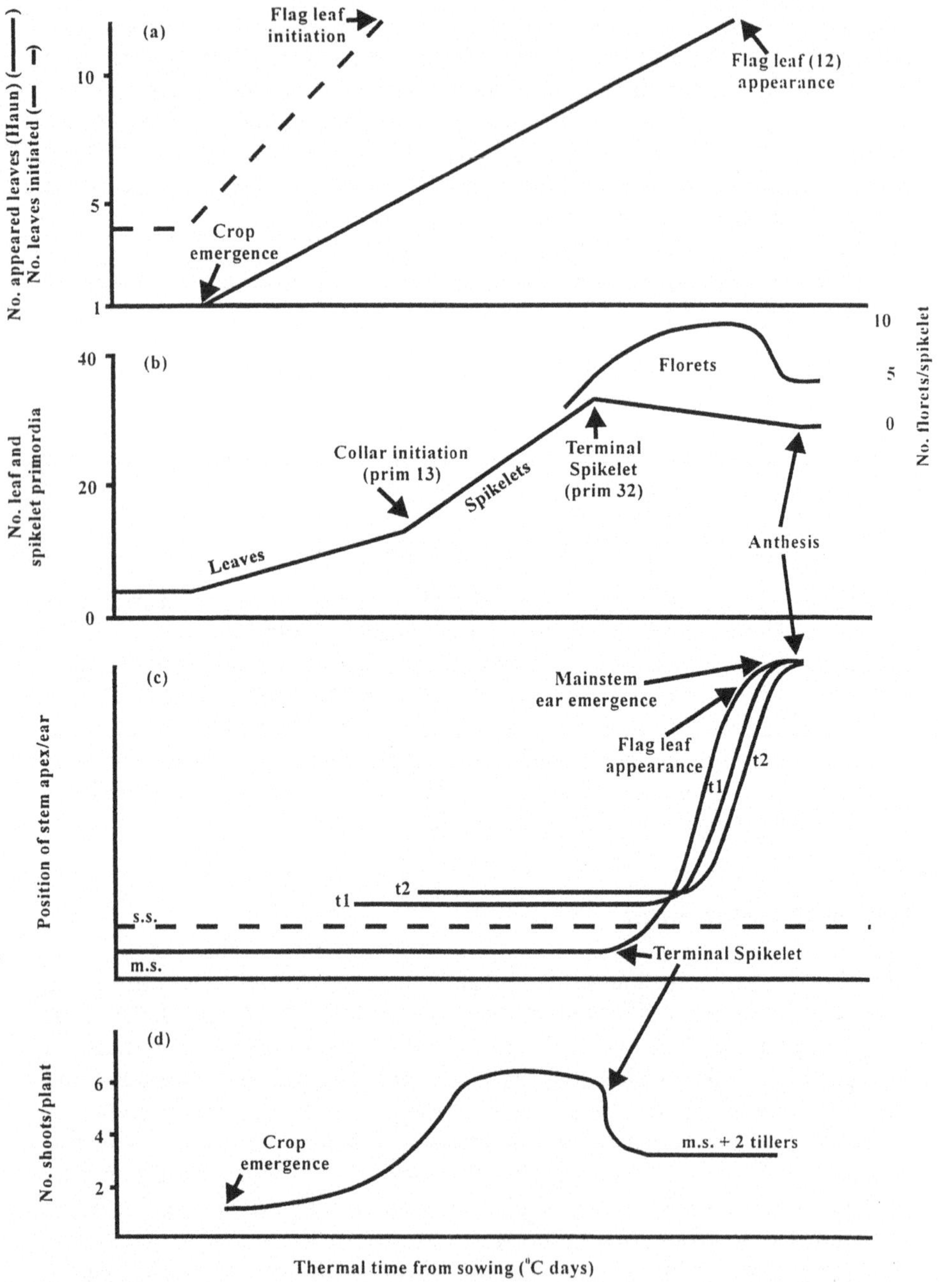

Fig. 1a–d. Schematic diagram of the thermal time course of development of a model winter wheat plant (12 mainstem leaves, 19 spikelets per mainstem ear, 4 grains per spikelet), including the relationships between events. **a** Mainstem leaf initiation and appearance. **b** Mainstem leaf, spikelet and floret initiation and death (with a change in scale for leaf initiation from **a**. **c** Relative positions of the stem apices of the mainstem (*m.s.*) and first two tillers (*t1*, *t2*), in relation to the soil surface (*s.s.*). **d** Appearance and death of the tillers produced by the mainstem. (Hay and Kirby 1991, with permission of Australian Journals of Scientific Research)

by Rickman and Klepper (1995). Quantitative aspects of wheat development are considered in more detail in Sections 4 and 7.

3.1 Leaf, Spikelet and Floret Initiation at the Mainstem Apex (Fig. 1a,b)

Mature wheat grains carry the primordia of three or four leaves, and a further two or three have been initiated by the time of crop emergence. The mainstem apex, located just below the soil surface, continues to generate leaf primordia regularly at intervals (plastochrons) of around 50 degree-days up to the initiation of the collar of the future ear; this is the first floral primordium (Kirby and Appleyard 1984, 1987). The timing of this transition from vegetative to reproductive development depends upon photoperiod (perceived only after crop emergence, i.e., by the leaf 6 stage at the earliest), modified by low temperature vernalization (perceived from the imbibed grain stage onwards) in winter varieties. For example, the winter variety illustrated in Fig. 1 required vernalization, such that the collar was preceded by 12 leaf primordia.

Although there are no whole plant or apical signs of this important transition, the rate of initiation of primordia rises sharply at, or within a few primordia of, the transition (i.e., the plastochron for spikelets is shorter than for leaves); the characteristic "double ridges" on the apex, which can be taken to be a sign that the stem is irrevocably committed to flowering, normally appear before 50% of the maximum number of spikelets have been initiated (Kirby and Appleyard 1984, 1987; Delécolle et al. 1989). The higher rate of spikelet production is maintained up to the initiation of the readily recognized terminal spikelet, in a plane at right angles to its predecessors. The timing of this event has been shown to be dependent upon photoperiod, but it is also associated with the coordination of leaf, ear and stem development (Hay and Kirby 1991): initiation of the terminal spikelet coincides with the start of stem internode extension which carries the embryonic ear upwards from near the soil surface to emerge above the flag leaf in time for anthesis (Fig. 1c). Detailed study of the components of stem extension has emphasized the close coordination of stem extension and apical development (Kirby et al. 1994).

The initiation of florets (each a potential grain site) within spikelets begins around the mid-point of the embryonic ear, progressing towards both the tip and the base in such a way that development is convergent, and all of the spikelets are at the same stage by the time that the terminal spikelet is initiated (Hay and Kirby 1991). Floret initiation (up to 12 per spikelet) continues through the terminal spikelet stage up to around the time of flag leaf appearance when a high proportion die, leaving two to six potential grains per spikelet at anthesis (e.g., Table 2). Substantial mortality of whole spikelets can also take place at the base and tip of the ear around this time, and there can be further losses of potential grains during pollination.

3.2
Leaf Appearance and Expansion; Tiller Development (Fig. 1c,d)

By the time that the tip of the first true leaf appears (emerges) through the coleoptile, five to seven leaf primordia have been initiated by the mainstem apex, and successive leaves appear at intervals of around 100 degree-days (phyllochron measured between successive tips or ligules) up to the last or flag leaf (leaf 12 in Fig. 1); this regularity of appearance occurs in spite of the progressive increase in the above-ground distance that the leaf tip must travel before appearing above the sheath of the preceeding leaf (Fig. 2). Since the phyllochron is approximately twice the plastochron, leaf primordia and partly expanded leaves accumulate, with the result that the appearance of the flag leaf occurs long after the switch from vegetative to reproductive development (initiation of the collar).

Furthermore, since the duration of the expansion of each leaf exceeds the phyllochron (see Sect. 4), the tip of each leaf appears before the appearance of the ligule of its predecessor. Consequently, during the vegetative and early reproductive stages of a wheat crop, each mainstem carries three to five visible green leaves: one in the early stages of expansion, one fully expanded and one to three undergoing senescence (Sect. 4; Weigand et al. 1981; Hay and Tunnicliffe Wilson 1982).

Few wheat canopies are composed solely of mainstems; for example, temperate stands at commercial (N. European) plant population densities (2–300 plants/m^2) are normally made up of plants with up to three fertile primary tillers, giving stem populations of up to 900/m^2. Single, spaced plants can generate many more branches including secondary or higher-order tillers (Hay

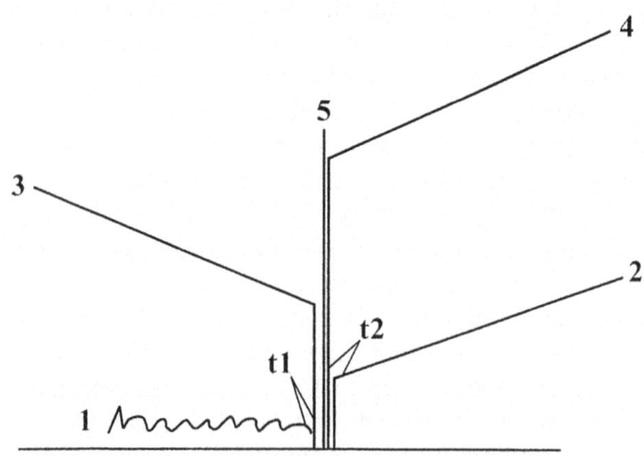

Fig. 2. Schematic diagram of a wheat plant shortly after the (tip) appearance of leaf 5. Note that leaf 1 has senesced and *t1*, *t2* indicate primary tillers in the axils of mainstem leaves *1* and *2*

and Kirby 1991). A primary tiller arises out of a bud in the axil of each lower leaf on the mainstem, and the emergence of the tiller from the first leaf axil is synchronous with the appearance of the tip of mainstem leaf 4. Subsequent primary tillers emerge at intervals of one phyllochron.

Although most wheat plants generate more than three visible tillers, around the time that the mainstem apex reaches the terminal spikelet stage (i.e., at the start of intense competition for resources among elongating stems), tillers begin to die in the reverse order of their initiation, leaving up to three to complete the course of development to bear mature ears (Hay 1986; Thorne and Wood 1988). Measurements by Davidson and Chevalier (1990) indicate that tillers which are destined to senesce around terminal spikelet grow more slowly from the start. Nevertheless, it is important to emphasize that the phases of "birth" and "death" of tillers are distinct, a feature which simplifies the simulation of tiller demography (see Sect. 4.6). There have been few quantitative studies of organogenesis by tillers but the pattern appears to be broadly similar to that of mainstems. Fewer leaves per stem are initiated, although they are of broadly the same dimensions as corresponding mainstem leaves, and the (shorter) duration of ear and stem development ensures that each tiller ear emerges and reaches anthesis within 2–3 days of the mainstem (Fig. 1c; Stern and Kirby 1979; Hay and Kirby 1991).

4 Quantitative Aspects of Development: Canopy Generation

Radiation interception by wheat crops is predominantly by leaf blades, although sheaths play a minor part after stem extension has begun, and ears contribute after anthesis. The blade area per plant (or per unit land area) at any point in the life of the crop can be calculated from the number and dimensions of (green) leaf blades carried by each plant, but since there is an ontogenetic trend in blade length and width (Gallagher 1979; Fig. 5), it is necessary to know the insertion as well as the number of the blades. It is also important to know the functional lifetime of each fully expanded leaf (i.e., from appearance to 50% senescence). To simulate/predict the crop leaf area available for the interception of solar radiation, it is, therefore, necessary to understand the factors determining:

germination and crop emergence,

and the following processes/features for both mainstems and tillers:

leaf appearance
final (flag) leaf number
leaf expansion (rate and duration)
leaf longevity.

4.1 Germination and Crop Emergence

In the absence of stress (drought, extreme temperature, waterlogging), the time elapsing from sowing to crop emergence is dependent upon temperature and sowing depth (Kirby 1993). Since this phase of crop development has been fully characterized under a range of growing conditions, the simulation of crop emergence is one of the most reliable components of crop models, depending upon a simple accumulation of thermal time, normally above a threshold of 0 °C.

4.2 Leaf Appearance

There is now a substantial body of evidence confirming that the number of appeared leaves [measured by tip appearance, ligule appearance or the Haun (1973) method] on each mainstem of a wheat crop increases linearly with accumulated temperature/thermal time from crop emergence, using a base temperature near 0 °C (Kirby 1995; Wilhelm and McMaster 1995). Although phyllochrons (in degree-days) measured in this way tend to remain constant up to flag leaf appearance for a given stand, their values vary with variety, latitude and season (e.g., Baker et al. 1986; Kirby and Perry 1987; McMaster et al. 1992; Slafer et al. 1994; Frank and Bauer 1995; Mosaad et al. 1995), usually within the range 70–160 °C days (tip or Haun method; threshold temperature 0 °C). In a number of investigations where there has been evidence of a bilinear response, with an increase in the rate of leaf appearance in the later stages of canopy development (e.g., Hay and Delécolle 1989; Boone et al. 1990; Cao and Moss 1991b; Fig. 3a), the transition appears to be associated with the switch to reproductive development at the mainstem apex.

The most striking variation is shown in sowing date experiments using winter varieties. Moving from autumn to spring sowings, with crop emergence dates from October to late March in the Northern Hemisphere, the rate of leaf appearance is progressively faster (shorter phyllochron), slowing only for crops emerging in late May and June (Baker et al. 1980; Kirby et al. 1985a; Hay and Delécolle 1989; Cao and Moss 1991b; Fig. 3a); Fig. 3b shows a similar pattern of declining phyllochron with delay in sowing from the Southern Hemisphere. This, and the progressive reduction in final leaf number (see Sect. 4.3) are, of course, integral components of the convergence of the different wheat crops of a sowing date experiment on a similar harvest date (Hay and Kirby 1991).

Two major conclusions can be drawn from the results of these field experiments: first, mainstem phyllochron is usually fixed early in the life of the crop, around crop emergence, presumably in response to an environmental signal which undergoes a regular annual variation. This need for regularity rules out most candidate factors, including temperature, because they can vary irregularly within and between seasons (although there is some interest in the

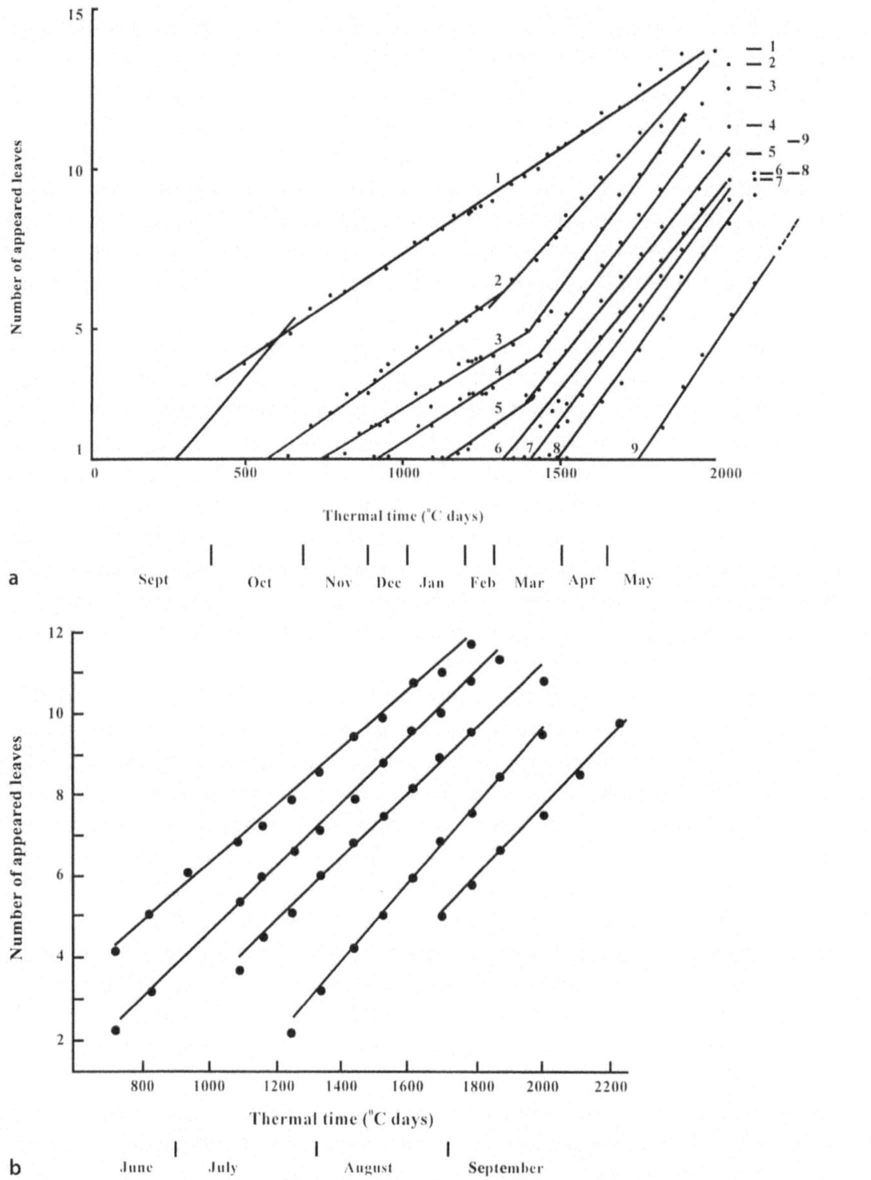

Fig. 3a,b. Thermal time courses of leaf appearance in **a** nine sowings of Maris Hustler winter wheat in southwest Scotland in 1982/83 (*numbered bars* indicate final leaf numbers per main-stem for each sowing); and **b** five sowings of Bencubbin wheat in Western Australia in 1983. Five of the sowings in **a** showed a resetting of the rate of leaf appearance but the "second phase" rates increased progressively with delay in sowing. The five sowings in **b** showed a similar, but less regular, trend with delay in sowing (see text). (After Hay and Delécolle 1989; Kirby and Perry 1987, with permission of Annals of Applied Biology, and Australian Journals of Scientific Research)

seasonal variation in the gradient in temperature between air and meristem; Brooking and Ritchie 1993). Secondly, for most crops, once the magnitude of the phyllochron is fixed, it remains constant up to flag leaf appearance in the face of considerable variation in temperature, day length, water relations and the availability of mineral nutrients. Exceptions include crops in which the phyllochron appears to be reset to a second constant rate around the start of reproduction (e.g., Fig. 3a). Furthermore, the process of vernalization, which contributes to the control of final leaf number (see Sect. 4.3) does not appear to influence the rate of leaf appearance (Cao and Moss 1991a; Mosaad et al. 1995; but see Cutforth et al. 1992).

These findings led Baker et al. (1989) to propose that the mainstem phyllochron is determined by the rate of change of day length at crop emergence, a factor which discriminates between autumn and spring day lengths if the sign (positive or negative) of the values is taken into account. Similarly close linear relationships between the rate of leaf appearance and the rate of change of day length were obtained in several, but by no means all, subsequent sowing date experiments (e.g., Hay and Tunnicliffe Wilson 1982; Delécolle et al. 1985; Kirby et al. 1985a; Kirby and Perry 1987). Other combinations of temperature and day length, and ontogenetic variation in the rate of leaf extension, have been invoked to explain the variation in the rate of leaf appearance with sowing (reviewed by Bindi et al. 1995).

Some attempts to explore the relationship between phyllochron and day length under controlled conditions have been undermined by the confounding of day length and the supply of photosynthetically active radiation (e.g., Cao and Moss 1989; Mosaad et al. 1995). Where day length has been extended using low-irradiance incandescent lamps, no evidence has been provided for the hypothesis that day length, or its rate of change, determines the rate of leaf appearance (e.g., in wheat and barley: Slafer et al. 1994; Kernich et al. 1995). Other investigations under controlled conditions have shown that suboptimal levels of other environmental factors, including water and nitrogen supply, can slow the rate of leaf appearance by slowing (preappearance) leaf extension (Sect. 4.4), but only when the plants are suffering severe stress (Krenzer et al. 1991; Cutforth et al. 1992; Longnecker and Robson 1994; Wilhelm and McMaster 1995).

There is, therefore, good experimental evidence for and against the role of rate of change of day length in determining the rate of leaf appearance in wheat mainstems. Furthermore, no mechanism for the detection of rate of change of day length, as opposed to its instantaneous value, has been identified in plants, although such systems do exist in animals. Cao and Moss (1994) have argued that the observed linearity may be an example of fortuitous compensation: as the season proceeds, the accelerating influence of increasing day length is opposed by the decelerating effect of higher temperature. This hypothesis depends upon responses to day length which are not generally accepted. By contrast, Jamieson et al. (1995a) propose that the use of correct temperatures (i.e., the temperature of the apex rather than the air) can bring all of the data for a given variety from a sowing date experiment together on the same linear response. It is difficult to reconcile this finding with previous work (e.g., Hay and Tunnicliffe Wilson 1982), and it may arise out of the particular climatic conditions in New Zealand.

The simulation of leaf appearance, therefore, poses several problems in addition to the fact that phyllochron varies among varieties. Phyllochron is set around crop emergence, under the influence of a factor, or factors, as yet unidentified, and once set, it usually remains constant under fluctuating environmental conditions. Although a number of ingenious models of wheat leaf appearance have been elaborated, recent tests show that those based upon rate of change of day length at crop emergence are as good as, if not better, than other simulations (Fig. 4; Bindi et al. 1995; Kirby 1995; McMaster and Wilhelm 1995).

4.3
Final Leaf Number

In the field, the final number of leaves per mainstem is determined at the initiation of the collar of the future ear (Fig. 1b), under the influence of day length in spring varieties, and day length and low temperatures in winter varieties requiring vernalization. The number of leaves produced under a given combination of day length and temperature varies among varieties (e.g., Kirby et al. 1985a; Slafer et al. 1994; Table 1), and it should be emphasized that, in the laboratory, unseasonable manipulation of environmental conditions can cause potential collar and spikelet primordia to revert to leaves (Brooking et al. 1995).

For varieties not requiring vernalization, there is a *minimum* leaf number because the day-length signal for collar initiation cannot be received by green leaf tissue until the crop has emerged. Since by that time the apex has already initiated five to seven leaves, the minimum leaf number per mainstem, even under the most favourable conditions for the switch to reproduction, cannot be less than six to eight, allowing for a lag of around 1 primordium (e.g., under 24-h day length, Table 1). The same holds for winter wheat varieties whose vernalization requirement (see below) has been satisfied by the time of crop emergence. There are striking differences between varieties in the pattern of increase in final leaf number as day length is reduced; for example, decrease in the (constant) day length from 24 to 9 h caused final mainstem leaf number to increase by between 1 (Sonora 64) and 8 (Justin, Thatcher) leaves (Table 1).

At a given day length, winter wheat varieties tend to produce more mainstem leaves than spring varieties because their requirement for low-temperature vernalization must be satisfied before the apex can respond to the day-length signal. (In practice, of course, most winter types do eventually initiate reproductive development without vernalization, but after producing many leaves). However, in contrast to the day-length signal, vernalizing low temperatures can be perceived (and accumulated) by the seedling before emergence, from the imbibed seed stage onwards. In spite of the importance of this

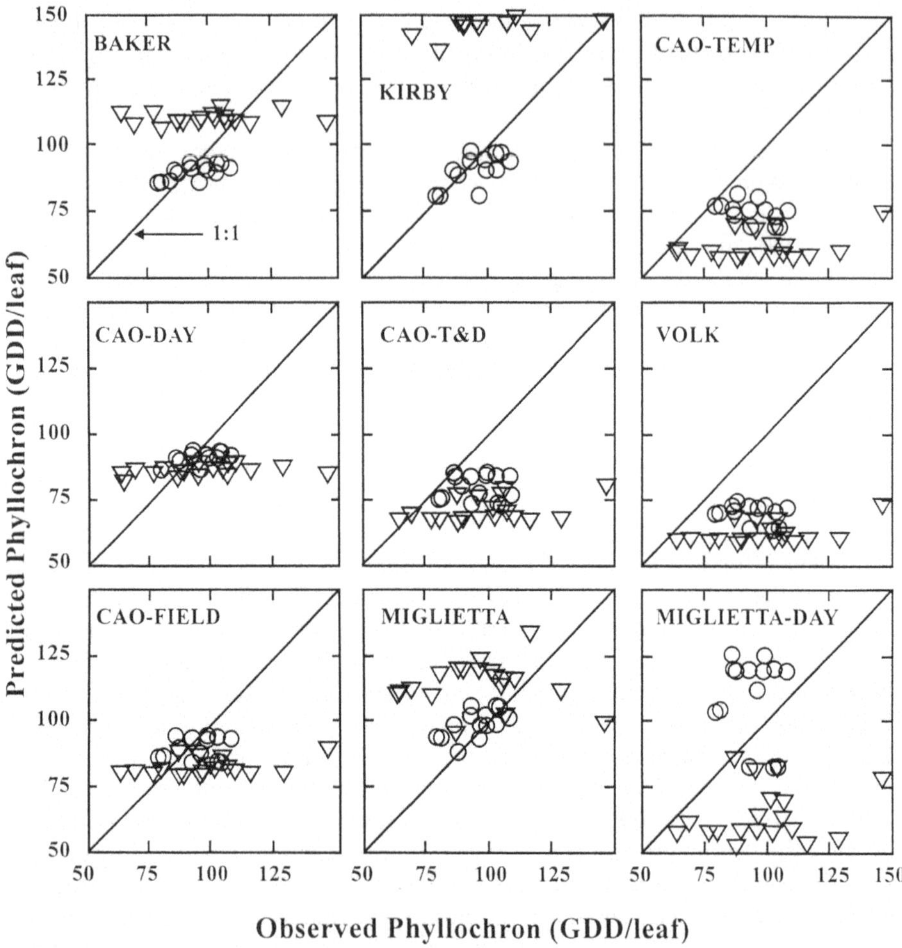

Fig. 4. Comparison of observed wheat mainstem phyllochron values (growth degree days per leaf) with values predicted by nine simulation models (using a range of data sets and appropriate cultivars). Each *square* refers to a single (named) model, and *points falling on the 1:1 line* indicate a precise prediction under the conditions imposed. *Circles* indicate spring cultivars and *triangles* winter cultivars. A few data points for the KIRBY and MIGLIETTA models fell outside the appropriate squares. (McMaster and Wilhelm 1995, with permission of the Crop Science Society of America)

process, there have, until recently, been few comprehensive quantitative studies of vernalization in wheat, and the data used in simulation models have tended to be a combination of values from wheat and rye experiments, even though it is known that responses vary among varieties (Flood and Halloran 1986; Hay and Kirby 1991; Craigon et al. 1995; Wang et al. 1995).

For example, when formulating a simulation model of Maris Hustler winter wheat development, Weir et al. (1984) used a vernalization response curve with

Table 1. Influence of day length on the final number of leaves on the mainstem of spring or vernalized wheat plants

Cultivar	Origin	Day length (h)													Reference
		7	8	9	10	11	12	13	14	15	16	17	20	24	
Sonora 64	Mexico			7.1		7.0		6.7		6.0		6.0			
Pitic 62				8.0		7.2		7.0		7.0		7.0			
Nadadores 63				11.9		11.0		9.0		7.0		6.0			Levy and
Justin	N. Dakota			14.5		12.8		9.7		7.0		7.0			Peterson
Galgalos	Russia			13.7		12.4		9.7		8.2		7.2			(1972)
Triumph	Oklahoma			13.7		12.5		10.0		7.7		7.0			
Pitic	Mexico		14.0				14.1				13.4		12.6	11.5	Holmes
Marquis	Canada		15.1				9.2				7.3		6.9	6.7	(1973)
Sonora 64	Mexico	12.8			10.5			7.8			7.0				Allison and
Siete Cerros		11.9			9.5			7.0			6.7				Daynard
Tokwe	Zimbabwe	9.3			7.7			6.0			6.0				(1976)
Zambesi II		12.4			9.6			7.0			6.6				
Marquis	Canada	16.2			14.4			9.5			7.0				
Triple Dirk	Australia		10.0	10.0	9.0				8.0					6.0	
Gabo			11.0	11.0	10.0				9.0					7.0	Rahman and
Kalyansona	India		10.0	9.6	9.2				8.0					6.2	Wilson
Thatcher	Canada		15.0	14.6	14.0				9.0					7.0	(1977)
Selkirk			12.0	12.0	12.0				9.6					8.4	
8–23	Afghanistan		13.0	12.6	12.0				9.0					7.0	
8–27			11.0	11.0	10.0				8.2					7.0	
Kogat	Alaska		10.0	10.0	9.0				7.6					6.2	
Condor	Australia						10.1		9.0	8.5					Slafer et al.
Thatcher	Canada						9.7		8.0	7.0					(1994)

a broad optimum between 3 and 10 °C, and minimum and maximum temperatures of −4 and 17 °C; the progress of vernalization could then be charted in terms of vernal days (equivalent to 24-h periods during which the temperature remains within the range 3–10 °C). It was estimated that Hustler wheat required 33 vernal days for full vernalization; this is in line with the normal experimental practice of exposing imbibed seeds to temperatures below 5 °C for up to 6 weeks (e.g., Cao and Moss 1991a). In contrast, Craigon et al. (1995), working on historic data for the variety Norin 27, generated a *bilinear* response curve with cardinal temperatures of −4.8, 5.2 and 26.6 °C; they estimated that a minimum of 275 vernalizing degree-days were necessary for full vernalization.

Most approaches to the vernalization of cereal plants have been based on measurements of the final number of mainstem leaves, from which can be estimated (in retrospect) the time of initiation of the collar and, by implication, the time of saturation of the vernalization requirement. It has been assumed that increasing the rate of saturation will inevitably lead to lower final leaf numbers. However, as pointed out by Brooking and his colleagues (Brooking 1996; Robertson et al. 1996), this analysis overlooks the fact that two independent processes are proceeding simultaneously at the stem apex: the progressive accumulation of vernalizing degree days, and the initiation of primordia, and that the rate of each process is determined by temperature. Thus, although the final number of mainstem leaves of a range of winter wheat cultivars increased with temperature from 5 to 14 °C, this was primarily because of higher rates of leaf initiation during vernalization; the greater response of initiation to temperature swamped the promotive influence of raised temperature on vernalization, with the result that more rapid vernalization at higher temperatures did not lead to lower leaf numbers. In disentangling these two effects, Brooking (1996) was able to show that the historic data reanalyzed by Craigon et al. (1995) (see above) actually showed a linear increase in the effectiveness of vernalization temperatures from 1 to a maximum around 11 °C, with lower effectiveness at 15 °C. Temperatures of −1 and 20 °C were ineffective. This useful advance in the understanding of vernalization has yet to be incorporated into models of wheat development.

Nevertheless, the modelling of final leaf number for a given variety is, generally, simpler than for leaf appearance. The factors affecting collar initiation (day length, vernalizing temperatures) are known and reasonably well characterised, although there is considerable variation in response curves between types and varieties, and some uncertainty about the interaction between daylength and vernalization. In addition to the submodels used in major simulations (see Sect. 4.7), there have been a number of recent models of final leaf number (e.g., Kirby 1992; Brooking et al. 1995; Craigon et al. 1995). The wide interest in phyllochron (Sect. 4.2) and final leaf number is understandable, since the appearance of the mainstem flag leaf is a major event in the life of the plant and crop, providing a valuable guide to the date of ear emergence and anthesis.

4.4 Leaf Expansion

Wheat plants show an ontogenetic trend in leaf size which, although characteristic of a variety, normally shows a progressive increase with insertion, decreasing only with the flag leaf and, in some varieties, its immediate predecessors (e.g., Fig. 5); for the mainstems of a single variety, with a given final leaf number (Sect. 4.3), there is a genetically determined upper limit to the size of the leaf blade at each node. (In experiments where sowing date varies, mainstems produce different numbers of leaves, and the ontogenetic trend can vary). The environmental conditions during leaf development determine to what extent each leaf achieves its potential size.

Studies of a range of temperate grasses and cereals have shown that, in the absence of moderate to severe stress, the rate of leaf expansion is linearly related to temperature if the temperature measurements are made close to the leaf expansion zone at the base of the leaf, near the stem apex (e.g., Gallagher and Biscoe 1979). It is therefore possible, as for other developmental processes, to chart the course of wheat leaf expansion (in practice *extension*, since expan-

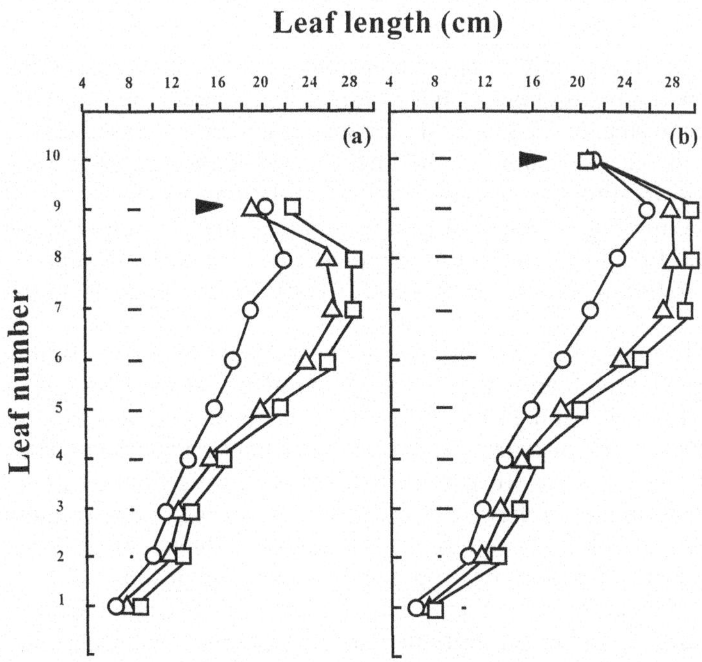

Fig. 5a,b. Ontogenetic trends in leaf length in three near-isogenic lines derived from the wheat cultivar, Maringa, grown in Argentina in **a** 1992 and **b** 1993. ○, △ and □ indicate dwarf (Rht1 Rht1 Rht2 Rht2), semi-dwarf (Rht1 Rht1 rht2 rht2) and standard height (rht1 rht1 rht2 rht2) lines. *Arrows* indicate flag leaf and *horizontal bars* represent SEM. (Calderini et al. 1996, with permission of the Annals of Botany Company)

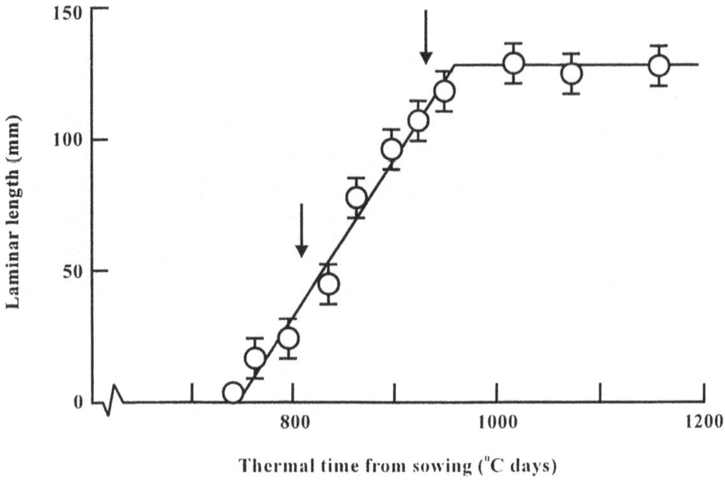

Fig. 6. Thermal time course of the laminar extension of mainstem leaf 7 of a wheat crop cv. Maris Huntsman growing in the Midlands of England in 1975. *Arrows* indicate the appearance of leaf tip and ligule, and *vertical bars* represent 2 × SEM. (After Gallagher 1979, with permission of Oxford University Press)

sion takes place predominantly in one dimension) in terms of thermal time, with a base temperature around 0 °C. For example, Fig. 6 shows the extension of the blade of mainstem leaf 7 of a field crop grown under unstressed conditions. Although the curve is sigmoid, the linear phase predominates and the errors involved in treating the process as a linear relationship, characterized by a rate and a duration, are small. It should be noted that the linear phase begins before leaf tip appearance and is complete shortly after ligule appearance (i.e., the duration of expansion of a leaf is considerably longer than the phyllochron).

Evidence is accumulating that both genetic and environmental factors influence leaf size predominantly through the rate rather than the duration of leaf expansion (Monteith and Elston 1983). This has been demonstrated in relation to plant water and nitrogen status, although osmotic adjustment tends to maintain the driving force (turgor) in field-grown plants (Tomos 1985). Particularly clear examples of the role of the rate of expansion in determining leaf blade area are provided by two recent studies of the influence of rht/Rht1, 2 and 3 dwarfing genes on canopy development in wheat (Tonkinson et al. 1995; Calderini et al. 1996): variation of these alleles had no influence upon leaf initiation or appearance (or on the initiation of reproductive primordia), but dwarf lines carrying Rht genes produced leaves with reduced blade length, surface area and dry weight compared with near-isogenic lines of normal height (e.g., Fig. 5). These changes, brought about by reduced rates of expansion but unchanged duration, were associated with reduced cell length rather than number.

In the simulation of radiation interception, modellers are interested principally in green/visible blade area per plant (i.e., the aggregate area of appeared, but non-senescent blades, plus the fraction of the youngest leaf blade which has extended beyond the subtending tissues of the last sheath). Several early mathematical models of wheat canopy development (e.g., reviewed by Hunt and Pararajasingham 1995), formulated with the simplifying assumption of low environmental stress, used empirical values for final leaf length, and estimated the proportion of the blade which is exposed by means of the thermal time from leaf tip appearance; other models deploy relationships in which final leaf length is dependent upon phyllochron, and some recognize the minor contribution to photosynthesis made by leaf sheaths after stem extension has begun. It is then possible to introduce a set of threshold values for individual environmental factors (e.g., water or nutrient supply) at which the rate of expansion or final length of each leaf blade is reduced (e.g., compare Porter 1984 with Porter 1993).

4.5 Leaf Senescence

It is becoming generally accepted that the ordered senescence of plant leaves is under genetic control (e.g., Thomas and Smart 1993; Gan and Amasino 1995) and can be treated in a way similar to that for other developmental processes. Thus, Hay and Kemp (1992) found that the rate of senescence of leek leaves, expressed in thermal time, was equal to the rate of leaf appearance, with the result that the number of fully expanded leaves per stem remained constant. The regularity of leaf death can, of course, be disturbed by early loss of leaves through mechanical damage, grazing, infection by pathogens, frost damage or other stresses.

There is a surprising dearth of detailed information on the time course of senescence of wheat leaves in the field, and most estimates of leaf longevity depend upon the observation that wheat mainstems carry three or four green leaves, at least up to anthesis (e.g., Weigand et al. 1981; Hay and Tunnicliffe Wilson 1982). For example, Hunt and Pararajasingham (1995) set the maximum longevity of each leaf in the early phases of canopy development at four phyllochrons from tip appearance, extending the duration for later leaves.

4.6 Tillers

The time course of emergence of mainstem branches is well understood, although the picture can become complex for widely spaced plants, where secondary and higher-order tillers arise (Hay and Kirby 1991). The emergence of the first primary tiller, from the axil of mainstem leaf 1, coincides with the appearance of the tip of mainstem leaf 4 (three phyllochrons after the appearance of the subtending leaf), and subsequent primary tillers appear at intervals of one phyllochron up to around the time of formation of the terminal spikelet (Hay 1986; Thorne and Wood 1988; Davidson and Chevalier 1990; Stapper and Fischer 1990) when they begin to die in the reverse order to their emergence. The emergence and contribution of a coleoptile tiller is more difficult to predict.

Although the intense competition amongst elongating stems, and between internodes of a given stem, must play a major role in the survival or death of tillers after the terminal spikelet stage, the details of this competition are poorly characterized. For example, under certain circumstances, survival can be influenced by the timing of nitrogen fertilizer application (Darwinkel 1983). However, survival can be modelled in a straightforward way (in a canopy development submodel) by allocating available assimilate (estimated by a photosynthetic production submodel) according to strict priorities (Hunt and Pararajasingham 1995), or by taking stem population density into account (Porter 1984).

Once tillers have emerged, their contribution to the leaf canopy can be evaluated in much the same way as for mainstems, using the, albeit limited, information on tiller leaf initiation, appearance, expansion and death from a series of experiments (e.g., Stern and Kirby 1979; Klepper et al. 1982; Kirby et al. 1985b; Longnecker et al. 1993). The regularity of tiller emergence, and the common pattern of development of tillers at each node, permit the modeller to treat the stems of a crop in groups, normally termed cohorts, of tillers of the same order or (calendar) age (Porter 1985).

4.7 Leaf Area Index

For an uncomplicated monocotyledonous species such as the leek (single stem, vegetative development only in the first season, simple ontogenetic trend in leaf size, simple leaf shape, predictable leaf senescence, single driving variable: temperature), the simulation of leaf area index (LAI: canopy leaf area per unit of soil surface, commonly expressed by modellers as GAI: green area index; see Fig. 7) involves a series of elementary arithmetic calculations (Hay and Kemp 1992). For wheat, whose more complex development is driven by temperature, day length and vernalization, modified by other environmental factors (water and nutrient relation; Sects. 4.1–4.7), there are still considerable areas of ignorance; nevertheless since the mid-1980s there has been sufficient understanding to permit the construction of a series of mathematical models, several of which evaluate leaf area index by the interaction of a phenology submodel (timing of crop emergence, terminal spikelet, anthesis etc.) with a leaf growth submodel.

The ARCWHEAT model, first formulated to describe the growth and development of unstressed Huntsman/Hustler winter wheat crops in the UK (Porter 1984; Weir et al. 1984; Hay and Walker 1989), and subsequently modified to incorporate the responses of Avalon wheat to water and nitrogen supply

(AFRCWHEAT2; Porter 1993), is a clear example of this approach. Figure 7 shows the results of an extensive test of the effectiveness of AFRCWHEAT2 in simulating the GAI of a series of Avalon crops in the South East of England (2 sites, 4 seasons, variation of sowing date, nitrogen supply, irrigation: 15 experimental treatments). Comparison of simulated GAI curves with observed values, over a range of peak values from 3 to 9, showed that AFRCWHEAT2 was extremely effective at simulating canopy development and duration under unstressed conditions (Fig. 7a,g,h,i), although the duration of the canopy was generally overestimated under more stressed conditions (e.g., Fig. 7b,c,d,f). Nevertheless, the scale of divergence was small, and would have resulted in differences in *interception* of no more than a few percent (Sect. 5), particularly in the more highly fertilized crops. Only where the application of nitrogen was high (250 kgN/ha), and applied in different seasonal patterns (Fig. 7m,n,o), did the model seriously overestimate GAI.

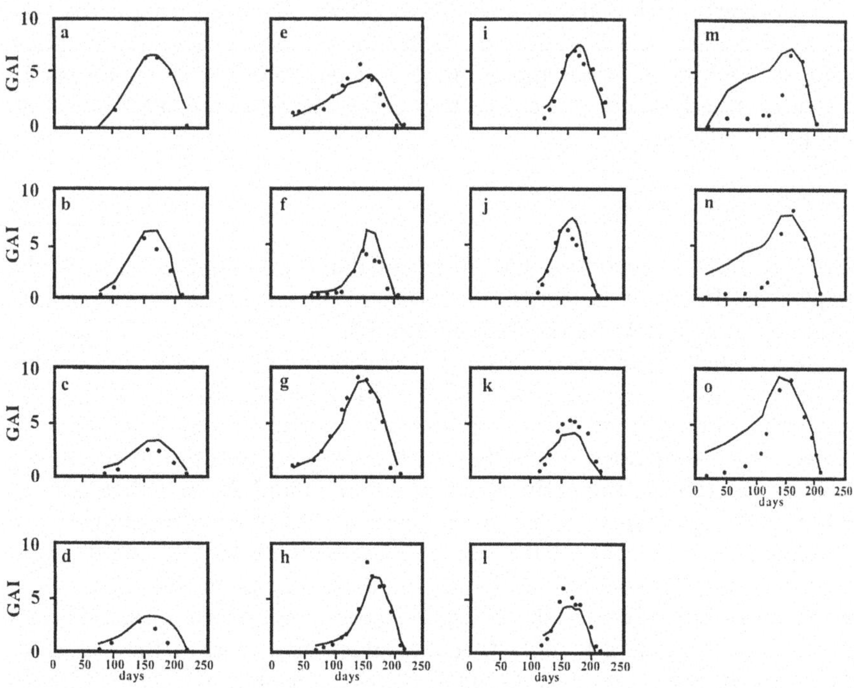

Fig. 7a–o. Observed (·) and simulated by AFRCWHEAT2 (*continuous curves*) time courses of green leaf area index (GA1) for a series of Avalon winter wheat crops grown in the southeast of England in 1985/86 (**a,i** with nitrogen fertilizer and irrigation; **b,j** with nitrogen fertilizer, no irrigation; **c,k** no nitrogen fertilizer, with irrigation; **d,l** no nitrogen fertilizer, no irrigation), 1982 (**e** low nitrogen fertilizer, early sowing; **f** low nitrogen fertilizer, late sowing; **g** high nitrogen fertilizer, early sowing; **h** high nitrogen fertilizer, late sowing), and 1983 [high nitrogen fertilizer in 1(**m**), 2(**n**) or 3(**o**) applications). (After Porter 1993, with the permission of the European Journal of Agronomy)

5
Leaf (Green) Area Index and the Interception of PAR

Since crop canopies consist of three-dimensional arrays of blades of different sizes, positions and orientations, the penetration of PAR into a canopy can be treated mathematically by analogy with the penetration of light into a solution of light-absorbing molecules i.e., by using a modification of Beer's Law:

$$I = I_o \exp\{-kL\}, \tag{4}$$

where I_0 is the irradiance above crop canopy; I is the irradiance at a point within the canopy above which there is a LAI (GAI) of L; and k is the extinction coefficient of the canopy. k, a dimensionless constant, is a composite expression of all the factors affecting penetration of PAR into the canopy (leaf angle, size, shape and thickness; the vertical stratification of leaf area; and leaf surface properties; Hay and Walker 1989). (More advanced treatments take account of the fact that PAR does not enter the canopy *vertically* throughout the day in the form of *direct* radiation). LAI (GAI) values, whether measured or calculated, can, therefore, be used to predict interception of PAR by a crop stand by the function:

$$F = 1 - \exp\{-kL\}, \tag{5}$$

where F is the % of incoming PAR intercepted and L is the LAI (GAI) of the canopy (e.g., Hipps et al. 1983). Most wheat crops reach 95% interception within the range of LAI (GAI) between 3 and 5.

In the simulation of wheat crops it is conventional to use a single value of k in generating the time course of PAR interception, in spite of the very considerable ontogenetic trends in canopy structure from emergence to anthesis. This is defended in the light of the small contribution made to dry matter production by the early stages of the crop, and the fact that, later in the life of the crop, most of the interception of PAR is by the youngest two or three leaves at the top of the canopy (Monteith 1994). In general, variation in leaf angle is probably the most important factor in determining the considerable differences in k which exist between wheat crops. In this context, it is instructive to note that the values of k used in the two most widely tested wheat models (AFRCWHEAT from the UK, and CERES-Wheat from the USA, the latter yet to be published in full; Rickman and Klepper 1991) are 0.44 and 0.85, respectively (Porter et al. 1993). This difference, which is consistent with field measurements (e.g., Gallagher and Biscoe 1978a; Hipps et al. 1983; Asrar et al. 1984), suggests that more PAR penetrates into the lower layers of the typically more erectophile wheat canopies of the UK.

6
Radiation-Use Efficiency

6.1
Photosynthetic Efficiency of Single Leaves and Canopies

A typical photosynthetic PAR response curve for a single wheat leaf held at right angles to the supply of PAR (Fig. 8) is curvilinear, reaching a plateau at irradiances varying between 100 and 250 W/m^2 (depending upon leaf age) owing to the saturation of the photosynthetic apparatus. Since, in practice, such curves can normally be resolved into bilinear responses without incurring undue errors (Monteith 1981), the leaf can be treated as if its photosynthetic efficiency (the slope of the response curve) is constant below the saturating irradiance. Efficiency falls sharply (but reversibly) if the leaf experiences irradiances above that necessary for saturation for short periods, but the photosynthetic apparatus can undergo irreversible damage if exposure is prolonged.

During most of the life of a wheat canopy, most of the interception of PAR does not take place at right angles to the leaf surface; solar angle varies diurnally, and the inclination of the upper leaves of the canopy means that the incident irradiance is lower, and a significant proportion of incident PAR pen-

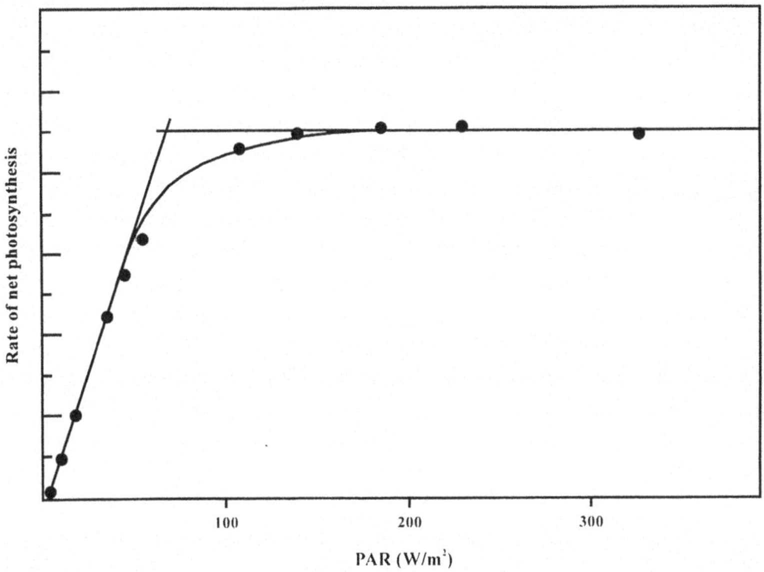

Fig. 8. Generalized photosynthetic PAR response curve for a single wheat leaf held normal to the supply of PAR

etrates more deeply into the canopy (Hay and Walker 1989). This serves to raise the saturating irradiance, for the canopy as a whole, typically to values over 300 W/m^2 (PAR). Radiation saturation is, therefore, uncommon or of short duration in temperate environments, and since mean leaf age remains steady up to anthesis, owing to the regular replacement of senescent leaves, the photosynthetic efficiency of unstressed wheat canopies tends to be constant at least up to anthesis (Gallagher and Biscoe 1978a; Monteith 1994).

6.2 Radiation-Use Efficiency of Wheat Canopies

Dry matter production depends not only upon the photosynthetic and respiratory activity of the leaf canopy, but also upon losses of fixed carbon by the respiration of the whole crop stand, including non-photosynthetic tissues. Building on the work of McCree (1974), it has become clear that whole crop respiration can be partitioned into two components:

$$R = aP_g + bW, \tag{6}$$

where R and P_g are the rates of respiration and gross photosynthesis respectively, W is crop dry weight (expressed in equivalents of CO_2 to conform with R and P_g), and a and b are constants (the value of a typically 10 or 20 times the value of b). Since, during active crop growth, bW (maintenance respiration) will be much lower than aP_g (growth respiration), the rate of crop respiration will be directly proportional to the rate of canopy (gross) photosynthesis. Together, the constancy of the photosynthetic efficiency of the canopy, and the dependency of the rate of respiration of the crop biomass on that of canopy photosynthesis explain why the radiation-use efficiency (ε) of unstressed cereal crops tends to remain constant at least up to anthesis. For example, Gallagher and Biscoe (1978a) established the constancy of ε in a range of cereal crops growing in the Midlands of England, over 14-day periods of peak growth and over the entire period from emergence to anthesis; furthermore the observed values of ε (around 3 g DM/MJ PAR) were similar for all of the wheat and barley crops in the study. When the analysis was extended to cover the lifetime of the crop up to harvest, by comparing crops, ε was again similar for all crops, but lower (around 2.2 g DM/MJ PAR) because of the lower efficiency of the ageing canopy after anthesis. It should be noted that these values of ε are per unit of PAR *absorbed* by the canopy, taking account of the fact that around 10% of intercepted radiation will be reflected back by, or pass through, the canopy. Furthermore, they refer to total, rather than above-ground biomass.

In comparing results of ε in the literature, it is important to note that consecutive measurements of the dry weight per unit area of a wheat stand are not strictly independent. Consequently, in evaluating ε, the plot should be of *rate* of increase of dry weight against *rate* of interception or absorption of PAR over a finite period (e.g., Fig. 9), rather than a simple plot of crop dry weight against the cumulative amount of intercepted/absorbed PAR (Russell et al. 1989; Monteith 1994).

The ε values measured by Gallagher and Biscoe (1978a) were similar to earlier measurements (around 2.8 g DM/MJ PAR intercepted), from a diverse range of temperate crop species (barley, potatoes, sugar beet, apples), assembled by Monteith (1977). However, any idea that there might be a universal radiation-use efficiency for all species was soon ruled out by other field measurements (e.g., Gosse et al. 1986; Azam-Ali et al. 1994): although temperate species rarely exceed 3 g DM/MJ PAR intercepted, tropical C_3 (e.g., rice) and many C_4 plants show much higher values. Nevertheless, the accumulating evidence from several wheat-growing areas (UK, USA, Australia, Mexico) confirms that an ε of 2.8 to 3.2 is typical of the preanthesis growth of unstressed wheat crops (Fig. 9; e.g., Fischer 1983, 1993; Garcia et al. 1988; Kiniry et al. 1989), although there must be some concern about observed variation in ε among varieties grown under the same conditions (e.g., Green 1989).

These findings have led to a widespread acceptance of the concept that, in the absence of significant stress, the environment influences wheat crop dry

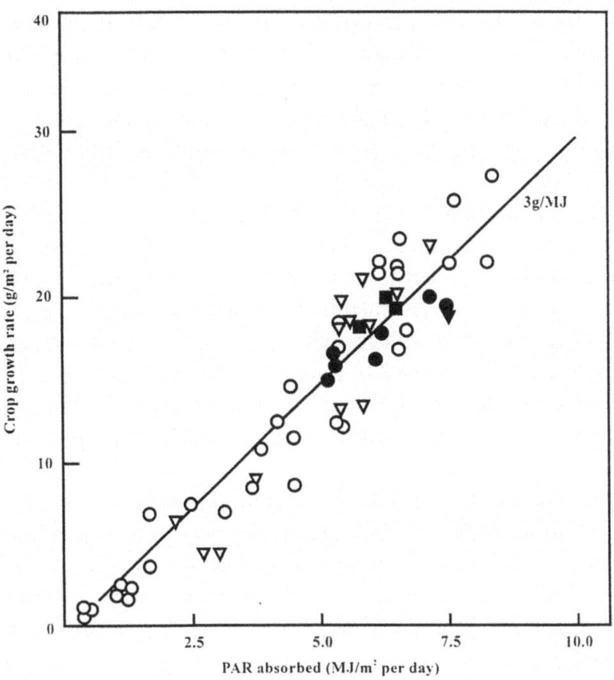

Fig. 9. Relationships between (preanthesis) crop growth rate and the rate of absorption of PAR by the canopy (here estimated as 0.9 × the rate of interception) for a range of wheat and barley crops, where the *line* indicates a radiation-use efficiency of 3 g DM/MJ PAR. (○, wheat UK; ▽ barley UK; • wheat Mexico; ■ wheat Australia; ▼ wheat Kansas, USA). (After Fischer 1983; with permission of the International Rice Research Institute)

matter production via canopy development and radiation interception rather than via radiation-use efficiency. Several widely applied simulation models of wheat are founded upon this principle, although CROPSIM-WHEAT (Hunt and Pararajasingham 1995) and CERES-Wheat allow for ontogenetic variation in ε, and AFRCWHEAT2 employs a canopy photosynthesis PAR response curve, making allowance for respiratory losses (Weir et al. 1984; Hay and Walker 1989). Once crop biomass production has been determined, then grain yield can be estimated simply by the use of an appropriate harvest index (Gallagher and Biscoe 1978a; Hay et al. 1986), by the apportioning of assimilate to the grains up to crop senescence (Weir et al. 1984), or by taking advantage of the fact that the rate of increase of harvest index during grain filling is relatively constant for a given cultivar (Moot et al. 1996; Wheeler et al. 1996).

6.3 Variation in Radiation-Use Efficiency: Stress and Ontogeny

Values of ε lower than 2.8g DM/MJ PAR can be anticipated for wheat crops in which the rate of gross photosynthesis (P_g) is depressed, for example, by radiation saturation of the canopy, by the disruption of leaf metabolism (low mineral nutrient levels, especially N; pathogenic infection), or by prolonged stomatal closure (in response to water stress or other factors such as air pollution). ε can also be impaired by factors which stimulate maintenance respiration (bW), including superoptimal temperatures, and the incidence of disease.

Depending upon the timing of their onset, each of these stresses can affect the time course of ε (i.e., disrupt the relative constancy up to anthesis), and it is generally accepted that ε falls after anthesis because of leaf ageing (lower photosynthetic potential and higher respiration of senescent tissues) (e.g., Fischer 1993). However, there are other factors which can affect the seasonal pattern of ε in a systematic way; in particular, a fall in efficiency can be predicted when a crop diverts its metabolism to the accumulation of products with a high energy cost of synthesis (e.g., postanthesis accumulation of lipids in oilseeds).

It is, however, important to emphasize that long-term exposure to stresses (water, nutrient, disease) will also tend to reduce crop growth rate via radiation interception (delay and curtailment of canopy expansion, and promotion of senescence). Some interactions of these factors can be explored by reviewing recent investigations of the influence of water stress, nitrogen supply and ontogeny on the ε of cereal crops.

An early indication that water stress can affect both canopy expansion and radiation-use efficiency in wheat was provided by Gallagher and Biscoe (1978a), who found that the data points for 1976 fell consistently below the straight line relationship, between crop growth rate and radiation absorbed, which was common to a range of crops and seasons in the Midlands of

England. Since 1976 was a year of extreme drought, this finding suggested that, in this environment at least, ε was sensitive to severe water stress, but less affected by the more moderate water stress of more normal years.

Support for this view has been provided by a more detailed study at Lincoln in New Zealand, in which water stress sufficient to cause reductions in biomass accumulation of up to 50% was imposed on irrigated *barley* crops (Jamieson et al. 1995b). Where the stress was imposed from crop emergence, ε was lower than for continuously irrigated crops, and the degree of depression was linearly related to the lowest observed soil-moisture deficit. This effect was in addition to the reductions in PAR interception caused by lower leaf area indices. By contrast, in those crops which experienced a more natural incidence of drought, starting around 40 days after emergence or later, water stress caused a reduction in PAR interception, owing to early senescence of the canopy, but ε remained unaffected.

Laboratory experiments with a range of crop species have shown that the rate of photosynthetic CO_2 fixation is dependent upon the nitrogen content of the leaves, particularly in plants of low to deficient nitrogen status. (Note that the SWHEAT model discussed in Sect. 7.2 assumes that photosynthetic rate is controlled by leaf nitrogen content; Van Keulen and Seligman 1987). It has proven difficult to confirm such reports in the field by fertilizer application experiments because the soil nitrogen contents of arable soils are rarely sufficiently low to give such low levels of leaf nitrogen. However, the application of fertilizer nitrogen can result in enhancement of the photosynthetic activity of cereal flag leaves, which can become deficient in nitrogen owing to translocation to the developing ear (reviewed by Hay and Walker 1989). Such findings formed the basis for the working hypothesis that the application of nitrogen influences wheat yield primarily by its effect on canopy development (survival of tillers, leaf size, rate of senescence), rather than its effects on ε (Biscoe and Gallagher 1978; Monteith and Elston 1983).

In a detailed examination of this hypothesis over two growing seasons in England, Green (1987) found that the application of nitrogen fertilizer to winter wheat crops gave the expected increases in peak leaf area index and in canopy duration (Fig. 10a). However, there were also increases in ε during canopy generation, which were more marked after the start of stem extension (Fig. 10b). Significant enhancement of radiation-use efficiency, which was not simply a sign of the relief of nitrogen deficiency, was maintained throughout the life of the crops studied. Nevertheless, care should be exercised in the interpretation of these data since some of the values of ε were unusually high (up to 4 g DM/MJ PAR absorbed), and the estimation of ε by "the regression of two accumulated indices" (Fig. 10b) may have introduced errors (Green 1987; Monteith 1994). In another major experiment at three sites (North Dakota, Kansas and Texas), involving 24 combinations of irrigation, nitrogen fertilization and variety, Garcia et al. (1988) also found modest, but not statistically significant, increases in ε with nitrogen application, but by far the predominant effect was on the interception of PAR. More recently, in an extensive experiment in Australia, Fischer (1993) has provided clear correlations between ε and the nitrogen content of the above-ground biomass of spring wheat crops.

Although Garcia et al. (1988) and Fischer (1993) provide evidence for and against the hypothesis that ε is relatively unaffected by changes in the environment, both record ontogenetic trends in ε which are difficult to reconcile with the concept that ε is relatively constant up to anthesis. Thus, the values of 2.8 to 3.2 g DM/MJ PAR (absorbed) observed by Garcia et al. (1988) for the full season were, in fact, means of around 1.5 between double ridge and terminal spikelet, 3.7–3.8 from terminal spikelet to anthesis, and 2.8–3.3 from anthesis to the grain "soft dough" stage. Similar trends were recorded for each of the three sites, but there are no coherent explanations of the very low values before the terminal spikelet stage. Fischer's (1993) values show a more modest range from 2.14 g/MJ PAR absorbed in the first few weeks of crop growth up to 2.92 around anthesis, thereafter falling to an estimated level of 2.12 during grain

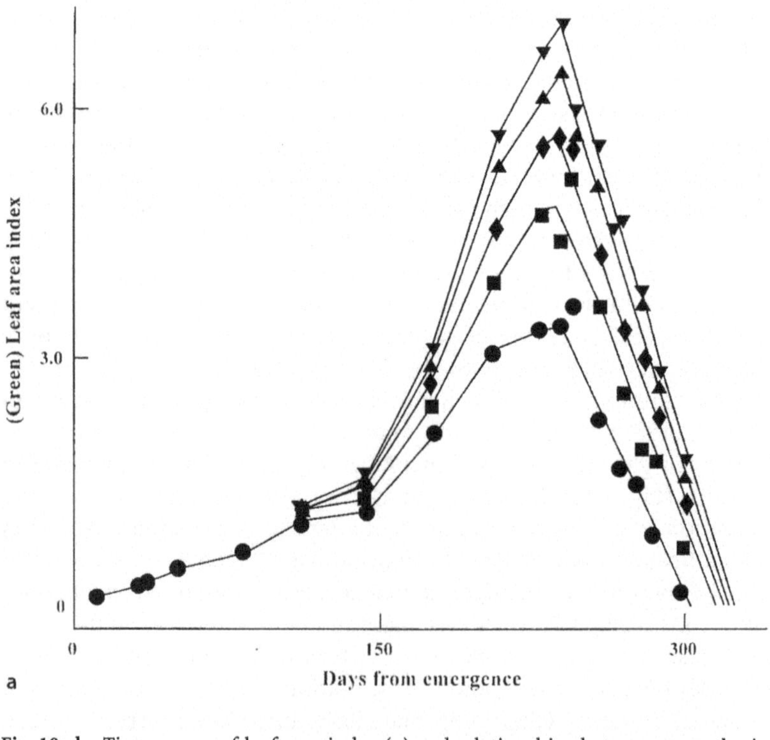

a

Fig. 10a,b. Time course of leaf area index (**a**) and relationships between preanthesis dry matter accumulation and absorption of total solar radiation (**b**) for crops of Armada winter wheat grown with different levels of nitrogen fertilizer application in 1980/81 in the Midlands of England. N_0 and ● indicate 0 kg N/ha; N_1 and ■ 50 kg N/ha; N_2 and ◆ 100 kg N/ha; N_3 and ▲ 150 kg N/ha; N_4 and ▼ 200 kg N/ha; *vertical arrows* indicate the start of stem extension. (After Green 1987, with permission of Elsevier Science – NL)

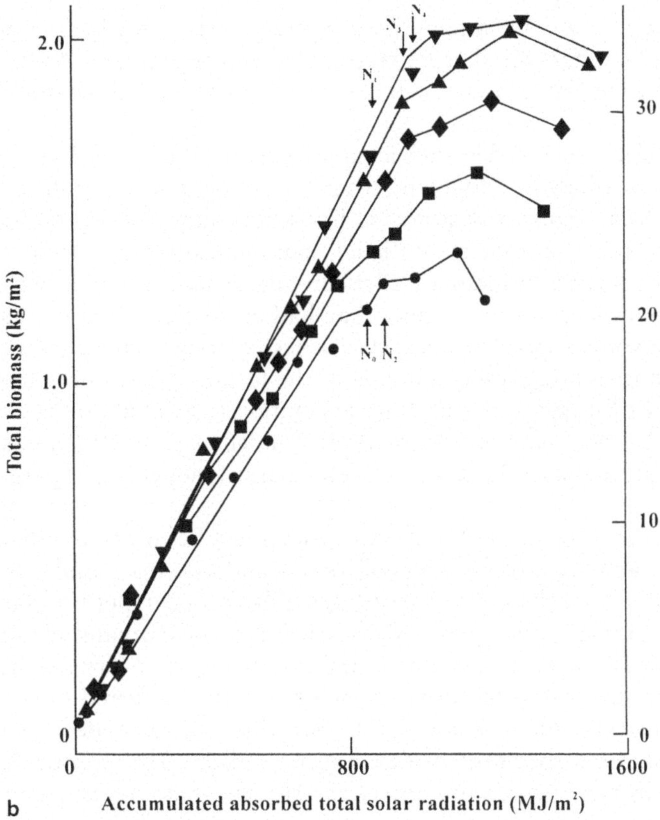

Fig. 10a,b. (*Continued*).

filling. There is, therefore, a need for more detailed monitoring of the time course of ε under a range of conditions; the valuable ontogenetic work of Gifford and Morison (1993) is a step in this direction, but carried out under controlled conditions, where enhanced levels of diffuse radiation can lead to higher values of ε (Sinclair et al. 1992).

7 Quantitative Aspects of Development: Generation of Grain Population Density

Grain yield can, as shown in Eq. (2), be expressed in terms of three conventional components: number of ears per unit area; number of grains per ear: and mean grain weight. Since, under productive conditions, mean grain weight is a relatively stable varietal characteristic, the product of the first two components: number of grains per unit area (i.e., the sink capacity of the crop), can been seen as a primary determinant of crop grain yield (Gales 1983; Hay and Walker 1989). This offers an alternative/complementary approach to the simulation of wheat yield (rather than via radiation interception, ε and harvest index).

However, the problems involved in such an approach are formidable. In particular, in contrast to canopy development, where all initiated leaves proceed to maturity (Sect. 5), the number of grains at harvest depends less upon the number of tillers, spikelets and florets initiated than upon the proportion of these organs which survives to maturity (Brisson and Delécolle 1991). Even under the most favourable conditions, more than half of the tillers and florets initiated, and 10–20% of the spikelets, are lost (Fischer 1983; e.g., Table 2); thus, although there is an increasing understanding of the factors determining the rate and duration of tiller, spikelet and floret generation (e.g., Siddique et al. 1989; Delécolle et al. 1989; Jedel and Hunt 1991; Youssefian et al. 1992), this information is of marginal usefulness in the mathematical modelling of grain production.

Instead, as shown by Fischer (1983), the approach must be to assume that each plant produces an excess of each type of organ, and that the number of ears, spikelets and florets surviving to harvest depends upon competition for limited resources (assimilate, nitrogen), with the earliest-formed organs taking precedence. For example, he proposed that, at anthesis, 10 mg of ear dry weight was required for the survival to maturity of each floret; the preferred florets would be in the spikelets at the mid point of the ear which differentiated first, and the first formed, proximal florets within a given spikelet. This approach, which is supported by the findings of more recent studies of assimilate partitioning between ear and stem (e.g., Siddique et al. 1989; Youssefian et al. 1992) and within the ear (Oscarson 1996), has been applied, with different degrees of success, in a series of wheat models.

Table 2. Production and survival of florets in mainstem spikelet 8 of a range of wheat cultivars grown at Perth, Western Australia, in 1986. (Siddique et al. 1989)

Variety	Date of introduction	Highest no. of florets	No. of florets at anthesis	No. of grains (% survival)
Purple Straw	1860	6.7	3.0	2.2 (32.8)
Gluyas Early	1894	7.8	3.0	2.9 (37.2)
Nabawa	1915	7.8	3.8	2.7 (34.6)
Bencubbin	1929	7.2	3.4	2.9 (40.3)
Gabo	1945	8.4	3.4	2.8 (33.3)
Insignia	1946	7.6	3.8	3.0 (39.5)
Gamenya	1960	8.0	4.0	3.3 (41.3)
Condor	1961	7.2	3.8	3.1 (43.1)
Tincurrin	1978	8.6	4.6	4.2 (48.8)
Kulin	1986	8.6	4.4	4.1 (47.7)

7.1 Simulation of Organ Survival

As indicated in Section 4.6, the first mainstem tiller appears at three phyllochrons after crop emergence and subsequent primary tillers appear regularly at intervals of one phyllochron. In several wheat models it is assumed that tiller production ceases at the (mainstem) double ridge stage, but by that time the plant already carries around twice the number of stems that will ultimately bear ears; the death of infertile stems tends to occur fairly abruptly after terminal spikelet, with little mortality thereafter. In constructing CROPSIM-WHEAT, Hunt and Pararajasingham (1995) dealt with this problem by estimating the quantity of assimilate necessary for each fertile stem after terminal spikelet and distributing the assimilate, according precedence to earlier-formed tillers. In ARCWHEAT, Porter (1984) applied a factor, derived from earlier work on the survival of apple buds, to ensure that the last-formed tillers died first and that the fertile stem population did not exceed a predetermined maximum level. In AFRCWHEAT2 (Porter 1993), this density-dependent approach was modified, by means of a water and nitrogen submodel, to allow for higher death rates of tillers under nitrogen shortage.

Modelling floret survival is more difficult because less is known about the fine control of floret death. Several studies have shown increased grain number per ear with improvement in assimilate and nitrogen supply to the ear (e.g., Siddique et al. 1989; Youssefian et al. 1992; Ayoub et al. 1994), and there are indications that floret death is lower in modern cultivars (Table 2), but it is not possible at present to convert such findings into rules for floret survival. Under these circumstances, the semiempirical rule of Fischer (1983) of one fertile floret per 10 mg of ear mass at anthesis has been employed in ARCWHEAT (Weir et al. 1984). A similar approach has been taken in CERES-Wheat, with some variation between varieties, but some more recent models have attempted to include analysis of spikelet and floret production (Van Keulen and Seligman 1987; Porter et al. 1993). Once the final grain number is known, grain filling can be simulated by partitioning the available assimilate to the fertile florets. The success of such approaches is considered in the next section (Fig. 11).

7.2 Simulation of Grain Yield and its Components

Crop simulation models are constructed and calibrated using information from a given, normally narrow, range of environmental conditions and cultivars. At the time of publication of a model, it is normally validated (i.e., its usefulness tested) by demonstrating its success at simulating the performance of similar varieties grown in similar environments. Such models are of great importance in stimulating the research needed to fill in gaps in our knowledge, and sensitivity analyses can be of great value in the identification of limiting factors. Nevertheless, in order to maximize the return of the considerable investment of research time on such local modelling (e.g., to generate useful information on the effects of global climatic change), it is important to work towards more universally applicable models. The first steps in this process – the testing of a range of crop models under different circumstances – is now well under way, as shown in the following two studies.

Three wheat models which had been developed for the Northern Hemisphere (AFRCWHEAT2, UK; CERES-Wheat, USA; SWHEAT, The Netherlands) were run using environmental data for three seasons (1984–86) at Lincoln, New Zealand (Porter et al. 1993). The results were then compared with measurements of unstressed crops of cv. Rongotea (a modern New Zealand spring wheat) and cv. Avalon (the winter cultivar used in the development of AFRCWHEAT2) during the same seasons. A modest degree of recalibration was carried out after the first run using 1984 data.

AFRCWHEAT2 emerged as clearly the best model for the simulation of the development/phenology of both cultivars, predicting the timing of events up to anthesis with a mean error (root mean square difference) of 5–8 days, compared with 10–16 days for CERES-Wheat and 26 days for SWHEAT. The results were less clear-cut for the interception of PAR since each model gave some degree of overestimation, although AFRCWHEAT2 tended to provide the most accurate estimates of dry matter production. The expression of these differences in terms of grain yield at harvest is presented in Table 3: both AFRCWHEAT2 and CERES-Wheat could simulate grain yield very accurately for some crops (e.g., both cultivars in 1984, 1986 (1)) but quite poorly for others (e.g., 1986 (2) & (3)). Results of this kind the important tools in investigating

Table 3. Observed grain yields (g/m^2) of crops of two wheat cultivars (Avalon, Rongotea) grown in three seasons at Lincoln, New Zealand, and the corresponding simulated grain yields (g/m^2) using three models (AFRCWHEAT 2, CERES-Wheat and SWHEAT). (Porter et al. 1993)

Season (sowing)	Cultivar	Observed yield (±SEM)	Simulated yield		
			AFRCWHEAT 2	CERES-Wheat	SWHEAT
1984	Avalon	874 ± 28	800	896	–
	Rongotea	780 ± 23	810	718	451
1985	Avalon	750 ± 29	920	913	–
	Rongotea	626 ± 146	700	737	276
1986 (1)	Avalon	1051 ± 30	1050	924	–
	Rongotea	990 ± 16	960	651	660
(2)	Avalon	1010 ± 25	740	883	–
	Rongotea	990 ± 86	760	611	631
(3)	Avalon	986 ± 32	780	871	–
	Rongotea	930 ± 38	680	567	589

the less robust components of each model. SWHEAT was generally unsuccessful at predicting the yield of Rongotea, and the values generated were subject to very large errors (equivalent to 30–50% of grain yield).

Figure 11 shows that although AFRCWHEAT2 and CERES-Wheat could give surprisingly accurate estimates of grain yield in an environment which was far from identical to that of the original modelling, there is still a long way to go in simulating the components of yield. For example, the precise prediction of the yield of Avalon in 1986(1) by AFRCWHEAT2 involved a 35% overestimate of ear population density and a 27% overestimate of grain population density, offset by a 23% underestimate in individual grain weight. These results confirm the difficulties of modelling organ survival (Sect. 7.1).

Recently, a more comprehensive comparison of eight wheat models has begun under the auspices of the GCTE Wheat Network (Global Change of Terrestrial Ecosystems, a component of the International Geosphere-Biosphere Programme; Goudriaan 1996). Instead of comparing the performance of the models against that of real crops, each was run using 30-year mean data for a typical winter wheat site in Northern Europe (Lelystad, The Netherlands) and a typical spring wheat site in the USA (Crookston, Minnesota). Even though the models were calibrated to the same phenology (dates of emergence, anthesis and maturity), the variation between models in predicted grain yield was very high: 2.5 to 8 t/ha for Crookston and 5.4 to 10.3 t/ha for Lelystad; forcing the models to generate the same time course of leaf area index did little to reduce the degree of variation (4.0 to 8.8 t/ha and 5.9 to 12.1 t/ha). Interpretation of these deviations from a common yield by the international GCTE Wheat Network offers a unique opportunity to improve the "portability" of crop models.

8 Conclusions

The results of extensive work on the control of the phenology and development of wheat crops in several countries over the past 20 years have been synthesized to give (locally applicable) mathematical models which predict the time course of phenology and leaf area index with a high degree of accuracy, in the absence of severe stress. Nevertheless, there are still considerable areas of uncertainty, notably the identity of the environmental factor(s) determining the phyllochron at crop emergence. Given relevant extinction coefficients, it is possible to use such time courses to estimate the quantity of PAR intercepted by the crop.

Measurements of radiation-use efficiency in several major wheat-growing areas of the world support the use of a universal factor (ε = 2.8–3.2 g DM/MJ PAR absorbed) in the modelling of wheat biomass yield, which can then be converted to grain yield using an appropriate harvest index, or by partitioning dry matter to grains. There is, however, sufficient evidence of variation of ε with ontogeny and stress to justify care in deploying the argument that the

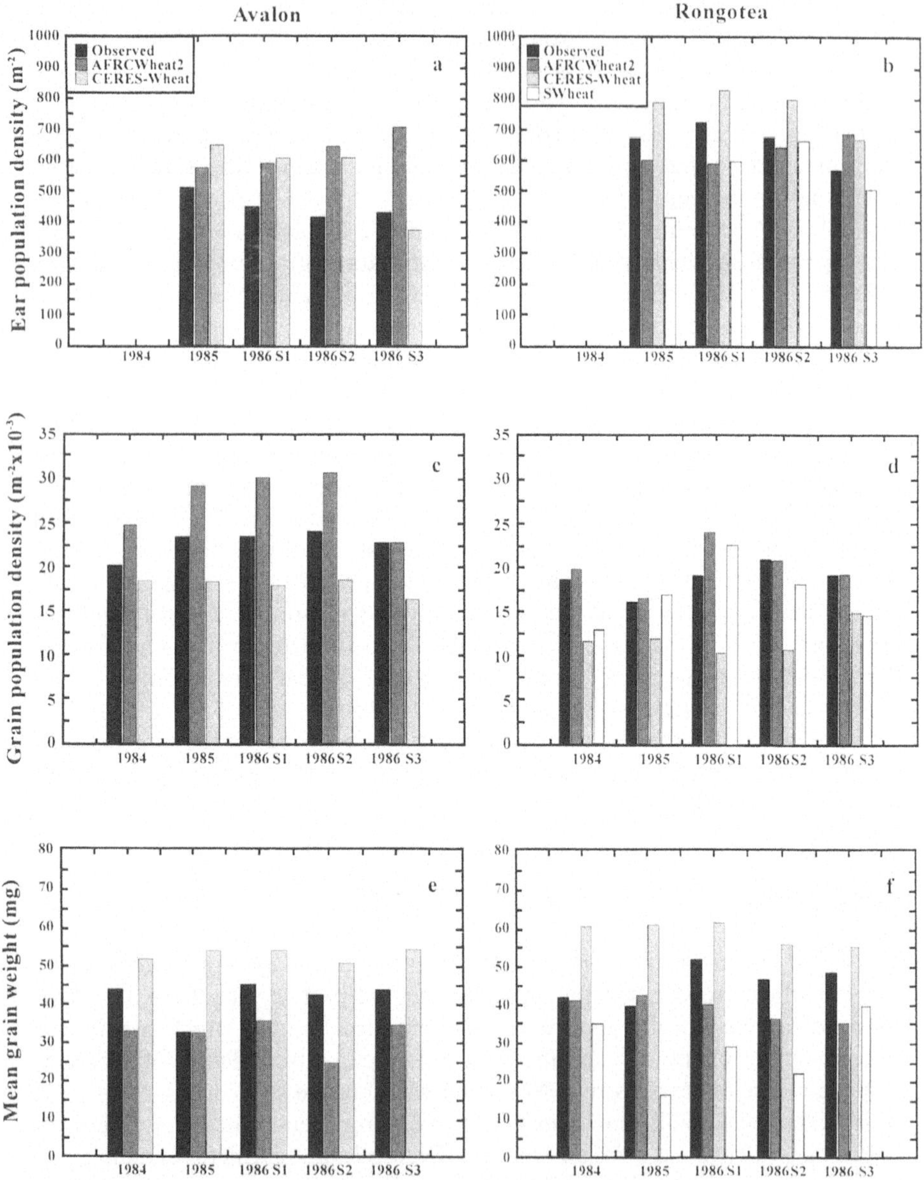

Fig. 11a–f. Observed and simulated values of components of grain yield for crops of Avalon winter wheat and Rongotea spring wheat grown at Lincoln, New Zealand, 1984–1986, where *S1* to *S3* refer to three dates of sowing. (After Porter et al. 1993, with permission of Elsevier Science – NL)

environment influences yield primarily via the interception of PAR. Indeed, under certain circumstances, values of ε can be used as indices of crop stress.

The modelling of grain population density is hampered by lack of understanding of the factors affecting survival of organs (tillers, spikelets and florets). Accurate prediction of grain yield can be associated with large errors in the prediction of ear population density, number of grains per ear and individual grain weight.

For the future, it is likely that progress in understanding the factors determining crop development and growth will come from a combination of conventional physiology and application of the methods of molecular genetics (note the ability to map genes for "days to heading"; Thomas et al. 1995; Tinker et al. 1996). Improved understanding of the genetic and environmental control of individual aspects of crop development and growth should contribute to improved accuracy of crop simulation models. However, it is likely that faster progress, and the improved "portability" of such models, will follow from comparisons of the performance of several models, and the relative success of their components, under uniform and contrasting conditions. Widely applicable crop models will be important tools in integrating the many potential effects of global climatic change on crop yield (notably effects on the rate of all aspects of development; on vernalization; on the rates of crop photosynthesis and respiration; and on the influence of stress, such as drought incidence). What has already been achieved by the modellers is a very positive answer to the pessimists of the 1970s who sought to discourage such approaches (Van Keulen and Seligman 1987).

Acknowledgement. It is a pleasure to thank Tony Fischer, Jan Goudriaan, Michael Kirby, John Porter and Don Smith for their help in providing material for this chapter, and reading the text in manuscript.

References

Allison JCS, Daynard TB (1976) Effect of photoperiod on development and number of spikelets of a temperate and some low-latitude wheats. Ann Appl Biol 83:93–102

Asrar G, Hipps LE, Kanemasu ET (1984) Assessing solar energy and water use efficiencies in winter wheat: a case study. Agric For Meteorol 31:47–58

Ayoub M, Guertin S, Lussier S, Smith DL (1994) Timing and level of nitrogen fertility effects on spring wheat in Eastern Canada. Crop Sci 34:748–756

Azam-Ali SN, Crout NMJ, Bradley RG (1994) Perspectives in modelling resource capture by crops. In: Monteith JL, Scott RK, Unsworth MH (eds) Resource capture by plants. Nottingham University Press, Nottingham, UK, pp 125–148

Baker CK, Gallagher JN, Monteith JL (1980) Daylength change and leaf appearance in winter wheat. Plant Cell Environ 3:285–287

Baker JT, Pinter PJ, Reginato RJ, Kanemasu ET (1986) Effects of temperature on leaf appearance in spring and winter wheat cultivars. Agron J 78:605–613

Bindi M, Porter JR, Miglietta F (1995) Comparison of models to simulate leaf appearance in wheat. Eur J Agron 4:15–25

Biscoe PV, Gallagher JN (1978) A physiological analysis of cereal yield. 1 Production of dry matter. Agric Prog 53:34–50

Boone MYL, Rickman RW, Whisler FD (1990) Leaf appearance rates of two winter wheat cultivars under high carbon dioxide conditions. Agron J 82:718–724
Brisson N, Delécolle R (1991) Développement et modè les de simulation de cultures. Agronomie 12:253–263
Brooking IR (1996) Temperature response of vernalization in wheat: a developmental analysis. Ann Bot 78:507–512
Brooking IR, Ritchie JT (1993) Leaf appearance and the phyllochron in wheat. In: Agronomy abstracts. American Society of Agronomy, Madison
Brooking IR, Jamieson PD, Porter JR (1995) The influence of daylength on final leaf number in spring wheat. Field Crops Res 41:155–165
Butterfield RE, Morison JIL (1992) Modelling the impact of climatic warming on winter cereal development. Agric For Meteorol 62:241–261
Calderini DF, Miralles DJ, Sadras VO (1996) Appearance and growth of individual leaves as affected by semidwarfism in isogenic lines. Ann Bot 77:583–589
Cao W, Moss DN (1989) Daylength effect on leaf emergence and phyllochron in wheat and barley. Crop Sci 29:1021–1025
Cao W, Moss DN (1991a) Vernalization and phyllochron in winter wheat. Agron J 83:178–179
Cao W, Moss DN (1991b) Phyllochron change in winter wheat with planting date and environmental change. Agron J 83:396–401
Cao W, Moss DN (1994) Sensitivity of winter wheat phyllochron to environmental changes. Agron J 86:63–66
Craigon J, Atherton JG, Sweet N (1995) Modelling the effects of vernalization on progress to final leaf appearance in winter wheat. J Agric Sci (Camb) 124:369–377
Cutforth HW, Jame YW, Jefferson PG (1992) Effect of temperature, vernalization and water stress on phyllochron and final main-stem leaf number of HY320 and Neepawa spring wheats. Can J Plant Sci 72:1141–1151
Darwinkel A (1983) Ear formation and grain yield of winter wheat as affected by time of nitrogen supply. Neth J Agric Sci 31:211–225
Davidson DJ, Chevalier PM (1990) Preanthesis tiller mortality in spring wheat. Crop Sci 30:832–836
Delécolle R, Couvreur F, Pluchard P, Varlet-Grancher C (1985) About the leaf-daylength model under French conditions. In: Day W, Atkin RK (eds) Wheat growth and modelling. Plenum Press, New York, pp 25–31
Delécolle R, Hay RKM, Guérif M, Pluchard P, Varlet-Grancher C (1989) A method of describing the progress of apical development in wheat, based on the time-course of organogenesis. Field Crops Res 21:147–160
Evans LT (1993) Crop evolution, adaptation and yield. Cambridge University Press, Cambridge
Evans LT, Wardlaw LF, Fischer RA (1975) Wheat. In: Evans LT (ed) Crop physiology. Cambridge University Press, Cambridge
Fischer RA (1983) Wheat. In: Symposium on Potential Productivity of Field Crops under Different Conditions. International Rice Research Institute, Philippines, pp 129–154
Fischer RA (1993) Irrigated spring wheat and timing and amount of nitrogen fertilizer. 2. Physiology of grain yield response. Field Crops Res 33:57–80
Flood RG, Halloran GM (1986) Genetics and physiology of vernalization response in wheat. Adv Agron 39:87–125
Frank AB, Bauer A (1995) Phyllochron differences in wheat, barley and forage grasses. Crop Sci 35:19–23
Gales K (1983) Yield variation of wheat and barley in Britain in relation to crop growth and soil conditions – a review. J Sci Food Agric 34:1085–1104
Gallagher JN (1979) Field studies of cereal leaf growth. 1. Initiation and expansion in relation to temperature and ontogeny. J Exp Bot 30:625–636
Gallagher JN, Biscoe PV (1978a) Radiation absorption, growth and yield of cereals. J Agric Sci (Camb) 91:47–60

Gallagher JN, Biscoe PV (1978b) A physiological analysis of cereal yield. 2. Partitioning of dry matter. Agric Prog 53:51–70
Gallagher JN, Biscoe PV (1979) Field studies of cereal leaf growth. 3. Barley leaf extension in relation to temperature, irradiance and water potential. J Exp Bot 30:645–655
Gan S, Amasino RM (1995) Inhibition of leaf senescence by autoregulated production of cytokinin. Science 270:1986–1988
Garcia R, Kanemasu ET, Blad BL, Bauer A, Hatfield JL, Major DJ, Reginato RJ, Hubbard KG (1988) Interception and use efficiency of light in winter wheat under different nitrogen regimes. Agric For Meteorol 44:175–186
Garfield E (1990) Journal citation studies. 53. Agricultural sciences. Curr Contents 51:3–15
Gifford RM, Morison JIL (1993) Crop responses to the global increase in atmospheric carbon dioxide concentration. In: International crop science I. Crop Science Society of America, Madison, Wisconsin, pp 325–331
Gosse G, Varlet-Grancher C, Bonhomme R, Chartier M, Allirand J-M, Lemaire G (1986) Production maximale de matière sèche et rayonnement solaire intercepté par un couvert végétal. Agronomie 6:47–56
Goudriaan J with contributions from Hunt LA, Ingram J, et al. (1996) Predicting crop yields under global change. In: Walker BH, Steffens WL (eds) Global change and terrestrial ecosystems. Cambridge University Press, Cambridge, pp 260–274
Green CF (1987) Nitrogen nutrition and wheat growth in relation to absorbed solar radiation. Agric For Meteorol 41:207–248
Green CF (1989) Genotypic differences in the growth of *Triticum aestivum* in relation to absorbed solar radiation. Field Crops Res 19:285–295
Haun JR (1973) Visual quantification of wheat development. Agron J 65:116–117
Hay RKM (1986) Sowing date and the relationships between plant and apex development in winter cereals. Field Crops Res 14:321–337
Hay RKM (1995) Harvest index: a review of its use in plant breeding and crop physiology. Ann Appl Biol 126:197–216
Hay RKM, Delécolle R (1989) The setting of rates of development of wheat plants at crop emergence: influence of the environment on rates of leaf appearance. Ann Appl Biol 115:333–341
Hay RKM, Kemp DR (1992) The prediction of leaf canopy expansion in the leek from a simple model dependent on primordial development. Ann Appl Biol 120:537–545
Hay RKM, Kirby EJM (1991) Convergence and synchrony – a review of the coordination of development in wheat. Aust J Agric Res 42:661–700
Hay RKM, Tunnicliffe Wilson G (1982) Leaf appearance and extension in field-grown winter wheat plants: the importance of soil temperature during vegetative growth. J Agric Sci (Camb) 99:403–410
Hay RKM, Walker AJ (1989) An introduction to the physiology of crop yield. Longman, London
Hay RKM, Galashan S, Russell G (1986) The yields of arable crops in Scotland 1978–82: actual and potential yields of cereals. Res Dev Agric 3:159–164
Hipps LE, Asrar G, Kanemasu ET (1983) Assessing the interception of photosynthetically active radiation in winter wheat. Agric Meteorol 28:253–259
Holmes DP (1973) Inflorescence development of semidwarf and standard height wheat cultivars in different photoperiod and nitrogen treatments. Can J Bot 51:941–956
Hunt LA, Pararajasingham S (1995) CROPSIM – WHEAT: a model describing the growth and development of wheat. Can J Plant Sci 75:619–632
Jamieson PD, Brooking IR, Porter JR, Wilson DR (1995a) Prediction of leaf appearance in wheat: a question of temperature. Field Crops Res 41:35–44
Jamieson PD, Martin RJ, Francis GS, Wilson DR (1995b) Drought effects on biomass production and radiation-use efficiency in barley. Field Crops Res 43:77–86
Jedel PE, Hunt LA (1991) A comparison of the development of multiflorous and standard wheats. Ann Bot 68:501–509

Kernich GC, Slafer GA, Halloran GM (1995) Barley development as affected by rate of change of photoperiod. J Agric Sci (Camb) 124:379–388
Kiniry JR, Jones CA, O'Toole JC, Blanchet R, Cabelguenne M, Spanel DA (1989) Radiation-use efficiency in biomass accumulation prior to grain-filling for five grain-crop species. Field Crops Res 20:51–64
Kirby EJM (1992) A field study of the number of main shoot leaves in wheat in relation to vernalization and photoperiod. J Agric Sci (Camb) 118:271–278
Kirby EJM (1993) Effect of sowing depth on seedling emergence, growth and development in barley and wheat. Field Crops Res 35:101–111
Kirby EJM (1995) Factors affecting rate of leaf emergence in barley and wheat. Crop Sci 35:11–19
Kirby EJM, Appleyard M (1984) Cereal development guide. 2nd edn. Arable Unit, National Agricultural Centre, Coventry, UK
Kirby EJM, Appleyard M (1987) Development and structure of the wheat plant. In: Lupton FGH (ed) Wheat breeding. Chapman & Hall, London, pp 287–311
Kirby EJM, Perry MW (1987) Leaf emergence rates of wheat in a mediterranean environment. Aust J Agric Res 38:455–464
Kirby EJM, Appleyard M, Fellowes G (1985a) Effect of sowing date and variety on main shoot leaf emergence and number of leaves of barley and wheat. Agronomie 5:117–126
Kirby EJM, Appleyard M, Fellowes G (1985b) Leaf emergence and tillering in barley and wheat. Agronomie 5:193–200
Kirby EJM, Appleyard M, Simpson NA (1994) Co-ordination of stem elongation and Zadoks growth stages with leaf emergence in wheat and barley. J Agric Sci (Camb) 122:21–29
Klepper B, Rickman RW, Peterson CM (1982) Quantitative characterization of vegetative development in small grain cereals. Agron J 74:789–792
Krenzer EG, Nipp TL, McNew RW (1991) Winter wheat mainstem leaf appearance and tiller formation vs. moisture treatment. Agron J 83:663–667
Levy J, Peterson ML (1972) Responses of spring wheats to vernalization and photoperiod. Crop Sci 12:487–490
Longnecker N, Robson A (1994) Leaf emergence of spring wheat receiving varying nitrogen supply at different stages of development. Ann Bot 74:1–7
Longnecker N, Kirby EJM, Robson A (1993) Leaf emergence, tiller growth and apical development of nitrogen-deficient spring wheat. Crop Sci 33:154–160
Lupton FGH (ed) (1987) Wheat breeding. Chapman & Hall, London
McCree KJ (1974) Equations for the rate of dark respiration of white clover and grain sorghum, as functions of dry weight, photosynthetic rate, and temperature. Crop Sci 14:509–514
McMaster GS, Wilhelm WW (1995) Accuracy of equations predicting the phyllochron of wheat. Crop Sci 35:30–36
McMaster GS, Wilhelm WW, Morgan JA (1992) Simulating winter wheat shoot apex phenology. J Agric Sci (Camb) 119:1–12
Monteith JL (1977) Climate and the efficiency of crop production in Britain. Philos Trans R Soc Lond B 281:277–294
Monteith JL (1981) Does light limit crop production? In: Johnson CB (ed) Physiological processes limiting plant productivity. Butterworths, London, pp 23–38
Monteith JL (1994) Validity of the correlation between intercepted radiation and biomass. Agric For Meteorol 68:213–220
Monteith JL, Elston J (1983) Performance and productivity in the field. In: Dale JE, Milthorpe FL (eds) The growth and functioning of leaves. Cambridge University Press, pp 499–518
Moot DJ, Jamieson PD, Henderson AL, Ford MA, Porter JR (1996) Rate of change in harvest index during grain filling of wheat. J Agric Sci (Camb) 126:387–395
Mosaad MG, Ortiz-Ferrara G, Mahalakshmi V, Fischer RA (1995) Phyllochron response to vernalization and photoperiod in spring wheat. Crop Sci 35:168–171
Oscarson P (1996) Transport of recently assimilated ^{15}N to individual spikelets in spring wheat grown in culture solution. Ann Bot 78:479–488
Porter JR (1984) A model of canopy development in winter wheat. J Agric Sci (Camb) 102:383–392

Porter JR (1985) Approaches to modelling canopy development in wheat. In: Day W, Atkin RK (eds) Wheat growth and modelling. Plenum Press, New York, pp 69–81
Porter JR (1993) AFRCWHEAT2: a model of the growth and development of wheat incorporating responses to water and nitrogen. Eur J Agron 2:69–82
Porter JR, Jamieson PD, Wilson DR (1993) Comparison of the wheat simulation models AFRCWHEAT2, CERES-Wheat and SWHEAT for non-limiting conditions of crop growth. Field Crops Res 33:131–157
Rahman MS, Wilson JH (1977) Determination of spikelet number in wheat. 1. Effect of varying photoperiod on ear development. Aust J Agric Res 28:565–574
Rickman RW, Klepper BL (1991) Environmentally driven cereal crop growth models. Annu Rev Phytopathol 29:361–380
Rickman RW, Klepper BL (1995) The phyllochron: where do we go in the future? Crop Sci 35:44–49
Ritchie JT (1991) Wheat phasic development. In: Hanks J, Ritchie JT (eds) Modelling plant and soil systems. American Society of Agronomy, Madison, Wisconsin, pp 31–54
Ritchie JT, NeSmith DS (1991) Temperature and crop development. In: Hanks J, Ritchie JT (eds) Modelling plant and soil systems. American Society of Agronomy, Madison, Wisconsin, pp 5–29
Robertson MJ, Brooking IR, Ritchie JT (1996) Temperature response of vernalization in wheat: modelling the effect on the final number of mainstem leaves. Ann Bot 78:371–381
Russell G, Jarvis PG, Monteith JL (1989) Absorption of radiation by canopies and stand growth. In: Russell G, Marshall B, Jarvis PG (eds) Plant canopies: their growth, form and function. Cambridge University Press, Cambridge, pp 21–39
Siddique KHM, Kirby EJM, Perry MW (1989) Ear:stem ratio in old and modern wheat varieties; relationship with improvement in number of grains per ear and yield. Field Crops Res 21:59–78
Sinclair TR, Shiraiwa T, Hammer GL (1992) Variation in crop radiation-use efficiency with increased diffuse radiation. Crop Sci 32:1281–1284
Slafer GA, Rawson HM (1994) Sensitivity of wheat phasic development to major environmental factors: a re-examination of some assumptions made by physiologists and modellers. Aust J Plant Physiol 21:393–426
Slafer GA, Connor DJ, Halloran GM (1994) Rate of leaf appearance and final number of leaves in wheat: effects of duration and rate of change of daylength. Ann Bot 74:427–436
Stapper M, Fischer RA (1990) Genotype, sowing date and plant spacing influence on high-yielding irrigated wheat in southern New South Wales. 1. Phasic development, canopy growth and spike production. Aust J Agric Res 41:997–1019
Stern WR, Kirby EJM (1979) Primordium initiation at the shoot apex in four contrasting varieties of spring wheat in response to sowing date. J Agric Sci (Camb) 93:203–215
Szeicz G, Monteith JL, dos Santos J (1964) Tube solarimeters to measure radiation among plants. J Appl Ecol 1:169–174
Thomas H, Smart CM (1993) Crops that stay green. Ann Appl Biol 123:193–219
Thomas WTB, Powell W, Waugh R, et al. (1995) Mapping genes of economic importance in spring barley. Annu Rep Scottish Crop Res Inst for 1994, Dundee, Scotland, pp 60–64
Thorne GN, Wood DW (1988) Contributions of shoot categories to growth and yield of winter wheat. J Agric Sci (Camb) 111:285–294
Tinker NA, Mather DE, Rossnagel BG, et al. (1996) Regions of the genome that affect agronomic performance in two-row barley. Crop Sci 36:1053–1062
Tomos AD (1985) The physiological limitations of leaf cell expansion. In: Baker NR, Davies WJ, Ong CK (eds) Control of leaf growth. Cambridge University Press, Cambridge, pp 1–33
Tonkinson CL, Lyndon RF, Arnold GM, Lenton JR (1995) Effect of the Rht3 dwarfing gene on dynamics of cell extension in wheat leaves, and its modification by gibberellic acid and paclobutrazol. J Exp Bot 46:1085–1092
Van Keulen H, Seligman NG (1987) Simulation of water use, nitrogen nutrition and growth of a spring wheat crop. Pudoc, Wageningen, 310 pp

Wang S-Y, Ward RW, Ritchie JT, Fischer RA, Schulthess U (1995) Vernalization in wheat. 1. A model based in the interchangeability of plant age and vernalization duration. Field Crops Res 41:91–100

Watson DJ, Thorne GN, French SAW (1963) Analysis of growth and yield of winter and spring wheats. Ann Bot 27:1–22

Weigand CL, Gerbermann AH, Cuellar JA (1981) Development and yield of hard red winter wheats under semitropical conditions. Agron J 73:29–37

Weir AH, Bragg PL, Porter JR, Rayner JH (1984) A winter wheat crop simulation model without water or nutrient limitations. J Agric Sci (Camb) 102:371–382

Wheeler TR, Hong TD, Ellis RH, Batts GR, Morison JIL, Hadley P (1996) The duration and rate of grain growth, and harvest index, of wheat (*Triticum aestivum* L.) in response to temperature and CO_2. J Exp Bot 47:623–630

Wilhelm WW, McMaster GS (1995) Importance of the phyllochron in studying development and growth in grasses. Crop Sci 35:1–3

Youssefian S, Kirby EJM, Gale MD (1992) Pleiotropic effects of the GA-insensitive *Rht* dwarfing genes in wheat. 2. Effects on leaf, stem, ear and floret growth. Field Crops Res 28:191–210

Growth and Development of Oat with Special Reference to Source–Sink Interaction and Productivity

P. Peltonen-Sainio[1]

1 Oat as a Subject for Crop Physiological Studies

Archeological surveys have revealed that oat (*Avena* sp.) was present quite early as a weed in barley (*Hordeum vulgare* L.) and emmer (*Triticum dicoccum* Schubl.) cultivations (Murphy and Hoffman 1992). As the culture of barley and wheat progressed northwards away from their center of origin, oat exploited these climates better and eventually achieved crop status. This was especially because oat (*A. sativa* L.) favors cool growing seasons, and the occurrence of high temperatures and drought results in markedly reduced grain set and fill (Coffman and Frey 1961).

Oat has not, however, managed to liberate itself from its historical past, and still bears the stamp of a secondary crop in many areas where it is grown. Oat is often cultivated on lands not deemed suitable for barley and wheat, and the status of oat as a cereal crop has clearly decreased since its golden days during the early half of this century. The world's total oat hectarage has decreased from over 50 million ha to close to 20 million ha during the last four decades (Anonymous 1993). This has been due mainly to a decline in the use of oat as animal feed, partly a result of mechanization of agriculture and the consequent decreased importance of workhorses. Nevertheless, oat still ranks among the 20 most commonly cultivated field crop species in the world (Anonymous 1993).

Such a drastic collapse of about 60% in oat hectarage has been reflected in reduced intensity of associated research. Basic agricultural management of oat is relatively well documented but interest in crop physiology, which was at the fore particularly in the early 1980s, has been modest compared with that taken in barley and wheat (*Triticum aestivum* L.). Fortunately, much of the knowledge pertaining to the physiology of those crops can also be applied to oat. Numerous differences exist among these species, however, and one must be careful in extending many of the generalizations. The form of the inflorescence is one of the most obvious differences that has implications for source–sink interaction. For example, the green area of the oat panicle is large and scattered when compared with that of spikes, which improves its ability to inter-

[1] Department of Plant Production, Section of Crop Husbandry, P.O. Box 27, FIN-00014 University of Helsinki, Finland

Crop Yield, Physiology and Processes
D. L. Smith, C. Hamel (Eds.)
© Springer-Verlag Berlin Heidelberg 1999

cept solar radiation. This results in a high contribution of the panicle to photosynthate production during the grain-filling period (Jennings and Shibles 1968).

In this chapter the physiological basis of productivity in spring oat is explained by describing the development and growth of individual plants and communities as a process of synergy and competition, i.e., source–sink interactions. In this connection, some of the impacts of cultural practices and plant breeding on expression of yield potential in both favorable and unfavorable environments are described – especially through alterations in yield-determining components including ability to intercept photosynthetically active radiation and partitioning of resulting dry matter between grain yield and vegetative nonyield components. Culture of oat for purposes other than production of grain per se, e.g., for forage (Lowe and Bowdler 1988), is excluded in this connection.

2 Phenostages and Phenophases of Oat with Agricultural Interest

2.1 Phenostages

It is useful to divide the life cycle of cereal plants into several developmental stages and time intervals. This allows management practices and occurrence of abiotic and biotic stresses (e.g., water deficit, temperature stress, invasion of pests and pathogens) to be recorded more precisely. The term growth is used here to describe approximate changes in gross morphology, in contrast to developmental scales which describe differentiation processes at the apex.

Several scales for the description of macromorphological growth stages visible to the naked eye have been introduced (e.g., Large 1954; Zadoks et al. 1974) and are applicable to oat with minor modifications (Landes and Porter 1989). For reasons of simplicity and sufficient accuracy, use of such scales has exceeded that of developmental scales. The growth scale introduced in Fig. 1 follows the main growth stages illustrated originally for wheat by Feekes (Large 1954). In Fig. 1 a uniculm growth habit in oat is depicted: tillering is inhibited and tillers emerging subsequently are unproductive and die. Such a tillering pattern characterizes dense plant stands and is particularly beneficial at high latitudes (Peltonen-Sainio and Järvinen 1995).

The only pictorial scales that comprehensively describe the stages of differentiation at the apical meristem of the main oat shoot with illustrations were published by Bonnett (1961) and Åfors et al. (1988). The developmental scale of Åfors et al. (1988) is presented in Fig. 2. Scales describing events at the apical meristem are generally developed to improve understanding of determination of yield components under different growing conditions and to promote greater accuracy when timing certain management practices. For example, in

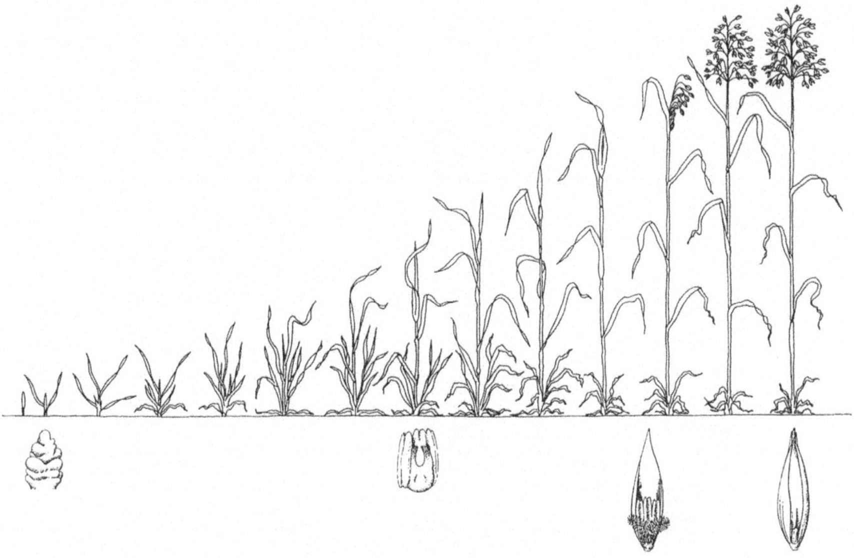

Fig. 1. Macromorphological growth stages of uniculm oat (*above*). Double ridge stage, start of flout abortion, pollination, and mature grain are indicated *below*

spring wheat, nitrogen fertilizer applied close to the initiation of double ridges (stages C-D, Fig. 2), and at stigmatic branch differentiation (stage Q) resulted more often in increased grain set and hence higher grain yield (Peltonen 1992a) than when applications were timed in accordance with external characteristics (Peltonen 1992b). Unfortunately, few reports exist on oat in which increase in productivity has been attempted by apical timed cultural practices. Developmental stages of the apex that are prone to herbicide injury through hormone-based effects can be avoided by assessing the developmental stage of the cereal apex (Andersen 1954; Hagsand 1954). Herbicide injuries such as fusion of spikelets, unilateral spikelet set, branched spikes, twisted rachis, and retarded differentiation have been reported, especially in wheat and barley (Andersen 1954; Loubser and Cairns 1989; Rajalahti and Peltonen 1993). Compared with cereals having spikes, the oat panicle is possibly better able to compensate for apex abnormalities occurring during early developmental stages and the consequent yield losses by producing additional spikelets in the lower branches of the axis.

As shown by Colville-Baltenberger and Frey (1987), within certain day lengths, development of oat is temperature-dependent. Therefore, calculation of cumulated thermal units is a considerably better predictor of development than merely the number of days between different phenostages. Effects of other factors such as agricultural management and the occurrence of stresses (e.g., moisture deficit) on the rate of cereal development conflict and reports for oat

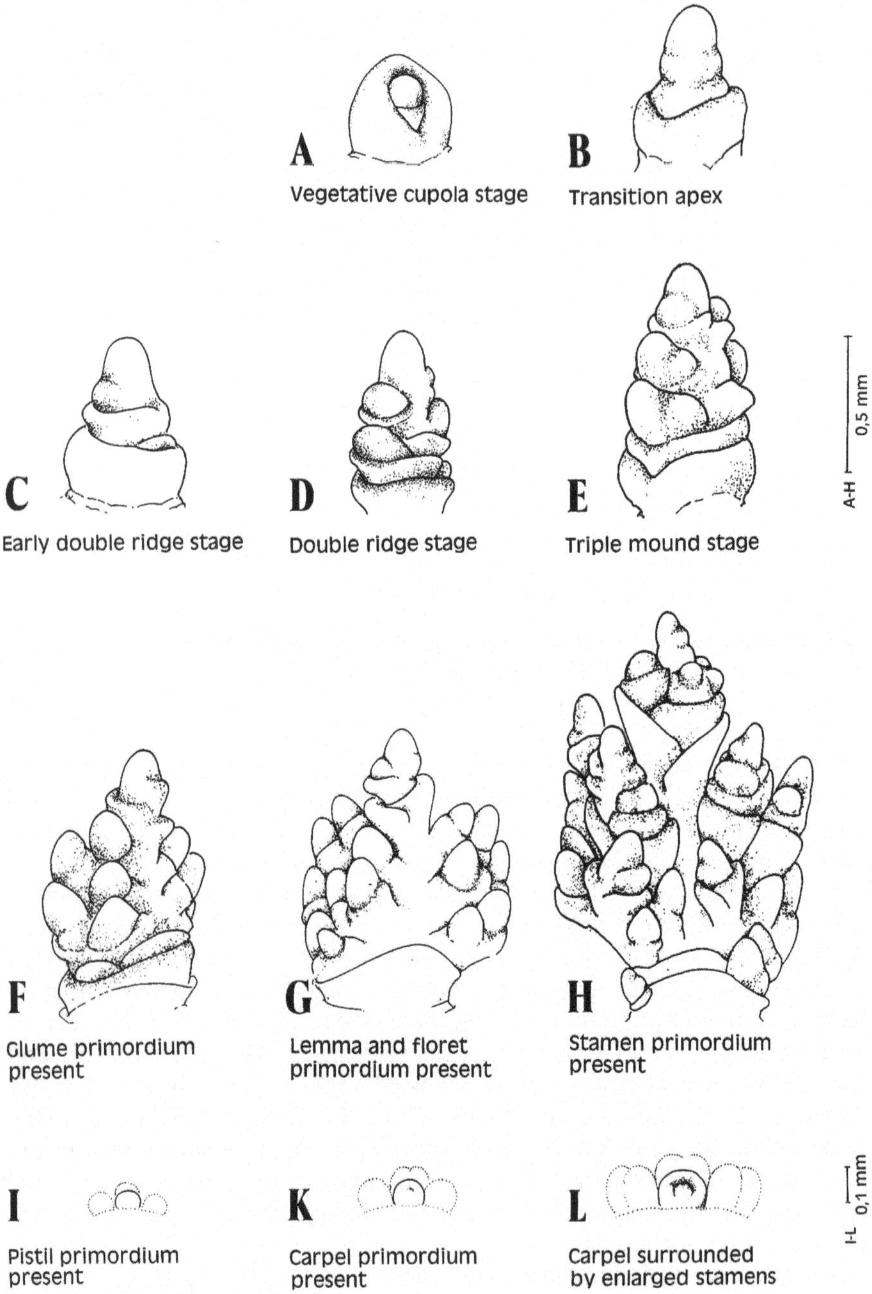

Fig. 2A–Y. Developmental stages of oat apex. Illustrations are from Åfors et al. (1988); descriptions are according to Waddington et al. (1986)

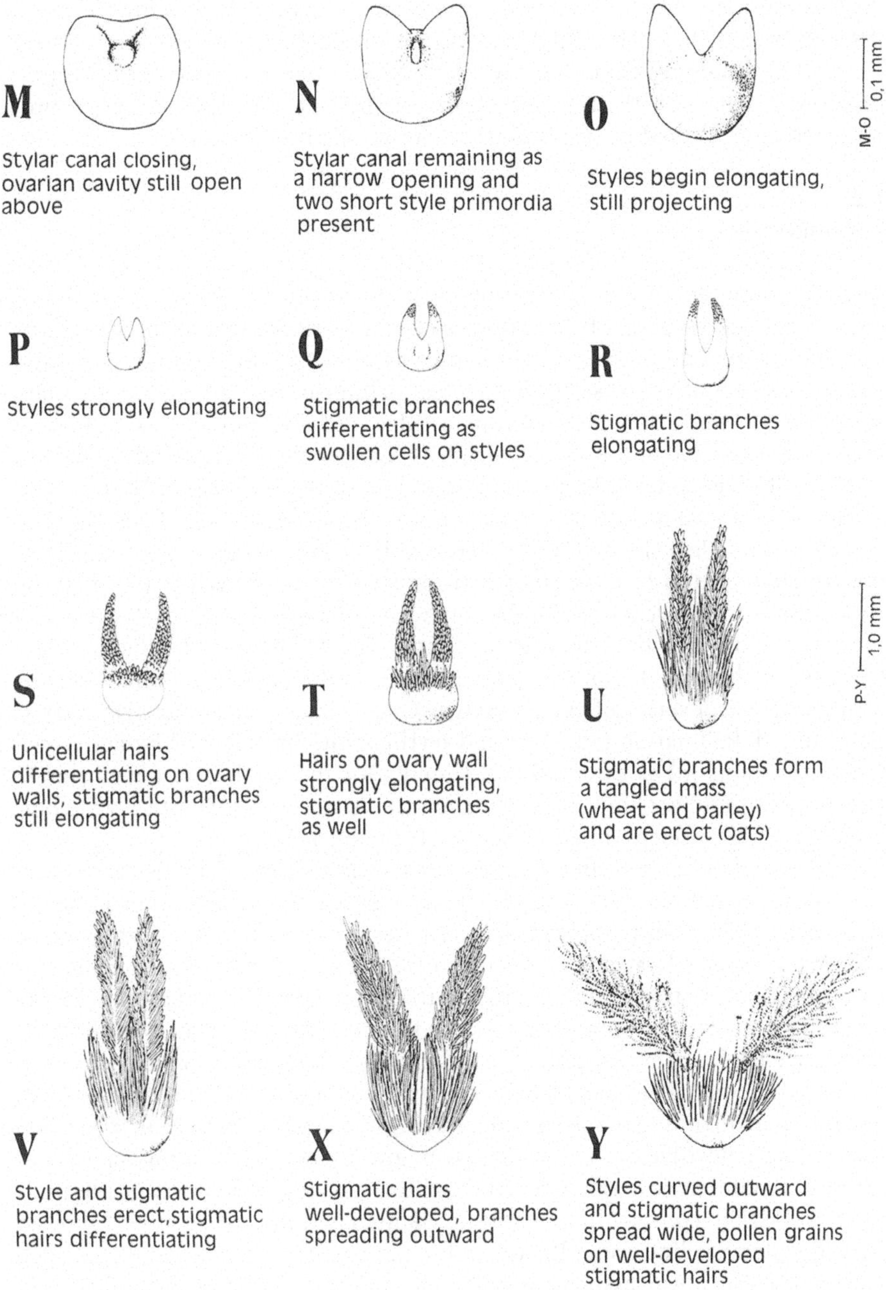

Fig. 2A–Y. (*Continued*)

are less common than for the other spring cereals. In a controlled environment, even severe moisture stress approaching the wilting point (water potential near −1.5 MPa) did not affect either the time of initiation of double ridges or pollination in oat, even though reductions in spikelet and floret set were pronounced and resulted in crop failure (Peltonen-Sainio 1991a).

2.2 Phenophases

Intervals between certain critical phenostages form the major phenophases such as vegetative, generative, and grain-filling periods. Two to three leaf primordia are already present in the embryos of oat grains (White 1995) and during the vegetative phase additional leaf primordia are initiated in the main shoot apex. After a few plastochrons a tiller bud also initiates as a ridge of meristematic tissue in the axil of each leaf (Langer 1979; Kirby and Appleyard 1984). Cell division leads to an increase in size of the leaf primordium so that it covers the apical meristem. This is followed by an expansion of cells in both leaf lamina and leaf sheath (White 1995), which continues throughout the generative phase. Up to 12 leaves may be produced per stem (White 1995). At the transition from the vegetative to the generative phase (stage B, Fig. 2) the apical meristem elongates to a cylindrical shape. The formation of double ridges (stage C) is the first step of the generative phase during which panicle organs differentiate as a series of events (stages C to Y) and thus storage capacity is determined. Pollination (stage Y) and fertilization are followed by active cell division. Grains fill as essential constituents for the synthesis of, e.g., starch, protein, and lipid are imported from source tissues and converted into components of the endosperm and embryo (Peterson 1988).

The duration of the different phenophases is influenced by genotype and environment, temperature and day length being the major environmental influences (Kirby and Appleyard 1984). Response to photoperiod prior to heading is especially strong in northern European oat cultivars, whereas cultivars originating from more southern areas responded less markedly to day length, especially at low temperatures (Bleken and Skjelvåg 1986). Modification of relative length of major phenophases through plant breeding may result in higher grain yield. The favorable directions of change in the duration of growth periods is evidently dependent on growing conditions. Takeda et al. (1987) established a positive association between grain yield and heading date in oat up to a certain critical value determined for growing conditions in the midwestern USA. In contrast to this finding, Salman and Brinkman (1992) showed that the later-heading oat cultivars had fewer postheading days and that such a growth pattern was associated with low grain yield. The possible direction of change that could be attempted when breeding oat for northern growing conditions (high latitudes and early summer drought followed by favorable soil moisture conditions) is presented in Fig. 3. It is suggested that the overall duration of the growth period remain unchanged, but the length of the vegetative and grain-filling periods be slightly shortened and, consequently, the period for spikelet set prolonged. This is supported by the finding that oat cultivars tested in southern Finland exhibited the following trend: the longer the vegetative period, the lower the grain–straw ratio, and the longer the period for maximum floret production (from stage C to Q, Fig. 2), the higher the panicle weight and the number of grains per panicle (Peltonen-Sainio 1994a). Therefore, such modifications in the relative length of the major growth phases are likely to result in higher yield potential, expression of which depends on post-anthesis growing conditions and the internal structure of the panicle, as discussed in the following paragraphs. In fact, in the northern extreme of oat cultivation (at ≥60°N), the vegetative period of oat is already exceptionally short due to long days, and the main shoot apex is in transition between the vegetative and generative phases (stage B, Fig. 2) when only some 140 °C days of effective temperature sum (base temperature T_b 5 °C) have accumulated after sowing (Peltonen-Sainio and Pekkala 1993). At this time, only two leaves have

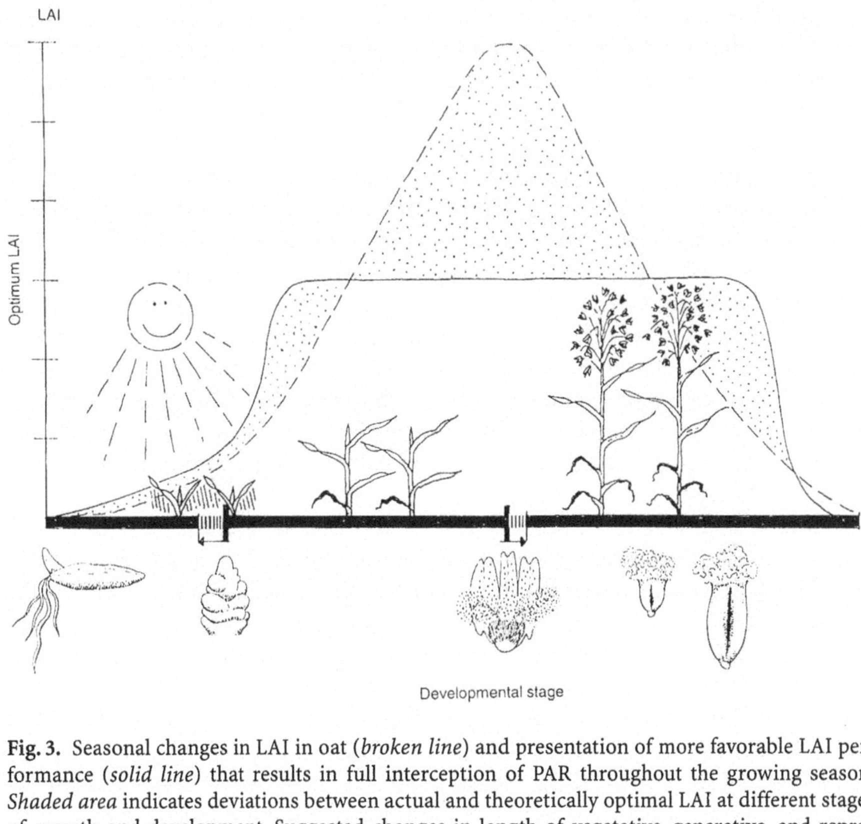

Fig. 3. Seasonal changes in LAI in oat (*broken line*) and presentation of more favorable LAI performance (*solid line*) that results in full interception of PAR throughout the growing season. *Shaded area* indicates deviations between actual and theoretically optimal LAI at different stages of growth and development. Suggested changes in length of vegetative, generative, and reproductive phases of oat grown under northern growing conditions are also indicated by *arrows*. Germination, double ridge stage, pollination, and grain filling are indicated *below*

unfolded in the oat plant. In spite of the speed of development, genotypic variation of about ±20 °C days of effective temperature sum was found in the duration of the vegetative period in oat material adapted for cultivation in these marginal regions (Peltonen-Sainio 1994a).

3 Canopy Architecture and Function in Relation to Dry-Matter Production

Solar radiation intercepted by the aboveground green area provides the energy for photosynthesis and is therefore the starting point for dry-matter production following exhaustion of the reserves contained in the seed. The availability of photosynthetically active radiation (PAR), the ability of the green area to capture it, the radiation-use efficiency (RUE – biomass accumulated per unit of intercepted radiation), and the partitioning of dry matter between harvestable organs and non-yield structures (harvest index, HI) define the major limitations to crop productivity. Leaf area index (LAI, m^2/m^2), i.e., the ratio of green leaf laminae (other photosynthetic areas often excluded to simplify measurement) and ground area, is commonly used to characterize the ability of the canopy to intercept PAR. Leaf area duration (LAD) takes into account the form of the LAI curve and the duration of green leaf area (Hunt 1982), when it is calculated by integrating LAI over time. LAD is, however, also commonly used as a 'qualitative' trait to describe solely days of green area until yellow ripening without providing any information about the quantity of the photosynthetically active leaf area (Peltonen-Sainio 1997a). Regardless of the interpretation of LAD, a positive relationship between LAD and productivity has been established for several crop species (Gardner et al. 1990; Sarma and Sarma 1992; Chau and Bhargava 1993; Jefferies and MacKerron 1993; Petr and Hradecka 1993; Peltonen-Sainio 1997a). Relationships between seasonal alterations in photosynthate-producing green leaf area and assimilate needs of harvestable yield components are seldom in perfect synchrony and, thus, it may be rewarding to improve this synchrony by cultivar choice and development of crop management.

Climatic factors, abiotic stressors, and crop management greatly influence the expression of genotype-dependent yield potential. One could say that this control is primarily through the environmentally caused changes in the ability to intercept PAR (Austin 1994). Peltonen-Sainio (1997a) showed that oat grown at high latitudes was not able to benefit from high main shoot LAD if drought occurred at grain fill, but in the absence of severe drought or if it occurred at preanthesis, high LAD favored yield formation. The contributions of growth stage-dependent, environment-dependent, and genotype-dependent changes in LAD and LAI, representing the main source of photosynthate, and differences in major competing sinks in demand for photoassimilates, are discussed in more detail in the following sections. The piecing together of this complex puzzle, termed *the crop physiological basis of oat productivity*, is possible even though some pieces are still missing and the picture remains incomplete.

3.1 Early Canopy Closure and Achievement of Optimum LAI During Formation of Yield Potential

Early canopy closure and the achievement of optimum LAI that enables full interception of PAR result in increased accumulation of aboveground biomass in terms of higher growth rate and, hence, possibly increased set of spikelets and florets. This is especially important in oat, because the harvest index (HI), which is one of the three yield-determining factors in addition to growth rate and growth duration, is already likely to be close to its optimum (Takeda and Frey 1987). Further increases in productivity require increased growth rate when breeding for higher HI is not advantageous and growth duration cannot be further prolonged, as it cannot be at high latitudes (Peltonen-Sainio 1993) and in the midwestern USA (Colville-Baltenberger and Frey 1987), where early maturity enables harvest prior to the occurrence of unfavorable growing conditions. The association between grain yield and vegetative growth rate was close in oats (Bregitzer et al. 1987), and selection for high vegetative growth rate has resulted in yield increases in oat lines (Takeda and Frey 1977; Takeda et al. 1979, 1980; Gupta et al. 1987). Oat breeders in the midwestern USA have not, however, changed the growth rate of oat cultivars released between 1914 and 1987 (Lynch and Frey 1993). Thus, further yield increases could possibly be attained by selecting for increased growth rate and, by this means, complete ground cover at early growth stages.

In addition to early canopy closure, reaching the optimum LAI rapidly and maintaining LAI as close to it as possible, most of the spikelet and floret initiation and grain-filling periods are likely to be favorable for grain production. The optimum value for LAI is likely to range from 3 to 5 in oat depending on leaf orientation (Hipps et al. 1983; Thorne et al. 1988). Under northern growing conditions, peak LAI of oat has varied from <2 to 12 depending on genotype and environment (Peltonen-Sainio 1997b; Peltonen-Sainio et al. 1997). LAI exceeding its optimum value only results in shading of the lower parts of the canopy and loss of self-supporting photosynthate production. Experiments carried out under northern growing conditions (60°N) indicated that an oat cultivar with peak LAI value ≥ 7 (due to the exceptionally high contribution of tillers to LAI – 34% at its highest) was one of the lowest yielding cultivars (Peltonen 1990). This causality may, however, be much more complex, and a negative impact of growth pattern on grain yield may result solely from deleterious effects of tillering under northern growing conditions, where auxin-derived apical dominance, maintained by the main shoot apex, is very strong. Therefore, retarded growth of tillers possibly resulted in an unbalanced input–output ratio in assimilate flow between tillers and the main shoot. The most productive oat cultivars under northern growing conditions had a peak value for LAI of about 5 (Peltonen 1990).

3.2 Maintaining Efficient Interception of PAR During Grain Growth

The role of the flag leaf as an assimilate source for grain filling is important, not least due to the small shaded area above it, the free capture of PAR, and the short distance between the source and sink tissues. Under northern European growing conditions, plant breeding has increased flag leaf area by up to 38% during the last 60 years, and concomitantly grain yield was increased by 20 to 30% (Peltonen-Sainio 1990). However, Jennings and Shibles (1968) indicated by using shading and removal of various plant parts that the green area in the panicle contributed more to grain filling than flag leaf, whereas in the experiments of Bregitzer et al. (1987) grain yield of oat lines was not related to the photosynthetic area above the uppermost leaf sheath. Small leaves are, however, often thicker, and populations characterized by thick leaves have better water-use efficiency, being better adapted to drought-prone environments (Austin 1994). Furthermore, small leaf size is associated with higher light saturated rates of leaf photosynthesis (Bhagsari and Brown 1986; Austin 1994). Bregitzer et al. (1987) found, however, that oat lines with small upper leaves had decreased floret length. They stated that, if a more open canopy is selected by choosing such genotypes, this may result in decreased grain size. Furthermore, as a clear negative effect of using plant growth regulators, Gendy and Höfner (1989) found that flag leaf area decreased, which is likely to reduce ability of this leaf, closest to the panicle, to capture PAR and support grain growth.

Salman and Brinkman (1992) found that an oat cultivar of high-yielding ability was superior to all other cultivars regarding growth rate and dry-matter accumulation during the postanthesis period. Such superiority requires better ability of the canopy to maintain efficient interception of PAR at the postanthesis period. Figure 3 illustrates highly schematic differences (shaded area) between theoretically optimal LAI versus LAI with high peak and rapid collapse of green area (see, e.g., Peltonen-Sainio 1997b). The shaded area could be further reduced by developing appropriate management practices or breeding for improved LAI. One could of course comment on the form of the optimal LAI curve by indicating that the peak value, and thereby the whole curve, must be temporarily higher than that already resulting in 95% capture of PAR. This is because otherwise it would be impossible to maintain optimal LAI given that senescence proceeds continuously and the potential of the green area to produce photoassimilates for grain fill decreases, resulting in incomplete realization of yield potential. This figure is, however, drawn to emphasize the crucial points of growth and development where the assimilate-producing source area could be further modified, i.e., the period of incomplete ground cover at the beginning and end of the growing season minimized.

Using modeling, Goudriaan and van Laar (1978) showed that differences in daily photosynthesis associated with following differences in leaf angle were relatively small. However, 10 years later the necessity to consider the effects of leaf-angle distribution in radiation models for canopy photosynthesis and energy exchange was pointed out and use of three leaf-angle classes (0–30°, 30–60° and 60–90°) was suggested (Goudriaan 1988). The importance of flag leaf angle, in particular, is likely to be connected to LAI. Austin (1994) postulated that wheat erectophile canopies would have higher rates of photosynthesis than planophile ones when LAI and leaf area density are high. Thus, leaf orientation and arrangement are likely to be crucial especially close to heading when peak LAI occurs rather than at early and late growth stages when LAI is low. Also, Goudriaan (1988) found that the influence of leaf angle disappeared under very low LAI. The importance of leaf angle varies depending on latitude. In addition, optimal leaf orientation varies also within one environment: for any particular position of the sun there is one optimal arrangement of leaves (Duncan 1971).

The actual requirement for such a persistent green area at the postanthesis period, as shown in Fig. 3, can be exceeded if stem and additional reserves compensate for the reduced LAI at late grain filling and, hence, contribute to grain growth. Such reserves comprise excess products of photosynthesis present in the plant prior to active growth of the grain endosperm (Bidinger et al. 1977; Blum et al. 1991). Their contribution to grain filling is estimated by comparing genotypes or altering the ratio of source to sink by using, e.g., defoliation, floret emasculation, and chemical desiccation techniques (Blum et al. 1983a,b; Peterson 1988; Hossain et al. 1990). The contribution of stem reserves to yield formation of oat has not, however, been studied as intensively as in wheat and barley. In the experiments of Peterson et al. (1975) 15–24% of carbohydrates used for grain growth in oat resulted from mobilized nonstructural carbohydrates in vegetative tissue. Mobilized carbohydrates in the flag leaf, all lower leaves, and the stem had equal contribution to grain fill.

In wheat, mobilization of plant reserves, relative to transient photosynthesis, was found to be especially important when stresses such as high temperature, which induces a heat-shock response, and drought occurred during grain filling (Blum et al. 1983a, 1989, 1991). Differences between cultivars in stem reserves mobilization were associated with differences in ability to sustain kernel growth in the absence of photosynthesis (Blum et al. 1983a; Hossain et al. 1990). However, losses of nitrogen from stems were not related to grain yield response to chemical desiccation (Hossain et al. 1990).

Stem reserves have also some yield-enhancing potential in non-stressed environments (Blum et al. 1991). For example, in the cool growing season of northern Europe late abortion of oat resulting in husks without groat at maturity could be avoided with an adequate continuous flow of photoassimilates for grain filling that follows mobilization of reserved carbohydrates. Stem reserves built up with excess carbohydrate could also decrease spikelet and floret death that occur prior to heading (Peltonen-Sainio and Peltonen 1995), because intensity of abortion is likely to be connected to severity of competition between several active, demanding sinks. This question of competition for photoassimilates between reproductive and vegetative structures is discussed further in the following sections. In general, achievement of high stem reserves as a breeding goal, especially under humid growing conditions, requires a better understanding of the benefits and costs of storage and mobilization of temporarily sequestered carbohydrates (Evans 1994). Furthermore, it is not yet clear if the need for stem reserves and the storage capacity of the straw (Blum et al. 1989) have decreased in modern short-strawed cultivars. It is possible that the stem of the dwarf lines is a weaker competitor for assimilate and the available supply is therefore for the requirements of generative organs. The dwarf lines are, however, characterized by higher tillering (Mäkelä et al. 1996) which possibly causes additional competition for photoassimilates. Yield may be reduced, however, not only by soil moisture and high temperature stress but also by stresses at critical points of oat growth and development such as insufficient nutrition, invasion of insect pests, and infection caused by plant pathogens, even when they occur only briefly. In such situations, stem reserves could ensure continuous growth with a minimum amount of abortion (Evans 1994).

Evans (1994) postulated that expanded duration of photosynthetic activity is typical for high-yielding cultivars of several crops including wheat, rice, maize, soybean, and cowpea. Furthermore, slower decline in photosynthetic activity has clearly contributed more to yield improvements than increase in the maximum CO_2-exchange rate (CER) (Evans 1994). Lynch and Frey (1993) indicated that this was also the case in oat cultivars in the midwestern USA. In addition, more recently developed cultivars had a slower rate of postanthesis leaf senescence. Similarly, under northern growing conditions delayed senescence, which is associated with rapid collapse of photosynthetic systems, is an attractive trait. Experiments comparing old and modern cultivars revealed that, as the flag leaf of the landrace cultivar emerged, its assimilation rate was higher than that of the modern cultivars (Peltonen-Sainio 1993). The high CER of the landrace cultivar possibly resulted from a smaller flag leaf. Such superiority in CER disappeared, however, during the very early part of the grain-filling period. This was possibly due to limited capacity of the panicle to utilize high amounts of photoassimilates (sink limitation), as the landrace cultivar had the smallest grain number of the cultivars screened (Peltonen-Sainio 1993).

Regarding the goal of increased canopy CER Day and Chalabi (1988), using a simple mathematical model, showed that a 10% increase in maximum photosynthetic rate (P_{max}) corresponded to a mere 2.5% increase in canopy photosynthesis. Even this marginal benefit would cease to exist if the increased P_{max} were to be associated with a decreased leaf area index following smaller leaf size, as has been observed in several crop species (Bhagsari and Brown 1986). This example indicates that the influences following manipulation of a particular physiological process are always dependent on many other processes interacting with each other. This interaction involves both immediate physiological responses and growth response during the growing season. Also, breeding for different subcomponents of crop photosynthetic CO_2 fixation has not been successful (Gifford et al. 1984). It could be stated that biochemical processes including photosynthesis and respiration have such a long evolutionary history that natural selection through competition has probably resulted in populations characterized by optimal photosynthetic efficiency.

Senescence processes are controlled in particular by the ratio between growth promoters (cytokinins, gibberellins, and auxins) and growth inhibitors (ABA and ethylene). Cytokinins and ABA are especially important, the former retarding aging processes and the latter causing stomata to close and chlorophyll to break down, thereby reducing photosynthetic potential (Goodwin and Mercer 1990). Zhi-Yi et al. (1988), however, found ABA to exercise synergistic control with cytokinins over these processes in oat exposed to a dark period. Reduced photosynthetic activity of the canopy during the postanthesis period is still evidently a limiting link in yield formation of modern cultivars and should be further improved in oat (Peltonen-Sainio 1993). This is because photosynthetic activity of green tissue frequently decreases much earlier than realization of maximum grain size (Evans 1994).

Nitrogen fertilizer has a well-demonstrated effect on LAI and LAD and consequently on productivity in oat (e.g., Peltonen-Sainio et al. 1993a, 1997). High rates of inorganic nitrogen fertilizer accelerated expansion of the leaf area, especially prior to stem elongation, and resulted in high preanthesis LAD due to enhanced tiller growth (Peltonen-Sainio et al. 1997). Use of a high nitrogen fertilizer application rate and green manuring also resulted in high postanthesis LAD. Such changes enabled higher rates of grain growth (Peltonen-Sainio et al. 1997). Effects of nitrogen fertilizer are likely to be controlled through increased production and improved translocation of cytokinins from root caps, an important site of cytokinin synthesis, to aboveground organs (Marschner 1986). Also gibberellic acid (GA), synthesized in the shoot apex and the expanding leaves, is involved but its effect is presumably indirect. When the supply of nitrogen is insufficient, cytokinin export from the roots is insufficient to maintain a high growth rate of the GA biosynthesis sites (Marschner 1986). Use of foliar application of a synthetic cytokinin, benzylaminopurine, as a management practice did not, however, prolong maturation, whereas nitrogen fertilizer applications increased the chlorophyll content of the flag leaf and delayed senescence (Peltonen-Sainio 1996). Sufficient available water in the soil also enables continuous translocation of cytokinins to leaves and other photosynthetic tissue, thereby maintaining a cytokinin–ABA ratio favorable for photosynthesis. This is emphasized in oat by the fact that its water-use efficiency (WUE) was lower than in many other crops (Ehlers 1989).

4 Vegetative Organs as Competing Sinks During Formation and Realization of Yield Potential

The life cycle of plants involves a catenary series of competing processes. The competition for photosynthate between plant organs of economic importance and the remaining vegetative structures, which not only compete but may also support floret set and grain fill, is especially important. Therefore, the role of vegetative nonyield organs varies from one growth stage to another. The major competing alternative sinks for photosynthate during the period when the majority of spikelets and florets are initiated in the main shoot apex (yield potential formed) are root growth, stem elongation, and tillering. Vegetative plant organs also have an important role in nitrogen dynamics. Peterson et al. (1975) showed that groat protein concentration correlated with the ratio of nitrogen to dry matter mobilized from the aboveground vegetative plant organs.

4.1 Root Growth and Function: The Hidden Component of Crop Physiology

The major part of the dry matter accumulated in the roots of small-grain cereals is produced during preanthesis (van Keulen and Seligman 1987). Therefore, it is likely that root growth competes more with formation of vegetative plant organs and yield potential than with grain filling. Varietal differences in structure and function of the root system may contribute more to the differences in yield than has been recognized, due to direct effects on water and nutrient uptake and their consequent influence on the ability to accumulate aboveground biomass. Also important is the maintenance of an appropriate balance between above- and belowground biomasses throughout the growth period. As compared to other crops, water-use efficiency of oat is low (Ehlers 1989), which emphasizes the role of root–shoot synchronization in yield formation. Varietal differences in root characteristics are very apparent when stresses are evident including drought and inadequate nitrogen. Léon and Schwarz (1992) found that, under such conditions, genotypic rank orders for grain yield and root characteristics were associated in oat, whereas differences in root performance of oat in nonstress environments were not associated with differences in biomass and grain yield (Schwarz et al. 1991; Léon and Schwarz 1992).

In addition to the obvious positive effects on water and nutrient uptake, a highly branched root system also provides better conditions for aboveground growth through improved hormonal control (Marschner 1986). The endogenous cytokinins synthesized in the numerous root caps play an important role in maintaining active photosynthesis, retarding senescence processes, and enhancing tillering by releasing tiller buds from apical dominance (Marschner

1986). Therefore, early and vigorous root growth evidently provides better possibilities for a high growth rate of aboveground biomass. Experiments of Mac Key (1988) with oat indicated that aboveground dry matter was positively correlated with root dry matter.

The control of root and shoot growth is evidently reciprocal. For example, inadequate nitrogen supply depressed shoot growth by 20–25% in oat and simultaneously stimulated growth of seminal and adventitious roots by 9–13% and 25–39%, respectively (Górny 1993). As another example, root signals control leaf expansion when plants are exposed to drought. In the experiments of Passioura (1988) roots of wheat seedlings sensed alterations in the water potential and hardiness of the soil. As the soil dried, roots sent signals to the leaves and by this means controlled leaf expansion independently of whether or not the leaves were highly turgid (Munns 1987; Passioura 1988).

Primitive genotypes and landrace populations characterized by low yield potential are often suggested to be better adapted to diverse environments. Peltonen-Sainio et al. (1993b) using traditional (Finlay and Wilkinson 1963; Eberhart and Russell 1966) and novel stability analyses (e.g., Kempton 1984) indicated that high grain yield in oat was associated with unstable yield performance under northern growing conditions. Differences in the root-shoot ratio between modern high-yielding and old oat cultivars may partly explain this finding. During the course of oat breeding, the root system has become concentrated in the uppermost layers of the soil profile. Root and straw length of a modern oat cultivar were up to 30 and 40% shorter, respectively, when compared with those of older cultivars (Mac Key 1988). Furthermore, the bulk of the root mass (53%) in an old cultivar occurred below 60 cm, whereas in a modern short-strawed cultivar only 6% of the root mass penetrated to such a depth and 42% was formed in the top 20 cm of soil (Mac Key 1988). Górny (1993) found no clear trend that indicated that old and modern oat cultivars differed in root growth in a low-nitrogen environment, even though improved nutrient-use efficiency has been an attractive goal for plant breeders. Evidently plasticity in root response to nitrogen supply, recorded in certain oat cultivars, is a valuable adaptive trait (Górny 1993).

4.2 Tillers: Waste or Economic Reservoirs of Assimilates?

Tillering is an important adaptive characteristic that considerably increases the morphological plasticity of a cereal plant. The role of tillers in yield formation and their contribution to grain yield in oat are genotype- and environment-dependent and can be manipulated markedly through crop management. Oat cultivars are, however, often unable to produce panicle-bearing tillers under long-day conditions, and high seeding rates are used, e.g., in Finland, to prevent postanthesis tillering which would result in uneven ripening of oat stands in the short growing season (Peltonen-Sainio and Järvinen 1995). The dwarf oat cultivar with the *Dw6* dwarfing gene, however, tillered more than the conventional height oat cultivars under northern growing conditions, but this increased number of tillers was unable to compensate for yield reduction resulting from low yield potential of the main shoot (Mäkelä et al. 1996).

Harrison and Kaufman (1980) showed that, in oat, panicle emergence from the leaf sheath induces tillering by releasing lateral tiller buds from apical dominance. During release, the tiller bud swells and then elongates at a slow rate (Harrison and Kaufman 1982). Contrary to the classical theory that the proposed apical dominance is controlled solely by auxin synthesized in the main shoot apex and transported to the tiller bud region, tiller bud release and tiller growth are regulated by several interacting hormones – mainly cytokinins and auxins with stimulatory and inhibitory effects, respectively. Fluctuations in relative amounts of cytokinins and auxins are due to transport and de novo synthesis (Harrison and Kaufman 1984). Results of experiments by Harrison and Kaufman (1980) on oat showed that hormonal control of apical dominance was similar in cereals to that earlier established for dicotyledonous crops. Furthermore, ABA is likely to have synergistic effects with auxins, and gibberellic acid (GA_3) with cytokinins (Harrison and Kaufman 1980). GA_3 is, however, only of secondary importance because, in contrast to cytokinins, when alone it is unable to promote tiller bud release from apical dominance. ABA is likely to inhibit tillering by preventing transport of cytokinin, but not of auxin, to the tiller bud (Harrison and Kaufman 1984). Furthermore, ethylene seems to stimulate tiller swelling during the onset of tiller release from apical dominance and may also promote tiller elongation in the presence of cytokinins (Harrison and Kaufman 1982). The mechanism of inhibition of tiller bud release induced by auxins, ABA, and ethylene is likely to be prevention of cytokinin transport and promotion of its catabolism (Harrison and Kaufman 1984).

Evans (1994) postulated that attempts of cereal breeders to produce cultivars with a low proportion of sterile tillers have resulted in assimilate savings that are likely to be invested in additional florets and spikelets as well as in grain mass. In the case of oat, this would clearly be an indication that production of nonpanicle-bearing tillers represents a waste of assimilates at the expense of economic plant organs. In oat cultivars adapted to northern growing conditions, tiller death coincided with floret abortion, indicating a lack of photosynthate for maintenance of multiple sinks simultaneously (Peltonen-Sainio and Peltonen 1995). Results of the experiments of Lauer and Simmons (1988) indicated, however, that tillers of barley paid the main shoot back with interest for the assimilates used for their growth prior to autotrophic growth and maintenance.

Crop management is able to affect tillering by influencing fluctuation of hormone balance in substrates. In a similar manner to their effect of delaying senescence processes, use of sufficient nitrogen fertilizer and providing adequate water increase tillering through stimulation of cytokinin synthesis in roots and continuous transport to the aboveground plant parts (Marschner

1986). Contrary to the findings in spring barley (Ma and Smith 1992), application of CCC – an antigibberellin – did not result in an increased number of panicles per m^2 and, furthermore, effects on grain yield were inconsistent (Leitch and Hayes 1990). According to Gendy and Höfner (1989), CCC used in mixture with dichloro-isobutyric acid (DCiB), however, reduced in particular upper and middle internodes of the straw, prevented lodging, and increased grain yield through formation of more panicles per m^2. Studies with barley (Ma and Smith 1992) have indicated that CCC treatment first inhibited tillering but that later the growth of the main shoot was retarded to such a degree that it enabled tiller growth. At high latitudes foliar application of CCC and ethephon tended to increase the proportion of LAD produced by tillers at the expense of grain yield in oat, which was attributed to lower postanthesis precipitation (Peltonen-Sainio 1997a).

4.3 Stem Elongation and Effects of Plant Height on Assimilate Competition

Shortening of oat straw and the consequent increase in HI has been one of the main reasons for genetically determined yield increases achieved with oat during this century (Lawes 1977; Wych and Stuthman 1983; Peltonen-Sainio 1990; Lynch and Frey 1993). Stem elongation occurring at the time of intensive floret set, together with competition caused by unproductive tillers, may be an unnecessary competitor for photoassimilates. Therefore, the introduction of short cultivars has possibly enabled the use of photosynthetic products for the set of spikelets and florets. Particularly at late preanthesis, reduced partitioning of photoassimilates to stem elongation in short stature cultivars may result in improved ability to reach a high yield potential, i.e., to maintain a high floret number through reduced abortion. Peltonen-Sainio and Peltonen (1995) indicated that in the oat panicle 75 to 90% of florets (both fertile and incompletely developed florets included) aborted, indicating an inability to realize a high yield potential. Within a spikelet, florets other than primary and secondary ones were aborted, in particular. Moreover, the proportion of aborted florets was higher at low application rates of nitrogen fertilizer even though the number of initiated florets was smaller. Whether floret abortion has been reduced by breeding for higher grain number per panicle is not certain. It could be hypothesized that introduction of oat cultivars with the dwarfing gene *Dw6* (Smart 1992) would be beneficial for building up yield potential, as dwarfing may result in considerably weakened competition for assimilates between straw and florets. This gene has pleiotropic effects, however. The panicle of dwarf cultivars has fewer florets and grains when grown at high latitudes, being a weaker competitor for assimilates per se and having a low panicle-filling rate and grain yield (Mäkelä et al. 1996).

In order to benefit from dwarf cultivars – especially through prevention of lodging – agricultural management must be fine-tuned by paying attention to the considerable differences in the dwarf canopy and plant stand structure compared with conventional cultivars, and by optimizing performance of LAI, tillering, biomass accumulation, and dry-matter partitioning for specific growing conditions (Anderson and McLean 1989; Mäkelä et al. 1994, 1996; Peltonen-Sainio and Järvinen 1995). In contrast to the possibility of a negative impact of straw length on grain yield due to increased competition, Payne et al. (1986) indicated that selection for high oat grain yield resulted in a longer vegetative growth period associated with taller plants, greater phytomass, and reduced HI. The role of stem reserves, and differences between short- and long-strawed cultivars in ability to build up these reserves, have not been investigated in depth for oat and may be important, especially when stresses occur.

Another possibility for lessening competition between stem and generative plant structures is to use plant growth regulators to reduce plant height. The results of Gendy and Höfner (1989) support this idea, because application of CCC with DCiB resulted in reduced straw length and an increased number of grains and panicles per m^2. Use of plant growth regulators does not only inhibit stem elongation but may also result in differences in stem characteristics. Clark and Fedak (1977) and Knapp et al. (1987) did not establish that CCC treatment had any effect on concentrations of structural carbohydrates and lignin of cereal culms, and in studies by Gendy and Höfner (1989), CCC and DCiB treatment did not change stem diameter. Knapp et al. (1987), however, indicated that concentrations of water-soluble carbohydrates increased following ethephon and CCC treatments. This may be an important indication of reduced partitioning of photoassimilates to the stem growth, and by this means production of greater stem reserves.

4.4 Grain Filling as a Final Process Affecting Realization of Yield Potential

4.4.1 *Increasing Grain-Filling Rate*

Large panicle weight (total weight of grains per panicle) is an indication of high yielding ability in oat (Brinkman and Frey 1978; Peltonen-Sainio 1991b; Salman and Brinkman 1992). However, in addition to being dependent on postanthesis growing conditions, the rate of grain and panicle filling is also dependent on the internal structure of the panicle and, thus, on the ability of the translocation system to import essential compounds for the synthesis of starch, protein, lipid, cell wall, and other components of the endosperm and embryo (Peterson 1988). Results from the experiments of Lesar and Peterson (1981) and Jenner (1984) have suggested that the ability of the translocation system to transport sucrose for grain fill, rather than sucrose availability, limits grain growth. Whether this operates at the level of sucrose loading into the phloem for export, transport, or unloading into the growing sink is, however, uncertain. Housley and Peterson (1982) and Peterson et al. (1982) found that a large number of spikelets per panicle and total weight of grains per panicle – but not a large number of grains per spikelet – correlated positively with vascular bundle traits in the peduncle (bundle number, bundle area, sieve tube number, and sieve tube area). Therefore, a panicle characterized by a large number of spikelets including only primary and secondary grains is likely to have a high filling rate. This is especially important when exceptionally intensive grain-filling processes are required to enable harvest before unfavorable growing conditions ensue. In fact, Peltonen-Sainio (1992) indicated that improved productivity in oat has been achieved under Finnish growing conditions, not through prolonged duration of green area but, rather, through development of cultivars with a shorter grain-filling period and an increased panicle-filling rate (increased by 30–35%). This was possible by increasing the number of grains per panicle (Peltonen-Sainio 1990), but not per spikelet. Takeda et al. (1987) indicated that oat grain yield was linearly and positively associated with grain number when grain weight was less than 23 mg per single grain but independent of number above that weight.

This concept is supported by the finding that one of the main reasons for smaller groat yield in naked oat cultivars, compared with husked ones after dehulling, is that naked cultivars produce a smaller number of spikelets per panicle associated with a larger number of grains per spikelet and consequently a lower grain-filling rate (Peltonen-Sainio 1994b, 1997b). A low single groat weight of naked oat cultivars was not, however, a result of insufficient main shoot LAD but tertiary grain was more often present in the spikelets of naked than of conventional cultivars and was invariably small (Peltonen-Sainio 1997b). Adjustment of nitrogen fertilizer application rate and seeding rate did not result in elimination of this undesirable trait of naked oat (Peltonen-Sainio 1997b). Tibelius and Klinck (1987) indicated that, even though secondary grain had greater potential for growth following removal or sterilization of the primary floret, weights of secondary grains in treated spikelets were never as great as those of primary grains in intact spikelets. In addition to reduced availability of assimilates due to limited capacity for translocation, the superiority of primary over secondary grain may be due to its earlier differentiation (Tibelius and Klinck 1987). The importance of capacity for carbohydrate transport is further emphasized by the finding that secondary grains at the top of a panicle were rarely as heavy as primary grains at the base, even though they reached anthesis earlier (Tibelius and Klinck 1987).

In addition to genotype, crop management offers an alternative to manipulating the internal structure of the oat panicle (Peltonen-Sainio et al. 1993a; Peltonen-Sainio and Järvinen 1995; Mäkelä et al. 1996; Peltonen-Sainio 1996, 1997b; Peltonen and Peltonen-Sainio 1997) and, by this means, indirectly to modifying the ability of the translocation system to transport essential compounds for grain growth. For example, the nitrogen fertilizer application rate, the seeding rate, and foliarly applied synthetic cytokinin have a marked impact on the number of grains and/or spikelets per panicle, which often correlates positively with the panicle filling rate (Peltonen-Sainio and Järvinen 1995; Peltonen-Sainio 1996).

Selection for high grain weight resulted in improved grain yield in Iowa (Pixley and Frey 1992), whereas under northern growing conditions no relationship was found between grain weight and yield (Peltonen-Sainio 1991b). Grain composition may have some impact on the accumulation of dry matter into grain. Oat grain is characterized by a high protein concentration and an exceptionally high oil concentration compared with barley and wheat. Penning de Vries et al. (1974) showed that the substrate required for synthesis of oil is more than that for synthesis of an equal weight of protein, while substrate use for carbohydrate synthesis is much less.

Schipper and Frey (1991) found, for oat lines rich in groat oil, that the substrate used for increased oil synthesis was not due to reduced groat protein but that such lines were characterized by differences in other physiological traits, including increased vigor during early growth, earlier heading, decreased rate of biomass accumulation during later growth phases, and smaller grain weight. Therefore, increased photosynthate production per se is suggested to be the primary means by which such cultivars achieved both high grain yield and high grain oil concentration (Forsberg et al. 1974; Gullord 1980; Thro and Frey 1984; Schipper and Frey 1991). Furthermore, McFerson and Frey (1992) suggested that further increases in groat protein yield may result from improved efficiency in nitrogen uptake and translocation. Nitrogen uptake was, however, independent of nitrogen harvest index (NHI) which indicates efficiency of nitrogen partitioning to the grain (Fawcett and Frey 1982), and NHI of oat declined as soil N increased (Fawcett and Frey 1982; Rattunde and Frey 1986; Kairudin and Frey 1988).

4.4.2
Increasing Harvest Index?

HI is the most frequently used indicator to show the success of the plant to partition dry mater into grain growth rather than nonyield structures (roots often excluded). Marked increases in HI have resulted from breeding for short-stature cereals (Sims 1963; Hanson et al. 1985; Gent and Kiyomoto 1989), but Evans (1994) pointed out that it remains unclear how the savings on stem growth are used to promote grain filling. Salman and Brinkman (1992) found that the correlation between HI and grain yield was positive and no trend of decline in grain yield was observed in their oat material, for which the highest HI was close to 54%. A similar positive association was observed by McMullan et al. (1988) who recorded an HI of about 46% at its highest. However, in Iowa the association between grain yield and HI was curvilinear and the maximum grain yield occurred when HI was only about 45% (Takeda et al. 1987). Similarly, in northern European oat material, optimum HI was likely to be close to 55%, and an HI as high as 60% recorded for some oat lines was associated with a slow accumulation rate of vegetative phytomass, which was possibly unable to satisfy the increased demands of grains for photosynthate as the postanthesis period proceeded (Peltonen-Sainio 1991b). Therefore, further increases in productivity of oat might be realized by introducing cultivars with increased total aboveground biomass, while maintaining HI at its current optimum level (Peltonen-Sainio 1991b,c).

Bregitzer et al. (1987) showed that selection resulting in a 14% increase in grain yield increased total aboveground biomass by 16%, i.e., HI was not actually increased but there was a proportional increase in the weights of both vegetative and generative plant structures. Furthermore, at high latitudes selection for higher grain yield in oat has not resulted solely from increased partitioning of aboveground dry matter into harvestable yield; also, vegetative phytomass per plant was higher in modern cultivars, even though old cultivars had up to 20% longer straw (Peltonen-Sainio 1991c). Further selection for high HI in oat has resulted in low grain yield (Takeda and Frey 1987) due to reduced vigor. However, intensive selection for growth rate index or unit straw weight resulted in high biological yield associated with low HI (Takeda and Frey 1987), i.e., dry-matter accumulation was enhanced for nonyield components of the plant stand at the expense of grain yield. Even though HI has become, since its introduction by Sims (1963) in the 1960s, one of the most important indicators of the success in synchronization between sources and economic sinks, its fundamental essence is still incompletely established when the recorded variations in HI associated with different growing conditions are considered (Evans 1994). For example, only in an unfavorable growing season is there an effect of the superior capacity of the modern oat cultivars to partition the assimilates of photosynthesis to the sinks of economic importance (Peltonen-Sainio 1992).

5
The Need for Description of Oat Ideotypes?

The development and growth of oat in relation to source–sink interaction and productivity have been shown to be a complex subject. The author will refrain from describing any productive oat ideotype because the characteristic composition of such oat models varies from one environment to another, e.g., different oat models are needed for low and high latitudes, for low- and high-input environments, and for growing conditions characterized by different stresses (Shabana et al. 1980; Atlin and Frey 1990; Peltonen-Sainio 1992; Young and Frey 1994). In fact, this is in agreement with the original definition of the ideotype concept developed by Donald (1968), according to whom ideotype is a biological model expected to perform or behave in a predictable manner within a defined environment. Figure 4 shows the most conspicuous differences in an oat plant when grown under favorable conditions (left-hand side) versus drought stressed conditions (right-hand side). Some of the explanatory changes in hormone balance are also shown. This figure clearly indicates the need for having separate crop models when aiming to improve

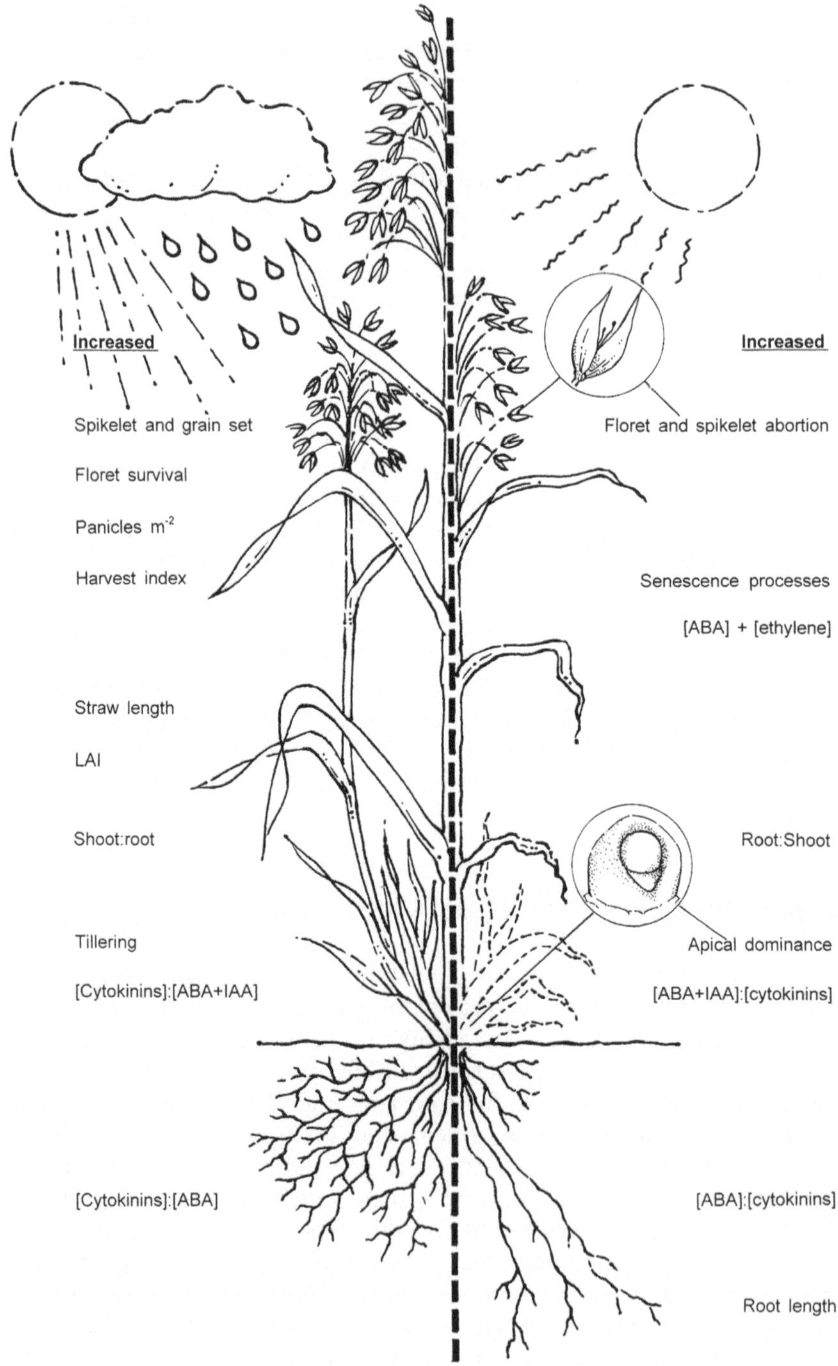

Fig. 4. Appearance of an oat plant when grown under favorable (*left*) and drought-stressed environments (*right*; sides separated by *vertical broken line*) with special reference to the changes in components and hormone balance contributing to yield formation

productivity in such contrasting environments; different morphological and, especially, physiological characteristics of the plant stand and individual plant must be emphasized, depending on their role in accumulation and the partitioning of dry matter between vegetative and generative sinks in that particular situation.

Ideotypes are traditionally used by plant breeders when superior types are searched for and selected in order to improve productivity further, but there is no reason to disparage their use when new techniques are developed that eliminate even temporary disharmony between sources and sinks. In light of this review of the recent literature devoted to the crop physiology of oat, it is necessary to mention that, even under certain very marginal growing conditions, a series of different stages of an ideotype are needed and not only one mature plant model.

Acknowledgement. Olli Parkkari provided figures and his work is gratefully acknowledged. Fredrik Stendahl kindly permitted reproduction of the apical developmental scale (Fig. 2) that originally appeared in Åfors et al. (1988).

References

Åfors M, Ohlander L, Stendahl F (1988) Stråsädens utveckling I. En litteraturstudie och beskrivning av en skala för bestämning av stråsädens ax- respektive vippanlag. Sveriges Lantbruksuniversitet, Institutionen för växtodlingslära, Uppsala, p 75

Andersen S (1954) Effect of 2,4-D on ear development in barley. Physiol Plant 7:517–522

Anderson WK, McLean R (1989) Increased responsiveness of short oat cultivars to early sowing, nitrogen fertilizer and seed rate. Aust J Agric Res 40:729–744

Anon ymous (1993) FAO production year book. FAO, Rome

Atlin GN, Frey KJ (1990) Selecting oat lines for yield in low-productivity environments. Crop Sci 30:556–561

Austin RB (1994) Plant breeding opportunities. In: Boote KJ, Bennett JM, Sinclair TR, Paulsen GM (eds) Physiology and determination of crop yield. ASA, CSSA, SSSA, Madison, Wisconsin, pp 567–586

Bhagsari AS, Brown RH (1986) Leaf photosynthesis and its correlation with leaf area. Crop Sci 26:127–132

Bidinger FR, Musgrave RB, Fischer RA (1977) Contribution of stored pre-anthesis assimilate to grain yield in wheat and barley. Nature 270:431–433

Bleken MA, Skjelvåg AO (1986) The phenological development of oat (*Avena sativa* L.) cultivars as affected by temperature and photoperiod. Acta Agric Scand 36:353–365

Blum A, Poiarkova H, Golan G, Mayer J (1983a) Chemical desiccation of wheat plants as a simulator of post-anthesis stress. I. Effects on translocation and kernel growth. Field Crops Res 6:51–58

Blum A, Mayer J, Golan G (1983b) Chemical desiccation of wheat plants as a simulator of post-anthesis stress. II. Relations to drought stress. Field Crops Res 6:149–155

Blum A, Golan G, Mayer J, Sinmena B, Shpileer L, Burra J (1989) The drought response of landraces of wheat from the northern Negev Desert in Israel. Euphytica 43:87–96

Blum A, Shpiler L, Golan G, Mayer J, Sinmena B (1991) Mass selection of wheat for grain filling without transient photosynthesis. Euphytica 54:111–116

Bonnett OT (1961) The oat plant: its histology and development. Univ Illinois Agric Exp Stn, Bull 672, Urbana, Illinois, p 112

Bregitzer PP, Stuthman DD, McGraw RL, Payne TS (1987) Morphological changes associated with three cycles of recurrent selection for grain yield improvement in oat. Crop Sci 27:165–168

Brinkman MA, Frey KJ (1978) Flag leaf physiological analysis of oat isolines that differ in grain yield from their recurrent parents. Crop Sci 18:69–73

Chau NM, Bhargava SC (1993) Physiological basis of higher productivity in rice. Indian J Plant Physiol 36:215–219

Clark RV, Fedak G (1977) Effects of chlormequat on plant height, disease development, and chemical constituents of cultivars of barley, oats, and wheat. Can J Plant Sci 57:31–36

Coffman FA, Frey KJ (1961) Influence of climate and physiologic factors on growth in oats. In: Coffman FA (ed) Oats and oat improvement. Agron Monogr 8, ASA, Madison, Wisconsin, pp 420–456

Colville-Baltenberger DC, Frey KJ (1987) Genotypic variability in response of oat to delayed sowing. Agron J 79:813–816

Day W, Chalabi ZS (1988) Use of models to investigate the link between the modification of photosynthetic characteristics and improved crop yields. Plant Physiol Biochem 26:511–517

Donald CM (1968) The breeding of crop ideotypes. Euphytica 17:386–403

Duncan WG (1971) Leaf angles, leaf area, and canopy photosynthesis. Crop Sci 11:482–485

Eberhart SA, Russell WA (1966) Stability parameters for comparing varieties. Crop Sci 6:36–40

Ehlers W (1989) Transpiration efficiency of oat. Agron J 81:810–817

Evans LT (1994) Crop physiology: prospects for the retrospective science. In: Boote KJ, Bennett JM, Sinclair TR, Paulsen GM (eds) Physiology and determination of crop yield. ASA, CSSA, SSSA, Madison, Wisconsin, pp 19–35

Fawcett JA, Frey KJ (1982) Nitrogen harvest index variation in *Avena sativa* and *A. sterilis*. Proc lowa Acad Sci 89:155–159

Finlay KW, Wilkinson GN (1963) The analysis of adaptation in a plant-breeding programme. Aust J Agric Res 14:742–754

Forsberg RA, Youngs VL, Shands HL (1974) Correlations among chemical and agronomic characteristics in certain oat cultivars and selections. Crop Sci 14:221–224

Gardner FP, Valle R, McCloud DE (1990) Yield characteristics of ancient races of maize compared to a modern hybrid. Agron J 82:864–868

Gendy A, Höfner W (1989) Stalk shortening of oat (*Avena sativa* L.) by combined application of CCC, DCiB and ethephon. Angew Bot 63:103–110

Gent MPN, Kiyomoto RK (1989) Assimilation and distribution of photosynthate in winter wheat cultivars differing in harvest index. Crop Sci 29:120–125

Gifford RM, Thorne JH, Hitz WD, Giaquinta T (1984) Crop productivity and photoassimilate partitioning. Science 225:801–808

Górny AG (1993) Differences in root and shoot response to limited N-supply in oat and spring barley. J Agron Crop Sci 171:161–167

Goodwin TW, Mercer EI (1990) Introduction to plant biochemistry. Pergamon Press, Oxford, 677 pp

Goudriaan J (1988) The bare bones of leaf-angle distribution in radiation models for canopy photosynthesis and energy exchange. Agric For Meteorol 43:155–169

Goudriaan J, van Laar HH (1978) Calculations of daily totals of the gross CO_2 assimilation of leaf canopies. Neth J Agric Sci 26:373–382

Gullord M (1980) Oil and protein content and its relation to other characters in oats (*Avena* sp). Acta Agric Scand 30:216–218

Gupta SC, Frey KJ, Skrdla RK (1987) Selection for grain yield of oats via vegetative growth rates measured at anthesis and maturity. Euphytica 36:91–97

Hagsand E (1954) Hormonderivatens inverkan på stråsäd och lin efter besprutning vid olika tidpunkter. Kungl Lantbruksakademiens tidskrift, Årg 93, Almquist & Wiksells Boktryckeri AB, Uppsala, Sweden, pp 179–389

Hanson PR, Riggs TJ, Klose SJ, Austin RB (1985) High biomass genotypes in spring barley. J Agric Sci (Camb) 105:73–78

Harrison MA, Kaufman PB (1980) Hormonal regulation of lateral bud (tiller) release in oats (*Avena sativa* L.). Plant Physiol 66:1123–1127

Harrison MA, Kaufman PB (1982) Does ethylene play a role in the release of lateral buds (tillers) from apical dominance in oats? Plant Physiol 70:811–814

Harrison MA, Kaufman PB (1984) The role of hormone transport and metabolism in apical dominance in oats. Bot Gaz 145:293–297

Hipps LE, Aspar G, Kanemasu ET (1983) Assessing the interception of photosynthetically active radiation in winter wheat. Agric Meteorol 28:253–259

Hossain ABS, Sears RG, Cox TS, Paulsen GM (1990) Desiccation tolerance and its relationship to assimilate partitioning in winter wheat. Crop Sci 30:622–627

Housley TL, Peterson DM (1982) Oat stem vascular size in relation to kernel number and weight. I. Controlled environment. Crop Sci 22:259–263

Jefferies RA, MacKerron DKL (1993) Responses of potato genotypes to drought. II. Leaf area index, growth and yield. Ann Appl Biol 122:105–112

Jenner CF (1984) Control of the accumulation of starch and protein in cereal grains. In: Jeffcoat B, Hawkins AF, Stead AD (eds) Regulation of sources and sinks in crop plants, Proc Symp Br Plant Growth Regul Group, York, UK, pp 195–209

Jennings VM, Shibles RM (1968) Genotypic differences in photosynthetic contributions of plant parts to grain yield in oats. Crop Sci 8:173–175

Kairudin NMD, Frey KJ (1988) Soil N availability and nitrogen harvest index of oats. J Iowa Acad Sci 95:73–78

Kempton RA (1984) The use of biplots in interpreting variety by environment interactions. J Agric Sci (Camb) 103:123–135

Kirby EJM, Appleyard M (1984) Cereal development guide. Arable Unit, Warwickshire, UK, p 93

Knapp JS, Harms CL, Volenec JJ (1987) Growth regulator effects on wheat culm nonstructural and structural carbohydrates and lignin. Crop Sci 27:1201–1205

Landes A, Porter JR (1989) Comparison of scales used for categorising the development of wheat, barley, rye and oats. Ann Appl Biol 115:343–360

Langer RHM (1979) How grasses grow. The Inst Biol Studies in Biol no 34, Southampton, UK, p 66

Large EC (1954) Growth stages in cereals. Illustration of the Feekes scale. Plant Pathol 3:128–129

Lauer JG, Simmons SR (1988) Photoassimilate partitioning by tillers and individual tiller leaves in field-grown spring barley. Crop Sci 28:279–282

Lawes DA (1977) Yield improvement in spring oats. J Agric Sci (Camb) 89:751–757

Leitch MH, Hayes JD (1990) Effects of single and repeated applications of chlormequat on early crop development, lodging resistance and yield of winter oats. J Agric Sci (Camb) 115:11–14

Léon J, Schwarz K-U (1992) Description and application of a screening method to determine root morphology traits of cereal cultivars. J Agron Crop Sci 169:128–134

Lesar LE, Peterson DM (1981) Growth and composition of kernels developing on excised oat panicles in liquid culture. Crop Sci 21:741–747

Loubser JM, Cairns ALP (1989) Abnormalities of the growth point and ear of barley caused by 2,4-dichlorophenoxy acetic acid. South African J Plant Soil 6:103–107

Lowe KF, Bowdler TM (1988) Effects of height and frequency of defoliation on the productivity of irrigated cats (*Avena strigosa* cv. Saia) and perennial ryegrass (*Lolium perenne* cv. Kangaroo Valley), grown alone or with barrel medic (*Medicago truncatula* cv. Jemalong). Aust J Exp Agric 28:57–67

Lynch PJ, Frey KJ (1993) Genetic improvement in agronomic and physiological traits of oat since 1914. Crop Sci 33:984–988

Ma BL, Smith DL (1992) Modification of tiller productivity in spring barley by application of chlormequat or ethephon. Crop Sci 32:735–740

Mac Key J (1988) Shoot: root interrelations in oats. In: Mattsson B, Lyhagen R (eds) 3rd Int Oat Conf, Svalöv, Sweden, pp 340–344

Mäkelä P, Peltonen-Sainio P, Stuthman DD, Simmons SR (1994) Crop physiological evaluation of growing Minnesota-adapted dwarf oat in Finland. In: Borin M, Sattin M (eds) 3rd Cong of the European Soc for Agron, Colmar Cedex, France, pp 172–173

Mäkelä P, Väärälä L, Peltonen-Sainio P (1996) Agronomic comparison of Minnesota-adapted dwarf oat with semi-dwarf, intermediate, and tall oat lines adapted to northern growing conditions. Can J Plant Sci 76:727–734

Marschner H (1986) Mineral nutrition of higher plants. Academic Press, London, 674 pp

McFerson JK, Frey KJ (1992) Correlated response of selection for protein yield in oats after three cycles of recurrent selection. Plant Breed 108:149–161

McMullan PM, McVetty, PBE, Urquhart AA (1988) Dry matter and nitrogen accumulation and redistribution and their relationship to grain yield and grain protein in oats. Can J Plant Sci 68:983–993

Munns R (1987) Does leaf turgor determine leaf expansion rates in dry or saline soil? In: Cosgrove DJ, Knievel DP (eds) Physiology of cell expansion during plant growth, Am Soc Plant Physiologists, Rockville, pp 241–242

Murphy JP, Hoffman LA (1992) The origin, history, and production of oat. In: Marshall HG, Sorrells ME (eds) Oat science and technology. ASA, CSSA Publishers, Madison, Wisconsin, pp 1–28

Passioura JB (1988) Root signals control leaf expansion in wheat seedlings growing in drying soil. Aust J Plant Physiol 15:687–693

Payne TS, Stuthman DD, McGraw RL, Bregitzer PP (1986) Physiological changes associated with three cycles of recurrent selection for grain yield improvement in oats. Crop Sci 26:734–736

Peltonen J (1992a) Ear development stage used for timing supplemental nitrogen application to spring wheat. Crop Sci 32:1029–1033

Peltonen J (1992b) Developmental stages of the apex used as a basis for supplemental nitrogen application in spring wheat production. Univ Helsinki, Dept Plant Prod, Sect Crop Husb, publ 35, Yliopistopaino, Helsinki, p 51

Peltonen J, Peltonen-Sainio P (1997) Breaking uniculm growth habit of spring cereals at high latitudes by crop management. II. Tillering, grain yield and yield components. J Agron Crop Sci 178:87–95

Peltonen P (1990) Seasonal alterations in canopy architecture of oats (*Avena sativa* L.) and their relations to yield formation. Acta Agric Scand 40:221–230

Peltonen-Sainio P (1990) Genetic improvement in the structure of oat stands in northern growing conditions during this century. Plant Breed 104:340–345

Peltonen-Sainio P (1991a) Effect of moderate and severe drought stress on the pre-anthesis development and yield formation of oats. J Agric Sci Finl 63:379–389

Peltonen-Sainio P (1991b) Productive oat ideotype for northern growing conditions. Euphytica 54:27–32

Peltonen-Sainio P (1991c) High phytomass producing oats for cultivation in northern growing conditions. J Agron Crop Sci 166:90–95

Peltonen-Sainio P (1992) Description of a productive oat ideotype characterized by morpho-physiological traits associated with high grain yield. Univ Helsinki, Dept Plant Prod, Sect Crop Husb, publ 34, Yliopistopaino, Helsinki, p 47

Peltonen-Sainio P (1993) Contribution of enhanced growth rate and associated physiological changes to yield formation of oats. Field Crops Res 33:269–281

Peltonen-Sainio P (1994a) Growth duration and above-ground dry-matter partitioning in oats. Agric Sci Finl 3:195–198

Peltonen-Sainio P (1994b) Yield component differences between naked and conventional oat. Agron J 86:510–513

Peltonen-Sainio P (1996) Nitrogen fertilizer and foliar application of cytokinin affect spikelet and floret set and survival in oat. Field Crops Res 49:169–176

Peltonen-Sainio P (1997a) Leaf area duration of oat at high latitudes. J Agron Crop Sci 178:149–155

Peltonen-Sainio P (1997b) Groat yield and plant stand structure of naked and hulled oat under different nitrogen fertilizer and seeding rates. Agron J 89:140–147

Peltonen-Sainio P, Järvinen P (1995) Seeding rate effects on tillering, grain yield, and yield components of oat at high latitude. Field Crops Res 40:49–56

Peltonen-Sainio P, Pekkala T (1993) Numeric codes for developmental stages of oat apex in the growing conditions of southern Finland. Agric Sci Finl 2:329–336
Peltonen-Sainio P, Peltonen J (1995) Floret set and abortion in oat and wheat under high and low nitrogen regimes. Eur J Agron 4:253–262
Peltonen-Sainio P, Granqvist M, Säynäjärvi A (1993a) Yield formation in modern and old oat cultivars under high and low nitrogen regimes. J Agron Crop Sci 171:268–273
Peltonen-Sainio P, Moore K, Pehu E (1993b) Phenotypic stability of oats measured with different stability analyses. J Agric Sci (Camb) 121:13–19
Peltonen-Sainio P, Forsman K, Poutala P (1997) Crop management effects on pre- and post-anthesis changes in leaf area index and leaf area duration and their contribution to grain yield and yield components in spring cereals. J Agron Crop Sci 179:47–61
Penning de Vries FWT, Brunsting AHM, van Laar HH (1974) Products, requirements and efficiency of biosynthesis: a quantitative approach. J Theor Biol 45:339–377
Peterson DM (1988) Translocation and grain development in oats. In: Mattsson B, Lyhagen R (eds) 3rd Int Oat Conf, Svalöf, Sweden, pp 323–329
Peterson DM, Schrader LE, Cataldo DA, Youngs VL, Smith D (1975) Assimilation and remobilization of nitrogen and carbohydrates in oats, especially as related to groat protein concentration. Can J Plant Sci 55:19–28
Peterson DM, Housley TL, Luk TM (1982) Oat stem vascular size in relation to kernel number and weight. II. Field environment. Crop Sci 22:274–278
Petr J, Hradicka D (1993) The formation of biological yield in triticale. Cereal Res Communication 21:221–229
Pixley KV, Frey KJ (1992) Genetic interrelations among grain quality indicators and agronomic traits for oat. Euphytica 60:149–156
Rajalahti R, Peltonen J (1993) Mode of action of phenoxy and sulfonylurea herbicides and their implication in herbicide injuries of cereals. Univ Helsinki, Dept Plant Prod, Sect Crop Husb, publ 40, Yliopistopaino, Helsinki, Finland, p 22
Rattunde HF, Frey KJ (1986) Nitrogen harvest index in oats: its repeatability and association with adaptation. Crop Sci 26:606–610
Salman AA, Brinkman MA (1992) Association of pre- and post-heading growth traits with grain-yield in oats. Field Crops Res 28:211–221
Sarma BK, Sarma CM (1992) Association of some physiological determinants with seed yield in soybean (*Glycine max* (L.) Merill). Indian J Hill Farm 5:113–117
Schipper H, Frey KJ (1991) Growth analyses of oat lines with low and high groat-oil content. Euphytica 54:221–229
Schwarz K-U, Léon J, Geisler G (1991) Genetic variability in root development of spring barley (*H. vulgare* L.) and oats (*A. sativa* L.). In: McMichael BL, Persson H (eds) Plant roots and their environment. Elsevier, Amsterdam, pp 639–646
Shabana R, Bailey T, Frey KJ (1980) Production traits of oats selected under low, medium, and high productivity. Crop Sci 20:739–744
Sims HJ (1963) Changes in the hay production and the harvest index of Australian oat varieties. Aust J Exp Agric Anim Husb 3:198–202
Smart GF (1992) The effects of exogenous gibberellic acid (GA_3) on first leaf length of oat cultivars. Cereal Res Commun 20:151–154
Takeda K, Frey KJ (1977) Growth rate inheritance and associations with other traits in backcross populations of *Avena sativa* × *A. sterilis*. Euphytica 26:309–317
Takeda K, Frey KJ (1987) Improving grain yield in backcross populations from *Avena sativa* × *A. sterilis* matings by using independent culling for harvest index and vegetative growth index or unit straw weight. Theor Appl Genet 74:659–665
Takeda K, Frey KJ, Bloethe Helsel D (1979) Growth rate inheritance and associations with other traits and contributions of growth rate and harvest index to grain yield in oats (*Avena sativa* L.). Z Pflanzenzücht 82:237–249
Takeda K, Frey KL, Bailey TB (1980) Contribution of growth rate and harvest index to grain yield in F_9-derived lines of oats (*Avena sativa* L.). Can J Plant Sci 60:379–384

Takeda K, Frey KL, Bailey TB (1987) Relationships among traits in F_9-derived lines of oats. Iowa State J Res 62:313–327

Thorne GN, Pearman I, Day W, Todd AD (1988) Estimation of radiation interception by winter wheat from measurements of leaf area. J Agric Sci (Camb) 110:101–108

Thro AM, Frey KJ (1984) Relationship between groat-oil content and grain yield of oats (*Avena sativa* L.). Proc Iowa Acad Sci 91:85–86

Tibelius AC, Klinck HR (1987) Effects of artificial reduction in panicle size on weight of secondary seeds in oats (*Avena sativa* L.). Can J Plant Sci 67:621–628

van Keulen H, Seligman NG (1987) Simulation of water use, nitrogen nutrition and growth of a spring wheat crop. Pudoc, Wageningen, The Netherlands, 310 pp

While EM (1995) Structure and development of oats. Welch RW (ed) The oat crop, production and utilization. Chapman & Hall, London, pp 88–119

Wych RD, Stuthman DD (1983) Genetic improvement in Minnesota-adapted oat cultivars released since 1923. Crop Sci 23:879–881

Young C, Frey KJ (1994) Grain-yield characteristics of oat lines surviving uniform and shuttle selection strategies. Euphytica 76:63–71

Zadoks JC, Chang TT, Konzak CF (1974) A decimal code for the growth stages of cereals. Weed Res 14:415–421

Zhi-Yi T, Veierskov B, Park J, Thinmann KV (1988) Multiple actions of abscisic acid in senescence of oat leaves. Plant Growth Regul 7:213–226

Barley: Physiology of Yield

D.L. SMITH, M. DIJAK, P. BULMAN, B.L. MA, and C. HAMEL[1]

Physiology describes the operation of a plant at the biochemical, cell, tissue, organ and whole plant levels, recognizing that all of these are under the control of the plant genome. In the case of crop plants we are interested in how the combined operation of all plant's functions leads to the production of the harvestable component which, in cereal crops, is generally grain. Broadly speaking, the most important of the physiological activities in a crop plant with regard to yield are dry matter production, nutrient uptake, plant component partitioning and the processes of development (especially grain filling and the other aspects of yield component development). The developmental processes occur sequentially in time and involve progressive shifts in physiology. As barley (*Hordeum vulgare* L.) is a feed grain and a major source of malting grain for beer and spirits, aspects of physiology that affect the amount of grain produced and its quality (the amount of energy and protein) are of special interest.

1
Apex Development and Contribution to Yield

The components of barley grain yield include the number of spikes per unit area, the number of spikelets per spike and the mean grain weight. The number of spikelets per spike is determined by the duration and rate of apical development. The size of the spikes (or strength of sink) at anthesis substantially influences grain filling and final weight (Kirby 1977; Scott et al. 1983). Comparisons of grains varying in their position along a spike, and of lateral and centre grains at the same rachis node, have shown final weight to be strongly related to spikelet primordium width at the double ridge stage (Cottrell and Dale 1984), and to carpel size at anthesis (Scott et al. 1983). Therefore, apical development determines directly or indirectly the grain yield levels of barley (Ellis 1993).

[1] McGill University, Natural Resource Department, Macdonald Campus, 21, 111 Lakeshore Road, Ste. Anne de Bellevue, QC, Canada H9X 3V9

Crop Yield, Physiology and Processes
D. L. Smith, C. Hamel (Eds.)
© Springer-Verlag Berlin Heidelberg 1999

1.1 Inflorescence Development

1.1.1 *Structure of the Barley Spike*

The inflorescence of a barley plant is a spike (or ear) located at the tip of the culm. The axis of the spike, the rachis, is usually bilaterally symmetrical. The basal node of the rachis is referred to as a "collar". Typically, three spikelets arise at each node of the rachis, and these triplets alternate from side to side throughout its length. Each spikelet normally consists of one floret and two narrow glumes. Florets are composed of a lemma, a palea, a carpel and three stamens. Since the lemma and palea remain attached to the caryopsis after threshing in most cultivars, barley is referred to as a coarse grain, and the caryopsis is often called a "seed" or a "grain", but not a kernal. The central spikelet at each node is the largest with laterals ranging from fully fertile to completely sterile, in six- or two-rowed barleys, respectively (Reid and Wiebe 1979; Stoskopf 1985). The six-row phenotype is controlled by one pair of recessive genes, and there is evidence that six-row barley is derived from a two-row ancestor (Harlan 1975).

1.1.2 *Inflorescence Differentiation*

Barley inflorescence development consists of a sequential formation of components. The first discernible structure is the collar. After several leaves have been initiated, the apex lengthens, and double ridges develop and initiate the formation of each reproductive organ. Prior to heading, aspects of inflorescence development and spike growth greatly influence the production and survival of spikelets and ultimately grain number per spike (Waddingdon et al. 1983). The processes of shoot apex differentiation have been much investigated (e.g., Frank and Apel 1985; Ma and Smith 1992a). A widely used set of criteria and descriptions of development stages can be found in Kirby and Appleyard (1981), although a number of scales have been developed (e.g., Waddington et al. 1983). In these scales, the apex development stage is determined by the development of the most advanced primordium.

In the initiating apex, primordia located at the middle of the rachis develop more quickly and are dominant over primordia in the distal and basal sections of the rachis. Since large variations in stages of development exist among primordia, a scale that indicates the average developmental stage of the whole spike, by using a weighted stage of apical development of each main-stem apex, has recently been developed (Ma and Smith 1991a). This method accounts for variations in the rate of promordium development all along the spike and provides a sensitive way of directly determining the effects of treatments on apical development, especially in short- to mid-length growing season areas, where the duration of apical development, is usually brief and several development stages may overlap.

1.1.3 Factors Affecting Apical Development

Environmental factors that affect shoot apical development in cereal crops include temperature (Frank et al. 1992; Kernich and Halloran 1996), irradiance (Kirby and Appleyard 1980; Baker and Gallagher 1983), photoperiod (Kirby and Appleyard 1980; Kernich et al. 1995, 1996), photoperiod - temperature interactions (Tew and Rasmusson 1978; Barham and Rasmusson 1981) and vernalization duration (Chun 1993). Factors such as high plant population (Kirby and Faris 1970), soil moisture deficit (Nicholls and May 1963; Bergner and Teichmann 1993), nutrient stress (Frank et al. 1992), sowing date (Wright and Hughes 1987; Kirby 1995) and use of plant growth regulators (Petr et al. 1988; Ma and Smith 1991a,b) also affect apical development.

Kirby and Faris (1970) found that leaf number, leaf size, stem and internode length, apex development and primordium production were affected by plant density (50 to 1600 plants per m^2), the high density treatments leading to earlier floral initiation, earlier spikelet primordium abortion, decreased spikelet survival and shorter spikes. Nicholls and May (1963) reported that although the duration of spike initiation was somewhat longer if water stress occurred during the spike initiation period, the total number of spikelet primordia was reduced as the spikelets were initiated at a slower rate. Bergner and Teichmann (1993) found the largest yield reductions if water stress occurred during jointing and pre-anthesis.

The sowing date effectively changes the temperature and photoperiod regime experienced by a developing crop and has a large effect on apical development. Delayed sowing in the spring (i.e. increasing photoperiod length) is associated with elevated rates and shorter durations of leaf appearance and spikelet initiation, and with a shorter time to attainment of all stages of apical development in spring barley (Wright and Hughes 1987; Kernich et al. 1995, 1996). Total spikelet primordium production is a curvilinear function of thermal time, such that apical development occurs more rapidly in a short but warm crop-growing season than under cool, longer season conditions (Ma and Smith 1992a). Frank et al. (1992) found that, during apical development, grains per spike decreased by 1.5 for a six-rowed cultivar and by 0.6 for a two-rowed cultivar for each degree increase in mean daily maximum air temperature. In studying high temperature effects on development, Kernich and Halloran (1996) found that the length of the spikelet growth phase was increased in thermal time units by high temperature and that the percentage of aborted spikelets was greater at higher temperature; they concluded that length of spikelet growth phase was only one factor in spikelet abortion.

Agrochemicals (growth regulators, herbicides and fungicides) have a large impact on apical development. Hormone-based herbicides may induce inflorescence abnormalities in cereals when applied during apex development (Thompson et al. 1984), while application of plant growth regulators (PGRs) to promote tiller production or alter the crop in other ways may not be effective unless applied at the optimum apex development stage (Cartwright and Jaddoa 1985). There has been considerable interest in the hormonal manipulation of growth and developmental relationships between the main-stem and tiller shoots and/or within the spike (Williams and Cartwright 1980; Höfner and Kühn 1982; Hutley-Bull and Schwabe 1982; Petr et al. 1988). Cytokinin application can increase uniformity both among shoots within a plant (main spike and tiller spikes) and among spikelets within the ear (Williams and Cartwright 1980).

Apical development is influenced by early growth regulator treatment. Early application (Zadoks' Growth Stages 13, 30) (Zadoks et al. 1974) of either chlormequat (2-chloroethyl trimethylammonium chloride, CCC) or ethephon (2-choloethyl phophonic acid) delays the apical development stages (Ma and Smith 1991a), resulting in increased tillering and a greater number of ears per unit area (Ma and Smith 1992b). Early application (ZGS 30) of ethephon alone, or in combination with chlormequat, suppresses the development of main-stem spikes in spring barley while increasing the number of tiller-derived spikes (Ma and Smith 1991b). For winter barley, chlormequat application at the lemma primordium stage transiently retards development and elongation of the more advanced shoots and lessens their dominance so that growth and survival of higher order tillers are promoted (Cartwright and Jaddoa 1985). The retarding effect of early chlormequat is transient but the associated increase in 'sink size' is persistent. Chlormequat slightly delays main-stem apical development without deleterious effects on spikelet formation and can, ultimately, reduce floret abortion, while ethephon has deleterious effects on main-stem apices and, in extreme cases, the main-stem apex may have failed to produce a spike (Ma and Smith 1991b, 1992b). Application of chlormequat prior to stem extension did not significantly affect yield in the experiments of Jenkyn et al. (1992).

Early GA_3 treatments are very similar to those produced by intercalating periods of longer days in a standard photoperiodic regime (Hutley-Bull and Schwabe 1982), while the effects of chlormequat and other growth retardants resemble those brought about by the intercalation of shorter days: gibberellin treatment hastens development, advances spikelet differentiation, and reduces leaf, spikelet and tiller numbers, whereas retardants such as chlormequat and paclobutrazol ([2RS, 3RS]-1-[4-chlorophenyl]-4,4-dimethyl-2-[1,2,4-triazol-1-yl]-pentan-3-ol) have the opposite effect. Mecoprop (MCPA, Mecoprop/bromoxynil/ioxynil mixture) produced abnormalities when applied before the triple mound apex stage, but not after (Thompson et al. 1984). Therefore, application of agrochemicals for plant growth enhancement or crop yield promotion through weed control, inhibition of stem elongation, or reduction of apical dominance should be timed with regard to apical development stages.

1.2
Yield Components

In barley, grain yield is more strongly related to grain number than grain size, with individual grain weight being the most stable component of yield (Gallagher et al. 1975; Hadjichristodoulou 1990; Frank et al. 1992). The positive relationship between yield and grain number, and the stability of individual grain weight, suggest that sink size limits yield (Hay 1995). Grain number is largely controlled by the number of spikes per m^2 (plant population and tiller number) and grains per spike, and so is determined prior to anthesis (Gallagher et al. 1975; Frank et al. 1992; Garcia del Moral and Garcia del Moral 1996). Assimilate supply to the developing inflorescence plays a critical role in establishing final grain number and grain size (Hendrix et al. 1986). Defoliation treatments prior to awn emergence reduce yield by reducing grain number per spike and individual grain weight, while defoliation treatments during the grain-filling period generally have little effect on yield (Jenkyn and Anilkumar 1990). Mature grain dry weight was higher in an awned genotype than in an unawned genotype, associated with greater photosynthesis in the awned ears (Bort et al. 1994). Increased sowing density increased ear number per m^2; however, grain yield was not increased as grain size was smaller (Jenkyn et al. 1992). Individual grain weight can be negatively affected by drought, high temperature, plant growth regulators or foliar-applied post-anthesis N treatments, which may reduce grain growth duration or the growth rate of grains (Moes and Stobbe 1991a; Mogensen 1992; Bulman and Smith 1993a; Savin and Nicholas 1996). Individual grain weight, grain number per spike and grain protein concentration are higher in early- than in later-emerging, higher-order tillers (Cannell 1969a; Lauer 1991).

2
Tillering and Tiller Contribution to Grain Yield

2.1
Tillering

Tillers are secondary stems initiated at the coleoptilar node or in the axillary cavity formed between leaves and stems. Many tiller buds are formed but only a few survive to produce grain. Some tiller buds fail to emerge from the leaf sheath while others senesce prior to flowering (Kirby and Faris 1972). Tiller bud initiation ceases after main-stem anther primordia formation (Kirby and Faris 1970). Prior to emergence from the subtending leaf sheath, a tiller bud is dependent on the main stem or its parent tiller for photoassimilates until it has established sufficient leaf area (Simmons and Lauer 1986; Lauer and Simmons 1988).

Tillers are produced in a consistent pattern which closely follows leaf appearance (Cannell 1969a; Kirby et al. 1985b). In stressed plants, the appearance of some tillers may be suppressed or delayed. The coleoptilar tiller does not always elongate and is particularly sensitive to environmental and cultural conditions (Cannell 1969b; Kirby and Riggs 1978). Typically, tiller numbers increase during the vegetative phase, reach a peak 1 to 2 weeks before heading and then decline to a stable spike number (Kirby 1967; Laude et al. 1967; Simmons et al. 1982). Tillering ceases with the onset of rapid main-stem elongation and the early stages of flower development (Kirby et al. 1985a; Hay 1986). Lowered red/far-red light ratio within a crop canopy as the season progresses is believed to play a role in the cessation of tillering (Davis and Simmons 1994). Tillering patterns are similar for high- and low-tillering barley cultivars, although the rate of tiller appearance, the maximum tiller number, and the time when maximum tiller number is reached may vary (Cannell 1969a; Simmons et al. 1982). Tillering can resume after flowering if moisture is abundant (Kirby 1967; Kivisaari and Elonen 1974) and nutrient supply is high (Aspinall 1961); this is most likely to occur if main-stem growth has been suppressed by conditions such as lodging or the application of ethephon under stressful conditions (Ma and Smith 1991b, 1992c). Late tillering is apparently independent of earlier tiller production (Aspinall 1961).

With the exception of the coleoptilar tiller (Thorne 1962; Cannell 1969b; Kirby and Riggs 1978), tiller survival is generally higher for early-produced tillers than for late-produced ones (Kirby and Riggs 1978). Tiller death is less severe when nutrient levels are high (Thorne 1962) and can be eliminated altogether under artificial conditions where nutrient levels are constantly maintained at high levels (Aspinall 1961). Although tiller dieback is often attributed to shading and increased competition for light among a plant's shoots (Kirby and Faris 1972; Lauer and Simmons 1985), stem elongation does not increase shading within the canopy, and competition among the shoots of the plant is for assimilates rather than for light (Simmons and Lauer 1986). The development of an adequate root system may also be involved in tiller survival. Tillers form fewer nodal roots than the main stem, and late-emerging tillers have poor root systems (Anderson-Taylor and Marshall 1983). Senescing tillers export dry matter (El Alaoui et al. 1992) and N (Thorne 1962) to the main stem and other plant parts; however, the dry matter contribution of higher-order tillers is unlikely to be significant because of their small leaf areas (Chafai-Elalaoui and Simmons 1988).

Tiller production and dieback are affected by genotypic and environmental factors (Simmons et al. 1982; Garcia del Moral and Garcia del Moral 1996). Earlier planting increases maximum tiller number but also increases tiller dieback. Conversely, delayed planting of spring barley reduces tiller numbers per plant but increases the proportion of tillers that survive (Lauer and

Partridge 1990). Late autumn sowing of winter barley reduces tiller numbers (Walker and Matthews 1991) and ultimately final spike numbers (Green et al. 1985; Conry and Hegarty 1992). Total tiller number is related to total light energy received (Aspinall and Paleg 1964). Low irradiance reduces the number of coleoptilar and later appearing tillers (Aspinal and Paleg 1964). Increased photoperiod extends the tillering period and increases the maximum tiller number (Guitard 1960; Aspinall and Paleg 1964). Shorter photoperiods delay the appearance of the first and coleoptilar tillers (Cannell 1969b).

Drought stress during the tillering stage decreases maximum tiller number and the proportion of tillers that survive (Day et al. 1978; El Alaoui et al. 1992). Tiller production ceases during drought stress, but can resume upon irrigation, especially if the stress occurs early in the tillering stage (Aspinall et al. 1964). High temperatures reduce tiller numbers by shortening the duration of the tillering stage (Guitard 1960; Garcia del Moral and Garcia del Moral 1996). The coleoptilar tiller is particularly sensitive to high temperatures (Cannell 1969b). Low nutrient levels, especially N, lower the rate and shorten the duration of tillering (Aspinall 1961). High N supply at seeding increases maximum tiller number and tiller survival, especially for late tillers (Garcia del Moral et al. 1984; Birch and Long 1990). Post-heading N uptake appears related to the amount of tillering after anthesis (Bulman and Smith 1994).

Tiller production begins earlier and the rate of tillering is higher at low than at high plant densities (Kirby and Faris 1972). In dense stands, tiller production per plant decreases (Kirby 1967; Leakey 1971; Kirby and Faris 1972), and tiller dieback occurs earlier and more rapidly than in sparse stands (Kirby 1967). Tiller dieback increases with narrowing row width (Simmons et al. 1982). Differences in the rate of leaf emergence for main-stem and tiller culms are more pronounced at higher plant densities, especially with high-order tillers (Kirby and Faris 1972). Leaf sheath length decreases with higher-order tillers and with increasing plant density (Kirby and Faris 1972). This has implications for the determination of tiller populations since these are routinely counted as they emerge from the subtending leaf sheath. Plant density, for example, has a strong effect on the number of tillers that emerge from the leaf sheath but has a minor effect on tiller bud initiation (Kirby and Faris 1972).

Genotypic differences in maximum tiller number and tiller survival are usually attributable to variations in coleoptilar and higher-order tiller production (Kirby and Riggs 1978; Simmons et al. 1982). In the United Kingdom, newer cultivars produce more tillers and tend to have a greater proportion of surviving tillers than older ones (Riggs et al. 1981). Two-row cultivars produce more leaves and tillers than six-row cultivars (Kirby and Riggs 1978). Kaufmann and McFadden (1960) reported a positive effect of seed size on tiller number. Cultivar differences for tiller production and survival are more apparent at low than at high planting densities (Baker and Briggs 1982; Birch and Long 1990), since higher-order tillers are suppressed at higher densities.

2.2 Tillering and Yield

The usefulness of tillers has been questioned (Donald 1968). Several lines of evidence indicate that tillers compete with the main stem for fixed carbon and water (Jones and Kirby 1977; Kirby and Jones 1977):

1. Tillers import C from the main stem during the tillering stage and continue to do so during rapid main-stem extension, albeit in much smaller quantities (Simmons and Lauer 1986). Proportionately more C is translocated to tillers in high- than in low-tillering cultivars (Lauer and Simmons 1985).
2. In uniculms and oligoculms, main stems grow faster than in normal-tillering plants (Kirby 1973a). Uniculm genotypes are taller, produce leaves more quickly and reach anthesis and maturity earlier than normal-tillering genotypes.
3. Early tiller excision increases leaf area, spikelet number, grain number and individual grain weight of the main stem (Alaoui et al. 1988; Kirby and Jones 1977), suggesting that competition for assimilates by tillers occurs during the crucial stages of main-stem apical development (Kirby and Jones 1977).
4. Main-stem yields are greater at high densities than at low ones, providing indirect evidence of competition from tillers (Lauer 1991).

Tiller number and total leaf number per plant are positively associated, and genetic variability for the rate of leaf emergence and total leaf number suggests that selection for either of these traits could increase tiller numbers (Kirby and Riggs 1978; Kirby et al. 1985a,b). However, absolute shoot mortality tends to be greater in high-tillering cultivars (Simmons et al. 1982) and spike numbers are not necessarily increased (Thorne 1962). In two-row barley, maximum tiller number is inversely related to final spike number (Kirby 1967). Also, compensation among-yield components and increased lodging due to thinner tiller stems may limit yield increases (Simmons et al. 1982; Benbelkacem et al. 1984). The proposed use of uniculms (Donald 1968, 1979) has had limited success. Some yield increases have been reported (Donald 1979), but results have been inconsistent because of yield component compensation (McDonald 1990), or because of spike abnormalities appearing at higher plant densities and higher temperatures (Kirby 1973b; Badra and Klinck 1981).

2.3 Tillering: Hormonal Control and Plant Growth Regulators

Hormonal balances are implicated in the regulation of tiller bud growth and development. Bud growth is inhibited by auxins (Johnston and Jeffcoat 1977; Sharif and Dale 1980a,b) and inhibited or reduced by GA_3 (Woodward and

Marshall 1988; Ma and Smith 1992c). Anti-auxins (TIBA and ethephon), GA_3 biosynthesis inhibitors (chlormequat chloride, mepiquat chloride and ancymidol), and formulations of ethephon and chlormequat or mepiquat chloride promote tillering (Koranteng and Matthews 1982; Woodward and Marshall 1989; Lauer 1991; Moes and Stobbe 1991b). Auxins and GA_3 do not inhibit the growth of rapidly growing tiller buds (Sharif and Dale 1980a,b).

Ethephon increases the number of early (Foster et al. 1991; Moes and Stobbe 1991b; Taylor et al. 1991) and late emerging tillers (Foster et al. 1991; Lauer 1991), but the productivity of these tillers is often quite low and cultivar-dependent. Growth regulators that reduce stem growth in freely tillering cultivars have no effect on the development of tiller buds in uniculms (Woodward and Marshall 1989). The application of synthetic and natural cytokinins promotes tillering in some cases but not in others (Ruckenbauer and Kirby 1973; Johnston and Jeffcoat 1977; Sharif and Dale 1980b). Cytokinin is required for tiller growth once the bud has been released from inhibition (Sharif and Dale 1980b).

3 Grain Filling

Grain filling is divided into three stages: (1) an initial lag phase of about 2 weeks during which endosperm nucleus division, organelle formation and water accumulation in the grain occur; (2) a linear dry matter accumulation phase, during which starch and proteins are rapidly synthesized; and (3) a tapering-off phase during which dry matter accumulation ceases and grains begin to dehydrate. The rapid grain growth commences 7–10 days after anthesis and lasts 2–5 weeks (May and Buttrose 1959; Metzger et al. 1984). During the linear phase, grain growth rates vary from about 0.9 to 2.2 mg dry matter per grain per day (Scott et al. 1983; Ho and Jui 1989). Each stage is affected by external and internal factors which influence final grain dry matter accumulation.

Genotypic variation exists for both rate and duration of the grain filling, and individual grain weight is better correlated with rate of grain filling than with duration of grain filling (Metzger et al. 1984; Ho and Jui 1989). However, increased rate of dry matter accumulation is associated with shorter duration of grain filling, such the final grain dry weight is not increased by increased rate of grain fill (Cochrane et al. 1996). Variation in total grain dry weight is often better correlated with rate rather than with duration of the linear grain-filling phase, during which time about 85% of grain dry matter is accumulated (Walpole and Morgan 1971; Scott et al. 1983).

The embryo comprises 3–5% of the final grain weight (Tallberg 1982) and reaches its maximum dry weight about three quarters of the way through grain development (Duffus and Rosie 1975). Final grain weight is determined by conditions during inflorescence development and grain filling (Ellis and Kirby

1980; Scott et al. 1983). Dry matter accumulation varies with grain position along the spike. Individual grain weight is greatest for the middle spikelets of a spike and tapers gradually to the basal and distal ends (Scott et al. 1983). Middle spikelets have the fastest growth rates. The duration of grain growth is shortest for the distal grains and longest for basal grains (Koesmarno and Sedcole 1994). At each rachis node, centre grains are heavier than lateral ones (Ellis and Kirby 1980; Cottrell et al. 1985; Gu and Marshall 1988). The slower overall development (Ma and Smith 1991b) and poorer vascular development (Kirby and Rymer 1975) of the distal grains may explain their lower grain weights.

Cellular development of barley grains has been described previously (Duffus and Rosie 1975; Cochrane and Duffus 1979, 1981). Dry matter accumulation may be related to endosperm cell number and/or starch granule types (Olsen and Krekling 1980; Cochrane and Duffus 1983). Sucrose is the major assimilate translocated to the grain (Duffus and Cochrane 1992). It is rapidly absorbed by the endosperm (Felker et al. 1984) and is the predominant free sugar in developing grains (Chevalier and Lintel 1983). Energy requirements for synthesis reactions in the grain are met by normal respiratory processes within grain cells (Duffus and Rosie 1977; Nutbeam and Duffus 1978). Sucrose can be synthesized by the pericarp (Nutbeam and Duffus 1976). Sucrose content per grain is maximal 1–2 weeks after anthesis, during the period of endosperm call division (Baxter and Duffus 1973; Cerning and Guilbot 1973; Brooks et al. 1982), after which levels decline. Rapid starch synthesis begins about 2 weeks after anthesis (Baxter and Duffus 1973). The grain concentration of reducing sugars declines after the start of rapid starch synthesis (Baxter and Duffus 1973; Duffus and Rosie 1975).

Starch synthesis and total dry matter accumulation are limited by reactions involved in starch synthesis rather than by assimilate supply (Lingle and Chevalier 1984; MacLeod and Duffus 1988b). Changes in assimilate supply during grain filling brought about by shading or degraining spikes (Nosberger and Thorne 1965; Merbach et al. 1994), defoliation (Jenkyn and Anilkumar 1990) or by artificially increasing CO_2 concentrations (Natr and Apel 1974) have relatively little influence on grain weight. Grain yield is closely related to total irradiance interception, but the effect appears to operate on grain number rather than on grain weight (Gallagher et al. 1975; Williams and Hayes 1977).

Starch comprises 60–75% of final grain dry weight (Duffus and Cochrane 1992). In normal genotypes, about 25% of the starch is amylose and 75% is amylopectin (Duffus and Cochrane 1992). Waxy-type grains contain nearly 100% amylopectin, have a slightly lower starch concentration (4–9%), and much higher sucrose (33–44%) and maltose (111–176%) concentrations than normal grains (Newman et al. 1990). In non-waxy cultivars, the amylose content of the endosperm increases with increasing grain age (Kang et al. 1985). Barley starch is composed of large (10–20 μm in diameter), lenticular (A-type) and small (<10 μm in diameter), spherical (B-type) granules that are initiated at different times during grain development (MacGregor et al. 1971; Williams and Duffus 1977). A-type granules are initiated during endosperm cell division, and B-type granules after this. B-type starch granules are about ten times more numerous than A-type granules, but the latter make up about 90% of the starch volume in the endosperm (May and Buttrose 1959; McDonald et al. 1991). In non-waxy cultivars, A-type starch granules contain slightly more amylose than B-type granules (McDonald et al. 1991).

Termination of grain filling is related to anatomical changes in the grain (Lingle and Chevalier 1985). Cessation of starch accumulation involves the reactions converting sucrose to starch, rather than a lack of substrate (La Berge et al. 1973; Brooks et al. 1982). Enzymatic activity decreases towards the end of grain filling (Chevalier and Lingle 1983; Lingle and Chevalier 1984), even though sucrose is still being supplied to the endosperm (Cochrane 1983, 1985).

Grain maturation begins 30–40 days after anthesis (Duffus and Rosie 1975; Ma and Smith 1992a). It is characterized by loss of pericarp chlorophyll, rapid acidification of the starchy endosperm (Macnicol and Jacobsen 1992), attainment of maximum grain dry matter and onset of rapid grain dehydration. Grain dehydration proceeds along the spike from the distal to the basal end, and may be related to changes in vascular tissues along the rachis (Cochrane 1985).

Post-anthesis water deficits reduce final grain weight (Wells and Dubetz 1970; Day et al. 1978). Water stress does not alter the number of endosperm cells, but decreases the size of A-type starch granules without decreasing the total number of A- and B-type starch granules in the endosperm (Brooks et al. 1982). Water stress has no effect on endosperm cell volume, endosperm cell number or rate of grain filling, but shortens the duration of the grain-filling period (Brooks et al. 1982). Increased rainfall is associated with decreased grain protein (Correll et al. 1994) as it increases carbohydrate deposition more than protein deposition.

Air temperatures influence starch, protein and lipid contents of barley grains, and individual grain weight. Temperatures above 15 °C lower starch and protein content and individual grain weight, and increase lipid content; the size of A- and B-type starch granules and the number of B-type starch granules is decreased (MacLeod and Duffus 1988a; Tester et al. 1991). Temperatures above 35 °C are associated with increased grain protein (Correll et al. 1994). Amylose and amylopectin contents of normal and waxy starches respond little to increasing temperatures in the range of 10–20 °C. Sucrose synthase activity and starch deposition decrease in the endosperm when day/night temperatures are raised from 20/15 °C to 30/25 °C, even though the sucrose concentration increases (MacLeod and Duffus 1988b). Temperatures in the range of 15–30 °C do not affect endosperm cell number (MacLeod and Duffus 1988a). The duration of grain filling is negatively correlated with mean air temperature during this stage, while rate of grain filling is positively correlated (Gallagher et al. 1975). Total fatty acid content and composition are related to grain size, and change during growth and maturation of the grain (de Man and Cauberghe 1988; de Man and Vervenne 1988).

4 Carbon Metabolism

4.1 Photosynthesis and Radiation Use Efficiency

Barley is a C_3 plant, and its productivity is the result of a balance between CO_2 fixation via photosynthesis and CO_2 loss through dark respiration and photorespiration. A number of recent reviews discuss the processes involved in light harvesting and carbon metabolism (Douce and Neuburger 1989; Golbeck 1992; Geiger and Servaites 1994; Horton et al. 1996; Creelman and Mullet 1997). Crop biomass production is influenced by the duration of the growing period and the rate of growth (Jamieson et al. 1995a). The rate of growth is influenced by the amount of photosynthetically active radiation (PAR) intercepted by the plant (which is influenced by leaf area index) and the efficiency with which this radiation is used.

Radiation use efficiency (RUE) is defined as the proportion of dry plant biomass accumulated per unit of intercepted PAR (Kiniry et al. 1989). It may be expressed for the growing season or for specific phases of development, as values differ over the growing season (Monteith 1994). The term must specify the type of carbon (CO_2 uptake, shoot dry matter or total dry matter), the type of radiation (intercepted total, absorbed total, intercepted PAR or absorbed PAR), and the time scale over which the carbon is accumulated (Arkebauer et al. 1994). It is linked to physiological processes so it may change; for example, during grain filling of crops producing seeds containing high-energy compounds such as protein or oil (Kiniry et al. 1989). Radiation use efficiency is very stable in unstressed crops (Jamieson et al. 1995b), but is affected by water, nutrient and temperature stresses (Arkebauer et al. 1994). For example, when barley was drought-stressed from emergence, decreased growth rates and reduced biomass yield were caused mainly by reduced RUE, and the RUE remained depressed even when drought was relieved (Jamieson et al. 1995b).

Leaf area index (LAI) is a unitless index of leaf area per land area and, as such, is related to canopy closure, radiation use efficiency and water use efficiency. It is an indication of solar radiation interception efficiency of a crop. Leaf area index may be measured directly, by measuring total leaf area within a specific field area, or indirectly, as by reflectance measurements (Bouman et al. 1992; Kuusk 1995). The LAI of barley is generally at its maximum at anthesis (Ramos et al. 1995). Planting density influences barley LAI (Stoskopf 1985; Gontia and Khare 1992) with higher densities causing more rapid accumulation and, sometimes, higher final values. Nitrogen fertilizer applied at tillering was found to increase LAI (Ramos et al. 1995). Decreased LAI was found to occur with low K fertility (Andersen et al. 1992a,b), soil compaction (Lipiec et al. 1992) and application of chlorsulfuron (Lemerle et al. 1993). When barley plants were drought-stressed from 2 weeks before anthesis or later, decreased biomass yield was associated with reduced radiation interception due to more rapid leaf senescence (decreased LAI), while RUE remained stable and at near maximum levels (Jamieson et al. 1995b). Maximum yield is associated with an optimal LAI, which is not necessarily the highest LAI (Stoskopf 1985; Gontia and Khare 1992; Bulman and Smith 1993a).

From the management point of view, crop photosynthesis can be manipulated through sowing date, sowing rate, row spacing, row direction, nutrient level and application of plant growth regulators. Measurement of photosynthetic rates has shown that, in general, narrower rows result in greater light interception and increased grain yield, while increased N fertility increases penultimate leaf photosynthetic rates (Leibovitch et al. 1992). Ethephon application has little effect on barley photosynthesis during the first few days after application, although there is a tendency toward increased photosynthetic rate in treated plants thereafter (Ma et al. 1992). Application of ethephon to barley 1 week after spike emergence also delays leaf senescence and increases photosynthetic rates and grain yield (Ma and Smith 1992e). Ethephon treatment may also extend the grain-filling duration of spring barley (Ma and Smith 1991b, 1992f).

4.2 Water Use Efficiency

Water use efficiency (WUE) is defined as biomass produced per unit evapotranspiration, and depends on transpiration efficiency and soil evaporation. Transpiration efficiency parallels radiation use efficiency, while soil evaporation is related to canopy development. The slower the canopy development, the greater the soil radiation load which leads to greater evaporation from soil, and so the lower the WUE of the canopy (Thomas and Fukai 1995). Yield, WUE and photosynthesis are interrelated (Bort et al. 1994; Dakheel et al. 1994). Transpiration is directly associated with dry matter accumulation, as water vapour and carbon dioxide both pass through the stomata. Soil evaporation indirectly affects dry matter accumulation through loss of water which is made unavailable to the plant roots, and by modification of temperature and humidity in the canopy (Jamieson et al. 1995a). Relative WUE varies in value, depending upon the parameter on which it is based, i.e., aboveground dry weight, total root and shoot weight, or grain production (Richards 1992).

Water use efficiency has been shown to increase in barley under water stress, more so than in wheat, triticale and oat (Loboda 1993; Rekika et al. 1995). It is correlated with carbon isotope discrimination by C_3 photosynthesis, a reflection of internal leaf gas exchange (Apel et al. 1993; Araus and Buxo 1993).

It can also be positively correlated with yield (Dakheel et al. 1994; Simpson and Siddique 1994). For example, mature grain dry weight, photosynthetic rate and WUE during the grain-filling period were greater in awned than in awnless barley genotypes (Bort et al. 1994). Where soil water was limited, the higher grain and biomass yield of barley than for bread wheat, durum wheat, triticale and oat was associated with greater WUE via more rapid development of leaf area and earlier flowering than the other species; these factors appeared to minimize soil surface water loss and ensure rapid completion of growth (Lopez-Castaneda and Richards 1994a,b). In the experiments of Mogensen et al. (1994) net photosynthesis of barley decreased after two-thirds of available soil water had been utilized, and continued to decrease in relation to relative soil water content thereafter. However, while genotype differences in drought response of CO_2 assimilation rate, stomatal conductance and CO_2 intercellular concentration occurred, none of these were correlated with yield (Arnau and Monneveux 1995).

A number of factors have been shown to influence the WUE of barley. It was improved by addition of fertilizer N, P and K, or rotation of barley with *Vicia sativa* (Andersen et al. 1992a; Harris 1994) and shown to decrease in Fe-deficient plants (Sharma et al. 1994). Spraying of acetylsalicylic acid in aqueous solution on drought-stressed barley plants increased yield and WUE (Bergmann et al. 1994). Intercropping of *Pisum sativum*, *Vicia faba* or *Festuca rubra* with barley resulted in greater WUE; this was attributed to differences in canopy structure and plant biomass production (Izaurralde et al. 1994). Zero tillage was associated with greater WUE and better soil water conservation than were minimum or conventional tillage (McAndrew et al. 1994); WUE decreased with an increase in soil compaction (Arvidsson and Jokela 1995). Although WUE is not generally affected by irrigation (Carefoot and Major 1994), mulching, irrigation and irrigation design can increase WUE (Mandal and Mahapatra 1993; Wessolek and Rengar 1993).

Salinity is a consideration in barley production under irrigation, as growth and grain yield are reduced at relatively low salinities (0.1 M NaCl) (Gorham and Chapman 1992). Barley is considered to be the most salt-tolerant cultivated cereal, with tolerance associated with salt exclusion (Gorham and Chapman 1992; Stanca et al. 1992). In a study comparing barley and wheat with salt-tolerant species, it was found that barley and wheat could be as productive, or more productive, than the most salt-tolerant species, as they used all available water and had the highest WUE when grown under high density (Richards 1992). Dry matter partitioning is affected by salinity, with a higher harvest index under high salinity (Grattan et al. 1994). Photosynthetic rate, growth and grain yield are reduced even at low salinities (Gorham and Chapman 1992; Shen et al. 1994) although a causal relationship between effects of salinity on growth and leaf gas exchange has not been shown (Dunn and Neales 1993). Some studies show decreased plant growth without major photosynthesis effects, suggesting little salinity stress in barley at the photosynthetic level (Belkhodja et al. 1995).

4.3
Harvest Index

Harvest index is defined as the ratio of the dry weight of desirable crop material (e.g., grain, or shoots in the case of silage) to the total aboveground dry matter produced by the crop. It is most useful as an index of productivity for small grain cereal crops and pulses, with values for modern varieties of most intensively cultivated grain crops between 0.4 and 0.6 (Hay 1995). Increases in spring barley yields over the last century have been largely attributed to increased harvest indices; however, the harvest index increases for cereal crops appear to be reaching a plateau (Bulman et al. 1993; Jedel and Helm 1994; Hay 1995). Higher total dry matter production and increased lodging resistance have been other major factors in increased barley yields (Bulman et al. 1993; Jedel and Helm 1994; Lopez-Castaneda and Richards 1994a). Harvest index is highly heritable (Hay 1995). However, it was not useful as a selection criterion in breeding programmes to increase yield or as a predictor for barley silage grain content (Baron et al. 1992; Theoulakis et al. 1992).

A number of factors have been shown to influence harvest index. Ethephon treatment can result in a reduced barley harvest index. While ethephon reduces lodging, it also results in incomplete spike emergence and reduced tiller survival and, consequently, reduced grain weights (Bridger et al. 1995). Harvest index is influenced by interactions among rainfall patterns, N fertilization and grain-filling temperature (Gonzales-Ponce et al. 1992). High rainfall and moderate grain-filling temperatures were associated with increased harvest index and yield in response to N fertilization; low rainfall and high grain-filling temperatures were associated with decreased harvest index in response to N fertilization. High soil fertility, while increasing yield, decreased harvest index (Jedel and Helm 1992). Harvest index and yield were shown to decrease with increased water deficit (Salam et al. 1991), saline irrigation (Ahmed et al. 1993) and interference by quackgrass (Pageau and Tremblay 1995). Decreased yield generally occurs where harvest index is reduced (Bridger et al. 1995). However, growing conditions may be such that yield and total biomass production are increased but harvest index is reduced (Jedel and Helm 1992).

4.4
Dry Matter Partitioning

The export and distribution patterns of assimilates in barley vary according to age and the developmental stage of the plant (Zagal 1994). In young plants, assimilates are translocated rapidly towards the roots, but this movement slows down as the plant ages and translocation of C to the roots diminishes (Gregory and Atwell 1991; Jensen 1993). In the youngest mature leaf of uniculm barley, 56 and 93% of $^{14}CO_2$ is moved out within 7 and 24 h, respectively, to other leaves and roots in roughly equal proportions (Gordon et al. 1977). About 70% of assimilated C is allocated to the shoots at the reproductive stage (Zagal 1994). Carbon translocated to the roots may be lost through leaching, exudation and the sloughing off of cells (Jensen 1993). During the tillering stage, roots comprise from 40 to 48% of the total plant weight; however, their contribution to total plant weight by early stem elongation and by anthesis declines to about 30 and 15%, respectively (Gregory and Atwell 1991; Gregory et al. 1992). Barley inflorescences receive photoassimilates preferentially from the top two to three leaves of the plant (Lauer and Simmons 1985). During the grain-filling phase, these top leaves, together with the green tissues of the spike, are the major sources of photoassimilate production (Caley et al. 1990; Ma and Smith 1992d). The photosynthetic capacity of the spike is related to its total green area including awns (Thorne 1965).

The physiological determinants of yield are not well defined for barley plants treated with plant growth regulators (PGRs). Early application of either chlormequat or ethephon retards development of spring barley main-culm apices (Ma and Smith 1991a). This may reduce the degree of dominance exerted by these sinks, making relatively more assimilates available for the secondary sinks (tillers within a plant or tip and bottom spikelets within a spike). Thus, PGRs may decrease the differences in competitive ability among 'sinks', leading to more synchronized development of spikelets within a spike (Williams and Cartwright 1980; Höfner and Kühn 1982). Translocation or retranslocation of dry matter from non-grain plant tissues (leaves, sheaths, culm and chaff) to the grain is slower in PGR-treated plants (Ma and Smith 1992d). This may have occurred because of a decreased ability to translocate material to the sink, or because of an increased sink limitation. Decreased ability to translocate could result from decreased loading, translocation or unloading efficiency in phloem elements, or from alterations in vascular anatomy. Increased sink limitation could occur through increased source size, with no accompanying change in sink size or through decreased sink size while the source remained constant.

Ethephon treatment increases the number of sterile rachis nodes and florets per spike (Simmons et al. 1988; Ma and Smith 1991b; Moes and Stobbe 1991c; Bulman and Smith 1993d). This can reduce the total sink size, unless there is a compensatory increase in seed size. Differences in the weight of non-grain plant material between PGR-treated and untreated plants (Ma and Smith 1992d) may be related to differences in respiratory rates, retranslocation of carbohydrate to roots or in the degree of structural dry matter accumulation. Chlormequat and ethephon application increases the concentration of water-soluble carbohydrates in the culms of wheat (Knapp et al. 1987). The data of Ma and Smith (1992d) indicate that ethephon or chlormequat application to spring barley (1) does not alter total aboveground dry matter accumulation per shoot; (2) alters the partitioning of dry matter among mature plant parts within a shoot; (3) reduces the rate of photoassimilate translocation out of leaves, sheaths and chaff, or decreases the plants' ability to remobilize material from these structure; (4) positively or negatively affects grain yield, depending on the conditions in the crop year; and (5) is not reliable as a tool for increasing grain production.

5 Nitrogen Metabolism

5.1 Nitrogen Uptake

Induction and the rate of nitrate uptake are influenced by external nitrate concentration, light, pH, temperature, the concentrations of other ions, anaerobic conditions, metabolic inhibitors (Warner and Huffaker 1989; Aslam et al. 1990; Glass et al. 1990) and genotype (Bloom and Finazzo 1986). Increased nitrate uptake results if temporal or spatial N limitations occur within or around the plant (Drew and Saker 1975; Pettersson and Jensén 1978). Local N supply around the root system affects the size and production of lateral roots (Drew and Saker 1975; Samuelson et al. 1992). However, although dry matter distribution within the root system may be altered, the whole plant root/shoot dry matter ratio will remain unaltered (Öhlén and Larsson 1992). When external nitrate concentrations are limiting, the uptake capacity of roots increases significantly (Lee and Rudge 1986; Lee and Drew 1989).

Absorbed nitrate is partitioned between a small active metabolic pool, which regulates nitrate reductase induction and the conversion of nitrate to ammonia (Butz and Jackson 1977), and a large storage pool (Ferrari et al. 1973), from which N can be released when external N sources are limiting (Barneix et al. 1984). Vacuoles are the major temporary storage site for nitrate in leaves (Granstedt and Huffaker 1982). The rate of nitrate release from the storage pool is influenced by cellular carbohydrate status and by irradiance (Aslam et al. 1976, 1979; Aslam and Hufaker 1982, 1984).

Plants fed ammonium or ammonium-nitrate mixtures accumulate N more quickly than those fed nitrate only (Lewis and Chadwick 1983; Soares and Lewis 1986). During the grain-filling stage, newly assimilated N is translocated rapidly to the developing grain (Soares and Lewis 1986). Although plants retain the capacity to reduce nitrate during grain filling, ammonia fertilizers provide the grain with N more efficiently because of a decline in the nitrate absorption rate and in the capacity of senescing shoots to reduce nitrate during the grain-filling stage (Soares and Lewis 1986). Bloom et al. (1992) reported that the energy costs for the uptake and assimilation of nitrate were about 64% greater than those required for ammonium, suggesting that plant growth may be more energy-limited when nitrate is used as a fertilizer rather than ammonium (Bloom et al. 1989, 1992).

5.2 Nitrate and Nitrite Reductase Activity

Several reviews have dealt with the structure, function and regulation of nitrate reductase (NR) (Barber and Solomonson 1990; Hoff et al. 1992) and barley follows the general pattern in this regard. Reduction of nitrate to ammonia involves two steps, each catalyzed by a specific enzyme. In the first step NR catalyzes the reduction of nitrate to nitrite, which is reduced in turn to ammonia by nitrite reductase (NiR). Once N has been reduced to ammonium it is assimilated into organic compounds by the GS/GOGAT cycle (glutamine synthetase/glutamate synthase). The induction of NR and NiR has been reviewed by Rajasekhar and Oelmüller (1987). Nitrate reductase activity (NRA) is induced by nitrate and light (Warner and Huffaker 1989; Somers et al. 1983) and involves regulation by both de novo synthesis and degradation of the enzyme (Somers et al. 1983; Small and Gray 1984).

Nitrate reduction occurs mainly in leaves, but also in roots; roots export nitrate to the shoot via the xylem, and extensive recycling of N occurs between the roots and the shoot (Aslam and Huffaker 1982; Lewis et al. 1982; Gojon et al. 1986). Nitrate is the major N form translocated in the xylem (Lewis et al. 1982). Of the organic N found in the xylem, glutamine is the most abundant form, representing 50% of all amino acid N, followed by serine, asparagine and arginine. In leaves, the dark nitrate reduction rate is proportional to the level of available carbohydrate (Aslam and Huffaker 1982, 1984). Both nitrate reduction in the roots and retranslocation of assimilated N to the roots via the phloem contribute to the N nutrition of the roots (Gojon et al. 1986).

5.3 Nitrogen Assimilation

Nitrogen uptake and assimilation in plants have been extensively reviewed (Barber and Solomonson 1990; Warner and Kleinhofs 1992; Hunt and Layzell 1993; Udvardi and Day 1997). Nitrogen assimilation and C metabolism are linked within plants (Stulen 1986; de Cires et al. 1993; Mattsson and Schjoerring 1996). Nitrogen uptake and assimilation require fixed C for the production of amino acids (Dean-Drummond et al. 1979), and the nitrate storage pool level is affected by the carbohydrate status of a cell (Aslam and Huffaker 1982, 1984). Photosynthesis in barley plants is strongly influenced by leaf N concentration, which is in turn affected by overall nutrient status (Fritsch and Jung 1984; Groot 1988). Under conditions of N stress, carbohydrate partitioning within the plant favours the root system. This results in a lower shoot/root dry matter ratio (Brouwer 1962). Under N stress leaf expansion slows before a decrease in total dry matter or C accumulation occurs (Lee and Rudge 1986; Chapin et al. 1988).

5.4
Senescence

Ribulose-bisphosphate carboxylase (RUBPCase) is the most abundant leaf protein and, during senescence, photosynthetic rates and the levels of RUBPCase and its activity decline (Peterson and Huffaker 1975; Friedrich and Huffaker 1980). As RUBPCase is degraded, it becomes an important N source for developing grain (Peterson and Huffaker 1975). At this time leaf protease activity increases (Peterson and Huffaker 1975; Miller and Huffaker 1985), and leaf N concentration decreases (Corke et al. 1989; Carreck and Christian 1991). Additionally, chloroplast numbers (Martinoia et al. 1983) and the activity of NR, NiR and other N-assimilating enzymes decrease (Feller and Keist 1986).

5.5
Amino Acid Translocation

Most amino acid N is translocated as glutamine and glutamate (Tully and Hanson 1979; Winter et al. 1992). Serine, aspartate, and alanine are also found in significant quantities (Tully and Hanson 1979; Weibull et al. 1990). The relative proportions of amino acids in the phloem change with plant age (Weibull 1987) and do not seem to be affected by drought stress (Tully and Hanson 1979). The amount of glutamic acid and aspartic acid in the phloem increases as fertilizer N application levels increase (Weibull 1987). The high amide concentrations may be related to lower bioenergetic costs for the synthesis of these amino acids in leaves (Miflin and Lea 1977), and reduced costs for conversion of glutamine to other amino acids required for storage protein synthesis (Duffus and Rosie 1978). Barley grain protein contains high levels of amino acids of the glutamate family. Under conditions of N stress and during plant senescence, N translocated out of leaves originates from catabolized proteins rather than from newly assimilated N (Tully and Hanson 1979).

5.6
Pattern of N Uptake

Nitrogen and dry matter accumulation in plants are related (Perby and Jensén 1984; Bulman and Smith 1994). Delayed N application to seedling plants leads to reductions in the growth rate of the seedling, in their leaf area and dry weight and in the photosynthesis of the first leaf, leading to a 40% reduction in fixed carbon (Dale 1972). Reductions in photosynthesis are more closely related to reductions in leaf size than in photosynthetic activity. Total N content per plant increases until about anthesis, at which time there is often a short period during which total N decreases slightly before increasing again during grain filling (Corke et al. 1989; Pettersson 1989; Carreck and Christian 1991). This apparent loss of N does not appear to be related to losses in plant dry matter. In cereals, N provided to the developing grains comes from continued

N uptake and assimilation after anthesis and from the retranslocation of previously accumulated vegetative N; these are negatively correlated (Bulman and Smith 1994). Post-anthesis N uptake and N retranslocation from vegetative organs determine final plant N content. The amount of N accumulated by anthesis and ultimately retranslocated to the grain is variable, but generally estimated at 75–80%. Barley plants have the capacity to absorb and assimilate N even during grain filling and to rapidly incorporate this newly assimilated N into the grain (Soares and Lewis 1986). Post-heading N uptake is correlated with total dry matter accumulation after anthesis, and total plant and grain N per plant at harvest (Bulman and Smith 1994).

Total plant and leaf N concentrations decline as plants develop (Leigh and Johnston 1985; Carreck and Christian 1991). Plant N concentration increases with N supply (Leigh and Johnston 1985). Straw N concentrations also increase with increasing N fertilization rate (Kucey 1987; Bulman and Smith 1993b) and are higher in drier environments (Grant et al. 1991). Late foliar applications of N, capable of raising the N content per grain (Turley and Ching 1986a,b; Bulman and Smith 1993c), increase the activity of glutamine synthetase (GS) and glutamine 2-oxo-glutarate amino transferase (GOGAT) in barley leaves, as well as leaf N (protein and amino acids) and chlorophyll content (Turley and Ching 1986b). Nitrogen use efficiency decreases under low soil moisture conditions and decreases with increasing levels of available N (Grant et al. 1991).

5.7 Grain Nitrogen

Grain yield and grain protein concentration (GPC, $g\,kg^{-1}$) are generally negatively correlated (Torp et al. 1981). Two N gradients are apparent in the grains of a barley spike. Nitrogen content per grain is highest for the central grain at each rachis node, and for the grains situated in the middle of the spike., Drought and/or high temperature stress appear to influence the carbon – nitrogen balance within barley grains, with starch deposition in the grain often being more sensitive to post-anthesis stress than nitrogen deposition (Savin and Nicholas 1996). It has been reported that water deficits reduce GPC without reducing the protein and amino acid content (g per seed) of the grain (Brooks et al. 1982), and that drought stress may decrease the amount of carbohydrate in the grains without decreasing protein, resulting in a higher GPC (Ma et al. 1992).

In contrast to starch, protein accumulation under normal conditions is controlled by the amount of available substrate. For example, increasing the supply of available N in detached spike cultures greatly increased the amount of protein per grain (Giese et al. 1983; Giese and Hejgaard 1984). However, in field experiments, high-GPC cultivars did not accumulate more N after heading than low-GPC cultivars (Bulman and Smith 1994). If N was added to barley plants via peduncle perfusion, seed protein concentration and content were increased, even under conditions of high soil N fertility. This suggests that seed protein accumulation is more limited by the ability of roots to take up N from the soil than by the seed to take up N from the rest of the plant (Foroutan-pour et al. 1997).

Nitrogen harvest index (NHI) varies from about 0.50 to 0.70, depending on moisture conditions, level of fertilizer N and cultivar. NHI decreases with increasing fertilizer N, and with decreasing soil moisture (Bulman and Smith 1993b,c; Lewicki and Chery 1992; Grant et al. 1991). Nitrogen retranslocation and N harvest index are not correlated with GPC (Bulman and Smith 1994).

6 Protein Synthesis and Amino Acid Composition

Protein synthesis and its control in barley grain have been reviewed by Kreis and Shewry (1992). Proteins are stored in protein bodies that are believed to originate from vacuoles or the endoplasmic reticulum (Cameron-Mills and von Wettstein 1980; Miflin et al. 1981). Prolamines, or hordeins in barley, are the major storage proteins in the grain, and account for 35–50% of the total grain N (Kirkman et al. 1982). The albumin plus globulin fraction represents 15–30% of the total grain N and is located predominantly in the aleurone layer and the embryo (Newman and Newman 1992). Prolamines are generally rich in glu-amine and proline, but poor in lysine and tryptophan (Foster and Prentice 1987).

Hordeins are a complex mixture of polypeptides that can be divided into four groups, the B-, C-, D-, and gamma-hordeins. All hordeins are high in glutamine plus glutamic acid and proline (Kreis and Shewry 1992), and with the exception of the D-hordein are low in the essential amino acids lysine and threonine (Newman and Newman 1992). The C-hordeins are deficient in cyst(e)ine and methionine while the B-hordeins are rich in these sulphur-containing amino acids (Shewry et al. 1980).

Under normal field conditions, the B- and C-hordeins account for 70–80% and 10–20% of hordeins, respectively, while the D- and gamma-hordeins represent only minor fractions. The proportion of C-hordein increases under conditions of high N-fertility or sulphur stress (Kirkman et al 1982; Shewry et al. 1983).

7 Contribution of Preanthesis Carbon and Nitrogen to Yield

7.1 Retranslocation of Carbon During Grain Filling

Two carbohydrate sources exist for grain filling – current photosynthate and redistributed photosynthate (Schnyder 1993). Current photosynthesis, particularly that of the flag leaf and the ear, supplies most of the sucrose to the developing grain, with spike photosynthesis contributing as much as 30–40% of grain carbohydrate (Frey-Wyssling and Buttrose 1959; Thorne 1963; Ma and Smith 1992d). Sources of photosynthate for barley grains also include the awns, immature grains and the pericarp (Duffus and Rosie 1973; Kjack and Witters 1974; Watson and Duffus 1988).

The extent to which pre-anthesis dry matter accumulation contributes to grain yield is variable (Biscoe et al. 1975; Austin et al. 1980b; Bidinger et al. 1980) Estimates of the pre-anthesis assimilation contribution based on crop growth and ^{14}C analyses range from 2 to 74% (Gallagher et al. 1975; Bidinger et al. 1980). However, overestimations are likely to have occurred since vegetative dry matter loss during grain filling does not represent complete retranslocation towards the grain; contributions are thought to be more likely in the order of 2–54% (Austin et al. 1977). The contribution of pre-anthesis carbon to yield increases when crops are subjected to drought stress (Austin et al. 1980a,b; Bidinger et al. 1980) or heavy rust (*Puccinia hordei* Otth.) infection (Scott and Dennis-Jones 1976). Bonnett and Incoll (1992) demonstrated differences in the contribution of pre-anthesis assimilation to grain yield between growing seasons, which are presumably linked to climatic differences between years.

Plant organs differ in their relative contribution of retranslocated assimilate to the grain (Daniels et al. 1982; Bonnett and Incoll 1992). The shoot below the peduncle contributes approximately 90% of the total reserve capacity of the plant (Daniels et al. 1982). Stems provide temporary storage of assimilates before and after anthesis, and contribute a significant amount of assimilates to the developing grain. This is especially so during the latter portion of the grain-filling phase, when senescence decreases current photosynthate supply; they can also be a major contribution to yield in stressful years (Bonnett and Incoll 1992; Schnyder 1993). Retranslocation of assimilates occurs earlier when plants are subjected to shading (reduction in source size) and is delayed in de-eared plants (reduction in sink size), indicating that retranslocation of assimilates occurs when sink demand exceeds source capacity (Bonnett and Incoll 1993a). Fructose, sucrose and, in particular, fructans are the major water-soluble carbohydrates that are accumulated in the stems and retranslocated to the grain (Bonnett and Incoll 1993b). The short-term reserve pool exists mainly as sucrose rather than as starch, while the long-term reserve exists as fructan until after anthesis. Reserve pools may temporarily hold half or more of mature grain photosynthate and only if these reserve pools are depleted does manipulation of sink demand influence source leaf assimilation (Merbach et al. 1994).

7.2 Retranslocation of Nitrogen During Grain Filling

Little work has been done on the relative importance of pre- and post-anthesis N uptake on yield and grain protein accumulation in barley. Le Gouis

(1992) reported that about 80–90% of grain N was assimilated prior to anthesis. Post-anthesis N uptake and its contribution to grain N and total plant N accumulation were greater in a growing season associated with higher grain and straw dry matter yields. Apparent N retranslocation (measured as the difference between plant N content at anthesis and straw, including chaff, N content at maturity) is generally lower when water and N fertilizer levels are high (Bulman and Smith 1993b). In field experiments, apparent post-anthesis N uptake increases with increasing levels of N applied at seeding (Bulman and Smith 1993b). Genotypic and N treatment differences appear to be associated, at least in part, with differences in the amount of late tillering.

8
Phosphorus Uptake and Mycorrhizal Associations

The role of mycorrhizae in agriculture has been the subject of several reviews (Jeffries 1987; Bethlenfalvay and Linderman 1992; Hamel 1996). Like other members of the family Graminae, barley forms symbiotic associations with arbuscular mycorrhizal fungi. As with other mycorrhizal plants the bulk of the benefit derived from mycorrhizal symbiosis is the effective extension of the root system by the extramatrical hyphae which, by extending several centimetres beyond the root surface, facilitate the uptake of relatively immobile nutrients, principally phosphorus (Bieleski 1973). Positive correlations observed between barley yield, fungal abundance and nutrient uptake (Hamel et al. 1994; Wani et al. 1991) suggest that mycorrhizal symbiosis is a component of barley nutrient acquisition strategy.

Arbuscular mycorrhizal fungi are obligate biotrophs and production of non-host plants such as members of the Brassicaceae or Chenopodiaceae or fallowing prior to a barley crop reduces their spore populations and delays mycorrhizal colonization of barley roots (Black and Tinker 1979). This delay in mycorrhizal development may lead to poorer nutrition and reduced growth during early crop development. Responsiveness to mycorrhizal inoculation varies among plant species and, conversely, efficiency of different mycorrhizal strains differs. Barley is less responsive to mycorrhizal inoculation than crops such as leek, corn or most legumes; however, yield increases often result from inoculation (Owusu-Bennoah and Mosse 1979; Clarke and Mosse 1981; Powell 1981; Jakobsen 1983). Out of 11 tested mycorrhizal fungal species *Glomus epigaeus* and *G. macrocarpum* were most efficient on barley, increasing yield by 21 to 24% respectively (Jensen 1984). In another study, *G. constrictum* and *G. fasciculatum* increased grain yield by 63 and 53% repectively, while *G. margarita* had no effect (Jensen 1982). Mycorrhizal root colonization levels have been found to vary widely in barley. Boyetchko and Tewari (1990) have pointed out that selection of barley cultivars under high P fertility, as is usually the case, may select for mycorrhizal resistance rather than susceptibility as mycorrhizal development is deterred under high fertility conditions.

9 Potassium Uptake

Correlations between net potassium uptake and plant growth have been reported (Pitman 1972; Jensén and Pettersson 1980). Glass and Perley (1980) noted that cultivars exhibiting a high potassium uptake during the seedling stage tended to have elevated tissue potassium concentration during the later stages as well as high grain potassium concentrations. However, Siddiqi and Glass (1987) reported that while potassium influx was negatively correlated with root potassium concentration, shoot potassium concentration, shoot growth rate, and the shoot to root ratio had little influence on potassium uptake. Genotypic differences in net potassium uptake, the rate of potassium influx and potassium uptake efficiency (biomass produced per unit potassium concentration or per unit mg potassium absorbed) have been reported, but these are not well understood (Memon et al. 1985a,b).

10 Plant Lodging and Growth Regulators

Lodging in small grain cereals reduces yield and grain quality. Lodging can cause yield losses directly by reducing photosynthesis and indirectly by promoting conditions conducive to disease development and increased harvest losses (Stoskopf 1985). Among the small grain cereals, barley is the second most susceptible to lodging after oat (*Avena sativa* L.). Lodging near the time of heading is the most detrimental for barley (Sisler and Olson 1951; Stanca et al. 1979). The earlier lodging occurs during grain filling the more severe the reductions in grain yield and 1000-grain weight (Sisler and Olson 1951). Spring barley yields can be decreased by as much as 38% when lodging occurs near heading (Stanca et al. 1979). Lodging prior to anthesis is less serious because newly elongating internodes recover by bending upwards in a negative geotrophic

reduces yields much more than lodging to a 45° angle (Sisler and Olson 1951).

Several synthetic plant growth regulators (PGRs) are used commercially as lodging control agents. Among those most used in agricultural production are chlormequat and ethephon. Chlormequat is an anti-gibberellin agent that inhibits GA_3 synthesis, resulting in reduced levels of gibberellin in the plant (Jung 1984). While chlormequat is effective in reducing height and lodging in wheat, its poor absorption and rapid transport within the plant make it much less effective for barley (Skopik and Cervinka 1967). Chlormequat alters within-plant uniformity and increases the production of spike-bearing tillers in barley (Cartwright et al. 1985; Matthews et al. 1981).

Ethephon releases ethylene through a relatively slow pH-dependent reaction (Warner and Leopold 1969). The resulting ethylene inhibits the movement of auxin in stem tissues, thereby reducing auxin's ability to promote stem elongation (Sachs and Hackett 1972). Ethephon prevents lodging by reducing crop height (Sachs and Hackett 1972; Dahnous et al. 1982). Ethephon has been registered for barley in some regions of North America (Wiersma et al. 1986). However, yield increases following ethephon treatment are not consistent (Paterson et al. 1983), and this has caused some resistance to their large-scale adoption (Leary and Oplinger 1983). Yield reductions due to these PGRs have been reported for spring and winter barley (Paterson et al. 1983). Green et al. (1988) reported that mepiquat chloride (1,1-dimethyl-piperidinium chloride), which has effects similar to chlormequat, has the potential to increase barley yields only under conditions conducive to lodging.

Early application of chlormequat can increase barley yields (Matthews et al 1981; Koranteng and Matthews 1982) without affecting plant height or lodging (Waddington and Cartwright 1986), by increasing the number of spike-bearing tillers produced. This offers a potential strategy for the use of chlormequat as a yield promoter, rather than a yield protectant (Koranteng and Matthews 1982). Barley plants treated at the onset of stem elongation with ethephon alone or in combination with chlormequat increase the number of spike-bearing tillers but this has little effect on individual grain weight or yield (Ma and Smith 1991b). Early application of both PGRs (from three-leaf to stem-elongation stages) temporarily retards main-stem apical development (Ma and Smith 1991a), leading to intra-spike and intra-plant alterations in source-sink relationships. Changes in intra-plant relationships increase the production of spike-bearing shoots through enhanced tillering rather than enhanced tiller survival (Ma and Smith 1992c). Even post-anthesis application of ethephon can modify barley yield components and yield (Ma and Smith 1992e). In general, chlormequat application at the three-leaf stage increases the number of grains per spike due to increased grain number in the main-culm spike, whereas ethephon applied at ZGS 30 increased the number of spikes m^{-2} by increasing the production of tiller spikes (Ma and Smith 1992e,f). These changes occasionally increased grain yield. For barley, fewer detrimental side effects appear to be associated with chlormequat than with ethephon application. Plant height and lodging are reduced by ethephon application at the flag-leaf stage; however, this sometimes reduces grain yields by decreasing the main-stem spike yield (Ma and Smith 1992f).

In addition to reducing lodging, ethephon application may affect barley yields by altering individual yield components either positively or negatively (Moes and Stobbe 1991b). Ethephon treatment increases the number of spikes m^{-2} in spring (Simmons et al. 1988; Moes and Stobbe 1991b; Ma and Smith 1992f, 1991b) and winter barley (Hill et al. 1982). An increased spike number can result in improved yields even in the absence of lodging (Hill et al. 1982). However, ethephon often decreases grain number per spike, and/or the 1000-grain weight (Moes and Stobbe 1991c; Ma et al. 1992). The various responses to ethephon may be due to genotypic sensitivities (Caldwell et al. 1988), environmental conditions (Wiersma et al. 1986) or both (Caldwell et al. 1988). Under continental climatic conditions, such as those that occur on the Canadian Prairies, the limited potential for late tillering and/or for the risk of yield reduction restrict the use of ethephon even when the risk of lodging is high (Foster et al. 1991; Moes and Stobbe 1991a).

11 Conclusions

Yield in barley in dependent on developmental and physiological processes as affected by environment. Temperature, irradiance and photoperiod are environmental factors which can be manipulated in a limited fashion by altering sowing date. Field management practices which influence physiology and yield include sowing density, row spacing, fertility, soil cultivation and irrigation. Breeding programmes have considerably increased the harvest index over the past century, although this appears to be reaching a plateau. Yield is related to grain number, more than grain size, and so depends on grain-bearing spikes (dependent on the seeding rate and number of tillers per plant), seeds per spike (dependent on apical development) and assimilate supply to the developing grains. Plant growth regulators affect apical development and tillering; they must be applied at the optimum apex development stage. Assimilate supply to the grain is related to photosynthesis, nutrient uptake and water use efficiency. These processes are influenced by environmental and field management factors. Dry matter and nutrients are partitioned among plant parts during growth and retranslocated to the developing grains. Partitioning and retranslocation are affected by sink size, application of growth regulators and environmental factors.

References

Ahmed IA, Abo-Elenin RA, Ahmed WE, Lieth H, Al-Masoom AA (1993) Interactive effects of salinity and both forms and levels of nitrogen fertility of growth and yield of barley. In: Towards the rational use of high salinity tolerant plants, vol 2. Agriculture and forestry under marginal soil water conditions. Proc ASWAS Conf, 8–15 Dec 1990, Al Ain, United Arab Emirates. Kluwer, Dordrecht, pp 121–128

Alaoui ACE, Simmons SR, Crookston RK (1988) Effects of tiller removal on spring barley. Crop Sci 28:305–307

Andersen MN, Jensen CR, Losch R (1992a) The interaction effects of potassium and drought in field-grown barley. I. Yield, water-use efficiency and growth. Acta Agric Scand B. Soil Plant Sci 42:34–44

Andersen MN, Jensen CR, Losch R (1992b) The interaction effects of potassium and drought in field-grown barley. II. Nutrient relations, tissue water content and morphological development. Acta Agric Scand B. Soil Plant Sci 42:45–56

Anderson-Taylor G, Marshall C (1983) Root-tiller interrelationships in spring barley (*Hordeum distichum* (L.) Lam.). Ann Bot 51:47–58

Apel P, Maass I, Lau I (1993) Relationship between carbon isotope discrimination and grain yield in spring barley under drought conditions. Cereal Res Commun 21:345–347

Araus JL, Buxo R (1993) Changes in carbon isotope discrimination in grain cereals from the northwestern Mediterranean basin during the past seven millenia. Aust J Plant Physiol 20:117–128

Arkebauer TJ, Weiss A, Sinclair TR, Blum A (1994) In defense of radiation use efficiency: a response to Demetriades-Shah et al. (1992) Agric For Meteorol 68:221–227

Arnau G, Monneveux P (1995) Physiology and genetics of terminal water stress tolerance in barley. J Genet Breed 49:327–331

Arvidsson J, Jokela WE (1995) A lysimeter study of soil compaction effects on evapotranspiration in a barley crop. Swed J Agric Res 25:109–118

Aslam M, Huffaker RC (1982) In vivo nitrate reduction in roots and shoots of barley (*Hordeum vulgare* L.) seedlings in light and darkness. Plant Physiol 70:1009–1013

Aslam M, Huffaker RC (1984) Dependency of nitrate reduction on soluble carbohydrates in primary leaves of barley under aerobic conditions. Plant Physiol 75:623–628

Aslam M, Harbit KB, Huffaker RC (1990) Comparative effects of selenite and selenate on nitrate assimilation in barley seedlings. Plant Cell Environ 13:773–782

Aslam M, Huffaker RC, Rains DW, Rao KP (1979) Influence of light and ambient carbon dioxide concentration on nitrate assimilation by intact barley seedlings. Plant Physiol 63:1205–1209

Aslam M, Oaks A, Huffaker RC (1976) Effect of light and glucose on the induction of nitrate reductase and on the distribution of nitrate in etiolated barley leaves. Plant Physiol 58:588–591

Aspinall D (1961) The control of tillering in the barley plant. I. The pattern of tillering and its relation to nutrient supply. Aust J Biol Sci 14:493–505

Aspinall D, Paleg LG (1964) Effects of daylength and light intensity on growth of barley. III. Vegetative development. Aust J Biol Sci 17:807–822

Aspinall D, Nicholls PB, May LH (1964) The effects of soil moisture stress on the growth of barley. I. Vegetative development and grain yield. Aust J Agric Res 15:729–745

Austin RB, Edrich JA, Ford MA, Blackwell RD (1977) The fate of the dry matter, carbohydrates and ^{14}C lost from the leaves and stem of wheat during grain filling. Ann Bot 41:1309–1321

Austin RB, Bingham J, Blackwell RD, Evans LT, Ford MA, Morgan CL, Taylor M (1980a) Genetic improvements in winter wheat yields since 1900 and associated physiological changes. J Agric Sci 94:675–689

Austin RB, Morgan CL, Ford MA, Blackwell RD (1980b) Contributions to grain yield from preanthesis assimilation in tall and dwarf barley phenotypes in two contrasting seasons. Ann Bot 45:309–319

Badra A, Klinck HR (1981) The influence of inter-plant competition on the morphological development of spikes of non-tillering barley (*Hordeum vulgare* L. emend. Lam.). Can J Plant Sci 61:829–835

Baker RJ, Briggs KG (1982) Effects of plant density on the performance of 10 barley cultivars. Crop Sci 22:1164–1167

Baker CK, Gallagher JN (1983) The development of winter wheat in the field. 2. The control of primordium initiation rate by temperature and photoperiod. J Agric Sci 101:337–344

Barber MJ, Solomonson LP (1990) Assimilatory nitrate reductase: Functional properties and regulation. Annu Rev Plant Physiol Plant Mol Biol 41:225–253

Barham RW, Rasmusson DC (1981) Inheritance of photoperiod response in barley. Crop Sci 21:454–456

Barneix AJ, James DM, Watson EF, Hewitt EJ (1984) Some effects of nitrate abundance and starvation on metabolism and accumulation of nitrogen in barley (*Hordeum vulgare* L. cv. Sonja). Planta 162:469–476

Baron VS, Dick AC, Wolynetz MS (1992) Characterization of barley silage: maturity relationships for central Alberta. Can J Plant Sci 72:1009–1020

Baxter ED, Duffus CM (1973) Enzymes of carbohydrate metabolism in developing *Hordeum distichum* grain. Phytochemistry 12:1923–1928

Balkhodja R, Medrano H, Abadia A, Abadia J, Mathis P (1995) Effects of salinity on photosynthetic parameters of barley (*Hordeum vulgare* L.) grown under a triple line source sprinkler system in the field. In: Photosynthesis: from light to biosphere, vol IV. Proc Xth Int Photosynthesis Cong, Montpellier, France, 20–25 Aug, 1995. Kluwer, Dordrecht, pp 753–756

Benbelkacem A, Mekni MS, Rasmusson DC (1984) Breeding for high tiller number and yield in barley. Crop Sci 24:968–972

Bergmann H, Leinhos V, Machelett B, Geibel M, Treutter D, Feucht W (1994) Increase of stress resistance in crop plants by using phenolic compounds. Acta Hortic 381:390–397

Bergner C, Teichmann C (1993) A role for ethylene in barley plants responding to soil water shortage. J Plant Growth Regul 12:67–72
Bethlenfalvay GJ, Linderman RG (1992) Mycorrhizae in sustainable agriculture. ASA Spec Publ 54, American Society of Agronomy Publishers, Madison, Wisconsin
Bidinger FR, Musgrave RB, Fischer RA (1980) Contribution of stored pre-anthesis assimilate to grain yield in wheat and barley. Nature 270:431–433
Bieleski RL (1973) Phosphate pools, phosphate transport, and phosphate availability. Annu Rev Plant Physiol 24:225–252
Birch CJ, Long KE (1990) Effect of nitrogen on the growth, yield and grain protein content of barley (*Hordeum vulgare*). Aust J Exp Agric 30:237–242
Biscoe PV, Gallagher JN, Littleton EJ, Monteith JL, Scott RK (1975) Barley and its environment. IV. Sources and assimilate for the grain. J Appl Ecol 12:295–318
Black RLB, Tinker PB (1979) The development of endomycorrhizal root systems. II. Effects of agronomic factors and soil conditions on the development of vesicular-arbuscular mycorrhizal infection in barley and on the endophyte spore density. New Phytol 83:401–413
Bloom AJ, Finazzo J (1986) The influence of ammonium and chloride on potassium and nitrate absorption by barley roots depends on time of exposure and cultivar. Plant Physiol 81:67–69
Bloom AJ, Caldwell RM, Finazzo J, Warner RL, Weissbart J (1989) Oxygen and carbon dioxide fluxes from barley shoots depend on nitrate assimilation. Plant Physiol 91:352–356
Bloom AJ, Sukrapanna S, Warner RL (1992) Root respiration associated with ammonia and nitrate absorption and assimilation by barley. Plant Physiol 99:1294–1301
Bonnett GD, Incoll LD (1992) The potential pre-anthesis and post-anthesis contributions of stem internodes to grain yield in crops of winter barley. Ann Bot 69:219–225
Bonnett GD, Incoll LD (1993a) Effects on the stem of winter barley of manipulating the source and sink during grain-filling. I. Changes in accumulation and loss of mass from internodes. J Exp Bot 44:75–82
Bonnett GD, Incoll LD (1993b) Effects on the stem of winter barley of manipulating the source and sink during grain-filling. II. Changes in the composition of water-soluble carbohydrates of internodes. J Exp Bot 44:83–91
Bort J, Febrero A, Amaro T, Araus JL (1994) Role of awns in ear water-use efficiency and grain weight in barley. Agronomie 14:133–139
Bouman BAM, van Kasteren HWJ, Uenk D, Van Kasteren HWJ (1992) Standard relations to estimate ground cover and LAI of agricultural crops from reflectance measurements. Eur J Agron 1:249–262
Boyetchko SM, Tewari JP (1990) Effect of phosphorus and VA mycorrhizal fungi on common root rot of barley. In: Allen MR, Williams SE (eds) Proc 8th North American Conf on Mycorrhizae, Sept 5–8 1990, Jackson, Wyoming
Bridger GM, Klinck HR, Smith DL (1995) Timing and rate of ethephon application to two-row and six-row spring barley. Agron J 87:1198–1206
Brooks A, Jenner CF, Aspinall D (1982) Effects of water deficit on endosperm starch granules and on grain physiology of wheat and barley. Aust J Plant Physiol 9:423–436
Brouwer R (1962) Nutritive influences on the distribution of dry matter in the plant. Neth J Agric Sci 10:399–408
Bulman P, Smith DL (1993a) Yield and yield component response of spring barley to fertilizer nitrogen. Agron J 85:226–231
Bulman P, Smith DL (1993b) Accumulation and redistribution of dry matter and nitrogen by spring barley. Agron J 85:1114–1121
Bulman P, Smith DL (1993c) Grain protein response of spring barley to high rates and post-anthesis application of fertilizer nitrogen. Agron J 85:1109–1113
Bulman P, Smith DL (1993d) Yield and grain protein response of spring barley to ethephon and triadimefon. Crop Sci 33:798–803
Bulman P, Smith DL (1994) Post-heading nitrogen uptake, retranslocation, and partitioning in spring barley. Crop Sci 34:977–984

Bulman P, Mather DE, Smith DL (1993) Genetic improvement of spring barley cultivars grown in eastern Canada from 1910 to 1988. Euphytica 71:35–48

Butz RG, Jackson WA (1977) A mechanism for nitrate transport and reduction. Phytochemistry 16:409–417

Caldwell CD, Mellish DR, Norrie J (1988) A comparison of ethephon alone and in combination with CCC or DPC applied to spring barley. Can J Plant Sci 68:941–946

Caley CY, Dufus CM, Jeffcoat B (1990) Photosynthesis in the pericarp of developing wheat grains. J Exp Bot 41:303–307

Cameron-Mills V, von Wettstein D (1980) Protein body formation in the developing barley endosperm. Carlsberg Res Commun 45:577–594

Cannell RQ (1969a) The tillering pattern in barley varieties. I. Production, survival, and contribution to yield by component tillers. J Agric Sci (Camb) 72:405–422

Cannell RQ (1969b) The tillering pattern in barley varieties. II. The effect of temperature, light intensity and daylength on the frequency of occurrence of the coleoptile node and second tillers in barley. J Agric Sci (Camb) 72:423–435

Carefoot JM, Major DJ (1994) Effect of irrigation application depth on cereal production in the semi-arid climate of southern Alberta. Irrig Sci 15:9–16

Carreck NL, Christian DG (1991) Studies on the patterns of nitrogen uptake and translocation to the grain of winter barley intended for malting. Ann Appl Biol 119:549–559

Cartwright PM, Jaddoa KA (1985) Chemical manipulation of plant development and dominance in barley crops (*Hordeum sativum* JESS). Acta Univ Agric (Brno), Fac Agron XXXIII, 3:181–186

Cartwright PM, Jaddoa KA, Waddington SR (1985) Spike development stage in barley. Field trial methods and data handling. Assoc Appl Biol, Welleshourne, Warwick

Cerning J, Guilbot A (1973) Changes in the carbohydrate composition during development and maturation of the wheat and barley kernel. Cereal Chem 50:220–232

Chafai-Elalaoui A, Simmons SR (1988) Quantitative translocation of photoassimilates from non-surviving tillers in barley. Crop Sci 28:969–972

Chapin FS, Walter CHS, Clarkson DT (1988) Growth response of barley and tomato to nitrogen stress and its control by abscissic acid, water relations and photosynthesis. Planta 173:352–366

Chevalier P, Lingle SE (1983) Sugar metabolism in developing kernels of wheat and barley. Crop Sci 23:272–277

Chun JU (1993) Variations in rate of leaf emergence, initiation of ear primordium, stem elongation and heading time as affected by vernalization duration of barley with differing growth habits. Field Crops Res 32:159–172

Clarke C, Mosse B (1981) Plant growth responses to vesicular-arbuscular mycorrhiza. XII. Field inoculation responses of barley at two soil P levels. New Phytol 87:695–703

Cochrane MP (1983) Morphology of the crease region in relation to assimilate uptake and water loss during caryopsis development in barley and wheat. Aust J Plant Physiol 10:473–491

Cochrane MP (1985) Assimilate supply and water loss in maturing barley grains. J Exp Bot 36:770–782

Cochrane MP, Duffus CM (1979) Morphology and ultrastructure of immature cereal grains in relation to transport. Ann Bot 44:67–72

Cochrane MP, Duffus CM (1981) Endosperm cell number in barley. Nature 289:399–401

Cochrane MP, Duffus CM (1983) Endosperm cell number in cultivars of barley differing in grain weight. Ann Appl Biol 102:177–181

Cochrane MP, Paterson L, Duffus CM, Froud-Williams RJ, Harrington R, Hocking TJ, Smith HG, Thomas TH (1996) The effect of temperature on grain development and starch quality in barley. Aspects Appl Biol 45:139–146

Conry MJ, Hegarty A (1992) Effect of sowing date and seed rate on the grain yield and protein content of winter barley. J Agric Sci (Camb) 118:277–287

Corke H, Avivi N, Atsmon D (1989) Pre- and post-anthesis accumulation of dry matter and nitrogen in wild barley (*Hordeum spontaneum*) and in barley cultivars (*H. vulgare*) differing in final grain size and protein content. Euphytica 40:127–134

Correll R, Butler J, Spouncer L, Wrigley C (1994) The relationship between grain-protein content of wheat and barley and temperatures during grain filling. Aust J Plant Physiol 21:869–873
Cottrell JE, Dale JE (1984) Variation in size and development of spikelets within the ear of barley. New Phytol 97:565–573
Cottrell JE, Easton RH, Dale JE, Wadsworth AC, Adam JS, Child RD, Hoad GV (1985) A comparison of spike and spikelet survival in mainstem and tillers of barley. Ann Appl Biol 106:365–377
Creelman RA, Mullet J (1997) Structure and membrane organization of photosystem II from green plants. Annu Rev Plant Physiol Plant Mol Biol 48:641–671
Dahnous K, Vigue GT, Law AG, Konzak CF, Miller DG (1982) Height and yield response of selected wheat, barley, and triticale cultivars to Cerone. Agron J 74:580–582
Dakheel AJ, Peacock JM, Naji I, Acevedo E (1994) Carbon isotope discrimination and water use efficiency in barley under field conditions. Aspects Appl Biol 38:173–183
Dale JE (1972) Growth and photosynthesis in the first leaf of barley. The effect of time and application of nitrogen. Ann Bot 36:967–979
Daniels RW, Alcock MB, Scarisbrick DH (1982) A reappraisal of stem reserve contribution to grain yield in spring barley (*Hordeum vulgare* L.). J Agric Sci (Camb) 98:347–355
Davis MH, Simmons SR (1994) Tillering response of barley to shifts in light quality caused by neighboring plants. Crop Sci 34:1604–1610
Day W, Legg BJ, French BK, Johnston AE, Lawlor DW, Jeffers W de C (1978) A drought experiment using mobile shelters: the effect of drought on barley yield, water use and nutrient uptake. J Agric Sci (Camb) 91:599–623
de Cires A, De la Torre A, Delgado B, Lara C, De Cires A, De la Torre A (1993) Role of light and CO_2 fixation in the control of nitrate-reductase activity in barley leaves. Planta 190:277–283
de Man W, Cauberghe N (1988) Changes in lipid composition in maturing barley kernels. Phytochemistry 27:1639–1642
de Man W, Vervenne B (1988) Lipid composition of barley in relation to grain size. Phytochemistry 27:2037–2039
Dean-Drummond CE, Clarkson DT, Johnson CB (1979) Effect of shoot removal and malate on the activity of nitrate reductase assayed in vivo in barley roots (*Hordeum vulgare* cv. Midas). Plant Physiol 64:660–662
Donald CM (1968) The breeding of crop ideotypes. Euphytica 17:385–403
Donald CM (1979) A barley breeding programme based on an ideotype. J Agric Sci (Camb) 93:261–269
Douce R, Neuburger M (1989) The uniqueness of plant mitochondria. Annu Rev Plant Physiol Plant Molec Biol 40:371–414
Drew MC, Saker LR (1975) Nutrient supply and the growth of the seminal root system in barley. II. Localized, compensatory increases in lateral root growth and rates of nitrate uptake when nitrate supply is restricted to only part of the root system. J Exp Bot 26:79–90
Duffus CM, Cochrane MP (1992) Grain structure and composition. In: Shewry PR (ed) Barley: genetics, biochemistry, molecular biology and biotechnology. CAB International, Wallingford, pp 291–317
Duffus CM, Rosie R (1973) Some enzyme activities associated with the chlorophyll containing layers of the immature barley pericarp. Planta 114:219–226
Duffus CM, Rosie R (1975) Biochemical changes during embryogeny in *Hordeum distichum*. Phytochemistry 14:319–323
Duffus CM, Rosie R (1977) Carbohydrate oxidation in developing barley endosperm. New Phytol 78:391–395
Duffus CM, Rosie R (1978) Metabolism of ammonium ion and glutamate in relation to nitrogen supply and utilization during grain development in barley. Plant Physiol 61:570–574
Dunn GM, Neales TF (1993) Are the effects of salinity on growth and leaf gas exchange related? Photosynthetica 29:33–42
El Alaoui AC, Simmons SR, Crookston RJ (1992) Allocation of photoassimilates by main shoots and non-surviving tillers in barley. Crop Sci 32:1233–1237

Ellis RP (1993) Barley development from the viewpoint of a crop geneticist. Aspects Appl Biol 34:57–67
Ellis RP, Kirby EJM (1980) A comparison of spring barley grown in England and in Scotland. 2. Yield and its components. J Agric Sci (Camb) 95:111–115
Felker FC, Peterson DM, Nelson OE (1984) [^{14}C]-sucrose uptake and labeling of starch in developing grains of normal and *seg* 1 barley. Plant Physiol 74:43–46
Feller U, Keist M (1986) Senescence and nitrogen metabolism in annual plants. In: Lambers H, Neetson JJ, Stulen I (eds) Fundamental, ecological and agricultural aspects of nitrogen metabolism in higher plants. Nijhoff, Dordrecht, pp 219–234
Ferrari TE, Yoder OC, Filner P (1973) Anaerobic nitrite production by plant cells and tissues. Evidence for two nitrate pools. Plant Physiol 51:423–431
Foroutan-pour K, Ma BL, Smith DL (1997) Protein accumulation potential in barley seeds as affected by soil- and peduncle-applied N and peduncle-applied plant growth regulators. Physiol Plant 100:190–201
Foster E, Prentice N (1987) Barley. pp 337–396. In: Olson RA, Frey KJ (eds) Nutritional quality of cereal grains: genetics and agronomic improvement CSSA, Madison, Wisconsin
Foster KR, Reid DM, Taylor JS (1991) Tillering and yield responses to ethephon in three barley cultivars. Crop Sci 31:130–134
Frank AB, Bauer A, Black AL (1992) Effects of air temperature and fertilizer nitrogen on spike development of spring barley. Crop Sci 32:793–797
Frank, R, Apel P (1985) Early vegetative growth, shoot apex development and yield of spring barley under controlled environment and field conditions. Z Pflanzenzücht 94:50–63
Frey-Wyssling A, Buttrose MA (1959) Photosynthesis in the ear of barley. Nature 184:2031–2032
Friedrich JW, Huffaker RC (1980) Photosynthesis, leaf resistances, and ribulose 1,5-bisphosphate carboxylase degradation in senescing barley leaves. Plant Physiol 65:1103–1107
Fritsch H, Jung J (1984) Enzyme activities and leaf constituents in barley seedlings at different nutrient levels. Z Pflanzenphysiol 114:433–442
Gallagher JN, Biscoe PV, Scott RK (1975) Barley and its environment. V. Stability of grain weight. J Appl Ecol 12:319–336
Garcia del Moral LF, Ramos JM, Recalde L (1984) Tillering dynamics of winter barley as influenced by cultivar and nitrogen fertilizer: a field study. Crop Sci 25:179–181
Garcia del Moral MB, Garcia del Moral LF (1996) Tiller production and survival in relation to grain yield in winter and spring barley. Field Crops Res 44:85–93
Geiger DR, Servaites JC (1994) Diurnal regulation of photosynthetic carbon metabolism in C3 plants. Annu Rev Plant Physiol Plant Mol Biol 45:235–256
Giese H, Hejgaard J (1984) Synthesis of the salt-soluble proteins in barley. Pulse-labeling study of grain-filling in liquid-cultured detached spikes. Planta 161:172–177
Giese H, Anderson B, Doll H (1983) Synthesis of the major storage protein in barley. Pulse-labeling study of grain-filling in liquid cultured detached spikes. Planta 159:60–65
Glass ADM, Perley JE (1980) Varietal differences in potassium uptake by barley. Plant Physiol 65:160–164
Glass ADM, Siddiqi MY, Ruth TJ, Rufty TW (1990) Studies of the uptake of nitrate in barley. II. Energetics. Plant Physiol 93:1585–1589
Gojon A, Passama L, Robin P (1986) Root contribution to nitrate reduction in barley seedlings (*Hordeum vulgare* L.). Plant Soil 91:339–342
Golbeck JH (1992) Structure and function of photosystem I. Annu Rev Plant Physiol Plant Mol Biol 43:293–324
Gontia AS, Khare AK (1992) Optimum leaf area index of barley (*Hordeum vulgare* L.) genotypes under various planting patterns. Adv Plant Sci 5:43–52
Gonzalez-Ponce R, Mason SC, Salas ML, Sabata RJ, Herce A (1992) Environment, seed rate, and N rate influence on yield of winter barley. Fertil Res 34:59–65
Gordon AJ, Ryle GJA, Powell CE (1977) The strategy of carbon utilization in uniculm barley. J Exp Bot 28:1258–1269

Gorham J, Chapman GP (1992) Grass responses to salinity. In: Desertified grasslands: their biology and management. Papers presented at Int Symp, London, 27 Feb–1 March 1991, Linnean Society Symposium Series No 13, pp 165–180

Granstedt RC, Huffaker RC (1982) Identification of the leaf vacuole as a major nitrate storage pool. Plant Physiol 70:410–413

Grant CA, Gauer LE, Gehl DT, Bailey LD (1991) Protein production and nitrogen utilization by barley cultivars in response to nitrogen fertilizer under varying moisture conditions. Can J Plant Sci 71:997–1009

Grattan SR, Royo A, Aragues R (1994) Chloride accumulation and partitioning in barley as affected by differential root and foliar salt absorption under saline sprinkler irrigation. Irrig Sci 14:147–155

Green CF, Furmston DT, Ivins JD (1985) Time of sowing and the yield of winter barley. J Agric Sci (Camb) 104:405–411

Green CF, Chalmers IF, Packe-Drury-Lowe SJ (1988) Enhancing the performance of ethephon with mepiquat chloride barley using an adjuvant comprising acidified soyal phospholipid. Ann Appl Biol 113:177–188

Gregory PJ, Atwell BJ (1991) The fate of carbon in pulse-labelled crops of barley and wheat. Plant Soil 136:205–213

Gregory PJ, Tennant D, Belford RK (1992) Root and shoot growth, and water and light use efficiency of barley and wheat crops grown on a shallow duplex soil in a Mediterranean-type environment. Aust J Agric Res 43:555–573

Groot JJR (1988) Post floral nitrogen behaviour in grain crops: measurements and theory. In: Jorna ML, Slootmaker LAJ (eds) Cereal breeding related to integrated cereal production. PUDOC, Wageningen, pp 93–97

Gu J, Marshall C (1988) The effect of tiller removal and tiller defoliation on competition between the main shoot and tillers on spring barley. Ann Appl Biol 112:597–606

Guitard AA (1960) The influence of variety, temperature, and stage of growth on the response of spring barley to photoperiod. Can J Plant Sci 40:65–80

Hadjichristodoulou A (1990) Stability of 1000-grain weight and its relation with other traits of barley in dry areas. Euphytica 51:11–17

Hamel C (1996) Prospects and problems pertaining to the management of arbuscular mycorrhizae in agriculture. Agric Ecosyst Environ 60:197–210

Hamel C, Dalpé Y, Lapierre C, Simard RR, Smith DL (1994) Composition of the vesicular-arbuscular mycorrhizal fungi population in an old meadow as affected by pH, phosphorus and soil disturbance. Agric Ecosyst Environ 49:223–231

Harlan JR (ed) (1975) Crops and man. Am Soc Agron, Crop Sci Soc Am, and Soil Sci Soc Am Madison, Wisconsin, 295 pp

Harris HC (1994) Water use efficiency of crop rotations in a Mediterranean environment. Aspects Appl Biol 38:165–172

Hay RKM (1986) Sowing dates and the relationships between plant and apex development in winter cereals. Field Crops Res 14:321–337

Hay RKM (1995) Harvest index: a review of its use in plant breeding and crop physiology. Ann Appl Biol 126:197–216

Hayman DS (1974) Plant growth responses to vesicular-arbuscular mycorrhiza. VI. Effect of light and temperature. New Phytol 73:71–80

Haynes RJ, Goh KM (1978) Ammonium and nitrate nutrition of plants. Biol Rev 53:465–510

Hejnowicz Z, Wloch W (1980) Growth and development of shoot apex in barley. I. Morphology and in higher plants. Physiol Plant 84:616–624

Hendrix JE, Lindin JC, Smith DH, Ross CW (1986) Influence of assimilate partitioning during inflorescence development on grain numbers in winter wheat. In: Cronshaw J, Lucas WJ, Giaquinta RT (eds) Phloem transport AR Liss, New York, pp 557–559

Hill DH, Joice R, Squires NRW (1982) Cerone: its uses and effect on the development winter barley. In: McLaren JS (ed) Chemical manipulation of crop growth and development. Butterworth, London, pp 391–397

Ho KM, Jui PY (1989) Duration and rate of kernel filling in barley (*Hordeum vulgare* L.). Cereal Res Commun 17:69–76
Hoff T, Stummann BM, Henningsen KW (1992) Structure, function and regulation of nitrate reductase in higher plants. Phys Plant 84:616–624
Höfner W, Kühn H (1982) Effect of growth regulator combinations on ear development, assimilate translocation and yield in cereal crops. In: McLaren JS (ed) Chemical manipulation of crop growth and development. Butterworth, London, pp 375–390
Horton P, Ruban AV, Walters RG (1996) Regulation of light harvesting in green plants. Annu Rev Plant Physiol Plant Mol Biol 47:655–684
Hunt S, Layzell DB (1993) Gas exchange of legume nodules and the regulation of nitrogenase activity. Annu Rev Plant Physiol Plant Mol Biol 44:483–511
Hutley-Bull PD, Schwabe WW (1982) Some effects of low-concentration gibberellic acid and retardant application during early growth on morphogenesis in wheat In: McLaren JS (ed) Chemical manipulation of crop growth and development. Butterworth, London, pp 329–342
Izaurralde RC, Chanasyk DS, Juma NG (1994) Soil water under conventional and alternative cropping systems in cryoboreal subhumid central Alberta. Can J Soil Sci 74:85–92
Jakobsen I (1983) Vesicular-arbuscular mycorrhizae in field grown crops II. Effect of agronomic factors and soil conditions on the development of vesicular-arbuscular mycorrhizal infection in barley and on the endophyte spore density. New Phytol 83:401–413
Jamieson PD, Francis GS, Wilson DR, Martin RJ (1995a) Effects of water deficits on evapotranspiration from barley. Agric For Meteorol 76:41–58
Jamieson PD, Martin RJ, Francis GS, Wilson DR (1995b) Drought effects on biomass production and radiation-use efficiency in barley. Field Crops Res 43:77–86
Jedel PE, Helm JH (1992) Agronomic response of six-row barley cultivars to supplemental fertilization and late-season fungicide treatments. Can J Plant Sci 72:1121–1130
Jedel PE, Helm JH (1994) Assessment of western Canadian barleys of historical interest. I. Yield and agronomic traits. Crop Sci 34:922–927
Jeffries P (1987) Use of mycorrhizae in agriculture. CRC Crit Rev Biotechnol 5:319–357
Jenkyn JF, Anilkumar TB (1990) Effects of defoliation at different growth stages and in different grain-filling environments on the growth and yield of spring barley. Ann Appl Biol 116:591–599
Jenkyn JF, Carter N, Darby RJ, Gutteridge RJ, Mullen LA, Plumb RT, Ross GJS, Todd AD, Widdowson FV, Wood DW (1992) Effects of seven factors on the growth and yield of winter barley grown as a third consecutive take-all susceptible crop and of growing the barley after oats or a fallow. J Agric Sci 119:303–333
Jensen A (1982) Influence of four vesicular-arbuscular mycorrhizal fungi on nutrient uptake and growth in barley (*Hordeum vulgare*). New Phytol 90:45–50
Jensen A (1984) Response of barley, pea, and maize to inoculation with different vesicular-arbuscular mycorrhizal fungi in irradiated soil. Plant Soil 78:315–323
Jensen B (1993) Rhizodeposition by $^{14}CO_2$-pulse-labelled spring barley grown in small field plots on sandy loam. Soil Biol Biochem 25:1553–1559
Jensén P, Pettersson S (1980) Varietal variation in uptake and utilization efficiency of potassium (rubidium) in high-salt seedlings of barley. Physiol Plant 48:411–415
Johnston GFS, Jeffcoat B (1977) Effects of some growth regulators on tiller bud elongation in cereals. New Phytol 79:239–245
Jones HG, Kirby EJM (1977) Effects of manipulation of number of tillers and water supply on grain yield in barley. J Agric Sci (Camb) 88:391–397
Jung J (1984) Plant bioregulators in cereal crops: Action and use. In: Ory RL, Ritig FR (eds) Bioregulators, chemistry and uses. Am Chem Soc, Washington, DC, pp 29–43
Kang MY, Sugimoto Y, Sakamoto S, Fuwa H (1985) Developmental changes in the amylose content of endosperm starch of barley (*Hordeum vulgare* L.) during the grain-filling period after anthesis. Agric Biol Chem 49:3463–3466
Kaufmann ML, McFadden AD (1960) The competitive interaction between barley plants grown from large and small seeds. Can J Plant Sci 40:623–629

Kernich GC, Halloran GM (1996) Temperature effects on the duration of the spikelet growth phase and spikelet abortion in barley. J Agron Crop Sci 176:23–29

Kernich GC, Halloran GM, Flood RG (1995) Relative effects of photoperiod and irradiance on pre-anthesis development in spring barley (*Hordeum vulgare* L.). Cer Res Commun 23:425–431

Kernich GC, Halloran GM, Flood RG (1996) Constant and interchanged photoperiod effects on the rate of development in barley (*Hordeum vulgare*). Aust J Plant Physiol 23:489–496

Kiniry JR, Jones CA, O'Toole JC, Blanchet R, Cabelguenne M, Spanel DA (1989) Radiation-use efficiency in biomass accumulation prior to grain-filling for five grain-crop species. Field Crops Res 20:51–64

Kirby EJM (1967) Effect of plant density upon growth and yield of barley. J Agric Sci (Camb) 68:317–324

Kirby, EJM (1973a) The control of leaf and ear size in barley. J Exp Bot 24:567–578

Kirby EJM (1973b) Effect of temperature on ear abnormalities in uniculm barley. J Exp Bot 24:935–947

Kirby EJM (1977) The growth of the shoot apex and apical dome of barley during ear initiation. Ann Bot 41:1297–1308

Kirby EJM (1995) Factors affecting rate of leaf emergence in barley and wheat. Crop Sci 35:11–19

Kirby EJM, Appleyard M (1980) Effects of photoperiod on the relation between development and yield per plant of a range of spring barley varieties. Z. Pflanzenzuecht 85:226–239

Kirby EJM, Appleyard M (1981) Cereal development guide. In: Barron A (ed) Cereal unit. National Agric Centre, Stoneleigh, Kenilworth, Warwickshire, UK

Kirby EJM, Faris DG (1970) Plant population induced growth correlations in the barley plant main shoot and possible hormonal mechanisms. J Exp Bot 21:787–798

Kirby EJM, Faris DG (1972) The effect of plant density on tiller growth and morphology in barley. J Agric Sci (Camb) 78:281–288

Kirby EJM, Jones HG (1977) The relations between the main shoot and tillers in barley plants. Agric Sci 88:381–389

Kirby EJM, Riggs TJ (1978) Developmental consequences of two-row and six-row ear type in spring barley. II. Shoot apex, leaf and tiller development. J Agric Sci (Camb) 91:207–216

Kirby EJM, Rymer JL (1975) The vascular anatomy of the barley spikelet. Ann Bot 39:205–211

Kirby EJM, Appleyard M, Fellowes G (1985a) Effect of sowing date and variety on leaf emergence and number of leaves of barley and wheat. Agronomie 5:117–126

Kirby EJM, Appleyard M, Fellowes G (1985b) Leaf emergence and tillering in barley and wheat. Agronomie 5:193–200

Kirkman, MA, Shewry PR, Miflin BJ (1982) The effect of nitrogen nutrition on the lysine content and protein composition of barley seeds. J Sci Food Agric 33:115–127

Kivisaari S, Elonen P (1974) Irrigation as a method of preventing detrimental late tillering of barley. J Sci Agric Soc Finl 46:194–207

Kjack JL, Witters RE (1974) Physiological activity of awns in isolines of Atlas barley. Crop Sci 14:243–248

Knapp JS, Harms CL, Volenec JJ (1987) Growth regulator effects on wheat culm nonstructural and structural carbohydrates and lignin. Crop Sci 27:1201–1205

Koesmarno HK, and Sedcole JR (1994) A method for the analysis of barley kernel growth data from designed experiments. J Agric Sci (Camb) 123:25–33

Koranteng GO, and Matthews S (1982) Modifications of the development of spring barley by early applications of CCC and GA_3 and the subsequent effects on yield components and yield. In: McLaren JS (ed) Chemical manipulation of crop growth and development. Butterworth, London, pp 343–357

Kreis M, Shewry PR (1992) The control of protein synthesis in developing barley seeds. In: Shewry PR (ed) Barley: genetics, biochemistry, molecular biology and biotechnology. CAB International, Wallingford, pp 319–333

Kucey RMN (1987) Nitrogen fertilizer application practices for barley production under southwestern Canadian prairie conditions. Commun Soil Sci Plant Anal 18:753–769
Kuusk A (1995) A Markov chain model of canopy reflectance. Agric For Meteorol 76:221–236
Laberge DE, MacGregor AW, Meredith WOS (1973) Changes in the free sugar content of barley kernels during maturation. J Inst Brew Lond 79:471–477
Laude HM, Ridley JR, Suneson CA (1967) Tiller senescence and grain development in barley. Crop Sci 7:231–233
Lauer JG (1991) Barley tiller response to plant density and ethephon. Agron J 83:968–973
Lauer JG, Partridge JR (1990) Planting date and nitrogen effects on spring malting barley. Agron J 82:1083–1088
Lauer JG, Simmons SR (1985) Photoassimilate partitioning of main shoot leaves in field-grown spring barley. Crop Sci 25:851–855
Lauer JG, Simmons SR (1988) Photoassimilate partitioning by tillers and individual tiller leaves in field-grown spring barley. Crop Sci 28:279–282
Leakey RRB (1971) The effect of changing plant density on floral initiation and development of barley (cv. Sultan). J Agric Sci (Camb) 77:135–139
Leary WP, Oplinger ES (1983) Effect of plant growth regulators and nitrogen on the agronomic performance of small grains. In: Cooke AR (ed) Proc Plant Growth Regul Soc Am, vol 10, East Lansing, Michigan, 19–23 June, pp 277–286
Lee RB, Drew MC (1989) Rapid, reversible inhibition of nitrate influx in barley by ammonium. J Exp Bot 40:741–752
Lee RB, Rudge KA (1986) Effects of nitrogen deficiency on the absorption of nitrate and ammonium by barley plants. Ann Bot 57:471–486
Le Gouis J (1992) A comparison between two- and six-row winter barley genotypes for above-ground dry matter production and distribution. Agronomie 12:163–171
Leibovitch S, Ma BL, Maloba WE, Smith DL (1992) Spring barley responses to row spacing and fungicide triadimefon in regions with a short crop growing season. J Agron Crop Sci 169:209–215
Leigh RA, Johnston AE (1985) Nitrogen concentrations in field-grown spring barley: an examination of the usefulness of expressing concentrations on the basis of tissue water. J Agric Sci (Camb) 105:397–406
Lemerle D, Fisher JA, Hinkley RB (1993) Radiometry accurately measures chlorsulfuron injury to barley. Aust J, Agric Res 44:13–21
Lewicki S, Chery J (1992) Étude de l'accumulation et de la remobilisation de l'azote chez l'orge (*Hordeum vulgare* L.): comparaison de variétés possédant ou non le gène de seminainisme (*sdw*). Agronomie 12:235–245
Lewis OAM, Chadwick S (1983) A ^{15}N investigation into nitrogen assimilation in hydroponically-grown barley (*Hordeum vulgare* L. cv. Clipper) in response to nitrate, ammonium and mixed nitrate + ammonium nutrition. New Phytol 95:635–646
Lewis OAM, Watson EF, Hewitt EJ (1982) Determination of nitrate reductase activity in barley leaves and roots. Ann Bot 49:31–37
Lingle SE, Chevalier P (1984) Movement and metabolism of sucrose in developing barley kernels. Crop Sci 24:315–319
Lingle SE, Chevalier P (1985) Development of the vascular tissue of the wheat and barley caryopsis as related to the rate and duration of grain filling. Crop Sci 25:123–128
Lipiec J, Szustak A, Tarkeiwicz S (1992) Soil compaction: responses of soil physical properties and crop growth. Zesx Probl Post N Roln 398:113–117
Loboda T (1993) Gas exchange of different spring cereal genotypes under normal and drought conditions. Photosynthetica 29:567–572
Lopez-Castaneda C, Richards RA (1994a) Variation in temperate cereals in rainfed environments. I. Grain yield, biomass and agronomic characteristic. Field Crops Res 37:51–62
Lopez-Castaneda C, Richards RA (1994b) Variation in temperate cereals in rainfed environments. III. Water use and water-use efficiency. Field Crops Res 39:85–98

Ma BL, Smith DL (1991a) Apical development of spring barley in relation to chlormequat and ethephon. Agron J 83:270–274
Ma BL, Smith DL (1991b) The effects of ethephon, chlormequat chloride and mixtures of ethephon and chlormequat chloride applied at the beginning of stem elongation on spike-bearing shoots and other yield components of spring barley (*Hordeum vulgare* L.). J Agron Crop Sci 166:127–135
Ma BL, Smith DL (1992a) Apical development of spring barley under field conditions in northeastern North America Crop Sci 32:144–149
Ma BL, Smith DL (1992b) A new method of supplying substances to cereal inflorescences. Crop Sci 32:191–194
Ma BL, Smith DL (1992c) Modification of tiller productivity in spring barley by application of chlormequat or ethephon. Crop Sci 32:735–740
Ma BL, Smith DL (1992d) Growth regulator effects on aboveground dry matter partitioning during grain fill of spring barley. Crop Sci 32:741–746
Ma BL, Smith DL (1992e) Post-anthesis ethephon effects on yield of spring barley. Agron J 84:370–374
Ma BL, Smith DL (1992f) Chlormequat and ethephon timing and grain production of spring barley. Agron J 84:934–939
Ma BL, Leibovitch S, Maloba WE, Smith DL (1992) Spring barley responses to nitrogen fertilizer and ethephon in regions with a short crop growing season. J Agron Crop Sci 169:151–160
MacGregor AW, LaBerge DE, Meredith WOS (1971) Changes in barley kernels during growth and maturation. Cereal Chem 48:255–269
MacLeod LC, Duffus CM (1988a) Temperature effects on starch granules in developing barley grains. J Cereal Sci 8:29–37
MacLeod LC, Duffus CM (1988b) Reduced starch content and sucrose synthase activity in developing endosperm of barley plants grown at elevated temperatures. Aust J Plant Physiol 15:367–375
Macnicol PK, Jacobsen JV (1992) Endosperm acidification and related metabolite changes in the developing barley grain. Plant Physiol 98:1098–1104
Mandal BK, and Mahapatra SK (1993) Mulch and irrigation effects on growth and water use of barley, lentil linseed grown as sole and intercrop. Indian J Plant Physiol 36:246–249
Martinoia E, Heck U, Dalling MJ, Matile Ph (1983) Changes in chloroplast number and chloroplast constituents in senescing barley leaves. Biochem Physiol Pflanzen 178:147–155
Matthews S, Koranteng GO, Thomson WO (1981) Tillering and ear production: opportunities for chemical regulation. In: Hawkins AF, Jeffcoat B (eds) Opportunities for manipulation of cereal productivity. Br Plant Growth Regul Group, Monogr 7, Wye College, University of London, Ashford, Kent, 2–3 Sept, pp 88–96
Mattsson M, Schjoerring JK (1996) Ammonia emission from young barley plants: influence of N source, light/dark cycles and inhibition of glutamine synthetase. J Exp Bot 47:477–484
May LH, Buttrose MS (1959) Physiology of cereal grain. II. Starch granule formation in the developing barley kernel. Aust J Biol Sci 12:146–149
McAndrew DW, Fuller LG, Wetter LG (1994) Grain and straw yields of barley under four tillage systems in northeastern Alberta. Can J Plant Sci 74:713–722
McDonald AML, Stark JR, Morrison WR, Ellis RP (1991) The composition of starch granules from developing barley genotypes. J Cereal Sci 13:93–112
McDonald GK (1990) The growth and yield of uniculm and tillered barley over a range of sowing dates. Aust J Agric Res 41:449–461
Memon AR, Saccomani M, Glass ADM (1985a) Efficiency of potassium utilization by barley varieties: the role of subcellular compartmentation. J Exp Bot 36:1860–1876
Memon AR, Siddiqi MY, Glass ADM (1985b) Efficiency of K^+ utilization by barley varieties: activation of pyruvate kinase. J Exp Bot 36:79–90

Merbach W, Schumann F, Romer W, Reining E (1994) Apparent CO_2 assimilation, ^{14}C incorporation, and ^{14}C translocation in spring barley influenced by sink manipulations. Photosynthetica 30:593–601
Metzger DD, Czaplewski SJ, Rasmussen DC (1984) Grain-filling duration and yield in spring barley. Crop Sci 24:1101–1105
Miflin BJ, Lea PJ (1977) Amino acid metabolism. Annu Rev Plant Physiol 28:299–329
Miflin BJ, Burgess SR, Shewry PR (1981) The development of protein bodies in storage tissues of seeds. J Exp Bot 32:199–219
Miller BL, Huffaker RC (1985) Differential induction of endoproteinases during senescence of attached and detached barley leaves. Plant Physiol 78:442–446
Moes J, Stobbe EH (1991a) Barley treated with ethephon. I. Yield components and net grain yield. Agron J 83:86–90
Moes J, Stobbe EH (1991b) Barley treated with ethephon. II. Tillering pattern and its impact on yield. Agron J 83:90–94
Moes J, Stobbe EH (1991c) Barley treated with ethephon: III. Kernels per spike and kernel mass. Agron J 83:95–98
Mogensen VO (1992) Effect of drought on growth rate of grains of barley. Cereal Res Commun 20:225–231
Mogensen VO, Mortensen G, Jensen CR (1994) Photosynthesis of flag leaves and ears of field grown barley during drought. Eur J Agron 3:111–116
Monteith JL (1994) Validity of the correlation between intercepted radiation and biomass. Agric For Meteorol 68:213–220
Natr L, Apel P (1974) The influence of the rate of photosynthate production on the rate of dry matter accumulation in barley kernels. Photosynthetica 8:53–56
Newman CW, Newman RK (1992) Nutritional aspects of barley seed structure and composition. In: Shewry PR (ed) Barley: genetics, biochemistry, molecular biology, and biotechnology. CAB International, Wallingford, pp 351–368
Newman CW, Newman RK, Graham H (1990) Influence of the waxy starch gene on the carbohydrate composition of covered and hull-less barleys. Cereal Foods World 35:335
Nicholls PB, May LH (1963) Studies on the growth of the barley apex. I. Interrelationships between primordium formation, apex length, and spikelet development. Aust J Biol Sci 16:567–571
Nosberger J, Thorne GN (1965) The effect of removing florets or shading the ear of barley on production and distribution of dry matter. Ann Bot 29:635–644
Nutbeam AR, Duffus CM (1976) Evidence for C_4 photosynthesis in barley pericarp tissue. Biochem. Biophys Res Commun 70:1198–1203
Nutbeam AR, Duffus CM (1978) Oxygen exchange in the pericarp green layer of immature cereal grains. Plant Physiol 62:360–362
Öhlén E, Larsson C-M (1992) Nitrate assimilatory properties of barley grown under long-term N limitation: effects of local nitrate supply in split-root cultures. Physiol Plant 85:9–16
Olsen OA, Krekling T (1980) Grain development in normal and high lysine barley. Hereditas 93:147–160
Owusu-Bennoah E, Mosse B (1979) Plant growth response to vesicular-arbuscular mycorrhizae XI. Field inoculation responses in barley, lucerne and onion. New Phytol 83:671–679
Pageau D, Tremblay GF (1995) Effect of quackgrass on the productivity of barley seeded at different seeding rates and row spacings. Can J Plant Sci 75:613–618
Paterson WGW, Blackett GA, Gill WD (1983) Plant growth regulator trials on spring and winter barley. Research and Development Note 16:24, The Scottish Agric. Colleges, Edinburgh, Scotland
Perby H, Jensén P (1984) Net uptake and partitioning of nitrogen and potassium in cultivars of barley during ageing. Physiol Plant 1:559–565

Peterson LW, Huffaker RC (1975) Loss of ribulose 1,5 diphosphate carboxylase and increase in proteolytic activity during senescence of detached primary barley leaves. Plant Physiol 55:1009–1015

Petr J, Cerny V, Hruska L (1988) Yield formation in the main field crops. Developments in crop science. Elsevier, Amsterdam, 331 pp

Pettersson R (1989) Above-ground growth dynamics and net production of spring barley in relation to nitrogen fertilizer. Swed J Agric Res 19:135–145

Pettersson S, Jensén P (1978) Allosteric and non-allosteric regulation of rubidium influx in barley roots. Physiol Plant 44:110–114

Pitman MG (1972) Uptake and transport of ions in barley seedlings. III. Correlation between transport to the shoot and relative growth rate. Aust J Biol Sci 25:905–919

Powell CLL (1981) Inoculation of barley with efficient mycorrhizal fungi stimulates seed yield. Plant Soil 59:487–489

Rajasekhar VK, Oelmüller R (1987) Regulation of induction of nitrate reductase and nitrite reductase in higher plants. Physiol Plant 71:517–521

Ramos JM, De-la-Morena I, Garcia-del-Moral LF, De-la-Morena I (1995) Barley response to nitrogen rate and timing in a Mediterranean environment. J Agric Sci 125:175–182

Reid DA, Wiebe GA (1979) Taxonomy, botany, classification and world collection. In: Barley: origin, botany, culture, winter hardiness, genetics, utilisation, pests. USDA Agric Handb No 38, pp 79–104

Rekika D, Arnau G, El-Jaafari S, Monneveux P, Mathis P (1995) Photosynthetic gas exchange parameters as predictive criteria for drought resistance in durum wheat and barley. In: Photosynthesis: from light to biosphere, vol IV. Proc Xth Int Photosynthesis Congr, Montpellier, France, 20–25 Aug 1995. Kluwer, Dordrecht, pp 721–724

Richards RA (1992) Increasing salinity tolerance of grain crops: is it worthwhile? Plant Soil 146:89–98

Riggs TJ, Hanson PR, Start ND, Miles DM, Morgan CL, Ford MA (1981) Comparison of spring barley varieties grown in England and Wales between 1880 and 1980. J Agric Sci (Camb) 97:599–610

Ruckenbauer P, Kirby EJM (1973) Effects of kinetin on the growth and development of barley and its interaction with root size. J Agric Sci (Camb) 80:211–217

Sachs RM, Hackett WP (1972) Chemical inhibition of plant height. Hortic Science 7:440–447

Salam A, Al-Tahir OA, Al-Tahir MA (1991) Soil moisture regime effects on productivity of some barley (*Hordeum vulgare* L.) genotypes. Ann Agric Sci 36:121–127

Samuelson ME, Eliasson L, Larsson C-M (1992) Nitrate-regulated growth and cytokinin responses in seminal roots barley. Plant Physiol 98:309–315

Savin R, Nicolas ME (1995) Effects of short periods of drought and high temperature on grain growth and starch accumulation of two malting barley cultivars. Aust J Plant Physiol 23:201–210

Schnyder H (1993) The role of carbohydrate storage and redistribution in the source-sink relations of wheat and barley during grain filling – a review. New Phytol 123:233–245

Scott RK, Dennis-Jones R (1976) The physiological background of barley. J Natl Inst Agric Bot 14:182–187

Scott WR, Appleyard M, Fellowes G, Kirby EJM (1983) Effect of genotype and position in the ear on carpel and grain growth and mature grain weight of spring barley. J Agric Sci (Camb) 100:383–391

Sharif R, Dale JE (1980a) Growth regulating substances and the growth of tiller buds in barley; effects of IAA and GA_3. J Exp Bot 31:1191–1197

Sharif R, Dale JE (1980b) Growth regulating substances and the growth of tiller buds in barley: effects of cytokinins. J Exp Bot 31:921–930

Sharma PN, Alka-Tripathi, Tripathi A (1994) Water relations and photosynthesis in barley plants grown under conditions of deficiency and toxicity of iron. Indian J Plant Physiol 37:85–88

Shen ZG, Shen QR, Liang YC, Liu YL (1994) Effect of nitrogen on the growth and photosynthetic activity of salt-stressed barley. J Plant Nutr 17:787–799

Shewry PR, Field JM, Kirkman MA, Faulks AJ, Miflin BJ (1980) The extraction, solubility and characterization of two groups of barley storage polypeptides. J Exp Bot 31:393–407

Shewry PR, Franklin J, Parmar S, Smith SJ, Miflin BJ (1983) The effects of sulphur starvation on the amino acid and protein composition of barley grain. J Cereal Sci 1:21–31

Siddiqi MY, Glass ADM (1987) Regulation of K^+ influx in barley: evidence for a direct control of influx by K^+ concentration of root cells. J Exp Bot 38:935–947

Simmons SR, Lauer JG (1986) Shoot photoassimilate partitioning patterns during the tillering phase in spring barley (*Hordeum vulgare* L.). In: Cronshaw J, Lucas WJ, Giaquinta RT (eds) Phloem transport Liss, New York, pp 519–526

Simmons SR, Rasmusson DC, Wiersma JV (1982) Tillering in barley: genotype, row spacing, and seeding rate effects. Crop Sci 22:801–805

Simmons SR, Oelke EA, Wiersma JV, Lueschen WE, Warnes DD (1988) Spring wheat and barley responses to Ethephon. Agron J 80:829–834

Simpson PG, Siddique KHM (1994) Soil type influences relative yield of barley and wheat in a Mediterranean-type environment. J Agron Crop Sci 172:147–160

Sisler WW, Olson PJ (1951) A study of methods of influencing lodging in barley and the effect of lodging upon yield and certain quality characteristics. Sci Agric 31:177–186

Skopik P, Cervinka M (1967) Sudy of the effect of CCC (2-chloroethyl trimethylammonium chloride) on cereals. 1. Movement and changes of CCC after application to wheat the barley leaves. Rost/Vyroba 13:547–558

Small IS, Gray JC (1984) Synthesis of wheat leaf nitrate reductase de novo following induction with nitrate and light. Eur J Biochem 145:291–297

Soares MIM, Lewis OAM (1986) An investigation into nitrogen assimilation and distribution in fruiting plants of barley (*Hordeum vulgare* L. cv. Clipper) in response to nitrate, ammonium and mixed nitrate and ammonium nutrition. New Phytol. 104:385–393

Somers DA, Kuo T-M, Kleinhofs A, Warner RL, Oaks A (1983) Synthesis and degradation of barley nitrate reductase. Plant Physiol 72:949–952

Stanca AM, Jenkins G, Hanson PR (1979) Varietal responses in spring barley to natural and artificial lodging and to a growth regulator. J Agric Sci (Camb) 93:449–456

Stanca AM, Terzi V, Cattivelli L (1992) Biochemical and molecular studies of stress tolerance in barley. In: Shewry PR (ed) Barley: genetics, biochemistry, molecular biology and biotechnology. CAB International, Wallingford, pp 277–288

Stoskopf NC (1985) Cereal Grain Crops. 516 pp. Reston Publishing Co. Reston, Virginia

Stulen I (1986) Interactions between nitrogen and carbon metabolism in a whole plant context. In: Lambers H, Neetson JJ, Stulen I (eds) Fundamental, ecological and agricultural aspects of nitrogen metabolism in higher plants. Nijhoff, Dordrecht, pp 261–278

Tallberg A (1982) Characterization of high-lysine barley genotypes. Hereditas 96:229–245

Taylor JS, Foster KR, Caldwell CD (1991) Ethephon effects on barley in central Alberta. Can J Plant Sci 71:983–995

Tester RF, South JB, Morrison WR, Ellis RP (1991) The effects of ambient air temperature during the grain-filling period on the composition and properties of starch from four barley cultivars. J Cereal Sci 13:113–127

Tew TL, Rasmusson DC (1978) Scoring photoperiod sensitivity in barley. Crop Sci 18:897–989

Theoulakis N, Iconomou E, Bladenopoulos K (1992) Harvest index as a selection criterion for improving grain yield in segregating populations of barley. Rachis 11:3–6

Thomas A, Fukai S (1995) Growth and yield response of barley and chickpea to water stress under three environments in southeast Queensland III. Water use efficiency, transpiration efficiency and soil evaporation. Aust J Agric Res 46:49–60

Thompson WJ, Scragg EB, Mathews S (1984) The tolerance of winter barley to hormone-based herbicides in relation to apical and morphological development. In: Proc Crop Protection in

northern Britain, Scottish Crops Research Institute Invergowrie, Dundee. Conf held of Dundee University, Dundee, UK, 19–22 March, pp 31–37
Thorne GN (1962) Survival of tillers and distribution of dry matter between ear and shoot of barley varieties. Ann Bot 26:37–54
Thorne GN (1963) Varietal differences in photosynthesis of ears and leaves of barley. Ann Bot NS 27:155–174
Thorne GN (1965) Photosynthesis of ears and flag leaves of wheat and barley. Ann Bot 29:317–329
Torp J, Doll H, Haahr V (1981) Genotypic and environmental influence upon the nutritional composition of barley grain. Euphytica 30:719–728
Tully RE, Hanson AD (1979) Amino acid translocated from turgid and water stressed barley leaves. I. Phloem exudation studies. Plant Physiol 64:460–466
Turley RH, Ching TM (1986a) Storage protein accumulation in "Sico" barley seed as affected by late foliar applications of nitrogen. Crop Sci 26:778–782
Turley RH, Ching TM (1986b) Physiological responses of barley leaves to foliar applied urea-ammonium nitrate. Crop Sci 26:987–993
Udvardi MK, Day DA (1997) Metabolic transport across symbiotic membranes of legume nodules. Annu Rev Plant Physiol Plant Mol Biol 48:493–523
Waddington SR, Cartwright PM (1986) Modification of yield components and stem length in spring barley by the application of growth retardants prior to main shoot stem elongation. J Agric Sci (Camb) 107:367–375
Waddington SR, Cartwright PM, Wall PC (1983) A quantitative scale of spike initial and pistil development in barley and wheat. Ann Bot 51:119–130
Walker KC, Matthews S (1991) Effect of autumn nitrogen on the growth and yield of winter barley in the north of Scotland. J Agric Sci (Camb) 117:279–285
Walpole PR, Morgan DG (1971) A quantitative study of grain-filling in three cultivars of *Hordeum vulgare* L. Ann Bot 35:301–310
Wani SP, McGill WB, Tewari JP (1991) Mycorrhizal and common root-rot infection, and nutrient accumulation in barley grown on Breton loam using N from biological fixation or fertilizer. Biol Fert Soils 12:46–54
Warner RL, Huffaker RC (1989) Nitrate transport is independent of NADH and NAD(P)H nitrate reductases in barley seedlings. Plant Physiol 91:947–953
Warner RL, Kleinhofs A (1992) Genetics and molecular biology of nitrate metabolism in higher plants. Physiol Plant 85:245–252
Warner HL, and Leopold AC (1969) Ethylene evolution from 2- chloroethyl phosphonic acid. Plant Physiol 44:156–158
Watson PA, Duffus CM (1988) Carbon dioxide fixation by detached cereal caryopses. Plant Physiol 87:504–509
Weibull J (1987) Seasonal changes in the free amino acids of oat and barley phloem sap in relation to plant growth stage and growth of *Rhopalosiphum padi*. Ann Appl Biol 111:729–737
Weibull J, Ronquist F, Brishammar S (1990) Free amino acid composition of leaf exudates and phloem sap. A comparative study in oats and barley. Plant Physiol 92:222–226
Wells SA, Dubetz S (1970) Reaction of two barley cultivars to a period of soil water stress during heading. Can J Plant Sci 50:701–704
Wessolek G, Rengar M (1993) Influence of irrigation design on water balance and water use efficiency. Kulturtech Z Landentwick 34:9–18
Wiersma DW, Oplinger ES, Guy SO (1986) Environmental and cultivar effects on winter wheat response to ethephon plant growth regulator. Agron J 78:761–764
Williams JM, Duffus CM (1977) The development of endosperm amyloplasts during grain maturation in barley. J Inst Brew 84:47–50
Williams RH, Cartwright PH (1980) The effect of applications of a synthetic cytokinin on shoot dominance and grain yield in spring barley. Ann Bot 46:445–452
Williams RH, Hayes JD (1977) The breeding implications of studies on yield and its components in contrasting genotypes of spring barley. Cereal Res Commun 5:113–118

Winter H, Lohauss G, Heldt HW (1992) Phloem transport of amino acids in relation to their cytosolic levels in barley leaves. Plant Physiol 99:996–1004
Woodward EJ, Marshall C (1988) Effects of plant growth regulators and nutrient supply on tiller bud outgrowth in barley (*Hordeum distichum* L.). Ann Bot 61:347–354
Woodward EJ, Marshall C (1989) Effects of plant growth regulators on tiller bud outgrowth in uniculm cereals. Ann Appl Biol 114:597–608
Wright D, Hughes LG (1987) Relationships between time, temperature, daylength and development in spring barley. J Agric Sci 109:365–373
Zadoks JC, Chang TT, Konzak CF (1974) A decimal code for the growth stages of cereals. Weed Res 14:415–421
Zagal E (1994) Carbon distribution and nitrogen partitioning in a soil-plant system with barley (*Hordeum vulgare* L.), ryegrass (*Lolium perenne* L.), rape (*Brassica napus* L.) grown in a $^{14}CO_2$-atmosphere. Plant Soil 166:63–74

Rice

A.R. SHARMA[1] and D.P. SINGH[2]

1 Introduction

Rice is the second most important cereal in the world, occupying a land area of nearly 150 million ha with an average productivity of 3.5 t/ha. It is grown largely in Asia (>90% of the total area) with smaller areas in Latin America, Africa and parts of Europe. The cultivation of rice extends over a wide range of climatic, soil and hydrological conditions – from wet tropical to semi-arid and warm temperate regions, from heavy clay to poor sandy soils and from dryland to swampland in fresh or brackish water. Geographically, rice is grown between the latitudes of 53°N (in north-east China) and 35°S (in New South Wales, Australia). However, it grows best between the latitudes of 45°N to 40°S (Grist 1986). Rice cultivation extends from below sea level to an altitude of more than 2500 m in the Himalayas. Topographical conditions vary from uplands (drylands) in hilly or plateau regions, with problems of deficit soil moisture, to medium lands with efficient water management and lowlands with excess water up to a depth of 5 m (Yoshida 1977). In tropical countries rice is cultivated mostly during the wet season, coinciding with the period of active monsoon rains. However, in some south and south-east Asian countries, two to three crops are harvested in a year by extending cultivation to the dry season with the help of irrigation.

Rice belongs to the genus *Oryza*, tribe Oryzeae and family Gramineae. Of its more than 20 species, only *Oryza sativa* Linn. and *Oryza glaberrima* Steud are cultivated. *Oryza sativa* is the dominant species and probably originated in south-east Asia, particularly in India and Indo-China, from where the richest diversity of cultivated forms has been reported (Vishnu-Mittre 1977). Now, its cultivation extends throughout the rice-growing regions of the world. *Oryza glaberrima* originated in West Africa, where it is still grown in small areas under upland conditions but is gradually being replaced by *O. sativa*. Although the two species have only marginal morphological differences in ligule size and glume pubescence, their hybrids are mostly sterile. However, success in hybridization between *O. sativa* and *O. glaberrima* has recently been achieved (WARDA 1996; Dingkuhn et al. 1998). *Oryza sativa* is divided into two main subspecies, *indica* and *japonica*, which have some distinct morphological,

[1] Division of Agronomy, [2] Division of Soil Science and Microbiology Central Rice Research Institute, Cuttack 753006, India

Crop Yield, Physiology and Processes
D. L. Smith, C. Hamel (Eds.)
© Springer-Verlag Berlin Heidelberg 1999

physiological and genetic characteristics. A third group called *javanica* is also added to designate *bulu* and *gundil* varieties of Indonesia (Sharma and Sampath 1985). The two cultivated species and their wild relatives are diploids ($2n = 24$), while most of the other wild species are tetraploids ($2n = 48$).

Rice-growing ecosystems are broadly classified into three major categories: irrigated, rain-fed uplands and rain-fed lowlands - including deep-water and tidal wetlands, occupying an area of 55, 13 and 32%, respectively (IRRI 1994a). The green revolution of the mid-1960s brought about large increases in the yield of irrigated rice, primarily due to the introduction of modern high-yielding varieties and increased use of chemical fertilizers. The increased yield of these semi-dwarf, photo-insensitive varieties has been mainly due to improvement in the harvest index, owing to their better physiological efficiency. On the other hand, productivity of rice in the diverse rain-fed ecosystems of south and south-east Asia has not shown much improvement, due to non-availability of suitable varieties and various biotic and abiotic stresses. We discuss here various aspects of production physiology and approaches for achieving a further breakthrough in the yield potential of rice (*O. sativa*), with emphasis on irrigated ecosystems.

2 Growth and Development

Rice plant, like any other cereal, consists of roots, stem, leaves and panicle-bearing spikelets (Fig. 1). Cultivated rice is generally considered a semi-aquatic annual plant, although in the tropics it is also capable of perennial growth. Ratooning is practised in some places for raising the second crop but it is not a standard practice. The growth duration of rice varies from 2–8 months, depending on the variety and environmental conditions. Its life cycle has two distinct phases - vegetative and reproductive (including ripening). Development of the plant from germination to maturity is characterized by a series of discrete periods, each identified by a process of change in the structure, size or mass of specific organs. Such periods are those required for the germination of seeds and the emergence of seedlings; the initiation of the root, leaf, tiller or spikelet primordia; growth of different vegetative organs; and flowering and filling of spikelets. A 120-day variety, when grown in a tropical environment, spends about 60 days in the vegetative phase and 30 days each in the reproductive and ripening phases (Fig. 2). Total dry-matter production increases progressively up to maturity (Fig. 3). Leaf weight increases up to flowering and then decreases due to drying and death of lower leaves. Rapid increase in the dry weight of leaf sheath and culm up to flowering is followed by a sharp decline thereafter due to translocation of accumulated plant reserves to the panicle.

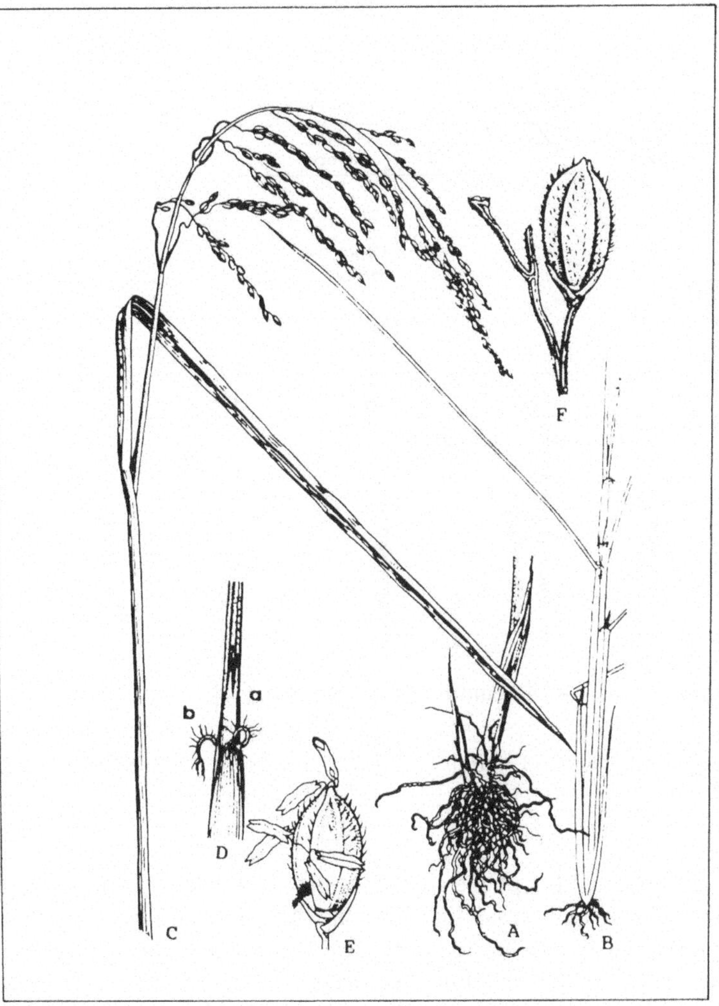

Fig. 1A–F. Different parts of the rice plant. **A** Plant base with roots ×1/2; **B** portion of the stem showing leaf ×1/4; **C** panicle ×1/2; **D** ligule (*a*) and auricle (*b*) ×1^1/$_2$; **E** spikelet ×5; **F** part of receme showing one mature spikelet attached and the reduced glumes remaining after the rest of another mature spikelet has fallen (E–F ×5). (Grist 1986)

2.1 Vegetative Growth

The vegetative phase is characterized by growth of roots, shoots and leaves, and is divided into two subphases, the active-vegetative and vegetative-lag phase. The active-vegetative phase lasts to maximum tillering and is accompanied by a rapid increase in plant height, tiller number and dry-matter

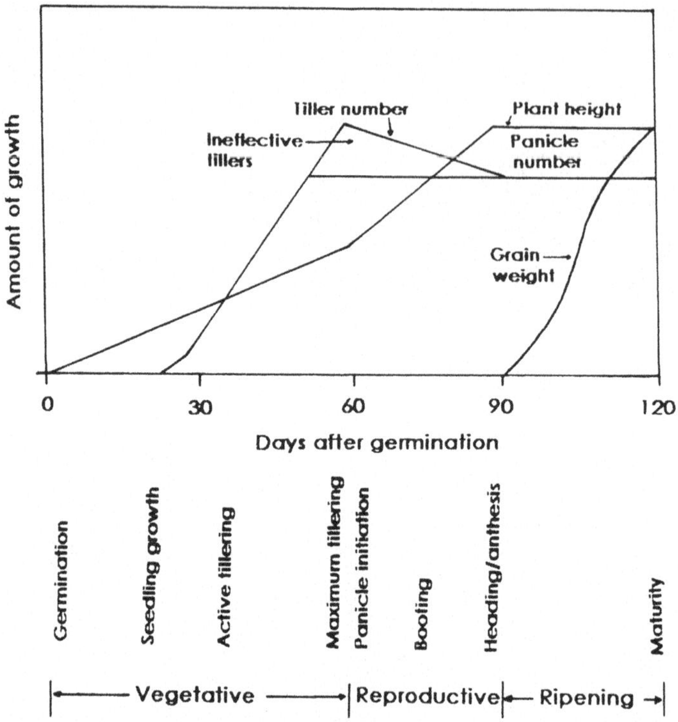

Fig. 2. Growth stages and amount of growth in different plant parts of a medium-duration variety. (IRRI 1993b)

production. The vegetative-lag phase continues up to panicle initiation. Plant height and dry matter increase less rapidly during this phase and tiller number decreases, primarily due to a reduction in the leaf N content and accumulation of less dry matter per tiller. Its duration depends on the variety and on climatic conditions, especially day length and temperature. During the vegetative-lag phase, maximum tillering, internode elongation and panicle initiation occur almost simultaneously in varieties of 105–120 days duration and successively in long-duration varieties (>140 days) (De Datta 1981).

The fibrous root system of rice consists of mainly two kinds of roots, the seminal and secondary adventitious roots. The seminal roots grow from the radicle and are subsequently replaced by the freely branched, secondary adventitious roots, produced from the lower stem nodes. Nodal roots may also develop from the aboveground nodes under deep-water conditions (Catling et al. 1988). The number and weight of the roots increase with the increase in tiller number, reaching a maximum between heading and anthesis. Development of the root system largely depends on the method of cultivation and the type of soil. Tillage and weeding facilitate lateral spread of the roots, while ridging favours their penetration deep into the soil (Grist 1986). Rice

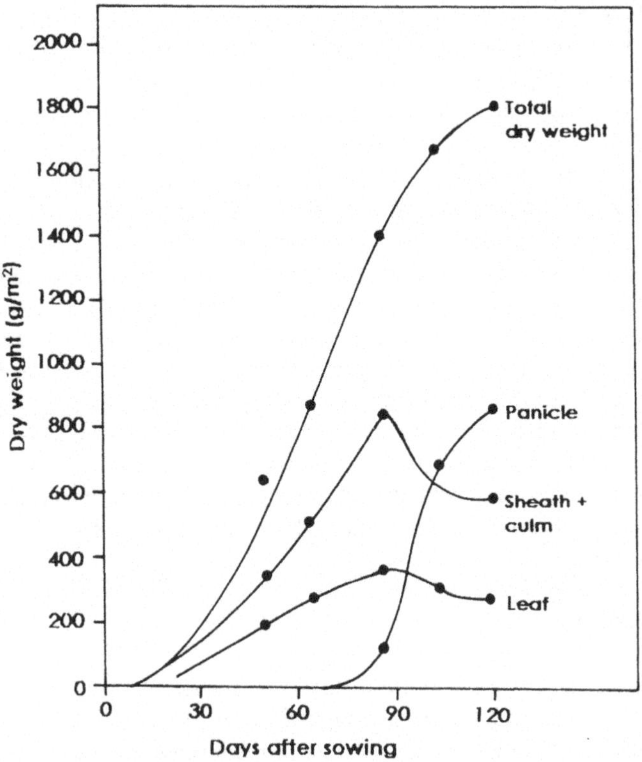

Fig. 3. Dry-matter production pattern of a modern high-yieîding rice variety. (IRRI 1987)

varieties exhibit a characteristic root system, short-duration varieties having a less developed root system than long-duration and coarse-grain varieties.

The leaves are arranged alternately along the stem and consist of a sheath enveloping the stem and the leaf blade or lamina. Auricles and ligules are generally present at the junction of the leaf sheath and blade (Fig. 1). The number of leaves on a stem equals the number of nodes. The physiological growth stage of a variety is indicated by the number of fully developed leaves on the main stem (De Datta 1981). Leaves develop consecutively and live for a short period. There are five to six active leaves in the beginning but only two to three during grain ripening. An increase in total leaf area per plant is caused mainly by increasing tiller number in a high-tillering variety and increasing the size of the successive leaves in a low-tillering variety. Application of N fertilizer has the maximum effect on leaf growth by increasing both the number of tillers and leaf size (Murata and Matsushima 1978).

The main stem is about 6 to 12 mm thick, more or less erect, cylindrical, smooth and hollow, except at the nodes. Its final height is a varietal characteristic and depends on the number and length of internodes. Leaves arise from just below the septum (sheath node) and axillary buds of the basal leaves give rise to tillers. Tillering starts at the four to five leaf stage, when the seedling becomes self-supporting. After the emergence of primary tillers, secondary tillers emerge at the age of about 30 days in semi-dwarf, photo-insensitive varieties. The tertiary tillers appear once the production of secondary tillers is completed. Tillering is synchronized with the development of leaves on the stem. Leaves on the tillers also develop in parallel with the development of leaves on the main culm. Tertiary and late-emerging primary and secondary tillers make only a marginal contribution to grain yield (Padmaja Rao 1994). Therefore, their production should be discouraged for achieving high yields.

Tillering is a varietal characteristic; the improved semi-dwarf varieties have a higher tillering capacity than the traditional tall types. High-tillering capacity enables the plants to make maximum use of the available space and is more useful under unfavourable conditions. Tiller development is greatly influenced by solar radiation, temperature and cultural practices such as plant population and supply of water and nutrients (Murata and Matsushima 1978).

2.2 Reproductive Growth

Reproductive growth begins just before or after the maximum tillering stage, depending on the variety and on environmental conditions (De Datta 1981). This phase is characterized by the culm elongation, emergence of the flag leaf, booting, heading and filling of the spikelets. During the period from panicle initiation to flowering, panicle primordia develop and plant height and dry matter increase more rapidly. The ripening phase lasts until maturity and is marked by a rapid increase in panicle weight accompanied by a decrease in straw weight.

Panicle initiation starts with formation of the first bract, which later develops into the neck node of the panicle (Murata and Matsushima 1978). This is immediately followed by the appearance of primordia on a rachis branch near the tip of the developing panicle and differentiation of other primary rachis branches. As the primary rachis branches start growing, the secondary rachis-branch primordia appear on their dorsal side alternately in two rows. The panicle is composed of a base axis, primary and secondary branches, rudimentary glumes and spikelets (Fig. 1). During panicle development, the spikelets become distinguishable and the panicle extends upwards inside the flag leaf sheath. After the panicle has grown to a length of about 5 cm, the spikelet primordia are differentiated. The spikelets are borne on a pedicel and contain a minute axis (rachilla), on which a single floret is carried in the axils of the two-ranked bracts, called glumes. Rice has only one fully developed floret (flower) per spikelet. The floret is enclosed by lemma and palea, which may be either awned or awnless. The floret consists of six stamens and a pistil, which is further composed of stigma, style and ovary. At the time of anthesis, the lodicules at the base of the flower become turgid and thrust the lemma and palea apart.

Booting, swelling of the flag leaf sheath, occurs in the later part of panicle development (Murata and Matsushima 1978). It is followed by emergence of the panicle out of the flag leaf sheath (heading) and protrusion of the first dehiscing anthers in the terminal spikelets (anthesis). Pollination normally occurs almost simultaneously with the opening of the flowers. The stamens elongate and the anthers move out of the flowering glumes as the pollen is shed. After the pollen falls on the stigma, the pollen tube emerges from the germ pore and grows rapidly. Many pollen tubes which have entered the stigmal tissues elongate downwards, pass through the space between the pericarp and integument, and finally enter the embryo sac through the micropyle. The fusion of a male gamete with two polar nuclei forms the primary triploid endosperm nucleus, which subsequently divides to form an endosperm. The fertilized diploid egg cell divides and forms the embryo. The lemma and palea close after the double fertilization of both the egg and the polar nuclei is completed. The development of the rice grain after fertilization is a continuous process, during which the grain undergoes several distinct changes in texture and colour before it finally matures. Grain ripening takes about 25–30 days in the tropics and 45–60 days in the temperate regions, regardless of the duration of the variety (Yoshida 1981).

3 Photosynthesis and Dry-Matter Production

Photosynthesis is the primary source of dry matter for production of total biomass and grain yield. Under a given set of climatic conditions, crop photosynthesis is determined by the total leaf area per unit land area (leaf area index), the potential photosynthetic rate per unit leaf area, the proportion of light transmitted to lower leaves to that of total light received above the canopy (light transmission ratio) and the longevity of the leaves (leaf area duration), which is the product of mean leaf area index (LAI) and total number of days from flowering to harvest (Murty 1977). The contribution to grain yield from the leaf area is much higher than that from the photosynthetic rate (Ishii 1993). Therefore, as a first step towards enhancement of bioproductivity, a high photosynthetic rate has to be superimposed over an optimum LAI.

Growth is dependent on the balance between photosynthesis and respiration, which is indispensable because it generates the energy needed for dry-matter production. A reduction in respiration by the organs not directly linked with growth, such as leaf sheath, culm and panicle, can lead to increased crop growth to a certain extent (IRRI 1989). A decrease in culm height and an increase in leaf weight ratio (leaf weight to total plant weight) at flowering lead to a more favourable balance between photosynthesis and respiration and thus result in higher dry-matter production (Peng et al. 1994). Total dry-matter production is lower under tropical than under the temperate conditions because of a less favourable balance between photosynthesis and respiration at higher temperatures (Yoshida 1981). Higher rice yields have been obtained in the temperate rice-growing regions of China, Japan, Australia and the USA mainly because low temperatures result in a longer duration of all growth stages, particularly grain filling, than the tropical environments of Asia, where high temperatures increase respiration losses and decrease the grain-filling duration (Norman et al. 1984; Williams 1992). However, because of shorter crop duration and the potential for year-round cropping, yields per year are higher in the tropics from three to four crops than from one crop in the temperate regions (Yoshida 1977).

3.1 Leaf Photosynthetic Rate

Photosynthetic rate is governed by the rate of CO_2 diffusion, the activity of carboxylating enzymes, chlorophyll and N content in leaf, interveinal distance, specific leaf weight and stomatal size and frequency (Janardhan and Murty 1977; Murty 1989a). Rice is basically a C-3 plant but its single leaf net photosynthetic rate per unit leaf area is relatively higher than that of other C-3 plants, such as wheat and barley (Yoshida 1983). This is due to the evolution of certain primary basic leaf characteristics in rice during its adaptation to the aquatic environment in hot and humid tropics. These include: (1) higher stomatal frequency, especially in *indica* varieties (800–1255/mm^2) as compared to wheat (117–273/mm^2); (2) smaller mesophyll cells with higher surface to volume ratio; and (3) lower stomatal resistance to CO_2 (0.7–0.8 s/cm) as compared to wheat (4.1 s/cm) (Tsunoda 1983). Besides, rice has a high CO_2 compensation point, exhibits photorespiration and lacks bundle sheath chloroplasts (Murata and Matsushima 1978). The maximum photosynthetic rate of 67.4 mg CO_2 $dm^{-2}h^{-1}$ recorded in Texas (USA) is nearly equal to that of C-4 plants such as maize (Tsunoda 1984). The leaf blade contributes most to photosynthesis, while the contribution of other green parts such as leaf sheath, culm and panicle is relatively low (5–8%) (Tsuno et al. 1975; Janardhan and Murty 1978a). On the other hand, dark respiration at heading is much lower in leaf (50 mg $CO_2g^{-1}Nh^{-1}$) than in culm (66 mg $CO_2g^{-1}Nh^{-1}$) or panicle (145 mg $CO_2g^{-1}Nh^{-1}$) (IRRI 1989).

Leaf photosynthetic rate increases linearly with increasing leaf N concentration (Fig. 4). Therefore, leaf N contributes to crop biomass production by affecting the rate of CO_2 assimilation (Kropff et al. 1994; Mae 1997). On the other hand, the respiration rate is also a function of N content; usually a 1% reduction in N concentration results in a 10% decline in growth respiration (IRRI 1989). In low N-responsive varieties, high inputs of N result in greater leaf area expansion and decrease the photosynthetic rate (De and Rao 1989). Nitrogen deficiency increases the stomatal resistance, particularly in the lower leaves, and thereby decreases the photosynthetic rate (Yoshida and Coronel 1976). Top dressing of N before flowering improves the leaf longevity, which is

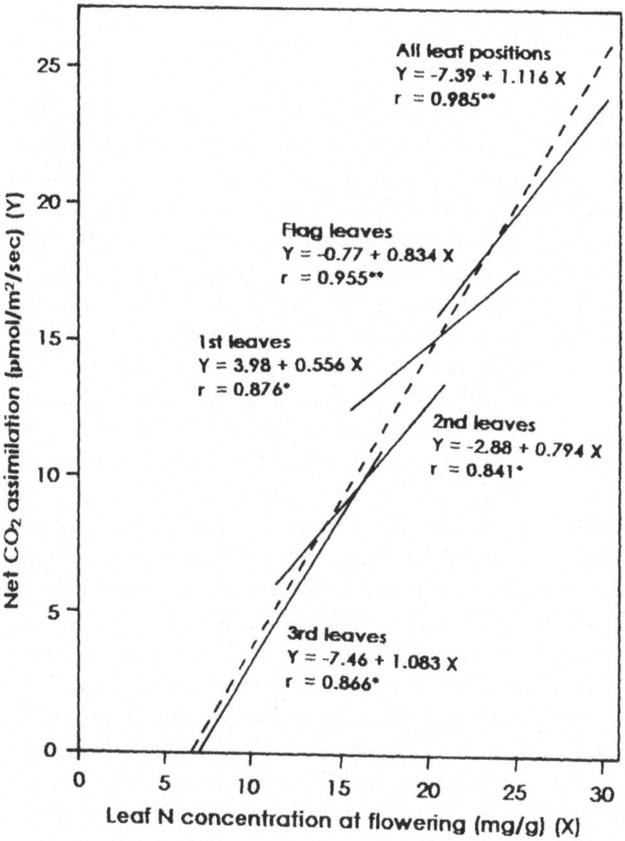

Fig. 4. Relationship between leaf N concentration at flowering and net CO_2 assimilation for individual leaf positions and across all leaf positions. (Dingkuhn et al. 1992b)

important for continued dry-matter production after heading through enhanced photosynthetic activity (Akita 1989). The net photosynthetic rate of individual leaves increases linearly with increasing light intensity up to about 50 klx (Deng et al. 1984). Thicker leaves with a higher N content show a higher light saturation point than thinner leaves with a lower N content. The light saturation level is nearly doubled when leaf N is increased from 11.5 to 17.5 mg/dm^2 and is about 83 klx at the maximum leaf N content (Tsunoda 1984). Photosynthetic rate under high light intensities is closely related to the chlorophyll content, which is directly influenced by the leaf N content. The photosynthetic rate is normally high in the dry season because of the availability of more solar radiation. However, in spite of high light intensity, a mid-day depression in photosynthetic rate occurs due to low air humidity, high temperature and decreased leaf water potential, which result in partial closure of stomata and enhanced photorespiration (Xu et al. 1984).

3.2
Leaf Orientation

Leaf orientation affects the photosynthetic rate by determining the light environment within the crop canopy. Erect leaves facilitate light penetration down the canopy and its even distribution and thereby increase crop photosynthesis, particularly under high light intensities (Yoshida 1972). The erectness of the flag leaf, from which much of the carbohydrate in the grain is derived, is considered an important desirable characteristic of high-yielding varieties. In a crop growth model, the erect-leaved canopy shows about 20% higher photosynthesis than the droopy-leaved canopy, when LAI is high (van Keulen 1976). Carbon assimilation of a leaf exposed to light on only one side is lower than when it is exposed on both sides, and this difference is greater for thick leaves with high N content (Peng et al. 1994). Therefore, a plant canopy with vertically oriented leaves gives better light penetration and higher assimilation per unit leaf area.

Distribution of light within the crop canopy is determined by the LAI and extinction coefficient, which in turn is influenced mainly by the leaf angle (Yoshida 1983). The extinction coefficient (k) is defined as the ratio between the light lost through the leaves and the light received at the top of the canopy, i.e., $\ln I/I_0 = -kF$, where I is the light intensity above the canopy, I_0 is the intensity in the crop stand and F is the LAI. The value of F is zero at the top of the canopy and highest at ground level. The higher the k value, the larger the shade produced by a unit leaf area (Murata and Matsushima 1978). The smaller the k value, the more even the light distribution within the canopy and the less the light saturation for photosynthesis of the individual leaves. As photosynthesis of rice leaves reaches light saturation level, k has a major influence on the dry-matter production. Mean k value for rice varieties ranges from 0.3 for those with erect leaves to 0.8 for those with droopy leaves (Yoshida 1983). The varieties with smaller k values show better response to high levels of N than those with greater k values (Yoshida 1981).

3.3
Leaf Area

The leaf area of rice plants increases as growth advances, reaches a maximum at around heading and thereafter decreases, due to death of the lower leaves. Significance of the leaf photosynthetic rate in determining the dry-matter production decreases as the leaf area increases (Murata and Matsushima 1978). With an increase in LAI, photosynthetic rate decreases due to increased mutual shading, but respiration increases. The LAI which supports the maximum crop growth rate is usually considered as the optimum. Assuming that gross photosynthesis increases asymptotically and respiration more or less linearly, the optimum LAI for rice is reported to be between 4 and 7 (Yoshida 1983).

However, nearly full light interception occurs at 6–8 LAI and increasing LAI up to this level is associated with linear increases in yield (Matsushima 1976). The crop planted closely and fertilized with high levels of N can attain a maximum LAI of 10 or more. In the modern varieties, increase in leaf area beyond the optimum level has no deleterious effect on grain yield, as the respiration is not a linear but a hyperbolic function of LAI and the net photosynthesis or dry-matter production remains constant even up to LAI 10 (Yoshida 1983). In a high N-solution culture experiment, gross photosynthesis (dark respiration + crop growth rate) increased with increasing LAI up to heading but thereafter this association showed hysteresis probably due to leaf senescence coupled with N remobilization and maintenance respiration (Fig. 5).

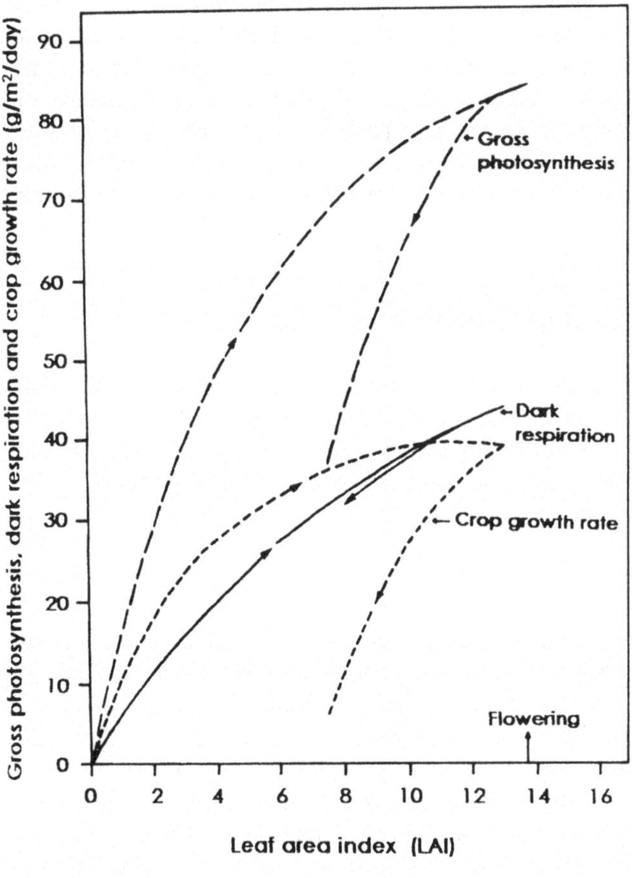

Fig. 5. Relationship between leaf area index and gross photosynthesis, dark respiration and crop growth rate in high N-solution culture. (IRRI 1989)

3.4 Varietal Differences

Evaluation of rice germplasm for photosynthetic efficiency per unit leaf area indicates large variations within and among different varietal groups (Nagamine 1990). The traditional tall and long-duration varieties often have higher leaf photosynthetic rate *per se* and total dry-matter production than the modern semi-dwarf and short-duration high-yielding varieties (Murty et al. 1976). However, their grain yields are low because of lower canopy photosynthesis during post-flowering stages, resulting from greater mutual shading, and lower harvest index, resulting from poor translocation efficiency due to premature lodging (Janardhan and Murty 1978b; Palit et al. 1979). The varietal differences in leaf photosynthesis are mostly attributed to carboxylation resistance, which is regulated by RuBP carboxylase activity (Ishii 1993). Varieties efficient in photosynthesis generally show higher photorespiration because the site of action for RuBP carboxylase and oxygenase is the same. Light intensity in many traditional rice-growing areas is low during the wet season. Therefore, there is a need to develop varieties with a high photosynthetic rate and low respiration losses under low light and high chlorophyll content, and increased dry-matter production with high reserve carbohydrates at flowering and harvest index (Murty 1989b).

4 Environmental Control of Growth and Yield

Rice is grown under a variety of diverse environmental conditions. Rainfall, temperature, light, CO_2 concentration, relative humidity and wind velocity influence its yield directly by affecting the growth, development and grain formation or indirectly by affecting the incidence of pests and diseases.

4.1 Rainfall

Rainfall is not very important for rice grown in the irrigated ecosystem but its amount and distribution affect crop growth and yield considerably in rain-fed ecosystems. Rice plant has a unique ability to adapt to deficit moisture stress as well as excessive waterlogging, which are of common occurrence in south and south-east Asia. In rain-fed uplands, the crop often experiences deficit moisture stress at different growth stages, depending on the intensity and frequency of rainfall. On the other hand, in rain-fed lowlands, a pre-flood period of 2–3 months is usually accompanied by drought, followed by a 3–5 month period of continuous submergence. These abiotic stresses influence plants' physiological functions, requiring special mechanisms for adaptation.

4.1.1 Drought

Drought occurs frequently in rain-fed uplands and limits crop growth and yield, which are directly proportional to the amount of water transpired (Yambao and Ingram 1988). Water uptake by rice decreases as soil matric potential drops below zero, leading to reduced leaf water potential. This reduces leaf expansion and CO_2 assimilation and finally causes leaf rolling and leaf-tip drying (O'Toole and Moya 1978; Dingkuhn et al. 1989). The adverse effects of drought depend on the timing, severity and duration of stress. Water stress at the vegetative stage reduces the plant height, tiller number and leaf area. However, the plants can recover once water is applied sufficiently before the onset of flowering and resume growth rates similar to those of non-stressed plants (Tuong et al. 1995). Although the vegetative phase is prolonged and the reproductive phase is delayed, grain yield is not reduced significantly due to delayed canopy senescence after recovery.

Rice plants are most sensitive to deficit moisture stress during the period from about 10 days before flowering to the end of flowering (Yoshida 1981). Water stress during this stage inhibits panicle exsertion and spikelet filling and causes high sterility, leading to decreased grain yield (Ekanayake et al. 1989). The sterility is irreversible and water supply at later stages does not help the crop. Pre-anthesis accumulation of carbohydrates in stem and leaf sheath contributes substantially to grain filling under water stress (Chaturvedi et al. 1996). High levels of sugar act as osmotic agent and help in maintaining higher leaf area and dry weight. Translocation is less affected by water stress than photosynthesis. However, carbohydrates are remobilized relatively faster in tolerant than in susceptible varieties. Therefore, high carbohydrate levels at anthesis and their faster translocation are desirable for tolerance to drought during the reproductive stage.

The two basic ways for the growth and survival of plants in dry habitats are drought escape and drought resistance. Drought escape can be achieved by matching variety duration with rainfall distribution, such that the crop passes critical developmental stages, especially panicle emergence, before or after the periods of probable water shortage. This is an important mechanism to ensure stable rice yields. Drought-resistant plants respond to dehydration by either avoidance or tolerance mechanisms. Dehydration avoidance is the ability of the plants to maintain a high water potential when soil available water is reduced, whereas dehydration tolerance is their inherent ability to sustain reduced internal water potential with least injury. Rice plants avoid dehydration either through increased water uptake by producing deep and extensive root systems or through reduced water losses by stomatal closure, leaf rolling, leaf waxiness and leaf abscission (Ingram et al. 1994).

4.1.2
Submergence

Submergence occurs during the periods of heavy rainfall in the catchment and coastal areas due to drainage congestion, resulting in partial to complete submergence of the crop. This causes plant mortality, reduced tillering and dry-matter production, leading to poor crop stand and grain yield (Pande and Reddy 1984; Sharma and Reddy 1992). The adverse effects of submergence are due to mechanical damage, siltation on leaves, reduced light, leaching of solutes from plant tissues, increased susceptibility to pests and diseases, and limited gas diffusion (Setter et al. 1995). Gases diffuse 10 000 times more slowly in water than in air (Armstrong 1979). Limited gas diffusion is the most dominant factor limiting survival of rice plants during submergence. Slow diffusion results in: (1) restricted CO_2 influx during photosynthesis; (2) restricted O_2 efflux from leaves during the day, causing enhanced photorespiration; (3) O_2 deficiency in shoots and also in roots during the night; and (4) increased accumulation of ethylene (Setter et al. 1989a). Limited gas diffusion under stagnant water results in the development of boundary layers around leaves, leading to increased resistance to gas diffusion and reduced availability of CO_2 for photosynthesis (Setter et al. 1989b; Waters et al. 1989). Further, limited O_2 supply to roots causes enhanced breakdown of carbohydrates and reduced nutrient uptake (Setter et al. 1987). Low O_2 levels lead to increased synthesis of ethylene in the plant tissues (Raskin and Kende 1984). Further, ethylene is trapped in the submerged tissues due to its slower diffusion in water. Accumulation of ethylene causes internode elongation and results in early senescence.

Adaptation of rice to O_2-deficient conditions (anoxia) is a complex phenomenon. It depends on the capacity of the plant cells to sustain and regulate alcoholic fermentation and their ability to reduce energy requirements for maintenance and membrane integrity (Greenway and Setter 1996). Additional maintenance respiration during submergence is partly for meeting the energy requirement for the synthesis of anaerobic proteins, which is considered an important adaptive feature of rice (Ricord and Pradet 1989).

Based on the depth and duration of submergence, the rain-fed lowlands are broadly classified into two groups: (1) deep-water (>50 cm depth) areas subjected to prolonged submergence (>1 month) and (2) flood-prone areas subjected to short-term (1–2 week) submergence. Rice plant requires different mechanisms for adaptation under these situations. Tall, elongating rice varieties possessing kneeing ability (stem floats horizontally on water surface and terminal section consisting of panicle bends upwards) are suited to deep-water areas and semi-tall, non-lodging types possessing submergence tolerance to flood-prone areas (Sharma and Reddy 1992).

4.1.2.1
Adaptation to Prolonged Submergence

Rice plants adapt to prolonged deep submergence mainly through internode elongation, together with lengthening of the terminal leaf sheath and blade, with rising water level (Catling et al. 1988). Sometimes, additional nodes may also contribute to stem elongation. Fast-elongating varieties grow up to 5–6 m tall to survive water depths of 3–4 m. Deep-water rices have thick culms with a massive lumen or hollow section and large air sacs that help in floating. Nodal roots develop from upper nodes and are capable of absorbing nutrients (Khan et al. 1982). The plants also produce tillers from the upper nodes but the late-emerging nodal tillers often bear infertile panicles. As the flood water recedes and part of the stem tends to float horizontally on the water surface, the terminal section bends upwards to keep the leaves and panicles above water. This kind of floating and bending characteristic is called kneeing ability, which is an essential adaptive feature of deep-water rice varieties.

The activities of various enzymes are increased during submergence (Smith et al. 1987). The enhanced activity of 1-aminocyclopropane-1-carboxylate synthase leads to increased production of ethylene. The activities of S-adenosylmethionine decarboxylase and arginine decarboxylase are related to an accelerated rate of cell division. An increase in the activity of α-amylase helps in the degradation of starch granules to support internode elongation. The latter three enzymes can be induced by application of ethylene and GA_3.

4.1.2.2
Submergence Tolerance

The plants in flood-prone areas should possess submergence tolerance because elongation may result in lodging and poor plant recovery during the post-flood period. Reduced elongation growth in such situations is beneficial, as it ensures greater availability of carbohydrates for maintenance processes during submergence (Greenway and Setter 1996). Submergence tolerance is generally associated with poor elongation ability and *vice versa* (Mazaredo and Vergara 1982; Jackson et al. 1987; Saha Ray et al. 1993), suggesting that the two characters are not compatible. There are large year to year variations in the flooding patterns with respect to the timing, duration, depth and flow of water at a given location. Therefore, a combination of moderate stem elongation ability with submergence tolerance, if achieved, would enable wider adaptability of the plant to prolonged deep-water and short-term flash-flood situations.

The amount of total non-structural carbohydrates (soluble sugars and starch) accumulated in rice plants before submergence is highly correlated with submergence tolerance (Palada and Vergara 1972; Emes et al. 1988; Chaturvedi et al. 1996). Therefore, adequate supply of carbohydrates is critical for plant survival under low O_2 concentrations, since photosynthesis is often limited during submergence. The decrease in carbohydrate levels following complete submergence is due to: (1) leakage from the tissues due to partial loss of membrane integrity; (2) decreased production through photosynthesis due to low light intensity and CO_2 concentration; (3) increased consumption in respiration due to low O_2 concentration in roots and shoots during the night; and (4) increased losses through photorespiration due to high O_2 concentration during the day (Setter et al. 1987). The relative dominance of these factors in curtailing growth depends on the characteristics of flood water, radiation levels and rice variety. Slow growth during submergence to maintain high carbohydrate levels for prolonged energy supply is an important mechanism for the tolerance of rice plants to submergence (Greenway and Setter 1996). The surplus energy can be used for maintenance processes for better plant survival during submergence, which is negatively correlated with α-amylase activity (Chaturvedi et al. 1996).

The development of suitable varieties and proper agronomic management are important for improving productivity of rain-fed lowland rice. Breeding efforts over the past two decades for developing high-yielding varieties adapted to prolonged submergence and short-term flash-flood situations have not been very successful. FR 13A, a traditional rice variety from India, is the best-known source for submergence tolerance, but it has poor combining ability. Molecular approaches, such as wide hybridization, are now being used to develop high-yielding varieties having tolerance to submergence (Brar and Khush 1995). Studies on agronomic management have shown that basal application of a moderate dose of N and P fertilizers together improves the initial plant vigour and dry-matter production, leading to increased survival during submergence (Reddy et al. 1985, 1991; Sharma 1995a). Aged, directly seeded plants withstand flooding better than young, transplanted plants that may experience excess water stress at or soon after transplanting (Sharma 1994). However, transplanting with tall, vigorous and N-fertilized seedlings (Panda et al. 1991; Sharma 1992) or vegetative tillers (Sharma 1994, 1995b) having more dry weight improves the crop stand and increases yield. Clonal propagation using tillers uprooted from the previously established crop or *Karuhan* seedlings (double transplanting technique) is an important risk-management strategy for alleviating rice productivity in flood-prone lowlands.

4.2 Temperature

Temperature greatly influences the duration and pattern of growth. Although rice is of tropical origin and can tolerate a wide range of temperatures from 10 to 50 °C, the optimum temperature for its growth is 25–31 °C (Table 1). The leaf photosynthetic rate is maximum at 25–30 °C (Tanaka 1976a) and translocation of assimilates occurs more efficiently at 17–29 °C (Nishiyama 1976).

The duration from germination to flowering and maturity in photoinsensitive varieties is determined mostly by the air temperature. When the

Table 1. Critical temperatures at different growth stages. (De Datta 1981)

Growth stage	Critical temperature (°C)		
	Low	High	Optimum
Germination	16–19	45	18–40
Seedling emergence	12–25	35	25–30
Rooting	16	35	25–28
Leaf elongation	7–12	45	31
Tillering	9–16	33	25–31
Panicle initiation	15	–	–
Panicle differentiation	15–20	30	–
Anthesis	22	35–36	30–33
Ripening	12–18	30	20–29

growing tips of vegetative parts are under water, its temperature greatly affects their growth and development (Yoshida 1983; Venkataraman 1987). However, air temperature is more important during the reproductive stages. Higher yield of rice in the temperate than in the tropical countries has been attributed mainly to an extended ripening period due to low temperatures during grain filling (Williams 1992). At high temperatures, the grain filling is faster but the grain-filling period is shorter. High day temperatures coupled with high solar radiation and low night temperatures are conducive to the production of more panicles without much reduction in spikelets/panicle (Yoshida 1983).

Temperatures above 35 °C cause different types of heat injury to rice plants, depending on the variety and growth stage (Yoshida et al. 1981). A maximum day temperature above 40 °C during hot months is not uncommon in the tropics, where a heat-susceptible variety may suffer from increased spikelet sterility (Satake and Yoshida 1977). Rice is most sensitive to high temperature around the flowering stage and its prevalence for even a few hours at anthesis has a decisive effect on spikelet sterility (Yoshida 1981). On the other hand, low temperature is a major constraint to rice production in the temperate and hilly areas of the tropical and subtropical regions. Depending on the growth stage, varying degrees of injury to rice plants are caused when the mean daily temperature drops below 20 °C (Yoshida 1983). Sensitivity to low temperature is maximum at the young micropore stage after reduction-division. The most common effects of low temperature are failure of seed germination, delayed emergence and discolouration of seedlings, stunted vegetative growth, delayed heading and partial panicle exsertion, degeneration of spikelets and high spikelet sterility, irregular maturity and the formation of abnormal grains (Kaneda and Beachell 1974).

4.3 Light

Rice growth is influenced by both intensity and duration of light. Low light during wet seasons is primarily responsible for poor yields in the tropics. Tripathy and Nayak (1995) reported a light intensity of 150–350 cal cm^{-2} day^{-1} with 4–6 h sunshine during the wet season and 400–450 cal cm^{-2} day^{-1} with 6–8 h sunshine during the dry season in eastern India. Light is generally not a growth-limiting factor in the early stages, but becomes progressively more critical with the advancing age of the crop (Fig. 6), primarily due to increased mutual shading. Artificial shading during the vegetative stage has only a marginal effect on grain yield, but yield is reduced significantly when shading is imposed during the reproductive stage (Akita et al. 1987). The most critical period for solar energy requirement is from panicle initiation to about 10 days before maturity. Grain yields are lower during wet seasons than during dry seasons, mainly because lower levels of solar radiation during the grain-filling and ripening stages reduce the percentage of filled spikelets (Murty and Sahu 1987).

Rice is basically a photoperiod-sensitive, short-day plant, but the development of photoperiod-insensitive varieties of less than 130 days duration has made its cultivation more flexible and suitable for multiple cropping. The photo-insensitive varieties mature within a fixed duration and can be planted any time during the year in the tropics. These have successfully replaced the photo-sensitive, long-duration varieties (>140 days) in irrigated areas.

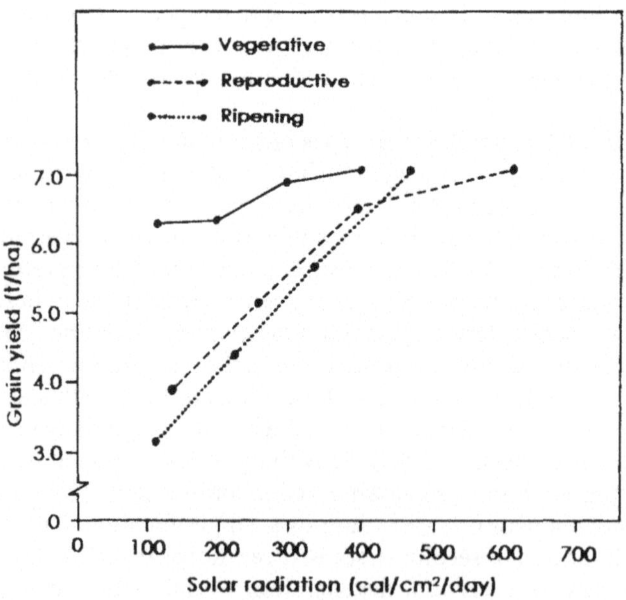

Fig. 6. Effect of solar radiation during different growth stages on grain yield. (Yoshida and Parao 1976)

However, in rain-fed lowlands, the photo-sensitive varieties are grown predominantly because of their greater tolerance to flooding than that of the photo-insensitive varieties. Their yields are fairly stable, despite the problem of lodging, damage due to typhoons, and poor fertilizer and weed management. Thus, insensitivity to day length is advantageous in some situations but disadvantageous in others.

4.4 CO_2 Concentration

An increase in yield due to CO_2 enrichment of different crops grown in the greenhouse or plastic enclosures is well established. However, increasing the CO_2 level under field conditions through additional supply is impractical and uneconomical because the CO_2-enriched air would diffuse rapidly into the atmosphere with little benefit to the crop. Nevertheless, this technique can be employed for analysing the yield-limiting processes.

Greenhouse experiments have shown that CO_2 enrichment during primordial initiation to flowering increases the sink size through enhanced photosynthesis (Yoshida et al. 1974). Dry-matter production and yield increase with increasing CO_2 concentration up to 2400 ppm under conditions of normal irradiance, temperature and adequate mineral nutrition (Yoshida 1976; Imai et al. 1985; Baker et al. 1990). However, an increase of 4 °C above ambient temperature eliminates the beneficial effects of CO_2 enrichment on yield, primarily due to a decreased grain-filling percentage (Fig. 7). Increasing both temperature and CO_2 concentration accelerates plant development but the grain yield is not altered and the quality is reduced (Zisaka et al. 1997). This shows that increasing CO_2 concentration may not be beneficial at high temperatures associated with global warming.

4.5 Relative Humidity and Wind Velocity

Relative humidity has not received much attention in our research programmes because its effects on rice are intermingled with those of temperature and light. The mean relative humidity during rice cultivation is generally negatively associated with solar radiation. Longer dew periods associated with high humidity and low night temperatures favour the incidence of diseases, particularly blast (Yoshida and Parao 1976; De Datta 1981). Photosynthesis of rice leaves is at a maximum at 50–60% relative humidity and decreases gradually with humidity beyond this level (Tsuno and Sato 1971). However, optimum humidity for photosynthesis may vary depending on the interrelationships between water absorption and transpiration.

A gentle wind causes turbulence in the crop canopy and increases photosynthesis, possibly because the air with reduced CO_2 supply in the plants' surroundings is blown out and replenished with fresh air (De Datta 1981).

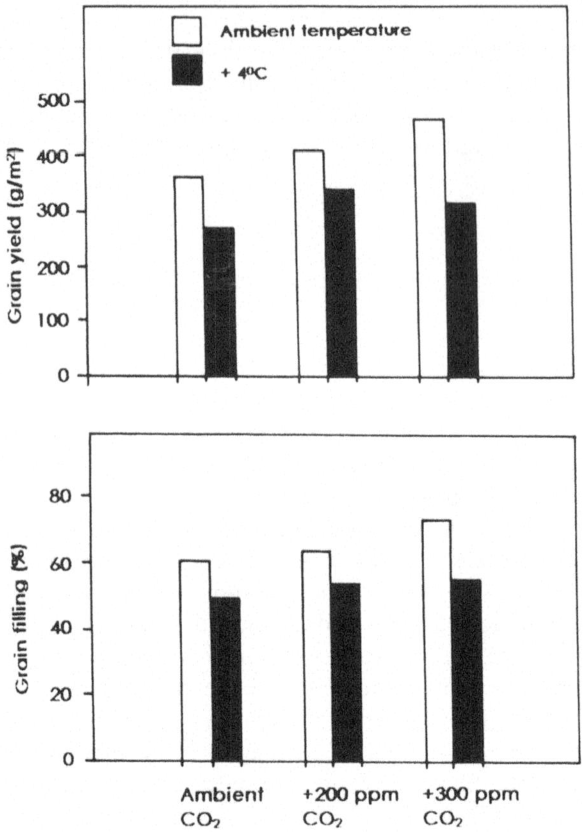

Fig. 7. Effect of CO_2 concentration and temperature on grain yield and grain filling. (IRRI 1995)

However, strong winds enhance the spread of bacterial leaf diseases and result in poor ascent of plant sap and the drying of top leaves, which are not favourable for absorption of nutrients and photosynthesis (Venkataraman 1987). It can mechanically damage the plants and cause severe lodging, thereby resulting in impaired translocation of nutrients to various plant parts, delayed ripening and shattering of maturing grains. Dry winds after flowering desiccate the panicles and increase floret sterility and the number of abortive endosperms (Vergara 1976).

5
Yield Determining Processes

Grain yield of rice broadly depends on the production of total dry matter and its partitioning to the grains. Total dry-matter production is attained in two distinct phases, pre-flowering and post-flowering. The proportion of dry matter accumulated during these phases varies with the duration of the variety. In varieties of 110–140 days duration, about 65–70% of the total dry matter is

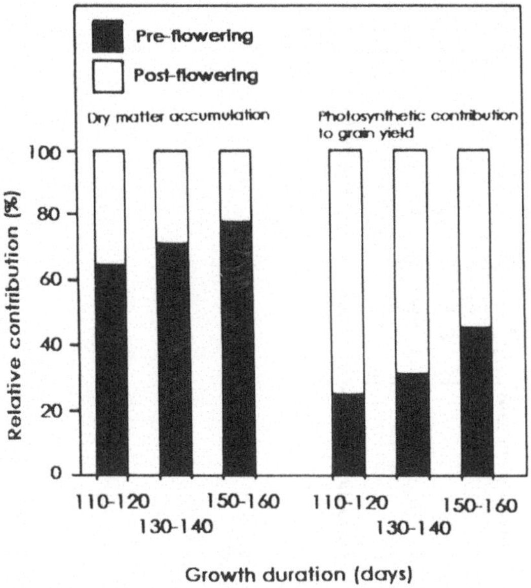

Fig. 8. Relative dry matter accumulation and photosynthetic contribution to grain yield during pre- and post-flowering stages in varieties of different maturity groups. (Venkateswarlu et al. 1977)

accumulated by the time of flowering and the remaining 30–35% is added during the post-flowering stages (Venkateswarlu et al. 1977). However, the contribution to grain yield of carbohydrates accumulated in different plant parts before flowering is only about 20–30% and the assimilates produced after flowering account for 70–80% of the total dry matter in the grain (Fig. 8).

5.1 Storage of Assimilates

Growth of vegetative parts slows down towards flowering and most of the assimilates (sugars and starch) begin to accumulate in the leaf sheath and the culm about 2 weeks before heading, reaching a maximum at around heading. The leaf sheath serves as a temporary storage site before flowering. The culm contains about 15% starch and 10% soluble carbohydrates, whereas their concentrations in the leaf blade are only 5 and 2%, respectively (Setter et al. 1994). After heading, the carbohydrate concentration in the leaf sheath and the culm decreases as the ripening proceeds and becomes lowest at about 3 weeks after heading, due to its mobilization to the panicle (Fig. 9). Removal of panicles at flowering prevents such a decrease in the carbohydrate levels, while their removal at about 20 days after flowering results in large increases in stem carbohydrates (Setter et al. 1994). The changes in leaf and stem dry weight occur almost parallel to those in the stored carbohydrates, the increase in panicle weight being nearly equal to the sum of increase in total plant dry weight and decrease in stored carbohydrates after heading. These observations suggest that carbohydrates stored in the vegetative parts are efficiently translocated to

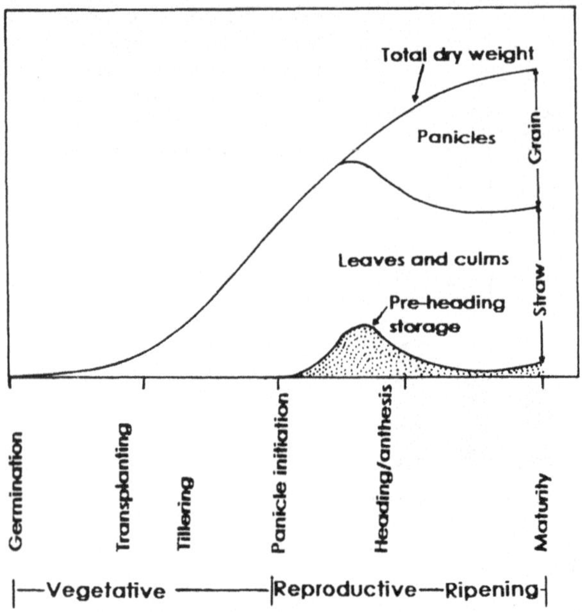

Fig. 9. Changes in the temporarily stored carbohydrates (pre-heading storage) and dry weight of various plant parts during different growth stages. (Murata and Matsushima 1978)

the panicle after anthesis. Pre-heading storage of assimilates is important for preventing the occurrence of abortive grains around 2–3 weeks after flowering, when the growth of the caryopsis is most active.

5.2 Determination of Yield Capacity

The number of panicles per unit land area, spikelets per panicle and grain weight are considered as the sink size or yield capacity. The panicle number is determined about 10 days after the maximum tillering stage and is greatly influenced by the supply of N and solar radiation (Murty 1977; Murata and Matsushima 1978). The number of spikelets per panicle is determined during the period from 32 to 5 days before heading in the temperate regions and about 10 days before heading in the tropical regions (Murata and Matsushima 1978). It is the difference between the number of spikelet primordia that have differentiated and the number of those that have degenerated, the former being strongly influenced by the supply of N during panicle differentiation and the latter by the amount of solar radiation and other environmental factors around the reduction-division stage. The size of the hull is decided 2 weeks before, and that of the kernel 3 weeks after, anthesis. There is not much variation in the size of rice grains because they are firmly enclosed by the glumes and cannot grow to a greater size than that permitted by the hull. This has been confirmed by placing an inert material inside the hull at the time of anthesis. The resultant grains at maturity are found to be indented with the shape of the enclosed object (Murata and Matsushima 1978).

5.3 Partitioning and Translocation of Assimilates

Partitioning of dry matter between the grain and vegetative parts is measured by the grain to straw ratio or the harvest index. This is determined by the amount of assimilates translocated to the panicle during grain filling, which in turn depends on the yield capacity, rates of photosynthesis and respiration, and efficiency of conducting tissues.

The presence of large amounts of starch and sugars in the leaf sheath and culm at harvest implies that either the translocation or the storage capacity is limiting. The translocation is generally poor during the wet season, at high levels of N fertilization, plant density and temperature, and in long-duration varieties with faster panicle senescence (Murty 1977). Leaf N above 2.5% and temperatures below 17 °C impair translocation. Increased sterility at high temperatures is believed to be due to increased losses of assimilates because of enhanced respiration, decreased LAI and shortened grain-filling duration (Murata and Matsushima 1978). Prevalence of cloudy weather during ripening restricts photosynthesis and, therefore, the carbohydrates accumulated in the vegetative parts before flowering are translocated more efficiently to facilitate grain filling. Varieties efficient in photosynthesis are not necessarily efficient in translocation. The traditional tall varieties have high photosynthesis but low translocation efficiency (Swain et al. 1987).

5.4 Contribution of Pre- and Post-Flowering Carbohydrates to Grain Yield

The contribution of pre- and post-flowering carbohydrates to grain yield is variable, depending on duration of the variety, level of N fertilization, method of cultivation and environmental conditions (Venkateswarlu et al. 1977; Yoshida 1983). Long-duration varieties utilise pre-flowering photosynthetic reserves more efficiently than short-duration varieties (Fig. 8). In the tropics, accumulation of carbohydrates in shoot is 10–15% (dry weight basis) during the wet season and 15–25% during the dry season (Setter et al. 1994). However, in the temperate regions, the total shoot carbohydrates are often 30–40%, presumably due to greater solar radiation (25 vs 20 MJ/day) and lower night temperatures (15 vs 25 °C) than in the tropics. Pre-flowering photosynthetic reserves support grain formation and stabilize yield, particularly under stress conditions such as low light, deficit moisture and incidence of insects and diseases. However, post-flowering photosynthates act as a major determinant of yield in varieties with less biomass at flowering (Venkateswarlu and Prasad 1980).

The contribution to the grain carbohydrates of carbohydrates stored in the vegetative parts before flowering can be calculated as follows:

$$\text{Contribution of carbohydrates accumulated in vegetative parts before flowing (\%)} = \frac{\text{Carbohydrates present in vegetative parts at flowering} - \text{Carbohydrates present in vegetative parts at maturity}}{\text{Carbohydrates present in grains at maturity}} \times 100$$

Based on this estimation, the contribution of accumulated carbohydrates under different conditions ranges from 20–40%, which implies that photosynthesis during the ripening period contributes 60–80% of the grain carbohydrates (Yoshida 1983). However, this method overestimates the percent contribution by presuming that all the carbohydrates lost from the vegetative parts during the grain filling are translocated to the grain. This assumption does not seem to be true because these carbohydrates are also partly used in respiration. Using ^{14}C-labelling, Cock and Yoshida (1972) observed that 68–72% of the stored carbohydrates was translocated to the grain, 20% was used in respiration and 12% remained in the vegetative parts. The contribution of accumulated carbohydrates to grain yield is reported at 21% using the ^{14}C-technique and 34% using the previous method (Yoshida 1981).

Post-flowering photosynthesis makes a large contribution and is closely related to grain yield (Yoshida 1972). However, increased growth duration, higher N application and greater dry-matter production at flowering decrease the contribution of post-flowering photosynthates (Venkateswarlu and Prasad 1980). The contribution of current photosynthates to grain yield also depends on the potential photosynthetic activity of leaves and their longevity during ripening, and the light environment within the crop canopy. Photosynthesis in the flag leaf continues up to maturity and has been used as an important criterion for breeding high-yielding varieties (Ishii 1993). The increase in dry weight during grain filling is correlated positively with crop photosynthesis under high levels of solar radiation and negatively with LAI under low levels of solar radiation (Yoshida 1983). The post-flowering dry-matter production being poor when solar radiation is low, a major contribution to grain yield comes from pre-flowering carbohydrate reserves. Therefore, varieties with greater dry-matter production at flowering are better suited for such conditions.

5.5 Source–Sink Relationship

Grain yield is controlled by dry-matter production which, in turn, is determined by the potential ability of the crop to photosynthesize (the source) and the capacity of spikelets to accept the photosynthates (the sink). The source is the supply of assimilates and includes both that produced before flowering and stored in the vegetative parts and that produced during post-flowering stages. The sink is the number of sub-units within the yielding parts (Squire 1990).

Grain yield is dependent on both the source and the sink capacity, which are interdependent. The sink depends on the source for carbohydrates and the efficiency of the source in synthesizing carbohydrates is triggered by the sink activity. Translocation of photosynthates from shoot to panicle depends on the sink capacity, the nature of the conducting vessels and the gradient of sugar concentration between the source and the sink (Murata and Matsushima 1978). Low yields in wet seasons are attributed to the inability of the source to ensure an adequate supply of carbohydrates to the panicle, reduced sink size, unfavourable source to sink ratio and poor translocation efficiency (Venkateswarlu 1976). It is not always possible to ascertain whether the source or the sink is limiting grain yield. The sink is considered to be limiting if a reduction in the source by removal of leaves does not affect the yield. Similarly, the source is considered to be limiting if the yield remains unchanged after a reduction in the sink by removal of spikelets. Yoshida (1981) compared the effects on rice yield of CO_2 enrichment before and after flowering to determine if the grain yield was limited by the source or the sink. The yield increased by 28.9% due to increased spikelets/m^2 when CO_2 enrichment was given before flowering and by 21.1% due to better grain filling when it was given after flowering. This suggests that both the source and the sink limit potential yield. Nonetheless, the source is reported to be limiting under a wide range of tropical environments, while the sink limitation occurs mostly in varieties whose reproductive sink is more sensitive to the environment (Squire 1990).

Venkateswarlu (1976) studied the source–sink relationship by maintaining varied plant populations (1 to 100 hills/m^2). The dimension of the source varied from 1 to 8 LAI and that of the sink size in terms of panicle number from 53 to 687/m^2. An increase in LAI beyond a certain level did not increase the grain number, which was far lower than the spikelet number (Fig. 10). The wide gap between spikelet number and grain number (20–45%) indicates the possibility of obtaining higher yield, had there been proper support from the source. Removal of panicles at anthesis has no effect on the relationship between the rate of photosynthesis and levels of major biochemical participants in photosynthesis, suggesting that the sink capacity of panicles does not limit the source capacity during grain filling (Mae 1997). Sterility can be reduced by 23% by removing 50% of the spikelets from a panicle. Although a certain amount of sterility may be inevitable due to plants' functional constitutions and climatic constraints, the considerable gap between the number of spikelets and grains clearly suggests that a greater capability of the source is essential for filling all the spikelets. Therefore, the lack of source capacity appears to be the main constraint to increased rice yield potential when the grain-filling percentage of spikelets is not high inspite of favourable climatic conditions (Mae 1997).

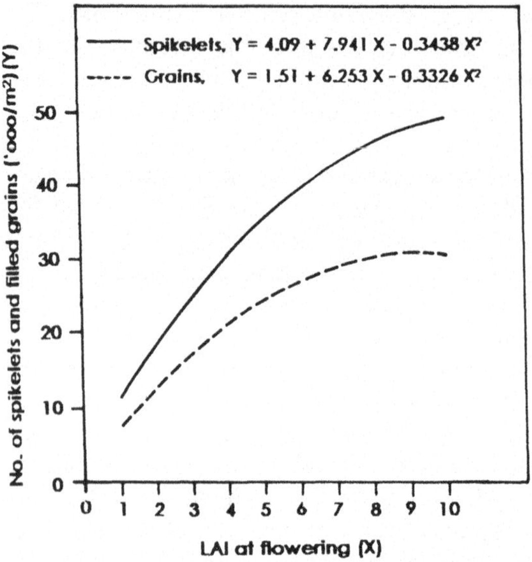

Fig. 10. Relationship between LAI at flowering and number of spikelets and filled grains. (Venkateswarlu 1976)

6 Analysis of Yield Components

Grain yield is the final product of a combination of different yield components, such as the number of panicles per unit land area, the number of spikelets/panicle, the percentage of filled spikelets and 1000-grain weight. An analysis of these components helps not only to identify the components limiting yield but also to suggest ways of increasing the yield. Of different yield components, panicle number is the most closely associated with grain yield. However, there is often a negative correlation between the number of panicles per unit land area and spikelets/panicle and between spikelets per unit land area and filled-grain percentage or 1000-grain weight (Yoshida 1983). The variability among rice varieties is the highest for panicle number, followed by that for spikelets/panicle, the percentage of filled grains and grain weight.

The relationship between grain yield and yield components is expressed as follows:

$$\text{Grain yield (t/ha)} = \text{Panicles/m}^2 \times \text{spikelets/panicle} \times \text{filled spikelets (\%)} \times \text{1000-grain weight (g)} \times 10^{-5}$$

or

$$\text{Spikelets/m}^2 \times \text{filled spikelets (\%)} \times \text{1000-grain weight (g)} \times 10^{-5}$$

Each component differs not only with respect to the growth stage at which it is determined but also in its relative contribution to grain yield, which can be

Table 2. Yield components of medium-duration rice varieties during wet and dry seasons. (Murty and Sahu 1987)

Yield component	Wet season	Dry season
Total dry matter (t/ha)	7.57	9.10
Grain yield (t/ha)	3.52	5.41
Panicles/m^2	411	346
Filled grains/panicle	50	77
Sterility (%)	31.6	13.5
1000-grain weight	21.2	22.1
Harvest index	0.47	0.60

calculated by the multiple regression analysis. Such an analysis has revealed that all the components together account for 81% of the yield variation, of which 60% is explained by the spikelets/m^2 alone and 21% by the filled spikelet percentage and 1000-grain weight together (Yoshida and Parao 1976). When the contribution of all the factors is taken as 100, which should be the case if there were no errors in measurement, the contribution of spikelets/m^2 becomes 74% and that of the filled spikelet percentage and grain weight together, 26%.

The relative importance of different components to grain yield varies with the location, season, crop duration and land situation. During wet seasons, grain yields are poor due to low grain number per panicle in short-duration (<100 days) varieties in rain-fed uplands, high spikelet sterility in medium-duration varieties in irrigated lands, and low panicle number/m^2 in long-duration (>140 days) varieties in rain-fed lowlands (Murty and Venkateswarlu 1978; Singh et al. 1989). The higher yields during dry seasons, when only medium-duration varieties are grown, are attributed to more grains per panicle due to lower sterility and higher harvest index (Table 2). Thus, identification of the components responsible for yield variation under a given situation and improvement in these specific components are important for achieving higher yields.

6.1 Spikelet Number

The number of spikelets per unit land area is the product of number of panicles and spikelets per panicle. Panicle number depends on the total number of tillers formed and the percentage of productive tillers. Grain yield has a linear relationship with the panicle number up to about 400/m^2, beyond which no such relationship is evident because of a negative correlation between the panicle number and grains per panicle (Venkateswarlu et al. 1984).

Under subnormal cultural practices, tillers/m^2 can be increased by increasing the seeding density. However, some of the late-formed tillers may die subsequently, particularly at low levels of N (Reddy et al. 1986). Optimum plant density varies with tillering capacity and the growth duration of the variety and level of N application, the density being higher for low-tillering and short-duration varieties under low N levels (Yoshida 1983). At an optimum combination of spacing and N level, the spikelets/m^2 are strongly influenced by solar radiation and temperature during the reproductive growth stages. High solar radiation combined with relatively low temperature is favourable for the production of spikelets (Venkateswarlu and Visperas 1987). The number of spikelets per panicle is also associated with the amount of N absorbed per unit dry weight of vegetative parts up to the flowering stage (Murty 1977). The number of spikelets produced per unit nitrogen absorbed or dry weight is much higher in cool regions than in warm regions (Yoshida 1983). Top dressing of N fertilizer at the beginning of the reproductive stage increases the number of spikelets per panicle (Akita 1989). The number of spikelets is strongly related to the increase in total crop dry weight between panicle initiation and flowering (Fig. 11).

6.2 Percentage of Filled Spikelets

The number of filled grains per panicle is the product of spikelets per panicle and percentage of filled spikelets, which is determined by the source capacity, ability of the spikelets to draw carbohydrates and translocation efficiency

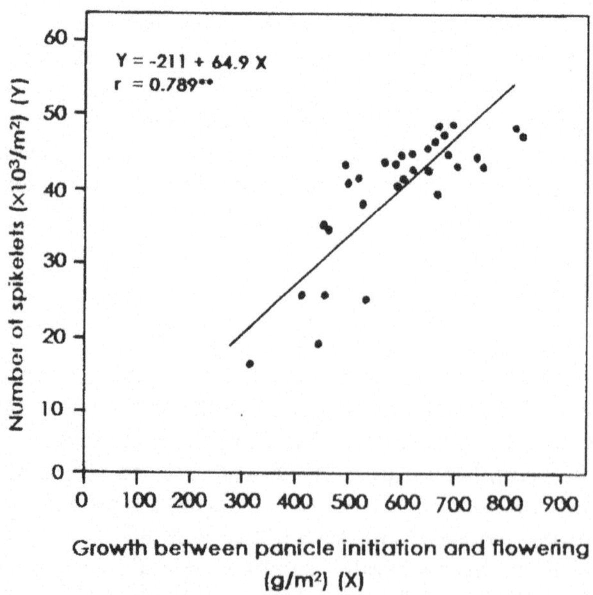

Fig. 11. Relationship between crop growth from panicle initiation to flowering and number of spikelets. (IRRI 1994b)

(Akita 1989). Poor spikelet filling is due to several factors, such as inadequate nutrition, moisture stress, incidence of insects and diseases and unfavourable weather conditions, particularly solar radiation and temperature during the reproductive growth stages (Yoshida 1983).

The number of filled spikelets per panicle is an expression of the varietal interaction with N status and photosynthetic activity of the plant from panicle initiation to heading. There is a considerable variation in filled-grain percentage, particularly at high levels of N. Nitrogen fertilization during vegetative stage increases the spikelet number but fails to sustain the survival of differentiated spikelets. The number of filled grains per panicle increases with increasing levels of N, despite a greater proportion of spikelet degeneration (CRRI 1992; Senanayake et al. 1996). A higher N content in the plant tissues favours spikelet differentiation and N absorbed after panicle initiation ensures an adequate supply of photosynthates to minimize their degeneration (IRRI 1987). Spikelet degeneration decreases linearly with increasing N concentration in the leaf at flowering (Fig. 12). Therefore, the percentage of filled spikelets can be increased by maintaining high rates of N assimilation during early grain filling and delaying leaf senescence in the later stages by additional N application at 10–15 days after flowering (Kropff et al. 1994).

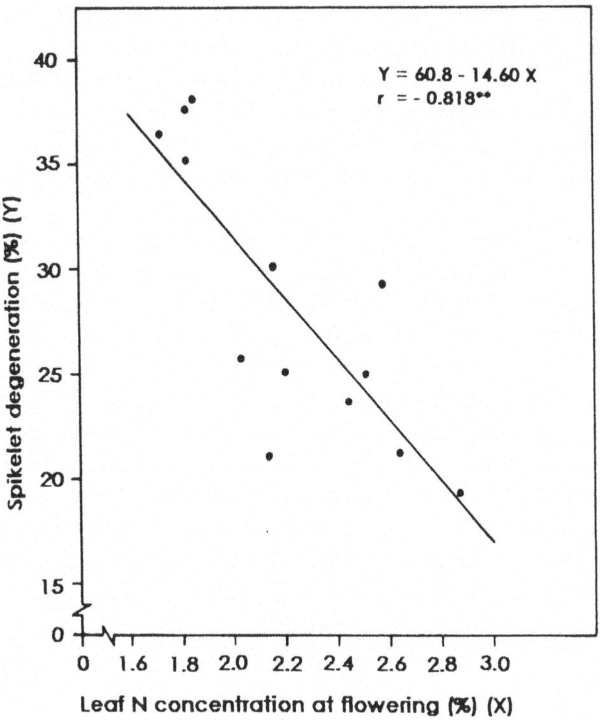

Fig. 12. Relationship between leaf N concentration at flowering and spikelet degeneration. (IRRI 1987)

Solar radiation during post-flowering stages affects spikelet filling mainly by controlling the source activity (Oldeman et al. 1987). Under a given solar radiation, the sink size relative to the source activity determines the filled spikelet percentage. This has been confirmed by the increased filled-grain percentage with partial removal of the spikelets (Venkateswarlu 1976). The increase in the number of unfilled spikelets under low solar radiation is due to inadequate supply of carbohydrates for their growth. Fertility of the spikelets depends on the light intensity during anthesis, while grain filling depends on the solar radiation, leaf area, photosynthetic rate and translocation of assimilates (Murty and Venkateswarlu 1978). Temperature affects the filled-grain percentage mainly by controlling the capacity of grains to accept carbohydrates and the length of the ripening period, which is inversely correlated with the mean daily temperature (Yoshida 1983). The persistence of cloudy weather conditions is more detrimental to grain filling at high temperatures because of a shortened ripening period.

6.3 Sterility

The extent of sterility varies from a few empty glumes to the entire panicle having unfilled grains, as a result of damage caused by certain insects and diseases, unfavourable weather conditions and genetical defects in the variety. The sterility is generally high during wet seasons, particularly in short-duration varieties whose flowering coincides with the low intensity and duration of light. The fact that pollen fertilization is not much affected under low light intensity indicates that high sterility is mostly due to post-fertilization problems and poor grain filling caused by limited source capacity (Venkateswarlu 1976).

Unfavourable climatic conditions, such as high rainfall (>200 mm), high night temperature (>25 °C) and low solar radiation ($<300\,\mathrm{cal\,cm^{-2}\,day^{-1}}$) during ripening are responsible for high sterility (Yoshida and Parao 1976; Murty and Murty 1981). Temperatures below 20 °C or above 35 °C and radiation lower than $200\,\mathrm{cal\,cm^{-2}\,day^{-1}}$ at anthesis are highly detrimental and result in up to 40–60% sterility. High sterility under low light at flowering results from: (1) reduced carbohydrate availability; (2) impaired protein synthesis and increased accumulation of soluble N; (3) poor mobilization of protein from shoot to panicles; and (4) a low cytokinin to gibberellin ratio in spikelets soon after anthesis (Murty and Sahu 1987). Strong and dry winds during flowering increase sterility by desiccating the plant and impairing grain development by causing mechanical damage and premature lodging (Vergara 1976). High N fertilization, moisture deficiency and lodging during post-flowering stages enhance the spikelet sterility (Murty and Murty 1982). Soil salinity is also responsible for high sterility.

6.4 Grain Weight

Wide variations in potential grain weight have been observed among rice varieties, with 1000-grain weight as high as 73 g (Takita 1983). However, varieties with high yield potential have grain weights in the range of 20–30 g. Unlike wheat and corn, the size of rice grain is physically restricted by the hull and its weight under most conditions is a very stable varietal character. However, there are large variations in the weight of individual grains within a panicle (Mahadevappa and Nandisha 1987; Venkateswarlu et al. 1988). Grain weight generally follows the order of maturity within a panicle, the first maturing grain being the heaviest. Extremely adverse weather conditions, such as high temperature and heavy shading, during ripening reduce grain weight (Yoshida 1981). High temperature increases the grain growth rate but decreases the grain-filling period (Akita 1989). However, the increased growth rate is not enough to compensate for the reduced grain-filling period, so that the grain weight declines. Variation in grain weight is relatively small under normal temperature ranges, which indicates that the developing grain will be filled to its capacity once ripening percentage is determined.

7 Response to Cultural Factors

Agronomic practices, mainly the method of crop establishment and N fertilization, influence the growth and yield of rice by affecting different physiological processes.

7.1 Method of Crop Establishment

Direct seeding is gradually replacing transplanting in many rice-growing areas of the tropics, due to the shortage and increasing cost of labour. However, semidwarf, short to medium duration and photo-insensitive high-yielding varieties released for irrigated areas during the past two to three decades are more suited to transplanted conditions. These varieties, when directly seeded, require greater N inputs to produce the same yield (Dingkuhn et al. 1990). The growth pattern of directly seeded rice is different from that of transplanted rice. Distribution of total aboveground biomass at heading in different plant parts varies considerably with the method of crop establishment and N fertilization (Fig. 13). Directly sown rice produces more dry matter than transplanted rice. The proportion of dry matter in green leaves is increased and that in dead leaves is decreased due to N application. Directly seeded rice is characterized by early production of large vegetative biomass, leaf area and tillers. Another important factor is the absence of transplanting shock and consequently the initial lag phase of growth (Dingkuhn et al. 1990). The absence of

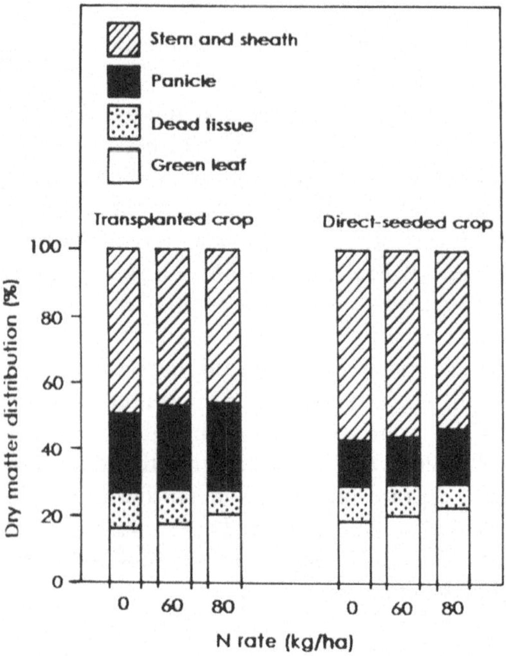

Fig. 13. Relative dry matter distribution in different plant parts at anthesis as influenced by method of crop establishment and N application. (Dingkuhn et al. 1992b)

transplanting shock and dense plant populations result in more vigorous vegetative growth, which improves the light interception during early crop establishment. Although a high plant population ensures a sufficiently large leaf area index (LAI) at early growth stages for maximum photosynthesis, excessive biomass production impedes the yield formation processes through foliar N dilution, reduced chlorophyll content, premature senescence and high maintenance respiration (Debata and Murty 1986).

Growth and yield of directly seeded rice are mostly constrained by foliar N deficiency during the reproductive growth stages caused by the supra-optimal production of tillers and leaf area. The lower leaf N concentration in directly sown rice than in transplanted rice (Fig. 14) reduces net C assimilation per unit leaf area and leads to early canopy senescence. This can be corrected by foliar application of N late during the season. The benefits of late-season N fertilization are comparatively higher in direct-seeded than in transplanted rice, because it restricts the early vegetative growth and increases the foliar N concentration (Dingkuhn et al. 1991a, 1992a). Therefore, the grain yield of directly seeded rice can be improved by limiting leaf area and increasing foliar N concentration during the reproductive growth stages. Vegetative biomass should not be channelled for further expansion of the foliage but instead should be allocated to storage tissues in stem and leaf sheath. This concept of assimilate and N partitioning may increase gross photosynthesis and growth efficiency, delay senescence and thus ensure high harvest index and grain yield.

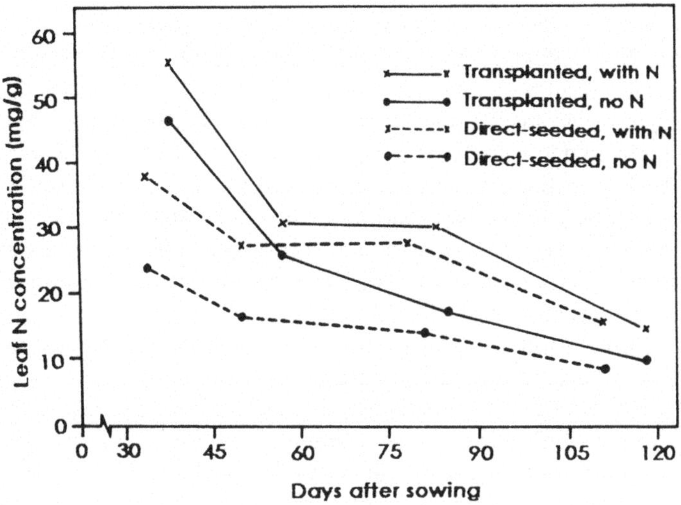

Fig. 14. Changes in leaf N concentration of transplanted and direct-seeded crop grown at 0 and 220 kg N/ha. (IRRI 1994b)

7.2
Nitrogen Fertilization

Application of N increases the LAI and photosynthetic rate due to increased chlorophyll synthesis in response to the increase in leaf N concentration (Akita et al. 1987). Adequate N supply increases the sink size and improves grain filling, resulting in a higher percentage of filled grains. Generally, more N is required for vegetative growth than for reproductive development. At moderate yield levels, more than 90% N required by the rice crop is taken up by the flowering stage (Murata and Matsushima 1978). However, at high yield levels relatively more N is required for panicle development and the crop absorbs up to 30–40% of the total N uptake during the grain-filling period.

Deep placement of N at panicle initiation improves the leaf and root activity, leaf N content and crop growth rate in the later stages and increases the number of panicles/m^2 and spikelets/panicle (Akita 1989). It enables the plants to maintain a high N concentration for a longer time, which improves the differentiation of spikelets and prevents their degeneration (Dingkuhn et al. 1992a). Nitrogen application during the late phase of panicle formation contributes to sink size by decreasing the number of degenerated spikelets and increasing the hull size (Mae 1997). It contributes to grain filing by increasing specific leaf weight and N content of leaves, which leads to the enhancement of photosynthetic activity and promotion of carbohydrate accumulation in culms and leaf sheaths (Miah et al. 1997). Combined application of N and K fertilizers around panicle initiation enhances the translocation of assimilates from flag leaf to panicle during ripening (Wang and Zhang 1995; Wang and Zou 1997). However, N fertilization at panicle initiation may result in poor crop response during wet seasons, when the solar radiation decreases towards crop maturity and higher N uptake occurs at the early growth stages. Application of more N at early stages results in excessive vegetative growth, LAI and culm elongation, leading to a less favourable balance between photosynthesis and respiration and thereby to decreased carbohydrate reserves at anthesis (Murata and Matsushima 1978). On the other hand, more N applied at panicle initiation increases the number of spikelets, leading to severe competition for carbohydrates among them, as a result of which weak spikelets in the lower portion of the panicle remain partially filled. Therefore, optimum N supply at the proper time is desirable for obtaining high yields.

Heavy N fertilization accompanied by high solar radiation and temperature can cause an acute shortage of carbohydrates at heading (Murata and Matsushima 1978). This would limit the development of mechanical tissues for active culm elongation and hinder the absorption and translocation of nutrients by inducing lodging. Consequently, photosynthesis of the lower leaves would be greatly reduced due to severe mutual shading and decreased root activity. The decreased assimilate supply to the panicle, further decreased under low light conditions, impairs grain filling and reduces yield. Therefore, varieties with large carbohydrate reserves in the vegetative parts before flowering and showing greater increase of photosynthesis but a relatively smaller increase in respiration are considered to be physiologically more efficient. In a high N-responsive variety, the photosynthesis should increase linearly with increasing levels of N application but less assimilates should be used for vegetative growth and respiration. Ishii (1993) reported that heavy N application increases leaf photosynyhesis more than respiration in N-tolerant varieties, leading to a higher ratio of photosynthesis to respiration.

Organic manures, including green manures and bio-fertilizers, release N at a slower rate than chemical fertilizers and maintain N supply to rice plants until the later growth stages (Sharma and Mittra 1988; Singh and Singh 1990). A regulated N supply increases the growth of roots and allows their penetration into deeper soil layers (Abe et al. 1995). It ensures optimum vegetative growth and prevents lodging, leading to an increased grain yield. Ensuring a continuous N supply through chemical N fertilizer alone is difficult, particularly under adverse soil and hydrological conditions. Therefore, integrated use of organic manures and chemical fertilizers is desirable for improving soil fertility and sustaining high yields.

8
Improving Yield Potential

Yield potential is defined as the maximum grain yield obtained when crop growth is not limited by water, nutrients and pests. It is determined by the interaction between varieties and climatic conditions, mainly temperature and solar radiation, and therefore varies with the location, year and cropping season. Yield potential can be used as a benchmark for its possible improvement through optimization of various agronomic practices. It can be expressed as the highest yield ever recorded at a given location or the theoretical value of maximum efficiencies of the biophysical and biochemical processes involved in the conversion of intercepted solar irradiance into plant dry matter, integrated over the life cycle of the crop.

The yield potential of modern rice varieties released since the mid-1960s has remained virtually constant. Maximum grain yields of 10–11 t/ha in the tropical regions and 13–15 t/ha in the temperate regions have been achieved (Yoshida 1981; Williams 1992; Ying et al. 1998a). The highest yield of 16.6 t/ha has been recorded with a *japonica* F_1 hybrid in the Yunnan province of south China (Amano et al. 1996). Although the gap between the present average yield under irrigated conditions and the potential yield is fairly large (about 5 t/ha), it will gradually diminish and might even disappear by the early part of the next century. With little scope for expanding the rice area, the challenge is to increase the yield from the existing land by breaking the present yield barrier.

8.1 Estimation of Yield Potential

The driving forces in grain yield formation include a source of carbohydrates and a sink for these carbohydrates. The source for grain production is determined by three components: (1) the amount of stem reserves allocated to the grains; (2) the rate of dry-matter production during the grain-filling period; and (3) the length of the grain-filling period or growth rate duration. Kropff et al. (1994) proposed a model for the grain yield formation i.e. $Y = S + (G \times D)$, where Y is the grain yield, S is the net amount of stem reserves translocated to the grain, G is the average daily growth rate during the grain-filling period and D is the length of the grain-filling period. The contribution of stem reserves to the grain yield has been estimated at 2.0–2.5 t/ha (Yoshida 1981). This implies that achieving grain yields of 10 t/ha would require a net dry-matter accumulation of about 8 t/ha (at 14% moisture content) during the 25 days of effective grain filling, which is equivalent to 284 kg/ha/day (oven-dry weight basis at 3% moisture content). It is an achievable estimate because rice in the tropical environments can produce 300–360 kg dry matter $ha^{-1} day^{-1}$ under conditions of high solar radiation and high LAI and leaf N content (Yoshida 1981; Akita 1989). These estimates for yield formation indicate that yields beyond 10 t/ha must result from increased accumulation of stem reserves and their mobilization to the grains, a prolonged grain-filling period or an increased growth rate during grain filling.

An increase in the contribution of stem carbohydrate reserves beyond the present level of 2.5 t/ha can be accomplished by increasing the net assimilation during the vegetative growth period through increased supply of N, but this may lead to lodging and the problem of diseases. The diseases can be checked by the use of fungicides or resistant varieties and lodging can be prevented by increasing the strength of stems, reducing plant height, lowering the height of the panicle and reducing the effects of wind through crop management (Setter et al. 1994). Considering that the dry-matter contribution from the stem reserves remains at 2.5 t/ha and the growth rate during the grain-filling period at 284 kg ha^{-1} day^{-1}, then an effective grain-filling period of 38 days would be needed to obtain a grain yield of 15 t/ha (at 14% moisture content). This means that a variety with a yield potential of 15 t/ha in the tropics should have the same grain-filling duration as that of varieties in the temperate regions.

8.2 Physiological Traits for Yield Improvement

Improvement in rice yield potential in the tropics is dependent on the ability to increase biomass production and sink size (Ying et al. 1998a). Increased photosynthetic efficiency coupled with decreased respiration and increased storage of carbohydrates in the grain are the major physiological processes which can be used to improve the yield potential. Plant parameters which need consideration for improving the source and sink capacity are listed in Table 3. The former is closely and positively related to the total biomass production and the latter to the harvest index.

8.2.1 Increasing Biomass Production

High biomass production can be achieved through increased canopy photosynthesis and reduced maintenance respiration, when the rice crop is grown

Table 3. Important parameters for improving the source and sink capacity of irrigated rice. (Freeman et al. 1993)

Source/sink capacity	Plant characteristics
Source capacity	Dwarfness
	Erect leaves
	Early vigour
	Leaf area index
	Photosynthetic efficiency
	Nutrient use efficiency
	Delayed senescence
	Healthy root system
Sink capacity	Panicle density
	Spikelets/panicle
	Panicle weight
	Grain weight
	Sterility
	Harvest index
	Crop growth rate prior to heading
	Reduced tillering
	Grain-filling duration

in a high solar radiation environment and supplied with adequate N (Akita 1989). The avenues for improving net photosynthesis are: (1) improving growth rate in the initial stages; (2) improving photosynthetic efficiency per unit N uptake; and (3) increasing the ratio between photosynthesis and respiration. Recently, incorporation of C-4 characteristics into rice has been proposed for increasing the net energy fixation and crop growth rate (Akita 1994). The likelihood of achieving this goal is not very high but, if successful, it could contribute significantly towards increasing biomass production and sustaining rice production.

Potential biomass production obtained in high N-solution culture using a high-yielding variety under the tropical conditions of the Philippines is about 22 t/ha (Fig. 15) and there is no further increase in biomass with increasing N levels, even when lodging is artificially prevented (Akita 1989). An increase in crop growth rate in the early stages, obtained by manipulating photosynthesis through increased N application, is associated with declining crop growth rate in the later stages. This is because increased biomass production is mostly accompanied by the inevitable enhancement of respiratory losses. Energy used in maintenance respiration cannot be manipulated without adversely affecting plant growth. However, one component of maintenance respiration that can possibly be reduced without much adverse effect is the energy consumption for remobilizing proteins. Such a reduction would help in maintaining net assimilation and root activity during later growth stages (Akita 1993).

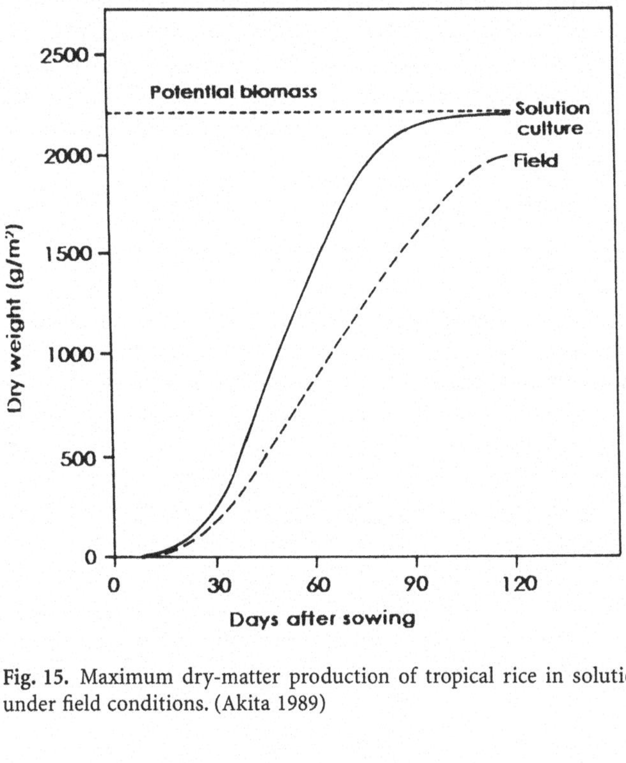

Fig. 15. Maximum dry-matter production of tropical rice in solution culture (80 ppm N) and under field conditions. (Akita 1989)

Achieving yields of 13–15 t/ha in the tropics without an increase in the rate of C assimilation would require a considerable delay in leaf senescence and maintenance of leaf N status during an extended grain-filling duration (Kropff et al. 1994). The rate of net C assimilation can be increased by increasing the flux of CO_2 from soil through root aerenchyma to the leaves for providing an additional source of CO_2 for photosynthesis (Penning de Vries 1991) or by increasing the vertical N gradient in the canopy so that a greater proportion of leaf N is displayed when light intensity is highest (Dingkuhn et al. 1991b). In view of these considerations, a strategy to increase net energy fixation without increasing N assimilation through manipulation of various physiological traits related to photosynthesis and respiration is important.

8.2.2
Increasing Harvest Index

The scope for improving biomass production being rather limited, another option for increasing the yield would be to increase the harvest index by manipulating the yield-determining processes. The higher total dry-matter production in long-duration varieties is generally accompanied by a sharp decrease in the harvest index (Venketeswarlu et al. 1977). There is ample scope for improving their harvest index by increasing the accumulation of non-structural carbohydrates and N compounds in grain. The key physiological parameters determining the harvest index are: (1) dry weight at heading, which is closely related to the sink capacity, and (2) crop growth rate around heading, which determines the ripening percentage. The two parameters are often inversely related and the ideal way to increase the harvest index is to maintain a higher crop growth rate and reduce the dry weight at heading without sacrificing the sink capacity.

8.2.2.1
Improving Sink Capacity

Variation in grain yield across seasons and under varying fertilizer levels is mainly attributed to the sink capacity. Increasing harvest index for improvement of yield potential involves increasing the sink size through increased sink formation efficiency (sink size per unit biomass at heading), greater partitioning of the assimilates for sink formation and faster N uptake. The sink formation efficiency is determined by the balance between the partitioning of N and photosynthates to the sink and the vegetative parts. High N content in the plant tissues favours differentiation of spikelets and greater supply of photosynthates prevents their degeneration. It is possible to decrease partitioning of photosynthates to the leaves by application of gibberellic acid (Katayama and Akita 1987). Short-duration varieties produce more spikelets/m^2 than long-duration varieties due to higher N content in the leaf sheath at panicle initiation

Table 4. Variation in yield components and highest yields achieved in temperate, subtropical and tropical regions. (Williams 1992; Freeman et al. 1993; Kropff et al. 1994; Ying et al. 1998a)

Yield components	Temperate	Subtropical	Tropical	
	Yanco (Australia)	Yunnan (China)	Manila (Philippines)	Hyderabad (India)
Grain yield (t/ha)	15.0	14.7	9.5	7.8
Panicles/m^2	700	433	560	489
Spikelets/panicle	100	148	80	96
Filled spikelets (%)	80	85	90	–
Grain weight (mg)	25	25	24	26
Variety	Amaroo (*japonica*)	Guichao (*indica*)	IR 72 (*indica*)	Jaya (*indica*)

(Akita 1989). Therefore, genetic improvement for high sink-formation ability in long-duration varieties has a great potential for increasing yield.

Analysis of 15 t/ha rice yield in the temperate or high altitude subtropical environments suggests plasticity in the yield components (Table 4). Under the tropical conditions the crop exhibits strong compensation mechanisms, particularly for panicle and spikelet number. Assuming a total biomass of 22 t/ha (at 3% moisture content) and harvest index of 0.6 for rough rice (0.5 for dehulled rice), a grain yield of 13 t/ha (15 t/ha at 14% moisture) can be expected (Akita 1989). Further, assuming that 30% of the grain weight is contributed by the translocation of carbohydrates stored in the vegetative parts, at least 9 t/ha additional dry weight may have to be obtained after heading. The key factor would then be whether the sink size is adequate to produce 13 t grain/ha from 13 t dry weight/ha at heading. This would require 65×10^3 spikelets/m^2, assuming a 1000-grain weight of 24 g and 85% ripening. However, currently the spikelet number is below 50×10^3/m^2. Achieving 13 t dry weight/ha at heading is possible, but to maintain 65×10^3 spikelets/m^2 seems to be impossible even at the highest spikelet formation efficiency. The present varieties do not have the capacity to produce sufficient assimilates to support the development of a larger sink. Therefore, the basic parameters that must be modified to achieve a major increase in yield potential of tropical rice include increased sink size, a longer period of effective grain filling and longer duration of green leaf area for active canopy photosynthesis to match the grain-filling duration.

8.2.2.2
Improving Ripening Percentage

Ripening involves two major processes, determination of the effective sink size and grain filling. It rarely limits the yield when sink size is low, but limits the yield greatly when sink size is high and the environmental factors during ripening are not favourable. Grain filling can be increased by efficient translocation of photosynthates, maintaining higher sink activity and a healthy root system, slower leaf senescence after heading and increased lodging resistance (Kato 1989; Khush 1993). Photosynthetic activity shortly before and after heading is critical for the development of spikelets, whereas supply of photosynthates during the entire ripening period determines the grain filling. Translocation of greater amounts of carbohydrates and N compounds from the leaf blade and sheath to the grains accelerates their senescence. In short-duration varieties, about half the grain N comes from the leaves which leads to early leaf senescence, decreased leaf area and consequently lower dry-matter production during ripening (Akita 1989). Therefore, a longer duration from flowering to ripening and delayed leaf senescence in varieties with adequate sink size are desirable for improving yield potential.

8.3 Yield Potential and Nitrogen Requirement

It is essential to match the pattern of N supply with N requirement of the crop at each developmental stage to realise the yield potential. Since the rice crop requires nearly 20 kg N for producing 1.0 t grain/ha, the N requirement for a potential yield of 15 t/ha is estimated at 300 kg N/ha (Setter et al. 1994). The soil provides about 50–80 kg N ha^{-1} $crop^{-1}$ in most double- and triple-cropped irrigated rice systems (Cassman et al. 1993) and the remaining N requirement for high yield levels must be met by supplying fertilizer N at the key growth stages. Assuming that the efficiency of fertilizer N is 50%, 440–500 kg fertilizer N/ha will be required for achieving this yield level. The fertilizer requirement can be reduced to 290–330 kg N/ha if the uptake efficiency of applied N is increased to 75%. This can be achieved by integrated use of different N fertilizers, including organic manures and slow release fertilizers, as well as by the proper timing and method of application in balance with crop N demand and soil N supply (Mae 1997). Following these practices, the grain yields of over 15 t/ha have been obtained with total N inputs of 290 kg N/ha in the subtropical environment at Yunnan (Ying et al. 1998b).

A declining trend in yield due to continuous double cropping of rice has been reported in recent years (Cassman et al. 1993; Sharma 1995c). This is partly due to reduced N supply from the soil, resulting in suboptimal leaf N content during the grain-filling period (Akita 1989). It has been shown in long-term experiments that the original yield level can be restored by higher fertilizer inputs (Cassman et al. 1993; Nambiar 1994). Therefore, adequate N supply throughout the growing season is crucial for realizing the yield potential, particularly in the tropics where high temperature leads to more gaseous losses of applied N from the soil-flood water system.

9 New Plant Types

The first high-yielding *indica* rice variety, Taichung Native 1, derived from a cross between Dee-geo-woo-gen (DGWG) and Tsai-yuan-chung (TYC), was released in 1956. Subsequently, the International Rice Research Institute (IRRI) developed IR 8, from a cross between DGWG and Peta, which gave an average yield of 6–8 t/ha in the wet season and 9–10 t/ha in the dry season. Its release in 1966 revolutionised rice cultivation in tropical Asia, where it is still grown in large areas under irrigated conditions. Today, more than 60% of the world's rice area is planted with varieties similar to IR 8, which account for 80% of the total production (Peng et al. 1994). The main characteristics of these varieties are: erect leaves, profuse tillering, semi-dwarf height (100 cm) with stiff culm, photo-insensitivity, high N-responsiveness and high harvest index. Tsunoda (1964), from his studies on crop photosynthesis, reported that thick, dark green leaves allow less light to be lost through reflection, while short, erect leaves permit more uniform light distribution within the canopy, reduce respiration and consequently increase dry-matter production. A short and sturdy culm minimizes lodging, thereby leading to higher grain yield. These observations along with the comparison of low- and high-yielding varieties led to the development of a plant type concept in rice similar to the wheat ideotype (Donald 1968). The term "plant type" refers to a specific combination of different morphological characteristics associated with the yielding ability of a variety. This concept has since attracted considerable attention from rice breeders and is widely accepted as a guideline for developing new high-yielding varieties.

9.1 Plant Type for Irrigated Ecosystems

For the past few years, the IRRI and many other institutes in the National Research Systems have been directing their major efforts towards breaking the present yield barrier. The "new plant type" aims at increasing the grain yield by 30–50% compared with that of the existing semi-dwarf varieties. This can be achieved by increasing the sink size, with some improvement in assimilate supply and its mobilization to grains and by reducing the number of unproductive tillers that are overexpressed in the current *indica* varieties. The additional assimilate supply is believed to come from greater carbohydrate storage in the thick culms, extension of the grain-filling period, delayed leaf senescence, lower panicle position to increase light interception by the top leaves and heterotic vigour. The important characteristics of the proposed new plant type for irrigated ecosystem are (Fig. 16a); (1) low tillering and absence of non-productive tillers; (2) heavy panicles with more primary branches, 200–250 grains per panicle and spikelets having large pedicellar vascular bundles; (3) short (90–100 cm) and sturdy culms bearing erect and thick leaves with dark green flag leaf sheath; (4) vigorous root growth and high photosynthetic rate even at low photosynthetically active radiation; and (5) medium growth duration (110–130 days) (Vergara et al. 1991; Peng et al. 1994). The new plant type should also possess multiple disease and insect resistance as well as acceptable grain quality.

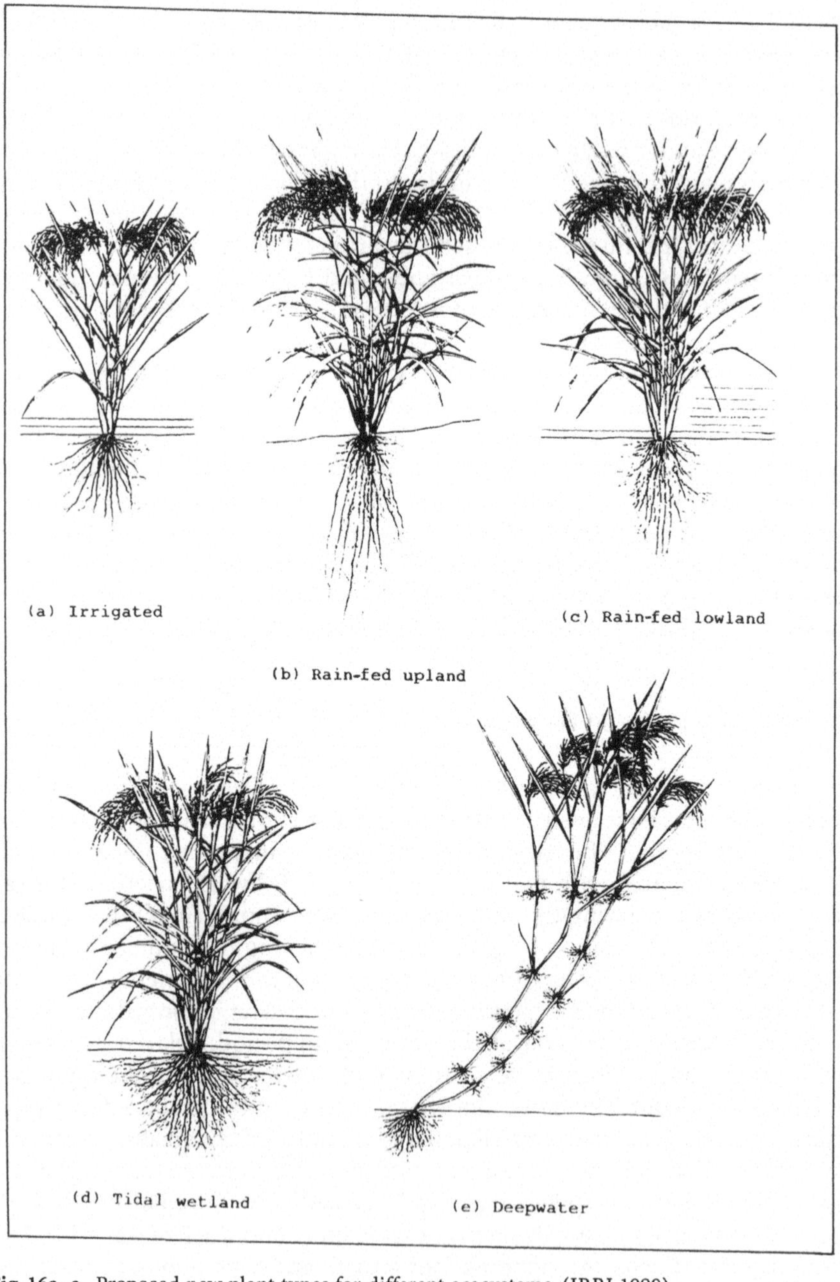

Fig. 16a–e. Proposed new plant types for different ecosystems. (IRRI 1990)

9.1.1
Reduced Tillering and Large Panicles

The modern high-yielding rice varieties tiller profusely under favourable conditions and each hill (3–5 plants) produces 30–40 tillers, of which only about half bear panicles. The remaining unproductive tillers compete with the productive ones for space, solar energy and nutrients. Further, a dense canopy resulting from the excessive tillering creates a humid micro-environment favourable for the development of diseases.

Reduced tillering is thought to facilitate synchronous flowering and maturity, production of more uniform panicles and efficient use of horizontal space. A single, semi-dominant gene is reported to control the low-tillering trait and it has pleiotropic effects on the length and thickness of the culm and panicle size (Ise 1992). In the modern high-yielding varieties, panicle number is positively correlated with grain yield (Yoshida 1983; Venkateswarlu et al. 1984). Therefore, reduced panicle number in a low-tillering plant type has to be suitably compensated with increased panicle size for achieving higher grain yield.

Large panicles usually result in a higher sink to source ratio and higher sink capacity through increased spikelet number and percentage of filled spikelets (Choi and Kwon 1985). Further, more primary branches can be accommodated on the panicle. In contrast to the present high-tillering and panicle number-type varieties, the low-tillering plant ideotype would ensure more vascular bundles and a greater proportion of high-density grains, thus facilitating the production of heavy panicles (Hayashi 1980).

9.1.2
Grain Size and High-Density Grains

The weight of individual grains within a panicle varies considerably, although the overall grain weight in rice is considered to be a fairly stable varietal characteristic. Coarse grains generally have lower density than fine grains, which suffer less from the deficiencies in grain filling. Grains having a 1000-grain weight of more than 20 g have low density and are usually not completely filled (Takita 1986). High-density grains are those which remain submerged in a salt solution of more than 1.2 specific gravity. About 70–85% of such grains are borne on the primary branches of the panicle (Padmaja Rao 1987; Venkateswarlu et al. 1987). Plants having a greater proportion of high-density grains would have a high yield, particularly if they have few tillers, heavy panicles and thick culms (Vergara 1988).

The superior spikelets are the first to accumulate dry matter and N during grain filling, followed by the inferior spikelets (Iwasaki et al. 1992). Such an apical dominance is altered by removal of the superior spikelets, indicating that delayed filling of the inferior spikelets results from source limitation and regulation of assimilate allocation within the panicle. Ensuring adequate translocation of assimilates to all the spikelets in a panicle to make them high-density grains can increase the grain yield by up to 30% (Venkateswarlu et al. 1986). High-density grains also result in better milling quality and more head rice recovery. Further, use of these grains as seed ensures better germination and seedling growth than that of low-density grains, leading to more uniform crop establishment.

9.1.3 Canopy and Leaf Characteristics

Leaf erectness facilitates more efficient light utilization at high LAI, particularly during the later growth stages. A plant canopy with vertically oriented leaves gives better light penetration and higher CO_2 assimilation per unit leaf area (Tanaka 1976b). On the other hand, droopy or horizontal leaves reduce light penetration and air movement inside the canopy and thereby increase the relative humidity and decrease the temperature, leading to a more favourable micro-climate for the incidence of diseases and insects. However, the plants with such leaves may have an advantage with respect to light interception and competition with weeds during early growth stages. Thick leaves are generally narrow and erect and contain relatively more chlorophyll and photosynthetic enzymes per unit area (Ishii 1993). Therefore, leaf thickness is considered a desirable trait for the new plant type. Leaf longevity is also an important parameter because early leaf senescence during the grain-filling period is a major limitation to increasing yield potential. It can be achieved by suppressing ethylene synthesis (Horton 1994).

Most of the current high-yielding varieties have their panicles at the top of the canopy where they absorb much of the incident radiation, thus lowering the efficiency of light utilization. Plant types which have their panicles below the top of the canopy are desirable; this can be achieved by shortening the topmost internode of the stem.

9.1.4 Short and Stiff Culm

The main trait responsible for increased grain yield in modern high-yielding varieties is reduced plant height, which results in an increased harvest index because of reduced straw yield and greater lodging resistance. The shorter plants require less maintenance respiration and have a more favourable balance between photosynthesis and respiration (Peng et al. 1994). However, further reduction in plant height from the present level of most high-yielding varieties is likely to reduce the total dry-matter production as well. The thicker culms with more vascular bundles can accumulate more carbohydrates and help in offsetting the decreased panicle number through the increased panicle weight (Hayashi 1980).

Increase in dry-matter production requires adequate N fertilization under high solar radiation (Akita 1989). However, increased biomass production might result in lodging, increased disease incidence and thereby decreased grain yield. Therefore, a strong and thick culm is required to prevent lodging and support larger panicles (Vergara 1988). Further, lowering the height of the panicle may have a profound effect on increasing lodging resistance, besides improving the light interception by leaves during grain filling (Setter et al. 1994).

9.1.5 Crop Growth Duration

The total dry weight, the LAI and the contribution of accumulated carbohydrates to grain yield increase linearly with increasing crop growth duration (Venkateswarlu et al. 1977). However, there is a negative correlation with the harvest index and a quadratic trend with the grain yield. The maximum yield of 8–10 t/ha under irrigated conditions is obtained from varieties of 110–130 days duration. The short-duration varieties of 100 days may produce higher grain yield per day, but the medium-duration varieties produce more total yield and can still fit into the double crop rice or rice–wheat systems.

Variation in total growth duration among varieties occurs mostly due to the difference in their vegetative growth period. In early-maturing varieties, shorter duration of panicle growth is often associated with decreased grain yield. The long-duration and photo-sensitive varieties have a long vegetative lag phase, which allows for the accumulation of more carbohydrates in the culm before the reproductive stage and leads to the production of larger panicles with heavier grains. However, such varieties are suitable for the lowland areas which remain waterlogged for considerably long periods (Sharma et al. 1995). The optimum growth duration involves a balance of different growth processes throughout the life of the plant, which can result in the production of more panicles, greater filled spikelets and increased grain weight. In view of these considerations, the new plant type for irrigated ecosystems should have a growth duration of about 120 days.

9.2 Plant Types for Rain-Fed Ecosystems

Rain-fed ecosystems are mostly monocropped with traditional rice varieties having varying degree of tolerance to abiotic and biotic stresses. Their grain yields are low (1–2 t/ha) but fairly stable at little or no input management level. Plant types with desirable morphological characteristics have now been conceived for achieving stable and reasonably high yields in these ecosystems (Table 5, Fig. 16b–e). The upland plant type is envisaged to be about 130 cm tall with 5–8 tillers per hill and very few unproductive tillers, sturdy stems, large panicles with 150–200 grains, a thick root system, 100–110 days growth duration, durable resistance to blast and tolerance to drought and acidic soils. The plant type for rain-fed lowlands having a water depth of up to 50 cm is conceived to be 130 cm tall with sturdy stems, dark green and erect leaves, 6–10 productive tillers with very few unproductive tillers and panicles with 150–200 grains. It should also possess tolerance to drought and submergence, and mul-

Table 5. Characteristics of improved plant types for different ecosystems. (Khush 1993)

Ecosystem	Plant height (cm)	Tiller number/ hill	Grains/ panicle	Growth duration (days)	Yield (t/ha)	Resistance to biotic stresses	Tolerance to abiotic stresses
Irrigated	90	4–5	200–250	100–135	13–14	Multiple resistance	-
Rain-fed upland	130	5–8	150–200	100–110	3–4	Blast	Drought, low pH
Rain-fed lowland							
(i) Water depth up to 50 cm	130	6–10	150–200	120–150*	4–5	Multiple resistance	Drought, submergence
(ii) Water depth >50 cm (deep water)	Elongating type	5–7	150–200	Photo-sensitivity	4–5	Stem borer	Submergence
(iii) Tidal wetlands	130	5–7	150–200	Photo-sensitivity	3–4	Multiple resistance	Peatiness, acidity

* Weak photo-sensitivity.

tiple resistance to biotic stresses. The plant type for deep-water areas (>50 cm depth) must have elongation ability, 5–7 panicles each with 150–200 grains, upper nodal rooting and tillering ability, photoperiod sensitivity, grain dormancy, drought tolerance at seedling stage and resistance to stem borers. The yield potential of rice subjected to tidal inundation can be improved by incorporating traits such as large panicles, sturdy stems and wide and relatively shallow rooting. These varieties must have tolerance to submergence, salinity, acidity and mineral deficiencies/toxicities.

9.3
Present Status of New Plant Type Development

The breeding efforts for the development of the new plant types began at the IRRI in the late 1980s and many promising germplasm from tropical *javanicas* with large panicles, low tillering and sturdy culms were identified. New breeding lines with semi-dwarf height (90–100 cm), sturdy culms, dark green and thick leaves, 8–10 panicles per hill without any unproductive tillers and 150–200 grains per panicle have been developed (Peng et al. 1994). Performance of these elite lines is being tested in the field under different environments (Martinez et al. 1995; Akita et al. 1996). The incorporation of improved grain quality and resistance to insects and diseases in the new plant type is also being attempted. It is hoped that varieties with the proposed ideotype will be released within a few years.

A variety called Lunishree, which has recently been released for the typical flood-prone lowlands by the Central Rice Research Institute (Cuttack, India), possesses some of the desirable morphological features of the proposed new plant type. It is a tall, photo-sensitive and long-duration variety with strong

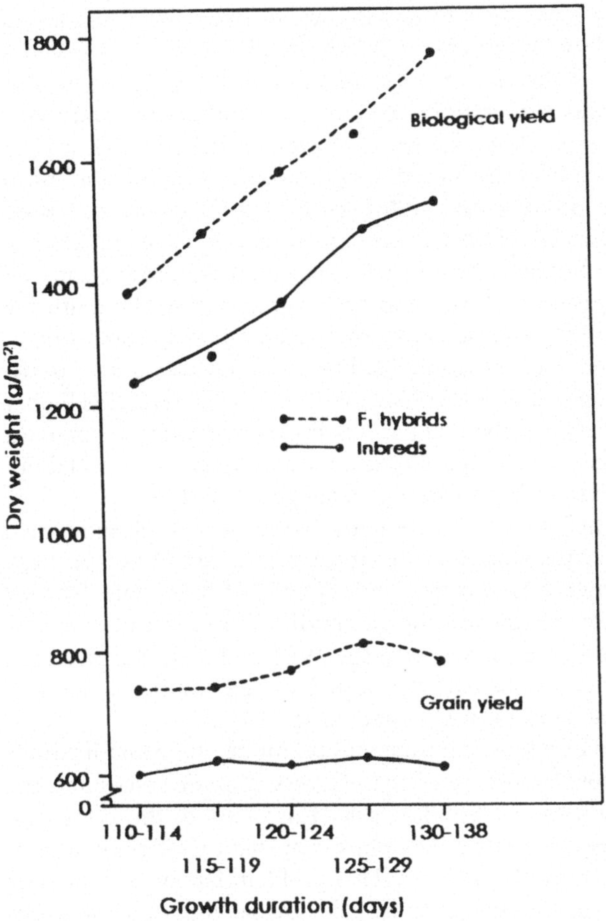

Fig. 17. Biological yield and grain yield of F_1 hybrids and inbred varieties with different growth durations. (Blanco et al. 1990)

culms, erect leaves, fewer tillers and larger panicles with a greater proportion of high-density grains. It produces up to 30% higher grain yield than the existing lowland varieties. The development of Lunishree is considered a major breakthrough in increasing the rice productivity of coastal areas with the accompanying problems of salinity and excessive waterlogging for prolonged periods.

10 Hybrid Rice

The yield potential of rice can be improved significantly through the exploitation of heterosis or hybrid vigour. The idea of utilizing this phenomenon was first conceived by Professor L.P. Yuan of China in the early 1960s, although the existence of cytoplasmic male sterility was known in India much earlier. In China, hybrid rice was introduced commercially in 1979 and by now it has covered more than 60% of the rice area (Virmani et al. 1993). Subsequently, hybrid rice technology was successfully exploited in South Korea and many other countries of south and south-east Asia. On average, hybrids have a yield advantage of about 1.0 t/ha over the inbred varieties under tropical conditions (Fig. 17). However under subtropical conditions, the yields are much higher and the potential of the hybrids is further exploited (Table 6). The higher yield of hybrids is mainly attributed to heterosis in total biomass production and harvest index through greater sink size and better grain filling (Ponnuthurai et al. 1984; Blanco et al. 1990). Hybrids have lower maintenance respiration in vegetative organs after panicle initiation, leading to higher assimilate partitioning to panicles (Virmani et al. 1995). The magnitude of heterosis is higher when genetic diversity between the two parents is greater. *Indica* × *japonica* hybrids have shown higher physiological efficiency and heterosis for total dry matter and grain yield than *indica* hybrids (Chuan'gen et al. 1995).

The development of hybrid rice is currently based on cytoplasmic male sterility (CMS) and fertility restoration systems, using a three-line approach involving CMS, maintainer and restorer lines. The CMS lines and hybrids developed in China are not suitable for tropical conditions because of their susceptibility to diseases and insects. Recently, the IRRI and National Research Systems in several Asian countries have developed a large number of promising CMS lines for tropical conditions (Virmani et al. 1995).

The three-line method of hybrid development is cumbersome and its use is constrained due to the low frequency of maintainers in *indica* rices and the lack of restorers in *japonica* rices. Besides, lack of diversity in the cytoplasm of hybrids because of the availability of few sources of stable CMS poses a great danger of susceptibility to abiotic stresses. This problem has been overcome by transferring male sterile cytoplasm from wild species into cultivated rice

Table 6. Performance of inbred cultivar (IR 72) and F_1 hybrid (Shanyou 63) in the tropical and subtropical environments. (Ying et al. 1998a,b)

Trait	Manila (Tropical)		Yunnan (Subtropical)	
	IR 72	Shanyou 63	IR 72	Shanyou 63
Grain yield (t/ha)	7.6	8.6	12.9	15.2
Panicles/m^2	494	369	614	490
Spikelets per panicle	81	99	108	112
Spikelet filling (%)	76	75	79	90
Grain weight (mg)	23.3	26.8	22.6	27.9
Harvest index (%)	45	45	49	47
Total biomass (t/ha)	16.2	16.5	24.1	29.7
LAI at flowering	5.7	7.5	6.7	9.1
Grain N (%)	1.46	1.54	1.12	1.21
Total N accumulation at maturity (kg/ha)	194	201	217	254
Total growth duration (days)	121	118	157	169

(Virmani et al. 1993). Another option is through reversible male sterility governed by a two-gene system, one gene causing male sterility and the other blocking the action of the first gene. This approach offers greater flexibility in hybrid rice production, provided male fertility and sterility characteristics are stable under diverse rice-growing environments. The development of thermosensitive genic male sterile (TGMS) and photo-sensitive genic male sterile (PGMS) lines in recent years has led to a two-line method of hybrid development, using a TGMS/PGMS line and a pollen parent, which need not be a restorer (Yuan 1993). This method does not require a maintainer line. The male sterility in these lines is genetically controlled and has no relation to cytoplasm. The two-line hybrids are likely to show higher heterosis compared to the three-line hybrids because of broader choice in selecting pollen parents and the lack of negative effects of the sterility-inducing cytoplasm (Siddiq 1995). Work on the development of two-line hybrids and interspecific hybrids with higher heterosis is in progress.

The large-scale adoption of hybrid rice technology is handicapped by the need to use fresh seeds in every season and their high cost as well as lack of adequate seed production and distribution infrastructure. These constraints can be overcome by developing true breeding hybrids having permanently fixed heterosis by using apomixis (Setter et al. 1994). Hybrid seed production is encountered with problems of low seed yields and poor outcrossing in CMS lines, which can be overcome by developing parental lines with long and exerted stigma, extended duration of floret opening and long anthers. The application of GA_3, flag leaf cutting and supplementary pollination increase the seed yields considerably (Virmani 1993).

11
Biotechnological Approaches

The yield potential of rice is conditioned by a large number of genes, whose expression is decisively influenced by the environment. Therefore, quantitative traits such as yield are difficult to manipulate by biotechnological approaches. However, recent developments in molecular techniques, such as wide hybridization, protoplast fusion, anther culture, molecular marker-aided selection, DNA fingerprinting and plant transformations through genetic engineering, have proved to be of immense use towards the enhancement of resistance to biotic and abiotic stresses and thereby to improve yield. Use of these techniques offers scope of attaining those objectives which are otherwise difficult through conventional breeding approaches, such as for transferring yield, adaptation and resistance genes, apomictic genes for the exploitation of F_1 hybrids and developing an associative biological nitrogen fixation system in rice (Singh 1993). The recent exciting developments is rice biotechnology include successful regeneration from protoplasts of *japonica* and *indica* rices, production of transgenic plants, development of a comprehensive molecular map consisting of more than 1500 DNA markers, tagging of genes of economic importance with RFLP markers, molecular characterization of pathognic and ulations based on restriction fragment pattern and transfer of genes from wild relatives (Khush 1995).

Wild rices are a valuable source of genes for resistance to biotic and abiotic stresses but transfer of these genes to cultivated rice through conventional breeding is handicapped by the abortion of hybrid embryos and poor recombination. Wide hybridization using embryo rescue technique has enabled the accessing of genes from all the wild species of genus *Oryza* and thus facilitated the development of their hybrids. Repeated backcrossing with cultivated rice followed by embryo rescue helps to restore fertility and chromosomal stability and results in the integration of fragments of wild genomes into the chromosomes of elite lines (Jena et al. 1992). Using this technique, genes for resistance to brown and white-backed planthoppers, bacterial blight and blast have been transferred from wild species to cultivated rice (Bennet et al. 1994).

Protoplast fusion has been used to produce somatic hybrids between sexually incompatible species. It is very useful for the speedy transfer of cytosterility and the production of cybrids and organelle recombinants. Somatic hybrids between cultivated and wild rices have been produced (Hayashi et al. 1988) and efforts are being made to transfer salt tolerance from *Sclerophyllum coarctata* to cultivated rice.

Anther culture enables the attainment of rapid homozygosity after hybridization and improves the selection efficiency, particularly for traits with low heritability. Besides, it is an important means of producing dihaploid population for gene mapping. The efficiency of anther culture is genotype-dependent and *indica* rices are less amenable to it than *japonica* rices (Guiderdoni et al. 1992). This technique has been successfully used to introduce salinity resistance into elite *indica* varieties (Zapata et al. 1991) and to transfer BPH biotype-1 resistance from *indica* to *japonica* rices (Toenniessen et al. 1993).

The discovery of DNA or restriction fragment length polymorphism (RFLP) markers has paved the way for tagging economically important genes and transferring them from one genetic background to another. Several genes for resistance to biotic and abiotic stresses have been mapped on the rice genome along with the most closely linked DNA markers (Bennet et al. 1994). RFLP analysis using PCR-based techniques ensures speedy screening of progenies and is particularly useful in breeding for traits for which true conditions do not exist at a given location. The application of RFLP and RAPD (randomly amplified polymorphic DNA) analysis to the genomes of rice insects and pathogens helps in understanding their evolution and in the development of resistant varieties (Levy et al. 1991; Leach et al. 1992).

Genetic engineering has been used to transfer alien genes to rice from organisms with which it normally does not hybridise and also from modified rice genes with altered properties (Christou et al. 1991; Shimamoto 1992; Shimamoto et al. 1993). Unlike conventional hybridisation, this technique avoides the introduction of undesirable genes and eliminates the need of extensive backcrossing. Studies on the transfer of foreign genes to rice are currently focussed on bacterial genes, mainly from *Bacillus thuringiensis* (B.t. genes) and the soybean trypsin inhibitor gene for insect resistance; chitinase and ribosome inactivating protein genes from barley for resistance to fungal diseases; the bacterial mannitol 1-phosphate dehydrogenase gene for salt tolerance; and genes regulating alcoholic fermentation for submergence tolerance (IRRI 1994b). The coat-protein gene of the rice tungro bacilliform virus has also been cloned and efforts are being undertaken to introduce it into rice (Toenniessen et al. 1993). Further, transgenic rice plants with useful foreign genes have been developed. Genetic transformation is also being attempted for resistance to sheath blight, apomixis in hybrid rice (Khush 1993) and root nodulation (Cocking et al. 1993). The identification of sources of resistance to sheath blight and stem rot diseases that do not exist in rice germplasm and their transferral to rice would be a significant contribution from molecular biology to the goal of higher yield potential in the tropical environments.

Gene mapping and marker-aided selection enables the achievement of durable stress resistance through the pyramiding of multiple resistance genes. Its application to quantitative characters such as yield requires the development of large mapping populations using many markers (Paterson et al. 1992) and mapping for these characters over a wide range of environments. However, this technique is likely to facilitate genetic transformation to improve the fundamental processes of yield formation involving many genes in the long term. Marker-aided analysis of yield components has been carried out in tomato (Paterson et al. 1991) and maize (Edwards et al. 1992; Stuber et al. 1992). In rice, the alteration of starch metabolism by introducing antisense-RNA constructs has resulted in reduced expression of the waxy gene, leading to lower amylose content in grains (Shimamoto 1992; Shimada et al. 1993). Work on mapping quantitative trait loci for resistance to bacterial blight, blast and drought is in progress.

12 Future Directions

Improving the physiological efficiency of the new rice varieties is contemplated for achieving a breakthrough in the yield potential. The opportunities for increasing the efficiency of potential photosynthetic capacity for better plant growth and yield are now available. This need not necessarily involve increasing the photosynthetic capacity *per se*, but removing the internal limitations to carbohydrate metabolism and increasing the functional life span of leaves. Physiological studies to identify the processes that can be manipulated to increase the net photosynthetic rate, particularly during the reproductive growth stages, in order to increase the total biomass production are important. The specific targets for such investigations should include: (1) increasing specific leaf N and the photosynthetic rate per unit specific leaf N; (2) increasing leaf photosynthesis at each stratum in the canopy by manipulating the canopy characteristics; (3) increasing panicle photosynthesis; (4) reducing respiration; (5) increasing leaf area duration and delaying leaf senescence and (6) incorporating C-4 characteristics into rice plant. Further, sink limitation needs to be quantified under a wide range of environments. Efforts should be made to improve the harvest index through increased sink size and greater partitioning of dry matter to grains, particularly in long-duration varieties.

The concept of the new plant type should be based not only on the morphological features but also on the physiological and biochemical processes. The recent advances in genetic transformation should be used for exploring and validating the yield-limiting processes and also in removing the constraints they impose on yield. Among the conventional approaches, improving agronomic practices for realizing yield potential should be an important strategy. This would include the proper timing and method of crop establishment, optimization of space utilization within the crop stand, adequate supply of nutrients and water and effective weed control.

Research on hybrid rice should be pursued more vigorously for development of: (1) genetically diverse CMS, maintainer and restorer lines; (2) heterotic hybrids with higher yield advantage, wider adaptability, resistance to insects and diseases and acceptable grain quality; (3) two-line hybrids using TGMS/PGMS lines; and (4) increasing the outcrossing potential of male sterile lines.

Biotechnological research should aim at: (1) improving yield potential through genetic transformation and heterosis aided by apomictic seed production and new sources of male sterility; (2) introducing tolerance to various biotic and abiotic stresses; (3) increasing nutrient uptake and assimilation efficiencies; (4) improving nutritive value and digestibility of storage proteins and enhancing biosynthesis of vitamin A precursors in grain; and (5) increasing starch biosynthesis and its translocation to endosperm. The transfer of "nif" genes to rice is also worth attempting.

References

Abe J, Songmuang P, Harada J (1995) Root growth of paddy rice with application of organic materials as fertilizers in Thailand. Jpn Agric Res 29:77–82

Akita S (1989) Improving yield potential in tropical rice. In: Progress in irrigated rice research. Int Rice Res Inst, Manila, Philippines, pp 41–73

Akita S (1993) Respiration: variation and potential for manipulation. In: Buxton DR, Shibles R, Forsberg RA, Blad BL, Asay KH, Paulsen GM, Wilson RF (eds) International crop science I. Crop Sci Soc Am, Madison, pp 799–805

Akita S (1994) Eco-physiological aspects of raising the yield plateau of irrigated rice in the tropics. In: Cassman KG (ed) Breaking the yield barrier. Int Rice Res Inst, Manila, Philippines, pp 85–89

Akita S, Parao FT, Coronel VP (1987) Physiological response of rice to light and nitrogen. In: Weather and rice. Int Rice Res Inst, Manila, Philippines, pp 81–91

Akita S, Ishikawa T, Fuzisawa H (1996) Growth characteristics of IRRI-developed new rice plant type breeding lines in Japan. Int Rice Res Notes 21(1):44–45

Amano T, Shi C, Qin D, Tsuda M, Matsumoto Y (1996) High-yielding performance of paddy rice achieved in Yunnan province, China. I. High yielding ability of Japonica F_1 hybrid rice, Yu-Za 29. Jpn J Crop Sci 65:16–21

Armstrong W (1979) Aeration in higher plants. In: Woolhouse HWW (ed) Advances in botanical research. Academic Press, London, pp 226–232

Baker JT, Allen LH Jr, Boote KJ, Jones P, Jones JW (1990) Rice photosynthesis and evapotranspiration in subambient, ambient and superambient carbon dioxide concentrations. Agron J 82:834–840

Bennet J, Brar DS, Khush GS, Huang N, Setter TL (1994) Molecular approaches. In: Cassman KG (ed) Breaking the yield barrier. Int Rice Res Inst, Manila, Philippines, pp 63–76

Blanco LC, Casal C, Akita S, Virmani SS (1990) Biomass, grain yield and harvest index of F_1 rice hybrids and inbreds. Int Rice Res Newsl 15(2):9–10

Brar DS, Khush GS (1995) Wide hybridization for enhancing resistance to biotic and abiotic stresses in rain-fed lowland rice. In: Fragile lives in fragile ecosystems. Int Rice Res Inst, Manila, Philippines, pp 901–910

Cassman KG, Kropff MJ, Gaunt J, Peng S (1993) Nitrogen use efficiency of irrigated rice: What are the key constraints? Plant Soil 155/156:359–362

Catling D, Puckridge DW, HilleRisLambers D (1988) The environment of Asian deepwater rice. In: Proc 1987 Int Deepwater Rice Worksh. Int Rice Res Inst, Manila, Philippines, pp 11–34

Chaturvedi GS, Ram PC, Singh AK, Ram P, Ingram KT, Singh BB, Singh RK, Singh VP (1996) Carbohydrate status of rain-fed lowland rices in relation to submergence, drought and shade tolerance. In: Singh VP, Singh RK, Singh BB, Zeigler RS (eds) Physiology of stress tolerance of rice. Proc Int Conf Stress Physiol in Rice, Lucknow, India, pp 103–120

Choi HC, Kwon KW (1985) Evaluation of varietal difference and environmental variation for some characters related to source and sink in the rice plants. Korean J Crop Sci 30:460–470

Christou P, Ford TL, Kofron M (1991) Production of transgenic rice (*Oryza sativa* L.) plants from agronomically important *indica* and *japonica* varieties via electric discharge particle acceleration of exogenous DNA into immature zygotic embroys. Biotechnology 9:957–962

Chuan'gen L, Fulin G, Jiangshi Z (1995) Biological characteristics of *indica-japonica* hybrids. Intl Rice Res Notes 20(4):6–7

Cock JH, Yoshida S (1972) Accumulation of ^{14}C-labelled carbohydrate before flowering and the subsequent redistribution and respiration in the rice plant. Proc Crop Sci Soc Jpn 41:226–234

Cocking EC, Kothari SL, Flinch RP, Lynch PT, Davey MR (1993) Genetic manipulations using rice protoplasts. In: Muralidharan K, Siddiq EA (eds) New frontiers of rice research. Directorate of Rice Res, Hyderabad, India, pp 141–147

CRRI (1992) Annual report for 1992–93. Cent Rice Res Inst, Cuttack, India

De SK, Rao Ch N (1989) Influence of nutrient deficiency on photosynthesis and productivity of early rice varieties. Oryza 26:317–319

Debata A, Murty KS (1986) Influence of plant density on leaf and panicle senescence in rice. Indian J Plant Physiol 29:281–285

De Datta SK (1981) Principles and practices of rice production. Wiley, New York

Deng X-Q, Lu Z-R, Li M-Q (1984) Effect of light intensity and temperature on the photosynthesis and photorespiration of rice. J South China Agric College 5:32–38

Dingkuhn M, Cruz RT, O'Toole JC, Dorffling K (1989) Net photosynthesis, water use efficiency, leaf water potential and leaf rolling as affected by water deficit in tropical upland rice. Aust J Agric Sci 40:1171–1181

Dingkuhn M, Jones MP, Johnson DE, Sow A (1998) Growth and yield potential of *Oryza sativa* and *O. glaberrima* upland rice cultivars and their interspecific progenies. Field Crops Res 57:57–69

Dingkuhn M, Schnier HF, De Datta SK, Dorffling K, Javellana C, Pamplona R (1990) Nitrogen fertilization of direct seeded vs. transplanted rice. II. Interactions among canopy properties. Crop Sci 30:1284–1292

Dingkuhn M, Schnier HF, De Datta SK, Dorffling K, Javellana C (1991a) Relationships between ripening phase productivity and crop duration, canopy photosynthesis and senescence in transplanted and direct-seeded lowland rice. Field Crops Res 26:327–345

Dingkuhn M, Penning de Vries FWT, De Datta SK, van Laar HH (1991b) Concept for a new plant type for direct-seeded flooded tropical rice. In: Direct-seeded flooded rice in the tropics. Int Rice Res Inst, Manila, Philippines, pp 27–31
Dingkuhn M, De Datta SK, Javellana C, Pamplona R, Schnier HF (1992a) Effect of late season N fertilization on photosynthesis and yield of transplanted and direct seeded tropical flooded rice. I. Growth dynamics. Field Crops Res 28:223–234
Dingkuhn M, De Datta SK, Pamplona R, Javellana C, Schnier HF (1992b) Effect of late season N fertilization on photosynthesis and yield of transplanted and direct seeded tropical flooded rice. II. A canopy stratification study. Field Crops Res 28:235–239
Donald CM (1968) The breeding for crop ideotypes. Euphytica 17:385–403
Edwards MD, Helentjaris T, Wright S, Stuber CW (1992) Molecular-marker-facilitated investigations of quantitative trait loci in maize. Theor Appl Genet 83:765–774
Ekanayake IJ, De Datta SK, Steponkus PL (1989) Spikelet sterility and flowering response of rice to water stress at anthesis. Ann Bot 63:257–264
Emes MJ, Wilkins CP, Smith PA, Kupkanchanakul K, Hawker K, Charlton WA, Cutter EG (1988) Starch utilization by deep water rice during submergence. In: Proc 1987 Int Deepwater Rice Worksh. Int Rice Res Inst, Manila, Philippines, pp 319–326
Freeman WH, Saxena NP, Siddiq EA (1993) Physiological approaches for genetic improvement of yield potential in irrigated rice. In: Muralidharan K, Siddiq EA (eds) New frontiers of rice research. Directorate of Rice Res, Hyberabad, India, pp 116–124
Greenway H, Setter TL (1996) Is there anaerobic metabolism in submerged rice plants? A view point. In: Singh VP, Singh RK, Singh BB, Zeigler RS (eds) Physiology of stress tolerance in rice. Proc Int Conf on Stress physiology of rice, Lucknow, India, pp 11–30
Grist DH (1986) Rice, 6th edn. Longman, London
Guiderdoni E, Galinato E, Luistro J, Vergara G (1992) Anther culture of tropical *japonica x indica* hybrids of rice (*Oryza sativa* L.). Euphytica 62:219–224
Hayashi H (1980) Studies on large vascular bundles in paddy rice plant and panicle formation. Bull Fukui Agric Exp Stn 17:31–35
Hayashi Y, Kyozuka J, Shimamoto K (1988) Hybrids of rice (*Oryza sativa* L.) and wild species obtained by cell fusion. Mol Gen Genet 214:6–10
Horton P (1994) Limitations to crop yield by photosynthesis. In: Cassman KG (ed) Breaking the yield barrier. Int Rice Res Inst, Manila, Philippines, pp 111–115
Imai K, Coleman DF, Yanagisawa T (1985) Increase in atmospheric partial pressure of carbon dioxide and growth and yield of rice (*Oryza sativa* L.). Jpn J Crop Sci 54:413–418
Ingram KT, Beuno FD, Namuco OS, Yambao EB, Beyrouty CA (1994) Genetic variation in rice root morphology, growth and water use. In: Kirk GJD (ed) Rice roots – nutrient and water use. Int Rice Res Inst, Manila, Philippines, pp 67–77
IRRI (1987) Annual report for 1986. Int Rice Res Inst, Manila, Philippines
IRRI (1989) Annual report for 1988. Int Rice Res Inst, Manila, Philippines
IRRI (1990) Planning for the 1990's. Int Rice Res Inst, Manila, Philippines
IRRI (1993a) Program report for 1992. Int Rice Res Inst, Manila, Philippines
IRRI (1993b) Rice almanac. Int Rice Res Inst, Manila, Philippines
IRRI (1994a) Rice facts. Int Rice Res Inst, Manila, Philippines
IRRI (1994b) Program report for 1993. Int Rice Res Inst, Manila, Philippines
IRRI (1995) Program report for 1994. Int Rice Res Inst, Manila, Philippines
Ise K (1992) Inheritance of a low tillering plant type in rice. Int Rice Res Newsl 17(4):5–6
Ishii R (1993) Leaf photosynthesis in rice in relation to grain yield. In: Abrol YP (ed) Photosynthesis – photoreactions to plant productivity. Kluwer, Dordrecht, pp 561–569
Ishii R, Samejima M, Murata Y (1977) Photosynthetic $^{14}CO_2$ fixation in the leaves of rice and some other species. Jpn J Crop Sci 46:97–102
Iwasaki Y, Mae T, Makino A, Ohira K, Ojima K (1992) Nitrogen accumulation in the inferior spikelet of rice ear during filling. Soil Sci Plant Nutr 38:517–525
Jackson MB, Waters I, Setter TL, Greenway H (1987) Injuries to rice plants caused by complete submergence – a contribution by ethylene (ethene). J Exp Bot 38:1826–1828

Janardhan KV, Murty KS (1977) Association of some leaf characters with photosynthesis in rice. Curr Sci 46:497–498
Janardhan KV, Murty KS (1978a) Photosynthetic carbon dioxide fixation by rice ears. Curr Sci 47:810–811
Janardhan KV, Murty KS (1978b) Variability in photorespiration and photosynthesis in rice varieties. Indian J Exp Biol 16:116–117
Jena KK, Khush GS, Kachen G (1992) RFLP analysis of rice (*Oryza sativa* L.) introgressive lines. Theor Appl Genet 84:125–128
Kaneda C, Beachell HM (1974) Response of *indica-japonica* rice hybrids to low temperatures. SABRAO J 6:17–32
Katayama K, Akita S (1987) Effect of exogenously applied gibberellic acid on initial growth of rice cultivars. Jpn J Crop Sci 56:173–174
Kato T (1989) Relationship between grain filling process and sink capacity in rice (*Oryza sativa* L.). Jpn J Breed 39:431–438
Khan MR, Ventura W, Vergara BS (1982) Uptake through aquatic roots and distribution of ^{15}N-tagged ammonium in deepwater rice. In: Proc 1981 Int Deepwater Rice Worksh. Int Rice Res Inst, Manila, Philippines, pp 321–326
Khush GS (1993) Varietal needs for different environments and breeding strategies. In: Muralidharan K, Siddiq EA (eds) New frontiers of rice research. Directorate of Rice Res, Hyderabad, India, pp 68–75
Khush GS (1995) Biotechnology approaches to rice improvement. In: Induced mutations and molecular techniques for crop improvement. Int Atomic Energy Agency, Vienna, Austria, pp 455–475
Kropff MJ, Cassman KG, Peng S, Matthews RB, Setter TL (1994) Quantitative understanding of yield potential. In: Cassman KG (ed) Breaking the yield barrier. Int Rice Res Inst, Manila, Philippines, pp 21–38
Leach JE, Rhoads ML, Vera Cruz CM, White FF, Mew TW, Leung H (1992) Assessment of genetic diversity and population structure of *Xanthomonas oryzae* pv. *oryzae* using a repetitive DNA element. Appl Environ Microbiol 58:2188–2195
Levy M, Romao J, Marchetti MA, Hamer JE (1991) DNA fingerprinting with a dispersed repeated sequence resolves pathotype diversity in the rice blast fungus. Plant Cell 3:95–102
Mae T (1997) Physiological nitrogen efficiency in rice: Nitrogen utilization, photosynthesis, and yield potential. Plant Soil 196:201–210
Mahadevappa M, Nandisha BS (1987) A review on the status of genetic analysis of characters important in harvest, post-harvest and seed technology of rice (*Oryza sativa* L.) in southern India. Seed Sci Technol 15:581–591
Martinez CP, Fischer A, Gonzalez D, Ramirez H, Mozika D (1995) Potential and limitations of the new rice plant type of the International Rice Research Institute. Preliminary observations. Arroz 44:15–20
Matsushima S (1976) High yielding rice cultivation. Univ Tokyo Press, Tokyo
Mazaredo AM, Vergara BS (1982) Physiological differences in rice varieties tolerant of and susceptible to complete submergence. In: Proc 1981 Int Deepwater Rice Worksh. Int Rice Res Inst, Manila, Philippines, pp 327–341
Miah MNH, Yoshida T, Yamamoto Y (1997) Effect of nitrogen application during ripening period on photosynthesis and dry matter production and its impact on yield and yield components of semidwarf *indica* rice varieties under water culture conditions. Soil Sci Plant Nutr 43:205–217
Murata Y, Matsushima S (1978) Rice. In: Evans LT (ed) Crop physiology. Cambridge Univ Press, Cambridge, pp 73–99
Murty KS (1977) Physiological aspects of production in rice. In: Padhi B (ed) Frontiers of plant sciences. Utkal Univ, Bhubaneswar, India, pp 79–86
Murty KS (1989a) Photosynthesis of rice plant. In: Gupta RP, Joshi RC, Ghildyal BP, Singh G (eds) Physiological environment of rice ecosystems. Agrophys Monogr No 1. Indian Agric Res Inst, New Delhi, India, pp 62–73

Murty KS (1989b) Physiological research on rice and future perspectives. In: Saxena HK, Srivastava RDL (eds) Proc Natl Seminar Plant Physiology Research Imperatives for the Nineties. CS Azad Univ Agric and Technol, Kanpur, India, pp 1-9

Murty PSS, Murty KS (1981) Note on the effect of climatic factors on spikelet sterility in rice. Indian J Agric Sci 51:803–805

Murty PSS, Murty KS (1982) Influence of stress conditions and N fertilization on spikelet sterility in *Oryza sativa* L. Geobios 9:193–196

Murty KS, Sahu G (1987) Impact of low-light stress on growth and yield of rice. In: Weather and rice. Int Rice Res Inst, Manila, Philippines, pp 93–101

Murty KS, Venkateswarlu B (1978) Physiological constraints on growth and development in rice during *Kharif* season. In: Proc Natl Symp Increasing rice yield in *Kharif*. Cent Rice Res Inst, Cuttack, India, pp 45–65

Murty KS, Nayak SK, Sahu G, Ramakrishnayya G, Janardhan KV, Rai RSV (1976) Efficiency of ^{14}C photosynthesis and translocation in local and high yielding rice varieties. Plant Biochem 3:63–71

Nagamine I (1990) Genetic diversity and inheritance of photosynthetic activity of rice (*Oryza sativa* L.) detected by oxygen evolution in leaves. Jpn Agric Res Q 23:249–254

Nambiar KKM (1994) Soil fertility and crop productivity under long-term fertilizer use in India. Indian Council Agric Res, New Delhi, India

Nishiyama I (1976) Effect of temperature on vegetative growth of rice plants. In: Proc Symp Climate and rice. Int Rice Res Inst, Manila, Philippines, pp 159–185

Norman MJT, Pearson CJ, Searle PGE (1984) Rice. In: The ecology of tropical food crops. Cambridge Univ Press, Cambridge, pp 82–119

Oldeman LR, Seshu DV, Cady FB (1987) Response of rice to weather variables. In: Weather and rice. Int Rice Res Inst, Manila, Philippines, pp 5–39

O'Toole J, Moya TB (1978) Genetic variation in maintenance of leaf water potential in rice. Crop Sci 18:873–876

Padmaja Rao S (1987) Panicle type: few structural considerations for higher yield potential in rice. Indian J Plant Physiol 30:87–90

Padmaja Rao S (1994) Partitioning pattern of biomass and yield components among different tillers of rice. Indian J Plant Physiol 37:277–278

Palada MC, Vergara BS (1972) Environmental effect on the resistance of rice seedlings to complete submergence. Crop Sci 12:209–212

Palit P, Kundu A, Mandal AK, Sircar SM (1979) Productivity of rice plant in relation to photosynthesis, photorespiration and translocation. Indian J Plant Physiol 22:66–74

Panda MM, Reddy MD, Sharma AR (1991) Yield performance of rain-fed lowland rice as affected by nursery fertilization under conditions of intermediate deep water (15–50 cm) and flash-floods. Plant Soil 132:65–71

Pande HK, Reddy BB (1984) Rice cultivation systems in waterlogged and flood-prone areas. In: Deepak A, Rao KR (eds) Applications of remote sensing for rice production. A Deepak Publ, Virginia, pp 127–140

Paterson AD, Damon S, Hewitt JD, Zamir D, Rabinowitch HD, Lincoln SE, Lander ES, Tanksley SD (1991) Mendelian factors underlying quantitative traits in tomato: comparison across species, generations and environments. Genetics 127:181–197

Paterson AH, Tanksley SD, Sorrells ME (1992) DNA markers in plant improvement. Adv Agron 46:39–90

Peng S, Khush GS, Cassman KG (1994) Evaluation of the new plant ideotype for increased yield potential. In: Cassman KG (ed) Breaking the yield barrier. Int Rice Res Inst, Manila, Philippines, pp 5–20

Penning de Vries FWT (1991) Improving yields: designing and testing VHYVs. In: Systems simulation at IRRI. Int Rice Res Inst. Res Pap Ser No 151, pp 13–19

Ponnuthurai S, Virmani SS, Vergara BS (1984) Comparative studies on growth and grain yield of some F_1 rice (*Oryza sativa* L.) hybrids. Philipp J Crop Sci 9:189–193

Raskin I, Kende H (1984) Regulation of growth in stem sections of deepwater rice. Planta 160:66–72

Reddy MD, Ghosh BC, Panda MM (1985) Flood tolerance of rice at different growth stages as affected by fertilizer application. Plant Soil 83:255–263

Reddy MD, Ghosh BC, Panda MM (1986) Effect of seed rate and application of N fertilizer on grain yield and N uptake of rice under intermediate deepwater conditions (15–50 cm). J Agric Sci (Camb) 107:61–66

Reddy MD, Sharma AR, Panda MM (1991) Flood tolerance of rice grown under intermediate deepwater conditions (15–50 cm) as affected by phosphorus fertilization. J Agric Sci (Camb) 117:319–324

Ricord B, Pradet A (1989) Anaerobic protein synthesis in different organs of germinating rice seeds. Plant Physiol Biochem 27:761–768

Saha Ray PK, HilleRisLambers D, Tepora NM (1993) Combination of stem elongation ability with submergence tolerance in rice. Euphytica 68:11–16

Satake T, Yoshida S (1977) Critical temperature and duration of high temperature-induced sterility in indica rice. Jpn Agric Res Q 11:190–191

Senanayake N, Naylor REL, De Datta SK (1996) Effect of nitrogen fertilization on rice spikelet differentiation and survival. J Agric Sci (Camb) 127:303–309

Setter TL, Waters I, Atwell BJ, Kupkanchanakul T, Greenway H (1987) Carbohydrate status of terrestrial plants during flooding. In: Crowford RMM (ed) Plant life in aquatic and amphibious habitats. Blackwell, Oxford, pp 411–433

Setter TL, Kupkanchanakul T, Greenway H (1989a) Submergence of rice. II. Adverse effects of low CO_2 concentrations. Aust J Plant Physiol 16:265–278

Setter TL, Waters I, Wallace I, Bhekasut P, Greenway H (1989b) Submergence of rice. I. Growth and photosynthetic response to CO_2 enrichment of flood water. Aust J Plant Physiol 16:251–263

Setter TL, Peng S, Kirk GJD, Virmani SS, Kropff MJ, Cassman KG (1994) Physiological considerations and hybrid rice. In: Cassman KG (ed) Breaking the yield barrier. Int Rice Res Inst, Manila, Philippines, pp 39–62

Setter TL, Ingram KT, Tuong TP (1995) Environmental characterization requirements for strategic research in rice grown under adverse conditions of drought, flooding or salinity. In: Ingram KT (ed) Rain-fed lowland rice – agricultural research for high-risk environments. Int Rice Res Inst, Manila, Philippines, pp 3–18

Sharma AR (1992) Effect of nursery fertilization and age of seedling on flood tolerance and productivity of rain-fed lowland rice (*Oryza sativa*). Indian J Agric Sci 62:657–663

Sharma AR (1994) Effect of different methods and timings of stand establishment on performance of rain-fed lowland rice under 0–50 cm water depth. J Agron Crop Sci 172:38–51

Sharma AR (1995a) Effect of basal and post-flood nitrogen fertilization on performance of rice (*Oryza sativa*) under conditions of intermediate deep-water and simulated flash-flooding. Indian J Agric Sci 65:399–404

Sharma AR (1995b) Direct seeding and transplanting for rice production under flood-prone lowland conditions. Field Crops Res 44:129–137

Sharma AR (1995c) Fertilizer use in rice and rice-based cropping system. Fertil News 40(5):29–41

Sharma AR, Mittra BN (1988) Effect of combinations of organic materials and nitrogen fertilizer on growth, yield and nitrogen uptake of rice. J Agric Sci (Camb) 111:495–501

Sharma AR, Reddy MD (1992) Technology for rice production in flood-prone and waterlogged areas of eastern India. Rice India 2(7):13–17

Sharma AR, Moorthy BTS, Rao KS, Venkateswarlu B (1995) Improvement of rice production technology in the conventional areas of eastern India. In: Deb DL (ed) Improving crop and animal productivity. Associated Publ Co, New Delhi, India, pp 157–178

Sharma SD, Sampath S (1985) Taxonomy and species relationship. In: Rice research in India. Indian Council Agric Res, New Delhi, India, pp 21–43

Shimada H, Tada Y, Kawasaki T, Fujimura T (1993) Antisense regulation of rice *waxy* gene expression using a PCR-amplified fragment of the rice genome reduces the amylose content in grain starch. Theor Appl Genet 86:665–672
Shimamoto K (1992) Genetic manipulation of rice: from protoplasts to transgenic plants. Jpn J Genet 67:273–290
Shimamoto K, Miyazaki C, Hashimto H, Izawa T, Itoh K, Tarada R, Inagaki Y, Iida S (1993) Transactivation and stable integration of the maize transposable element *Ds* cotransfected with the *Ac* transposage gene in transgenic rice plants. Mol Gen Genet 239:354–360
Siddiq EA (1995) Current status of hybrid rice development in India. In: Ilyas AM, Viraktamath BC (eds) Hybrid rice seed production technology: theory and practice. Directorate of Rice Res, Hyderabad, India, pp 1–12
Singh PK, Singh DP (1990) *Azolla* and rice cultivation. In: Somani LL, Bhandari SC, Saxena SN, Vyas KK (eds) Biofertilizers. Scientific Publ, Jodhpur, India, pp 39–65
Singh RB (1993) Research and development strategies for intensification of rice production in the Asia-Pacific region. In: Muralidharan K, Siddiq EA (eds) New frontiers of rice research. Directorate of Rice Res, Hyderabad, India, pp 25–44
Singh VP, De SK, Murty KS (1989) Growth and yield in relation to maturity duration in rice. Indian J Plant Physiol 32:344–349
Smith MA, Jacobsen JV, Kende H (1987) Amylase activity and growth of internodes of deepwater rice. Planta 172:114–120
Squire GR (1990) The physiology of tropical crop production. CAB Int, Wallingford
Stuber CW, Lincoln SE, Wolff DW, Helentjaris T, Lander ES (1992) Identification of genetic factors contributing to heterosis in a hybrid from two elite maize inbred lines using molecular markers. Genetics 132:823–839
Swain P, Nayak SK, Murty KS (1987) Photosynthesis and translocation of ^{14}C photoassimilates among rice varieties. J Nucl Agric Biol 16:18–21
Takita T (1983) Breeding of a rice line with extraordinarily large grains as a genetic source for high-yielding varieties. Jpn Agric Res Q 17:93–97
Takita T (1986) Studies on breeding for resistance to green leafhopper, varietal adaptability to local environments and high yield of rice in Malaysia. Trop Agric Res Cent, Tsukuba, Ibasaki, Japan, 56 pp
Tanaka I (1976a) Climatic influence on photosynthesis and respiration of rice. In: Proc Symp on Climate and rice. Int Rice Res Inst, Manila, Philippines, pp 223–247
Tanaka T (1976b) Regulation of plant type and carbon assimilation of rice. Jpn Agric Res Q 10:161–167
Toenniessen GH, Herdt RW, O'Toole TC (1993) Rockefeller Foundation's international rice biotechnology programme. In: Muralidharan K, Siddiq EA (eds) New frontiers of rice research. Directorate of Rice Res, Hyderabad, India, pp 125–133
Tripathy JN, Nayak SK (1995) Seasonal influence on growth and productivity of high-yielding varieties of rice (*Oryza sativa*). Indian J Agric Sci 65:356–357
Tsuno K, Sato K (1971) Relation between leaf water content and photosynthetic capacity. Proc Crop Sci Soc Jpn 40 (extra issue 2):41–42
Tsuno Y, Sato T, Miyamoto H, Harada N (1975) Studies on CO_2 uptake and CO_2 evolution in each part of the crop plants. II. Photosynthetic activity in the leaf sheath and ear of the rice plant. Proc Crop Sci Soc Jpn 44:293–300
Tsunoda S (1964) A developmental analysis of yielding ability in varieties of field crops. Nihon-Gakujitsu-Shinkokai, Maruzen Publ Co, Tokyo
Tsunoda S (1983) Photosynthetic strategy in rice and wheat. In: Genetics – new frontiers. Oxford and IBH, New Delhi, India, pp 225–266
Tsunoda S (1984) Biology of rice. Japan Scientific Society Press, Tokyo
Tuong TP, Ingram KT, Siopongco TP, Confesor RB, Boling AA, Singh U, Wopereis MCS (1995) Performance of dry-seeded rain-fed lowland rice in response to agrohydrology and N fertilizer management. In: Ingram KT (ed) Rain-fed lowland rice - agricultural research for high risk environments. Int Rice Res Inst, Manila, Philippines, pp 141–155

van Keulen H (1976) A calculation method for potential rice production. Contrib No 21, Cent Res Inst Agric, Bogor, Indonesia, 26 pp
Venkataraman S (1987) Agrometeorological aspects of growth, yield and water relations with special reference to rice. In: Weather and rice. Int Rice Res Inst, Manila, Philippines, pp 47–54
Venkateswarlu B (1976) Source-sink inter-relationships in lowland rice. Plant Soil 44:575–586
Venkateswarlu B, Prasad GSV (1980) Pre- and post-flowering photosynthetic contribution to grain yield in rice (*Oryza sativa* L.). Indian J Plant Physiol 23:300–308
Venkateswarlu B, Visperas RM (1987) Solar radiation and rice productivity. Res Pap Ser No 129, Int Rice Res Inst, Manila, Philippines
Venkateswarlu B, Rao JS, Rao AV (1977) Relationship between growth duration and yield parameters in irrigated rice (*Oryza sativa* L.). Indian J Plant Physiol 20:69–76
Venkateswarlu B, Rao SP, Acharyulu TL (1984) Studies on yield potential and inter-relationships to sink components in rice. Indian J Plant Physiol 27:313–321
Venkateswarlu B, Vergara BS, Parao FT, Visperas RM (1986) Enhanced grain yield potentials in rice by increasing the number of high-density grains. Philipp J Crop Sci 11:145–152
Venkateswarlu B, Parao FT, Vergara BS (1987) Occurrence of good and high density grains on rice panicles. SABRAO J 19:1–6
Venkateswarlu B, Vergara BS, Parao FT (1988) Appearance of different grades of grain during maturation of the rice (*Oryza Sativa* L.) panicle. Int Congr Plant Physiol, New Delhi, India, p 17
Vergara BS (1976) Physiological and morphological adaptability of rice varieties to climate. In: Proc Symp Climate and rice. Int Rice Res Inst, Manila, Philippines, pp 67–86
Vergara BS (1988) Raising the yield potential of rice. Philipp Tech J 13:3–9
Vergara BS, Venkateswarlu B, Janoria M, Ahn JK, Kim JK, Visperas RM (1991) Rationale for low tillering rice plant type with high-density grains. In: Direct seeded flooded rice in the tropics. Int Rice Res Inst, Manila, Philippines, pp 39–53
Virmani SS (1993) Global status of hybrid rice research and development. In: Barwale BR (ed) Hybrid rice - food security in India. Macmillan India, Madras, India, pp 3–22
Virmani SS, Prasad MN, Kumar I (1993) Breaking yield barrier in rice through exploitation of heterosis. In: Muralidharan K, Siddiq EA (eds) New frontiers of rice research. Directorate of Rice Res, Hyderabad, India, pp 76–85
Virmani SS, Khush GS, Pingali PL (1995) Hybrid rice for tropics: potential, research priorities and policy issues. In: Ilyas AM, Viraktamath BC (eds) Hybrid rice seed production technology: theory and practice. Directorate of Rice Res, Hyderabad, India, pp 13–30
Vishnu-Mittre (1977) Changing economy in ancient India. In: Reed CA (ed) Origin of agriculture. Mouton, The Hague, pp 570–588
Wang Y-R, Zhang Y-J (1995) Effect of nitrogen and potassium nutrition during late growth stage of rice on translocation of ^{14}C and grain yield. Int Rice Res Notes 20(2):16
Wang Y-R, Zou J (1997) Effects of N and K supply at the initial stage of panicle emergence on physiological traits in the flag leaf of hybrid rice Shanyou 63. Chinese J Rice Sci 11:165–169
WARDA (1996) Annual report for 1995. West Africa Rice Dev Assoc, Buoake, Côte d'Ivoire
Waters I, Armstrong W, Thomson CJ, Setter TL, Adkins S, Gibbs J, Greenway H (1989) Diurnal changes in radial oxygen loss and ethanol metabolism in roots of submerged and non-submerged rice seedlings. New Phytol 113:439–451
Williams RL (1992) Physiological comparison of Amaroo and YRL 39. In: Beecher HG, Dunn BW (eds) Annu Rep, Yanco Agric Inst, New South Wales, Australia
Xu D-Q, Li D-Y, Shen Y-G, Liang G-A (1984) On mid-day depression of photosynthesis of wheat leaf under field condition. Acta Physiol Sin 10:269–278
Yambao EB, Ingram KT (1988) Drought stress index for rice. Philipp J Crop Sci 13:105–111
Ying J, Peng S, He Q, Yang H, Yang C, Visperas RM, Cassman KG (1998a) Comparison of high-yield rice in tropical and subtropical environments. I. Determinants of grain and dry matter yields. Field Crops Res 57:71–84

Ying J, Peng S, Yang G, Zhou N, Visperas RM, Cassman KG (1998b) Comparison of high-yield rice in tropical and subtropical environments. II. Nitrogen accumulation and utilization efficiency. Field Crops Res 57:85–93
Yoshida S (1972) Physiological aspects of grain yield. Annu Rev Plant Physiol 23:437–464
Yoshida S (1976) Carbon dioxide and yield of rice. In: Proc Symp Climate and rice. Int Rice Res Inst, Manila, Philippines, pp 211–221
Yoshida S (1977) Rice. In: Alvim Paulo de T, Kozlowski TT (eds) Ecophysiology of tropical crops. Academic Press, New York, pp 57–87
Yoshida S (1978) Tropical climate and its influence on rice. Res Pap Ser No 20, Int Rice Res Inst, Manila, Philippines
Yoshida S (1981) Fundamentals of rice crop science. Int Rice Res Inst, Manila, Philippines
Yoshida S (1983) Rice. In: Symp Potential productivity of field crops under different environments. Int Rice Res Inst, Manila, Philippines, pp 103–127
Yoshida S, Coronel V (1976) Nitrogen nutrition, leaf resistance and leaf photosynthetic rate of rice plant. Soil Sci Plant Nutr 22:207–211
Yoshida S, Parao FT (1976) Climatic influence on yield and yield components of lowland rice in the tropics. In: Proc Symp Climate and rice. Int Rice Res Inst, Manila, Philippines, pp 471–491
Yoshida S, Coronel V, Parao FT, De los Reyes E (1974) Soil carbon dioxide flux and rice photosynthesis. Soil Sci Plant Nutr 20:381–386
Yoshida S, Satake T, Mackill DS (1981) High temperature stress in rice. Res Pap Ser No 67, Int Rice Res Inst, Manila, Philippines
Yuan LP (1993) Progress of two-line system in hybrid rice breeding. In: Muralidharan K, Siddiq EA (eds) New frontiers of rice research. Directorate of Rice Res, Hyderabad, India, pp 86–90
Zapata FJ, Alejar MS, Torrizo LB, Novero AU, Singh V, Senadhira D (1991) Field performance of anther culture-derived lines from F_1 crosses of *indica* rices under saline and non-saline conditions. Theor Appl Genet 83:6–11
Zisaka LH, Namuco O, Moya T, Quilang J (1997) Growth and yield response of field-grown tropical rice to increasing carbon dioxide and air temperature. Agron J 89:45–53

Physiology of Maize

M. TOLLENAAR[1] and L.M. DWYER[2]

1 Introduction

1.1 History, Production, and Utilization

Maize (*Zea mays* L.) or corn, as it is known in North America, is one of the three major cereal crops in the world. Although little consensus exists as to the origin and early evolution of maize, there is general agreement that maize was first domesticated 7000 to 10000 years ago in southern Mexico. After domestication, maize spread fairly quickly through North and South America, reaching the northeastern US and southern Canada prior to European colonization. Maize spread rapidly throughout Europe and, from Europe, to other parts of the world after the European discovery of the Americas.

The area in maize production in the world was 130 million ha and total world production was 507 million Mg in 1995 (Canada Grains Council 1995). Approximately 40% of the maize grain in the world in 1995 was produced in the US, followed by China (20%), Brazil (6.5%), and Mexico (3.25%). The average yield of maize in 1995 among the major maize-producing countries ranged from 7.78 and 7.60 Mg/ha in France and the US, respectively, to 2.36 and 2.20 in Brazil and Mexico, respectively (Canada Grains Council 1995). Commercial grain yields and maize grain production figures are usually expressed on a 15.5% grain-moisture basis.

The grain of maize has been traditionally used for direct human consumption, but the major use of maize in the world is as an animal feed. In the US, approximately 75% of the grain is used as an animal feed and approximately 20% of the maize is used as a source of industrial products. High fructose corn syrup and fuel ethanol consumed about 60% of the maize grain for industrial purposes in the US in 1992 (Duvick 1993) and the remaining 40% was used in consumer products such as snack foods, breakfast cereals, and degradable plastics. The major component of the grain is starch (Table 1) and starch is the basis for almost all industrial uses of the maize grain. With the advent of

[1] Crop Science Department, University of Guelph, Guelph, Ontario, Canada
[2] Agriculture and Agri-Food Canada, Eastern Cereal and Oilseed Research Centre, Ottawa, Ontario K1A 0C6, Canada

Crop Yield, Physiology and Processes
D. L. Smith, C. Hamel (Eds.)
© Springer-Verlag Berlin Heidelberg 1999

Table 1. Composition of component parts of maize kernels from seven Corn-Belt hybrids. (Watson 1987). Tabled values are means followed by a bracketed range

Composition	Endosperm	Germ	Pericarp	Whole kernel
		%		
Proportion of whole kernel	83 (82–84)	11 (10–12)	5.3 (5.1–5.7)	100
Starch	8.8 (8.6–8.0)	8.3 (5.1–10.0)	1.0 (0.7–1.2)	4.4 (3.9–5.8)
Fat	0.8 (0.7–1.0)	33 (31–35)	1.0 (0.7–1.2)	4.4 (3.9–5.8)
protein	8.0 (6.9–10.4)	18 (17–19)	3.7 (2.9–3.9)	9.1 (8.1–11.5)
Ash	0.3 (0.2–0.5)	10.5 (9.9–11.3)	0.8 (0.4–1.0)	1.4 (1.4–1.5)
Sugar	0.6 (0.5–0.8)	11 (10–13)	0.3 (0.2–0.4)	1.9 (1.6–2.2)
Unaccounted	2.7	8.8	8.7	9.8

biotechnology, it has been suggested that maize hybrids will be developed for specific industrial uses by changing the chemical composition of the kernel. In a more speculative scenario, the maize kernel will be used to produce and store completely different compounds, such as pharmaceuticals or industrial chemicals (Duvick 1993).

1.2 Taxonomy and Morphology

Maize is an annual plant belonging to the grass family (Graminea or Poaceae; Table 2). Maize is monoecious, with staminate flowers on the terminal branched end of a tall erect stem (i.e., the tassel) and pistillate flowers on axillary shoots (i.e., ear shoots). An ear shoot consists of a shank or cob, florets with long filamentous stigmas, the silk, and enveloping spathe leaves, the husks. In the beginning of floret development, staminate florets develop on the ear shoot and pistillate florets may develop on the tassel, but they usually remain rudimentary (Bonnett 1966). Occasionally, both staminate and pistillate florets will fully develop and become functional on the tassel (i.e., tassel seed), a phenomenon that occurs commonly on the tillers, which are elongated branches that sometimes develop from nodes lower on the stem. Axillary shoots develop in acropetal succession during early stages of plant development and the size

Table 2. The botanical classification of maize. (Porter 1967)

Class	Monocotyledonege
Subclass	Glumiflorae
Order	Poales
Family	Poaceae
Subfamily	Pinicoideae
Tribe	Tripsaceae
Genus	*Zea*
Species	*mays*

of the axillary shoots increases from the apex to the base of the stem. Later, when the ears begin to develop, the size sequence changes, so that the topmost shoot is the largest. The topmost shoot of fully expanded maize plants is usually located in the axis of the sixth or seventh leaf from the top in most commercial maize hybrids.

Maize is a tall plant with a fibrous root system. Corn Belt hybrids are about 3 to 3.5 m tall; inbreds are smaller; plant height tends to increase with increasing relative maturity. Long photoperiod also increases plant height, whereas drought stress may reduce plant height. Maize has a single leaf at each node, and two successive leaves along the stem of maize have an approximately opposite azimuthal orientation. Each leaf consists of a sheath surrounding the stem and a leaf blade that is connected to the leaf sheath at the collar or ligule. Number of leaves per plant varies from 7 for the short-season open-pollinated variety Gaspe Flint to more than 30 for tropical cultivars; leaf number for temperate-climate maize hybrids ranges from 16 to 23.

1.3 Agronomy and Physiology

The literature on the agronomy, production, genetics, structure and composition, and utilization of maize is extensive. The reader is referred to books such as *Corn and Corn Improvement* (Sprague and Dudley 1988) and *Corn: Chemistry and Technology* (Watson and Ramstad 1987), and the publication *How a Corn Plant Develops* (Ritchie et al. 1993) for more detailed information. In this chapter on maize we will concentrate on the interface between agronomy and the physiology of maize.

Physiology embraces functional processes ranging from gene translation and transcription to the formation of economic yield, transgressing levels of organization with a range in time scales from picoseconds to months. The focus of this chapter will be on integrating information on processes at lower levels of organization which can be used to explain yield formation of field-grown maize. In particular, we will discuss the impact of phenology or phasic development, which describes the progression of pivotal events during the life cycle from seed to seed, growth, expressed as dry matter accumulation, and dry matter partitioning on yield of maize. The chapter will conclude with a summary of physiological changes associated with grain yield improvement in maize during the past four to six decades in North America.

2 Phenology

2.1 Phasic Development

Maize plants come in various sizes and shapes depending on their origin, relative maturity, specific end use, or the environmental conditions under which they are grown. However, all maize plants go through the same phases of development from planting to maturity. The life cycle of maize can be divided into four distinct phases (Table 3): a period of predominantly vegetative growth (the leaf-growth phase), a period during which dry matter is allocated predominantly to the reproductive organs (the grain-filling period), a transition period between the two former phases (the flowering period), and the final phase of the life cycle during which no translocation occurs between the plant and the grain (the period of grain dry down).

The Leaf-Growth Phase. Maize seedlings can emerge in as few as 5 days after planting under warm moist soil conditions. Emergence can be delayed by up to 30 days in a cool soil. Maize seedlings are susceptible to disease organisms and insects, especially in a cold soil and stand establishment is often related to rate of seedling emergence (e.g., Eagles and Hardacre 1979; Mock and McNeill 1979). Application of a seed treatment and insecticides is important to ensure a uniform stand.

Even though leaves are emerging above ground, the growing point of maize is below ground until the four- to five-leaf stage. It is soil temperature rather than air temperature that influences rate of development during this phase. At about the four-leaf stage (four leaves visible) the plant has a short "growth crisis" (Deleens et al. 1984): the transition from heterotrophic to autotrophic

Table 3. Phasic development of maize

1. The leaf-growth phase
   - 1.1 Imbibition of the seed
   - 1.2 Plant emergence
   - 1.3 Transition from predominantly heterotrophic to predominantly autotrophic growth
   - 1.4 End of Juvenile phase
   - 1.5 Tassel initiation
   - 1.6 Initiation of topmost ear
   - 1.7 Emergence of topmost leaf
2. The flowering period
   - 2.1 Tassel emergence
   - 2.2 Anthesis
   - 2.3 Silking
   - 2.4 Fertilization of the florets
3. The grain-filling period
   - 3.1 Onset of lag phase of grain dry matter accumulation
   - 3.2 Onset of rapid grain dry matter accumulation
   - 3.3 End of rapid grain dry matter accumulation
   - 3.4 Half milk line
4. The period of grain dry down
   - 4.1 Black layer formation
   - 4.2 15.5% grain moisture

growth (until this point most of the resources for growth have been supplied by the reserves in the seed but, subsequently, photosynthesis of the seedling supports all growth). The seedling is particularly susceptible to unfavorable conditions during the "growth crisis" and effects of management, for example, low soil temperature associated with reduced tillage or soil compaction resulting from working soil when it is still wet, on seedling growth and development become apparent at this stage.

All leaves have been initiated by the time the plant reaches the eight- to ten-leaf stage, the stage of tassel initiation. Leaves continue to emerge and expand. The tassel will become visible soon after the topmost leaf has emerged from the whorl. The duration from planting to the emergence of the topmost leaf is directly related to the total number of initiated leaves (e.g., Dwyer and Stewart 1986a). Duration from planting to silking is longer in long-season than in short-season hybrids because long-season hybrids initiate more leaves.

Maize plants usually have one ear per plant, although up to eight ears may have been initiated. Axillary meristems are initiated at the same rate as the leaf primordia on the apical meristem, but with a delay of 5.6 to 7 plastochrons (i.e., interval between the initiation of two successive leaf primordia). The topmost axillary meristem is initiated on the same day as the tassel (Lejeune and Bernier 1996); hence, the topmost ear will be positioned five to seven leaves below the tassel. The transition of the topmost axillary meristem to an ear starts about 10 days after tassel initiation. Subsequently, the transition of vegetative axillary buds to ears in lower nodes will progress down the plant. Ears in these nodes will abort before silking, unless the environmental conditions for the plant are very favorable (e.g., plants in outside rows that receive high solar irradiance). When conditions are particularly unfavorable for corn growth and dry matter accumulation, for example, under extreme competition from surrounding plants at high population density, even the topmost ear may abort or fail to set grain (i.e., barren plants) (Tollenaar et al. 1992).

The Silking or Flowering Period. Maize attains its maximum height, pollen is shed from the tassel at the top of the plant, and silks are emerging from the earshoot during the flowering period. This period is critical for grain yield. Pollen production is abundant, as many as 10 million pollen grains are shed per day per plant, and fertilization of the florets on the ear is seldom limited by insufficient pollen. Silks that emerge from the earshoot should be pollinated within 8 days after emergence (Bassetti and Westgate 1994). After a pollen grain lands on a silk, the pollen grain germinates to produce a pollen tube which must grow down the entire length of the silk before fertilization can occur. Fertilization of the florets may not occur, or kernels that have been fertilized may abort during the first few days after fertilization, due to stress caused by drought, nutrient deficiency, pests, or high plant density. The result is incomplete seed set and reduced yields. The maize crop is prone to stress during the flowering period because crop growth and, consequently, the demand for resources for growth, is at a peak. In addition, flowering usually occurs during a period when daily air temperature and risk of water deficit are high.

The Grain-Filling Period. Rapid filling of the grain starts 2 to 3 weeks after silking. Grain grows at a rate of 3 to 4% of final yield per day during the period of rapid grain growth. The rate of grain growth is directly related to the prevailing temperature and is largely independent of crop dry matter accumulation (Duncan et al. 1965). If grain growth is more rapid than crop dry matter accumulation, dry matter for grain growth is remobilized from stalks, leaves, and husks (mainly stalks). Stress during this period can, therefore, result in increased lodging and premature leaf death.

The advance of grain maturity can be traced by following the progress of the "milk line" in the kernels (Afuakwa and Crookston 1984). This can be observed by breaking an ear in two when kernels have dented. The exposed face of the upper end of the ear will reveal the smooth surface of the kernels. The milk line should be visible near the top of the kernels. This line represents the boundary between the solid, starchy portion of the kernel and the milky, lower part. As the grain matures, the line moves downward towards the tip of the kernel. When the milk line is halfway down the kernel, the grain moisture content will be about 40%. The line will disappear at the tip of the kernel shortly before black layer formation, giving a good indication of when the crop is about to reach physiological maturity.

Maximum whole-plant dry matter is usually reached when the grain moisture is 40 to 45% (i.e., half milk line). Maximum maize silage yields can be obtained by harvesting at this stage. Grain will continue to accumulate dry matter, although at a much lower rate, until the formation of a distinct black layer at the base of the kernel (Daynard and Duncan 1969).

The Period of Grain Dry Down. The decline in moisture content of the grain is rapid during the grain-filling period when water in the kernel is replaced by starch and other constituents of the grain. Drying occurs more slowly after black layer formation. Moisture loss during this phase is exclusively a function of (1) the difference in moisture content of the kernel and the humidity of the air surrounding the plants and (2) physical barriers such as the kernel pericarp and the husks that restrict the flow of water vapor. Grain dry down is slow when the humidity of the air is high (e.g., when the temperature is low) and/or when the husks are tightly fitted around the ear (e.g., immature ears).

2.2 Relative Maturity and Rate of Development

Duration of each of the phases of development and, consequently, the duration of the life cycle is influenced by climatic factors, by the relative maturity of the genotype and, to a lesser extent, by soil and crop management factors. Maize hybrids are classified according to the duration of the life cycle (i.e., relative maturity). Relative maturity of a hybrid is determined by comparing its grain moisture percentage at harvest with that of a check hybrid; the higher the grain moisture relative to the check hybrid, the higher the relative maturity. Relative maturity can be expressed in terms of days from seeding to the date at which grain attains a set moisture percentage. Rate of development is most influenced by temperature and, consequently, duration of the life cycle will vary across locations and years.

In order to account for the influence of temperature on rate of development, heat unit systems have been used to classify the relative maturity of maize hybrids. The growing degree-day (GDD) system is used in the US. Growing degree-days accumulated during a day are calculated as:

$$GDD = (T_{max} + T_{min})/2 - T_{base}, \quad (1)$$

where T_{max} and T_{min} are daily maximum and minimum air temperatures respectively, and T_{base} is the base temperature (usually taken as 10 or 8 °C). Growing degree-days are frequently calculated with temperature limits of 30 and 10 °C (e.g., Plett 1992). Thus, if $T_{max} > 30$, $T_{max} = 30$ and if $T_{min} < 10$, $T_{min} = 10$. Relative maturity expressed in GDD is the sum of the daily GDDs from planting to harvest (at a set grain moisture). Another heat unit system that is used in Canada is the Crop Heat Unit system (CHU) (Brown and Bootsma 1993). Crop heat units accumulated during a day are calculated as:

$$CHU = (HUD + HUN)/2, \quad (2)$$

where HUD is heat units accumulated during the day time calculated as:

$$\begin{aligned} HUD &= 3.33(T_{max} - 10) - 0.084(T_{max} - 10)^2 \text{ for } T_{max} \geq 10°C; \\ HUD &= 0 \text{ for } T_{max} < 10°C, \end{aligned} \quad (3)$$

and HUN is heat units accumulated during the night time, calculated as:

$$HUN = 1.8(T_{min} - 4.4) \text{ for } T_{min} \geq 4°C; \; HUN = 0 \text{ for } T_{min} < 4.4°C. \quad (4)$$

Similarly, relative maturity expressed in CHU is calculated by summing daily CHU from planting to harvest (at a set grain moisture). The coefficients of variability for duration of maize development in accumulated GDD or CHU are substantially lower than those in calendar days across locations and years (Tollenaar et al. 1979).

The growing season for maize in temperate climates is delimited by subzero temperatures in early spring and late fall. In order to select maize hybrids of appropriate relative maturity for a particular location, long-term weather records are used to identify first and last day of the growing season with a 10% chance of frost.

Accumulated heat unit accumulation for the location can be estimated from daily maximum and minimum temperatures for each day between the first day in spring and last day in fall with a 10% chance of frost. Maps delineating seasonal heat unit accumulation facilitate hybrid selection. Examples of duration of development of two hybrids of different relative maturity grown at their respective heat unit locations in Ontario during an average year are depicted in Table 4.

Prediction of the duration of maize development stages was much improved with the introduction of heat unit systems in the 1960s, but large variations in heat unit accumulation among years and locations are still apparent, particularly for the period from silking to black layer formation (Major et al. 1983; Plett 1992). An accurate prediction of development is essential, particularly in mechanistic models of maize growth and development.

Most studies on effects of environmental influences on maize development have been carried out under controlled-environment conditions. In contrast, the heat unit systems (GDD, CHU) were developed through statistical manipulation of meteorological data and dates of silking and maturity of maize grown under field conditions. Controlled-environment conditions are appropriate for isolating the effects of a single environmental factor on development, but controlled-environment conditions do not entirely mimic field conditions. The improved understanding of maize development resulting from these studies may not lead instantly to a better prediction of development in the field because of differences between controlled environment and field conditions (e.g., absolute levels of environmental inputs such as incident solar radiation, interactions between environmental factors, continuously changing temperatures within and among days).

Most researchers define maize development in terms of the inverse of duration between two stages (e.g., planting and silking). This definition implies that

Table 4. Approximate dates and accumulated crop heat units (CHU) for various stages of corn development of a 2600-CHU and a 3200-CHU hybrid grown at their respective heat-unit locations during an "average" year

Stage of development	2600-CHU hybrid		3200-CHU hybrid	
	Date	CHU	Date	CHU
Seeding	15 May	0	5 May	0
Emergence [1.2][a]	29 May	200	17 May	200
4-leaf stage [1.3]	8 June	350	27 May	350
10-leaf stage	27 June	800	12 June	800
Tasselling [2.1]	20 July	1300	15 July	1550
Silking [2.3]	26 July	1450	21 July	1700
Start of grain filling [3.2]	11 Aug	1850	6 Aug	2100
Half milk line [3.4]	18 Sept	2450	11 Sept	2950
Black layer [4.1]	30 Sept	2600	21 Sept	3200

[a] Phasic development in square brackets according to Table 3.

the effect of any environmental factor on rate of development is equal throughout this period. However, this is often not true. For instance, in the planting to silking interval, the duration between anthesis and silking is particularly sensitive to environmental stresses (Dow et al. 1984). The duration of the leaf-growth phase (Table 3) is highly correlated with final leaf number, and final leaf number is modulated by environmental influences during a short period before tassel initiation (eight- to ten-leaf stage in temperate-climate hybrids). For instance, an increase in the photoperiod may result in an increase in final leaf number, but the impact of photoperiod on leaf number occurs only during a short period before tassel initiation (Kiniry et al. 1983; Tollenaar and Hunter 1983). Increasing temperature during a short period before tassel initiation also results in an increase in total leaf number and, consequently, the increased temperature will result in both an apparent decline in rate of development, due to increased total leaf number, and an increase in rate of development, due to a higher rate of leaf appearance (Tollenaar and Hunter 1983). Final leaf number is correlated with the relative maturity of maize genotypes (Chase and Nanda 1967) and environmental influences on leaf number would appear to be associated with changes in relative maturity. Therefore, quantification of environmental influences on rate of development should be phase-specific, and effects on final leaf number should be separated from other effects that influence duration of the vegetative phase.

Rate of Leaf Appearance. An accurate quantification of the effect of temperature on rate of development is the backbone of any phenological model. The response of maize to temperature has been reported in respect to seedling shoot extension (Lehenbauer 1914), radicle and shoot elongation (Blacklow 1972), and leaf appearance (Tollenaar et al. 1979). The response was remarkably similar in these three studies (Durand et al. 1982): a curvilinear relationship between rate of development and temperature with an optimum temperature of 31 °C and a basically linear response between 12 and 26 °C (Fig. 1). The response mimics the influence of temperature on enzyme kinetics, i.e., enzyme activation and denaturation. Rate of leaf appearance can be easily monitored and measured over short periods of time. Rate of leaf appearance as defined by Tollenaar et al. (1979) is the inverse of the phyllochron (i.e., the duration between the appearance of successive leaf tips), whereas others (e.g., Warrington and Kanemasu 1983a) have defined rate of leaf appearance as the inverse of the duration between the appearance of successive leaf collars or ligules. Rate of leaf-tip appearance is relatively stable across the leaf-growth phase (Tollenaar et al. 1979; Kiniry and Bonhomme 1991) whereas rate of leaf-collar appearance has been reported as stable (Dwyer and Stewart 1986a) or variable (Muldoon et al. 1984; Kiniry and Bonhomme 1991) across this period. In the following, we will refer to rate of leaf-tip appearance when discussing leaf appearance rates.

Rate of leaf appearance is influenced by the temperature of the growing point (Brouwer et al. 1973). The growing point of maize is below the soil surface

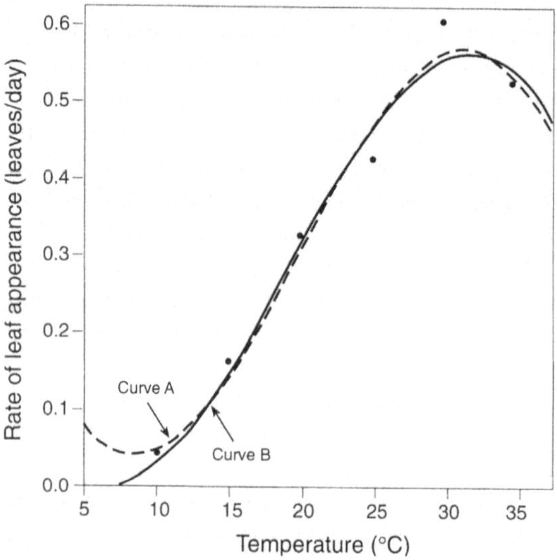

Fig. 1. Relationship between rate of leaf appearance and air temperature. *Circles* indicate mean rate of leaf appearance at six constant temperatures. *Curve A* is a polynomial equation of best fit to the data and *Curve B* is an alternate polynomial in which Y = 0 when X = 6

until the four- to five-leaf stage and, consequently, soil temperature rather than air temperature will influence rate of leaf appearance during the seedling phase. Above-ground plant tissue temperature is governed by the heat balance (incoming and outgoing radiation, sensible heat loss, and latent heat loss due to transpiration) and growing-point temperature could differ from air temperature at any stage of development. Jeppson and Crookston (1986) showed that reduced transpiration of defoliated maize seedlings resulted in an increase in growing-point temperature. However, growing-point temperature is usually not recorded in indoor and field studies. Rate of leaf appearance is affected by soil N, soil moisture, and incident PAR. McCullough et al. (1994) reported that rate of leaf appearance from the 3- to the 12-leaf stage was slower at low than at high N supply; the reduction ranged from 23 to 26% in the two hybrids tested. It has also been reported that soil-moisture deficit reduces rate of leaf appearance (Begonia et al. 1987; Albrecht and Carberry 1993). We have observed that the rate of leaf appearance of field-grown maize is consistently 15 to 20% higher than predicted from the equation reported by Tollenaar et al. (1979). An experiment was conducted by us under controlled-environment conditions with plants grown at 24.5 °C under a 10-h photoperiod (650 μmol $m^{-2}s^{-1}$), 10-h photoperiod plus 10-h extension with incandescent bulbs (40 μmol $m^{-2}s^{-1}$), or a 20-h photoperiod (Tollenaar 1998). Rates of leaf appearance from the 3- to 15-leaf stage were 0.46, 0.45, and 0.54 leaves/day, respectively, indicating that photoperiod did not affect rate of leaf appearance and that doubling incident photosynthetically active radiation (PAR) increased rate of leaf appearance by 17%. Increases in rate of leaf appearance associated with higher incident irradiance can also be inferred from results reported by Gmelig Meyling (1973) and Ellis et al. (1992).

Total Leaf Number. Total leaf number at the end of the leaf growth phase is determined by rate and duration of leaf initiation from imbibition of the seed to tassel initiation, plus the number of leaf initials present in the embryo. Assuming that five leaves are present in the embryo (Kiesselbach 1949), Warrington and Kanemasu (1983a) reported that leaf primordia initiation was apparently slow from imbibition of the seed to seedling emergence, but primordia initiation progressed at a constant rate from plant emergence to tassel initiation and rate of initiation showed near-linear increases as mean temperatures were increased from 15 to 28 °C. As previously mentioned, Tollenaar and Hunter (1983) showed that the total primordia number (or final leaf number) of a temperate-climate maize hybrid was influenced by photoperiod and temperature during a short period prior to tassel initiation: this period extended from the four- to seven-leaf stage for photoperiod and from the five- to seven-leaf stage for temperature. Final leaf number increased with increases in photoperiod and temperature (0.15 leaves/h and 0.17 leaves/°C, respectively). Leaf stage at tassel initiation was numerically equal to 50% of final leaf primordia number in these experiments, suggesting that the influence of photoperiod and temperature on final leaf number was mediated by their effects on the duration of leaf primordia initiation. This contention is confirmed (at least qualitatively) by results of other published research. Both Kiniry et al. (1983), for photoperiod, and Bonhomme et al. (1984), for temperature, reported a short sensitive or inductive period for leaf number just prior to tassel initiation. Gay (1980) reported a high correlation between leaf stage at tassel initiation and final leaf number ($r = 0.92$): the mean leaf stage at tassel initiation was 0.46 of final leaf number in a 3-year field experiment with various planting dates and maize hybrids representing a range of relative maturities. Warrington and Kanemasu (1983b) reported a linear increase in final leaf number with increase in mean temperature above 15 °C, but leaf number also increased when the temperature declined below 15 °C. Results of published research differ in respect to the quantitative aspects of the response of leaf number of temperature and photoperiod. For instance, the increase in leaf number when photoperiod is increased tends to be larger in later-maturing than in early-maturing genotypes, although a range of photoperiod responsiveness has been observed among genotypes of similar relative maturity (e.g., Kiniry et al. 1983; Russell and Stuber 1984).

Effects of factors other than photoperiod and temperature on total leaf number have not been well documented in the literature. Significant, but small, declines in final leaf number with increase in plant density, decrease in available soil N, and weed competition (Tollenaar et al. 1994a,b) have been reported. In our indoor study with the 10-h photoperiod, 10-h photoperiod plus 10-h extension with low PPFD, and 20-h photoperiod, we recorded final leaf numbers of 15.7, 17.3, and 18.4 leaves, respectively. All means were significantly different from each other, suggesting that incident PPFD also influences total leaf number.

Phenological Response to Temperature During the Grain-Filling Period. The simplicity of the thermal time concepts and their improvement over a day counter for prediction of development led to their widespread adoption, particularly for the vegetative period, but thermal time concepts have been less useful for the grain-filling period. Although Carter and Poneleit (1973) and Derieux and Bonhomme (1982) reported a close relationship between accumulated thermal units from planting to silking and from silking to maturity (with silking occurring at approximately 50 to 55% of accumulated heat unit accumulation from planting to maturity), relatively little is known about rate of development during the phases of the grain-filling period because of the logistical difficulty in carrying out research with fully expanded maize plants under controlled environment conditions. Thermal time required for specific genotypes to reach maturity has been found to vary with the thermal environment represented by mean daily air temperature; years or locations with higher mean air temperatures result in higher thermal unit accumulations to maturity for the same genotypes (Major et al. 1983; Plett 1992).

Dwyer et al. (1997) have recently shown that phenological response of maize to temperature is very different during the vegetative and grain-filling periods. During vegetative growth, phenological response to mean daily air temperature follows a sigmoidal curve beginning below 5 °C, with maximum response to temperatures between 25 and 30 °C (Fig. 2). During the grain-filling period, the temperature response function is flat from 0 to 12 °C and only rises significantly with mean daily temperatures above this range (Fig. 3). Both temperature-response functions (F_T) are described by a cubic polynomial

$$F_T = B_0 + B_1 T_A^2 + B_2 T_A^3, \tag{5}$$

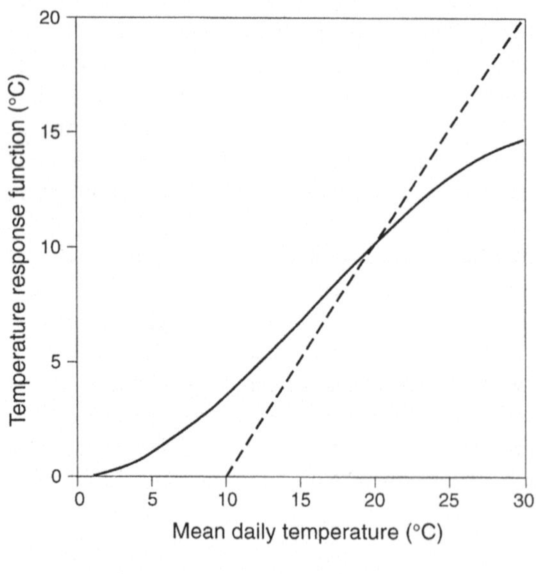

Fig. 2. Comparison of the temperature-response function fitted to Eq. (5) (*solid line*) and that assumed by GDD with a base of 10 °C and temperature limits of 30 and 10 °C (*dashed line*) for the planting to silking period

where T_A is mean daily air temperature and B_0, B_1 and B_2 are empirical coefficients. However, for the vegetative period (planting to silking) $B_0 = 0$. The sigmoidal response function for the vegetative period fitted by Dwyer et al. (1997) is consistent with the bulk of empirical evidence for this period summarized by Shaykewich (1995). The temperature response during the grain-filling period is very different from that during the vegetative period, and from that assumed by GDD (Fig. 3). However, the response is apparently consistent over a wide range of genotypes and growing conditions. Coefficients of Eq. (5) were fit to field measurements obtained on 28 Pioneer brand hybrids at 19 locations in north central and northeastern US and southern Ontario (latitudes 39 to 48°N) and produce fitted curves in Figs. 2 and 3. In addition, when growth room data on length of the grain-filling period under a range of temperature treatments (Tollenaar and Bruulsema 1988) were analyzed, they were found to produce a function of similar shape (Fig. 3). It should be noted that one of the treatments in the growth chamber study was a day/night temperature regime of 14/2 °C. The fact that grain filling was completed in a reasonable time provides evidence that development continues during the reproductive period at or below 10 °C. The fitted-temperature response functions reduced the standard error in estimating the duration of the grain-filling period (silking to physiological maturity) from 12 days using GDD to less than 6 days.

The significance of the temperature insensitivity of development from 0 to 12 °C depends on the distribution of mean air temperatures during the grain-filling period for particular location-year environments. The fitted temperature response function calculated faster development rates for cool days near the end of the growing season than did GDD. Incorporation of this

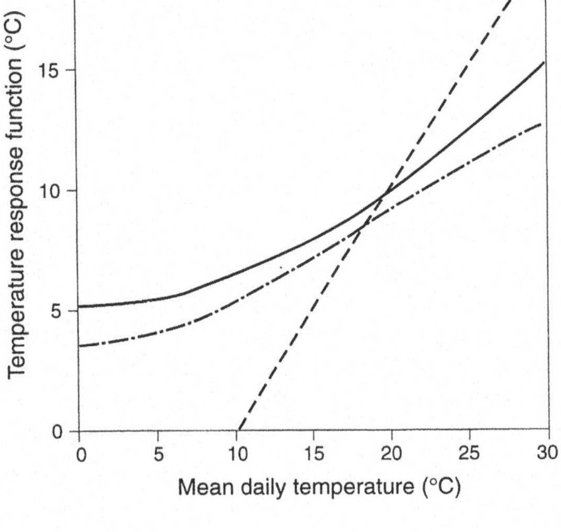

Fig. 3. Comparison of the temperature response function fitted to Eq. (5) using field data (*solid line*) and growth room data of Tollenaar and Bruulsema (1988) (*long and short dashed line*), and that assumed by GDD with a base of 10 °C and temperature limits of 30 and 10 °C (*dotted line*) for the silking to maturity period

response to low temperatures during grain filling improves estimation of the duration of the grain-filling period over that possible using current heat unit systems.

3
Dry Matter Accumulation

Green plants capture incident solar radiation by means of photosynthesis and convert the radiant energy into chemical energy which is stored in the carbon-to-carbon bonds of organic matter. Less than 6% of the dry matter of a maize plant is composed of minerals taken up from the soil. The energy required for uptake, translocation, and incorporation of minerals into dry matter is also directly or indirectly supplied by photosynthesis. Rate of dry matter accumulation of maize is the product of solar irradiance incident at the earth's surface (I), the absorption of solar irradiance by the maize canopy (α), and the efficiency of conversion of absorbed radiant energy into maize dry matter (ε). Total seasonal dry matter accumulation (DM) is the sum of daily rates of dry matter accumulation from planting to harvest:

$$DM = \int_{t=\text{planting}}^{t=\text{harvest}} (I\,\alpha\,\varepsilon)dt. \tag{6}$$

In the absence of limitations due to water, nutrients, or temperature, potential maize dry matter accumulation at any particular location is related to the duration of the growing season. The growing season may be delimited by low temperature, low soil moisture, or by the use of multiple-crop sequences. The components of rate of dry matter accumulation (I, α, and ε) vary throughout the growing season. Duration of the growing season, total incident solar radiation and its distribution, leaf area index (LAI) and canopy architecture, and leaf photosynthetic rate determine the rate of dry matter accumulation of the maize crop. Management practices and maize genotypes that produce the greatest dry matter in a particular environment can be predicted from the quantification of these factors and their interactions.

3.1
Incident Solar Radiation

Incident solar radiation is a function of solar height (i.e., latitude, day, and hour) and the attenuation of the radiation by the atmosphere (e.g., clouds). The incident radiation at the top of the atmosphere is about 1400 $W\,m^{-2}$ and maximum values at the Earth's surface are about 1000 $W\,m^{-2}$. Maximum average daily incident solar radiation in temperate climates is attained in early summer and declines rapidly during late summer (Fig. 4). It should be noted that daily incident solar radiation is highest on Earth during early summer at locations around 40°C latitude (due to the combination of long days and solar height).

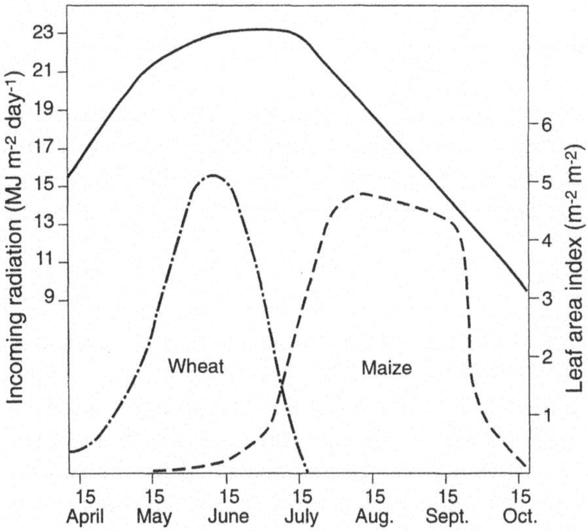

Fig. 4. Average daily incoming radiation and leaf area development of a wheat and a maize canopy during the growing season at Guelph, Ontario, Canada. (Tollenaar 1983)

About 50% of the incident solar irradiance can be used in photosynthesis. Photosynthetically active radiation (PAR, $W\,m^{-2}$) is the energy contained in wavelengths from 400 to 700 nm. The fraction of PAR as a proportion of global solar radiation is higher at low irradiance, due to the absorption of infra red radiation by the atmosphere and clouds, and the fraction is higher for diffuse than for direct radiation.

In photosynthesis, chlorophyll is excited by photons in the 400 to 700 nm range. The energy per photon is inversely related to its wavelength (i.e., $E = hc/\lambda$, where E is the energy per photon, h is the Planck constant, c is the speed of light, and λ is the wavelength). Photosynthetic photon flux density (PPFD, $\mu mol\,m^{-2}\,s^{-1}$) is the number of photons in the 400 to 700 nm range, and PPFD is the preferred measure of radiant energy in photosynthesis research. The conversion of PAR to PPFD varies with the spectral distribution of the radiant energy and ranges between 5 and 4.6 $\mu mol\,m^{-2}\,s^{-1}$ PPFD per $W\,m^{-2}$ PAR for incandescent bulbs and sunlight, respectively (Major 1987), i.e., 500 $W\,m^{-2}$ PAR of sunlight is approximately equivalent to 2300 $\mu mol\,m^{-2}\,s^{-1}$.

Photons of wavelengths smaller than 400 nm and larger than 700 nm are not used in photosynthesis but can indirectly influence maize canopy photosynthesis. The ratio of red (645 nm) and infrared (735 nm) radiation can affect internode length, and leaf width and length (Kasperbauer and Karlen 1994), and appears to influence leaf orientation (Girardin and Tollenaar 1993). UV-B radiation (290 to 315 nm) can result in damage to the leaf photosynthesis system, although maize appears to be relatively tolerant of UV-B radiation (van Doren 1995).

3.2 Absorption of Solar Irradiance

Incident solar radiation is absorbed by the maize canopy. The maize canopy consists of leaf blades, leaf stems, stalks, ear shoots, and tassels. However, only leaf blades are considered, unless noted otherwise. The transmitted radiation (I_t) at the bottom of a canopy can be estimated according to Monsi and Saeki (1953) as:

$$I_t = I_0 \exp(-k\,LAI), \quad (7)$$

where I_0 is incident irradiance at the top of the canopy, LAI is the leaf area index, and k is the extinction coefficient. The interception of incident radiation is the difference between incident radiation (I_0) and transmitted radiation (I_t). The Monsi and Saeki equation does not account for the reflection of incident radiation by the canopy. The irradiance that is absorbed by the canopy (I_A) can be estimated by taking the balance of incoming and outgoing fluxes as:

$$I_A = I_o[(1 - \rho_C)\exp(-k\,LAI)], \quad (8)$$

where ρ_C is the canopy reflection coefficient, which ranges from 4 to 5% for fully expanded maize canopies (Tollenaar and Bruulsema 1988). Equation (8) accounts for only one downwelling pass and ignores the flux of transmitted radiation that is reflected by the soil back into the crop canopy. Soil reflection coefficients range from 5 to 10% (Gallo and Daughtry 1986) and, consequently, the error in omitting this flux is minor for fully expanded maize canopies. The total daily absorption of incident radiation by a maize canopy is a function of LAI and k.

3.2.1 Leaf Area Index

The LAI of maize shows a sigmoidal response from seeding to around silking, when the topmost leaf is fully expanded, a slight decline during the grain-filling period and a rapid decline towards the end of the life cycle (Fig. 4). Both maximum LAI and the rate of change in LAI during the growing season vary substantially with maize genotype, environmental conditions, and cultural practices. The maximum LAI is smaller for inbreds than for hybrids and is, in general, substantially larger for late-maturing hybrids than for short-season hybrids (Major et al. 1972; Muldoon et al. 1984). Daynard (1972) reported that maximum LAI of a set of hybrids grown under nonlimiting field conditions was 33% larger in 1970 than in 1969, which appeared to be associated with the temperature regime during the leaf-growth phase. Maximum LAI increases when the plant density is enhanced; LAI is proportional to plant density at very low plant densities, but the change in LAI becomes increasingly smaller when plant density is enhanced from suboptimal to supraoptional densities for grain yield (Tollenaar 1992). In addition, hybrid by environment (or cultural practice) interactions for LAI are common.

Given the importance of LAI for rate of maize dry matter accumulation, various attempts have been made to predict LAI throughout the growing season. Methods range from the application of purely empirical equations (such as the Richards function) to simulations of leaf extension by estimations of leaf cell division and extension. In general, empirical methods give best results under a narrow range of conditions, but perform poorly when applied to conditions which deviate substantially from those for which the model was developed. In contrast, detailed knowledge of the physiological and biochemical processes is too scanty to formulate a mechanistic model that predicts leaf area development from processes that occur at the cell level. An entirely different approach is employed by the crop simulation group in the Netherlands established by the late C.T. de Wit (Goudriaan and van Laar 1994). They predict leaf area expansion by simulating dry matter partitioning to the leaves and multiplying leaf dry matter by specific leaf area (i.e., leaf area per gram of dry matter). Dry matter partitioning and specific leaf area vary across the life cycle of maize, among maize hybrids and environmental conditions (Tollenaar 1989a). Consequently, their method of simulating leaf area expansion requires empirical functions that relate dry matter partitioning and specific leaf area to genotypic and environmental variables. The current most satisfactory methodology for predicting LAI in maize is the formulation of empirical relationships between the area of each leaf in the canopy, its maximum leaf area expansion rate, duration of leaf area expansion, and leaf senescence (Stewart and Dwyer 1994a,b).

The LAI of a maize canopy can be predicted at any time during the growing season by estimating the area of each of the component leaves. Dwyer and Stewart (1986a) demonstrated a consistent relationship between leaf position and maximum area per leaf (Fig. 5). Exposed area of each leaf expands from the appearance of the leaf tip in the whorl until the appearance of the ligule (Thiagarajah and Hunt 1982). Rates of leaf tip and leaf ligule appearance have been documented fairly extensively (see Sect. 2.1) and, consequently, duration of leaf area expansion can be estimated. Leaf-area expansion follows a logistic pattern between the appearance of the tip and the ligule (Reid et al. 1990; Stewart and Dwyer 1994a), but rate of leaf-area expansion can be approximated by assuming a linear response (i.e., rate of area expansion = maximum area ÷ duration of area expansion). Hence, maize LAI can be predicted by this method if genotypic and environmental influences on maximum leaf area and leaf senescence can be quantified.

Leaf area per plant is correlated with total leaf number, but the relationship between leaf number and leaf area is not linear. For instance, Muldoon et al. (1984) reported that total leaf number was larger for late- than for early-maturing hybrids but leaf area of the late-maturing hybrids was disproportionately larger. A general relationship between total leaf number and maximum leaf area, similar to the relationship between leaf position and

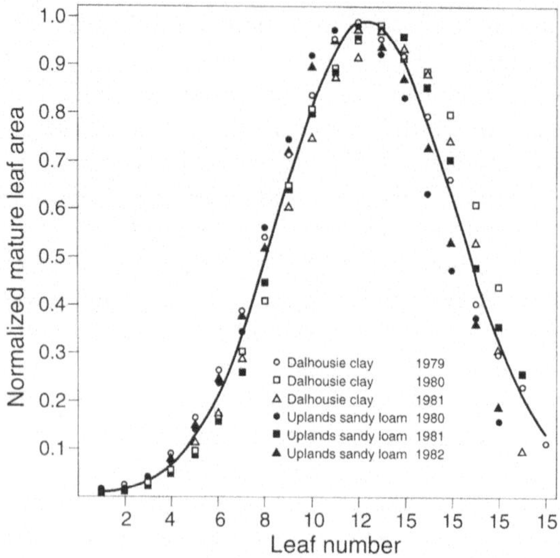

Fig. 5. Mature leaf area, normalized with respect to leaf area of the largest leaf per plant of individual years, as a function of leaf number. (Dwyer and Stewart 1986a)

maximum leaf area depicted in Fig. 5, could function as a framework for quantifying genotypic differences in maximum LAI. Such a relationship could also account for environmental factors that alter LAI by their influence on total leaf number (e.g., temperature, photoperiod, and incident radiation).

Leaf size is also modified by environmental influences such as temperature, incident radiation, and soil moisture and nutrient stress. Tollenaar (1989a) reported that the LAI of maize at the 12-leaf stage was highest at 19 °C (LAI = 1.87) and 23 °C (LAI = 1.85) when plants were grown at 4 °C intervals from 11 to 31 °C; LAI at 31 °C was 38% lower and LAI at 11 °C was 68% lower than that at 19 °C. Reid et al. (1990) also reported that the leaf area of fully expanded plants was higher at 19/14 °C than at 30/24 °C. The far-red: red ratio of irradiance is larger for light reflected by green leaves than for direct solar radiation, which can result in increased leaf length and a reduction in leaf width and leaf area (Kasperbauer and Karlen 1994). The reduction in leaf area per plant when plant density is increased may be, in part, attributable to this phenomenon. Water stress can reduce rate of leaf area expansion when leaf water potential declines below −0.7 Mpa (e.g., Boyer 1970), which results in prevalence of nighttime leaf expansion. Acevedo et al. (1971) and Barlow and Boersma (1972) found that rate of leaf area expansion can actually be higher immediately after a mild water stress than in unstressed plants. Effects of mineral nutrition on maximum leaf area in maize not have been well documented, in part, because effects of nutrient deficiency on total plant leaf area are confounded with leaf senescence.

Leaf senescence usually proceeds from the leaf at the lowest node upwards and leaf senescence is accelerated when plants are exposed to stress. High plant density, soil moisture deficit, and N stress have been reported to accelerate rate

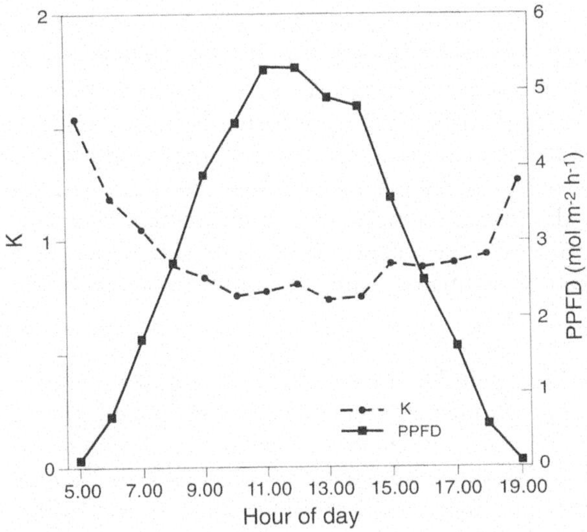

Fig. 6. Mean hourly incident PPFD and extinction coefficient (*K*) for maize canopy with LAI = 3.13 grown at Elora, Ontario, from 4 to 7 Aug. 1984. (After Tollenaar and Bruulsema 1988)

of leaf senescence (Begonia et al. 1987; Wolfe et al. 1988). Leaf senescence accelerates when plants approach physiological maturity. In this case, leaf senescence proceeds from leaves at the bottom of the plant to the leaf subtending the ear and from the topmost leaf downwards. The effect of aging on leaf senescence varies among maize hybrids but modern maize hybrids stay green longer than old hybrids (Crosbie 1982; Tollenaar and Aguilera 1992).

3.2.2 *Canopy Extinction Coefficient*

The incident irradiance transmitted or absorbed by $1\,m^2$ leaf area per $1\,m^2$ ground area (i.e., LAI = 1) is a function of the extinction coefficient k [i.e., from Eq. (7): $k = -\ln(I_t/I_0)/LAI$]. The extinction coefficient quantifies the influence of the arrangement of the leaves in the canopy and scattering of irradiance by the leaves on the attenuation of irradiance. The leaf-area arrangement in three-dimensional space above $1\,m^2$ of ground area is determined by (1) plant height, (2) plant spacing (i.e., row width vs. distance between plants in the row), (3) leaf length and width, (4) leaf angle with respect to the horizontal plane, and (5) leaf orientation with respect to north and south (i.e., azimuth angle). If all leaves in a canopy have the same leaf angle, the distribution of leaf azimuth is uniform, and leaf angle (α) < solar height, then $k = \cos\alpha$. The extinction coefficient is relatively constant during the middle of the daytime when close to 90% of the total daily incident PPFD is received, but k is greater during the 3-h period from after sunrise or before sunset (Fig. 6).

The effect of leaf angle on dry matter accumulation and grain yield is smaller in maize than in some other field crops due to the relatively high saturating PPFD for maize leaf photosynthesis. Leaf angle in US maize hybrids increased in the 1960s and 1970s (Crosbie 1982), but increases in leaf angle of short-season maize hybrids in Ontario were still not apparent in the 1980s (Tollenaar et al. 1994c). The effect of increased leaf angle on grain yield is difficult to quantify empirically because genetic yield improvement is associated with numerous factors. However, the magnitude of the impact of leaf angle on canopy photosynthesis can be estimated using a simplified representation of the maize canopy.

Canopy Photosynthesis – No Change in Plant Density. If we assume a uniform distribution of leaf area, canopy photosynthesis (P_c) can be estimated in four steps: sunlit leaf area (SLA), absorbed PPFD per unit leaf area (I_A), photosynthesis per unit leaf area (P_l), and canopy photosynthesis (P_c) (Table 5). Canopy photosynthesis at 2000 $\mu mol\,m^{-2}s^{-1}$ increased 14% in the short-season hybrid and 30% in the Corn-Belt hybrid when leaf angle was increased from 30° to 60°. The effect of increasing leaf angle is greater in Corn-Belt hybrids than in short-season hybrids because of the greater leaf area per plant (and greater LAI) of Corn-Belt compared to short-season hybrids. It should be noted that the actual increases in total crop dry matter and grain yield will be smaller than the increases in canopy photosynthesis in Table 5 because both incident PPFD and LAI will be smaller than maximum values in Table 5 during a large proportion of the growing season.

Canopy Photosynthesis – Increasing Plant Density. Plant density can be increased without affecting yield per plant when leaf angle is increased, if photosynthesis per plant remains constant. The new plant density (ΔPD) can be estimated because the new canopy photosynthesis is equal to both leaf photosynthesis at the 60° leaf angle times the new sunlit leaf area *and* canopy photosynthesis at the 30° leaf angle times times ΔPD, and all variables except ΔPD are known. In the example of Table 5, plant density could be increased to 129 and 139% of the plant density at the 30° leaf angle for the short-season and Corn-Belt hybrid, respectively. Again, these estimated increases in canopy photosynthesis due to an increase in leaf angle from 30° to 60° indicate the upper limit of yield improvement, as incident PPFD and LAI throughout the growing season will be frequently lower than values in Table 5 and, consequently, increases in canopy photosynthesis will be smaller.

3.3 Radiation-Use Efficiency and Total Dry Matter Accumulation

Rate of dry matter accumulation is a function of the interception of incident PPFD and the conversion of absorbed photons into plant dry matter [Eq. (6)], i.e., photosynthesis. The response of photosynthesis to PPFD is curvilinear

Table 5. Potential impact of mean leaf-angle distribution (i.e., 30° and 60°) on canopy photosynthesis of a short-season hybrid (S.S.) with LAI = 2.5 and a Corn-Belt hybrid (C.B.) with LAI = 4.5

Leaf angle treatment	Hybrid	Plant density %	Sunlit leaf area $m^2\,m^{-2}$	Absorbed PPFD $\mu mol\ m^{-2}(leaf)s^{-1}$	Leaf photosynthesi $\mu mol\ m^{-2}(leaf)s^{-1}$	Canopy photosynthesis $\mu mol\ m^{-2}(ground)s^{-1}$	Canopy photosynthesis relative to that of treatment 1
1[a]	S.S.	100	1.02	1740	35.45	36.13	1.00
	C.B.	100	1.13	1740	35.45	40.06	1.00
2	S.S.	100	1.43	1000	29.16	41.17	1.14
	C.B.	100	1.79	1000	29.16	52.17	1.30
3	S.S.	129	1.60	1000	29.16	46.66	1.29
	C.B.	139	1.91	1000	29.16	55.81	1.39

[a] Leaf angle treatment 1 is at 30°, treatment 2 is 60° with plant density equal to that of treatment 1, treatment 3 is 60° with plant density such that plant photosynthesis is equal to that of treatment 1. Incident PPFD is 2000 $\mu mol\ m^{-2}\ s^{-1}$ with no scattering of incident. PPFD by leaves is assumed; P_{max} is 50 $\mu mol\ m^{-2}\ s^{-1}$ and α = 0.07 mol CO_2/mol photon. See Appendix for steps involved in calculating canopy photosynthesis.

(Table 5, Appendix 1, Step 3). Photosynthesis per mol absorbed photons is highest at low PPFD and declines with increasing PPFD. Maximum photosynthetic efficiency of maize leaves is 0.062 mol CO_2 per mol absorbed photons (Ehleringer and Pearcy 1983). The positive effect of high leaf angle on canopy photosynthesis (Table 5) is a result of a reduction in mean PPFD per unit leaf area, but not in total absorbed PPFD per unit ground area, thereby increasing mean photosynthetic efficiency. In addition to PPFD, leaf photosynthesis is influenced by leaf age, leaf history, air temperature, mineral nutrition, and plant water status. Daily net canopy photosynthesis is also influenced by respiration, which is a function of chemical composition of the new biomass, and activity and size of the total biomass. Therefore, it is surprising that the efficiency of conversion of intercepted radiant energy into maize dry matter is relatively constant. A review of the literature revealed that the mean radiation-use efficiency (RUE) of maize grown under nonstressed conditions prior to the grain-filling period was 3.5 g accumulated dry matter per MJ of PAR intercepted by the canopy, with a 19% coefficient of variability of the data set (Kiniry et al. 1989). Tollenaar and Aguilera (1992) reported that RUE during the grain-filling period was 3 g/MJ for a recently released maize hybrid and 2.4 g/MJ for an old maize hybrid.

Total annual dry matter accumulation of a maize canopy is equal to the sum of the daily crop growth rates during the growing season [Eq. (6)]. In most cases, annual dry matter accumulation is primarily a function of the duration of complete (green) leaf-area expansion (Fig. 4) and, consequently, seasonal dry matter accumulation is positively associated with the relative maturity rating of a hybrid. Assuming mean incident PAR is $10\,MJ\,m^{-2}day^{-1}$ (i.e., short-wave irradiance is $20\,MJ\,m^{-2}day^{-1}$), interception of PAR is 0.85, and RUE is 3 g/MJ, then mean crop growth rate during the period of full leaf-area expansion is $25\,g/m^2$. If total dry matter at the onset of complete leaf-area expansion is 3 Mg/ha and the duration of complete leaf area expansion is 60 days, then total annual dry matter accumulation is 3 Mg/ha + 60 days $\times$ $250\,kg\,ha^{-1}day^{-1}$ = 18 Mg/ha.

Potential dry matter accumulation by maize canopies, however, is substantially greater than the aforementioned estimates. A farmer in Illinois consistently obtained grain yields ranging between 15 and 20 Mg/ha in commercial maize production during 1970s and 1980s (Warsaw 1985), which corresponds to total above-ground dry matter accumulation of 30 to 40 Mg/ha. Similar yields had been previously reported by a farmer in Michigan (Tollenaar 1983). Total dry matter accumulation of maize grown indoors is higher than that of field-grown maize although daily incident PAR was two times greater out side and, consequently, radiation-use efficiency was 2.5 times larger for the maize grown indoors (Tollenaar 1983). We speculate that very high RUEs and grain yields are associated with low abiotic stress levels and that the high rates are indicative of the current genetic potential for maize dry matter accumulation.

4 Dry Matter Partitioning

In mature leaves, starch and sucrose are major end products of photosynthesis. In maize, like most agronomic species, sucrose is the primary transport form of carbon. The import and utilization of soluble carbohydrates, along with amino acids as a source of reduced N, is essential to vegetative growth and reproductive development. Assimilate partitioning controls transfer of organic nutrients between grain and nongrain plant components.

Crop improvement through breeding has been based on increasing dry matter accumulation in the grain, that is, by either modifying partitioning of dry matter whose accumulation was limited by photosynthetic processes (Rocher 1988) or increasing total dry matter accumulation (Tollenaar 1991). Thus, strategies to improve grain yield include increased partitioning to the grain (harvest index, HI), delayed leaf senescence, and extended duration of the grain-filling period (Atkins 1993). Carbohydrates for grain originate from current photosynthesis and from translocation of reserves temporarily stored in stalks, leaves, cobs, and husks.

Nitrogen is supplied to storage organs as amino acids, produced after breakdown of proteins in vegetative tissue, or formed from recently absorbed nitrogen. In cereal crops generally, uptake of nitrogen after anthesis is a small fraction of the total absorbed (18–35%) (Penning de Vries and van Keulen 1993). In maize, nitrogen uptake after anthesis has been measured as 37 to 55% of total nitrogen absorbed (Ma and Dwyer 1998). Vegetative parts are mined efficiently, to an extent that the harvest index for nitrogen ranges from 50 to 64% (Penning de Vries and van Keulen 1993). The dilemma is that leaf N is also needed for photosynthesis, so reallocation to the grain undermines the capacity for assimilate supply (Prioul et al. 1990). Unless postanthesis uptake contributes significantly to the grain, final grain yield is determined by the size of the pool of available nitrogen at anthesis.

4.1 General Source-Sink Relationships

Partitioning of assimilates to the grain is influenced by the source (supply) and sink (demand for) photosynthate and is moderated by vascular connections, storage capacity of nongrain tissues and competing demands for tissue nitrogen. Empirical evidence is that phloem transport can occur over large distances, but a sink is generally supplied from a nearby source (Wardlaw 1968). In maize, leaves above the ear generally supply the developing kernels, while leaves below the ear support the lower stem and roots (Palmer et al. 1973; Fairey and Daynard 1978). Assimilate partitioning is also sensitive to environmental conditions. For example, water stress and low N fertility result in preferential allocation of dry matter to the roots during vegetative growth (e.g.,

Tolley-Henry and Raper 1986). Low temperature can also result in a change in the partitioning of dry matter and an accumulation of starch and carbohydrates at the source (Farrar 1988). Boote (1977) and Potter and Jones (1977) found that an increase in temperature lowered root: shoot ratios and increased partitioning of assimilates into new leaf area.

Timing of tissue development influences the relationships between photosynthetic sources and growing sinks. In cereals generally, there is short period following germination when the root is the major sink for seed reserves (Wardlaw 1987) but in maize this dominance is lost by the fourth leaf stage when the shoot is exposed to light and the leaves photosynthesize enough for the plant to convert to autotrophic growth. Brouwer (1962) postulated a generalized functional balance between carbon assimilation by the shoot and nutrient and water absorption by the root during vegetative development: resources (i.e., assimilates) are allocated according to the highest rate of return (i.e., rate of dry matter accumulation). Tollenaar (1989a) suggested that while this functional balance between root and shoot has contributed to a general understanding of partitioning, it has not lent itself to quantification.

The most economically significant source-sink relationship is that between photosynthesizing leaves and the grain. Following anthesis, grain growth is dominant. Even under drought conditions, evidence in wheat is that grain continues to grow although it requires a proportionally greater use of stored carbohydrates (Rawson and Evans 1971). However, there is a lower limit to the assimilation rate, below which grain fails to form (Edmeades and Daynard 1979), which Tollenaar et al. (1992) found was related to dry matter accumulation at anthesis and the threshold cost of producing an ear. Manipulation of source strength and sink size has sought to identify control mechanisms in grain filling. Recent evidence (Rajcan 1997) indicates that highest grain yields are associated with a balance between source and sink which results in neither a buildup of dry matter in the stalks, nor a mining of dry matter from the stalks during grain filling.

4.2 Source Strength

Leaf photosynthesis is the main source of carbon for growth and grain development. Although controls within the leaf can influence the supply of photosynthate, emphasis here is on the extent and duration of canopy leaf area. Extensive analysis of leaf photosynthesis and of canopy geometry in relation to light interception showed that photosynthetic capacities were rarely limiting but that leaf-area growth in relation to light interception was important (Evans 1975). Photosynthate production on an area basis can be increased by changing plant population density and leaf geometry without a change in the photosynthetic potential of individual leaves. A review of the yield advantage of variation in leaf angle (Trenbath and Angus 1975) concluded that more upright leaves gave no advantage in maize. However, Pendleton et al. (1968)

obtained increased yields in near isogenic maize lines with upright leaves and also demonstrated a yield increase when leaf angle was made more upright mechanically. More recent improvements in models to simulate leaf area distribution in space (e.g., Stewart and Dwyer 1993) permit comparison of light interception by genotypes with different architecture, and also for different plant population densities and planting arrangements.

Delayed senescence increases the duration of photosynthate production. This is particularly important to grain fill, as current photosynthate production is largely used in grain fill (Duncan 1975; Hageman 1986; Moutot et al. 1986). There is currently considerable interest in finding criteria to characterize leaf source capabilities. Leaf color and chlorophyll content have been used as simple indicators, although care must be taken in assuming a consistent relationship with rate of photosynthesis (Crafts-Brandner and Poneleit 1987). Carboxylase activities (PEPC and Rubisco) are also related to source strength, but there is no consistent association between these activities and leaf export or grain yield (Christy et al. 1986).

4.3 Export from Leaves

Assimilates are exported from leaves in the form of sucrose. Leaf nitrogen is also remobilized continuously throughout grain filling. Cliquet et al. (1990a) supplied $^{13}CO_2$ and ^{15}N-nitrate in nutrient solution during stalk elongation and found that carbon skeletons of soluble carbohydrates originated in almost equal proportions from current photosynthesis and from material stored before label exposure. This further supports the fact that mature and senescing leaves are exporters of new photosynthates and also of their cellular material. The soluble carbohydrate pool in the stalk was derived from many sources such as sucrose from the phloem and sucrose plus hexoses linked to in situ metabolism. Cliquet et al. (1990a) suggest that the gradient in soluble carbohydrates from lower to upper stalks reflects the successive contribution of leaves of decreasing age to the translocated sucrose.

Leaf carbohydrate metabolism is usually sensitive to source-sink variation (Prioul et al. 1990) and sucrose-phosphate synthase is the enzyme most representative of export capacity (Huber 1983; Rocher et al. 1989). A model of assimilate export from a maize leaf developed by Rocher and Prioul (1987) estimated the relative size of leaf sugar and starch compartments in the leaf through the day. Stewart et al. (1997) modified this model for nonsteady-state field conditions. In simple terms, carbon dioxide is taken up by the leaf and sucrose precursors (simple sugars) are formed. These precursors are converted to starch, which is stored in a leaf starch compartment, or to sucrose. Sucrose, in turn, is either stored in the leaf in one of two sucrose compartments or is exported from the leaf. Fits of this model to field data indicate that most of the leaf photosynthate from fully expanded maize leaves is exported directly from the leaves, and that export rates rise in proportion to increases in photosynthate production. This suggests that storage of carbohydrates in the leaf is not a significant factor when a sink exists, and net photosynthate is immediately available to the rest of the plant.

4.4 Temporary Storage

It has been known for several decades that soluble carbohydrates in corn stalks accumulate prior to anthesis and decline during grain filling (Daynard et al. 1969), especially in internodes below the ear (Hume and Campbell 1972). More recently, Swank et al. (1982) found that the dry weight and carbohydrate content of stalks were maintained until 34 days after anthesis, whereas stalk nitrogen content increased slightly up to 12 days after anthesis and decreased continuously thereafter. This suggested that through the first half of the grain-filling period leaves have the capacity to produce photosynthate that exceeds the needs of the ear and/or the transport system.

When the source of carbon and nitrogen in the kernels was examined it was found that maize reused more nitrogen reserves than carbon reserves for kernel filling (Below et al. 1981; Swank et al. 1982; Cliquet et al. 1990b). Results of Prioul et al. (1990) indicate that less than 10% of carbon in kernels came from remobilization. Ma et al. (1997) reported that between 20 and 63% of maize kernel nitrogen came from remobilization, with the proportion remobilized varying widely depending on the level and source of nitrogen amendment and the maize genotype. The stalk is a temporary carbon storage organ and carbon from the stalk is mainly remobilized to the cob, husk, and shank (Cliquet et al. 1990b). However, some of this carbon is translocated a second time; Moutot et al. (1986) showed by ^{14}C labeling that cob, husk, and shank serve as short-term storage reservoirs for kernel carbon feeding. Crawford et al. (1982) showed by ^{15}N labeling throughout grain fill that the cob, husk, and shank are first a nitrogen sink and then a nitrogen source for kernels. Cliquet et al. (1990b) confirmed that nitrogen incorporated at the beginning of stalk elongation and stored in vegetative organs was almost totally remobilized after anthesis, either directly to the kernel, or after temporary storage in cob, husk, and shank.

Whether or not carbon and nitrogen are remobilized depends on the stability of the structures in which they are incorporated. Nitrogen in amino acids and carbon in soluble carbohydrates and starch can be readily translocated to kernels, whereas cellulose is a permanent structure, not remobilizable under normal conditions (Cliquet et al. 1990a).

4.5 Competing Demands for Nitrogen

The reliance on photosynthetic carbon fixed after anthesis for grain fill results in competing demands for nitrogen. In spite of the high sink demand, leaf and canopy photosynthetic rates decline during grain filling (Christy et al. 1986; Dwyer and Stewart 1986b). Daynard and Kannenberg (1976) suggested that hybrids able to sustain longer source activity could attain higher yields. However, prolonged photosynthetic activity reduces nitrogen available for kernel protein. Whereas photosynthate produced after anthesis is the major source of kernel carbon, nitrogen uptake after anthesis tends to be restricted, and kernel nitrogen relies to a greater extent on remobilization. Although prolonged photosynthetic activity during grain filling requires maintenance of photosynthetic apparatus, Morita (1980) found that 90% of nitrogen lost from leaves during normal senescence is from the chloroplasts, and Ta and Weiland (1992) found that 45% of nitrogen remobilized to kernels is from leaves. Thus, it appears that remobilization of nitrogen to the kernel during grain fill prematurely reduces canopy photosynthesis.

Competition for nitrogen could be reduced if more of the kernel nitrogen requirement was taken up from the soil during grain fill. This is not a new idea, as Below et al. (1981) concluded that there was more flexibility for increasing the uptake of new nitrogen than for enhancing the remobilization of vegetative nitrogen. However, since that time we have learned more about the interrelationships between carbon and nitrogen demands of the plant. Tolley-Henry et al. (1988) found that nitrogen uptake during grain fill was associated with the availability of soluble carbohydrates to the roots. However, during grain filling, partitioning of carbohydrates is shifted from support of root activity to support of grain filling, resulting in the progressive decrease in nitrate uptake from the soil that has been noted (Below et al. 1981; Christensen et al. 1981). Recent experimental work to identify maize traits identified with higher nitrogen-use efficiency has determined that conditions that prolong the ability to take up nitrate from the soil during grain filling are associated with both higher nitrogen-use efficiency and higher grain yield. Postanthesis uptake from the soil is increased for "stay-green" plants (Ma and Dwyer 1998) and when high levels of soil nitrogen are available during the grain-filling period (Ma et al. 1997). These data suggest that competition for nitrogen within the plant may be reduced by selecting for plant types with high photosynthetic rates that are maintained during grain filling, to supply photoassimilates to the root as well as the grain, and by supplying nitrogen amendments that release nitrate nitrogen throughout the growing season and provide more available soil nitrogen during grain fill.

5 Genetic Improvement

Average maize grain yield in the US has increased from about 1.2 Mg/ha in the early 1930s to about 6.5 Mg/ha in the 1990s (Fig. 7), yields in Canada and France increased at rates of 100 to 150 $kg\,ha^{-1}\,yr^{-1}$ from 1950 to the 1990s (Tollenaar et al. 1994c), and average maize yield improvement in tropical countries has been 43 $kg\,ha^{-1}\,yr^{-1}$ (Edmeades and Tollenaar 1990). The yield improvement

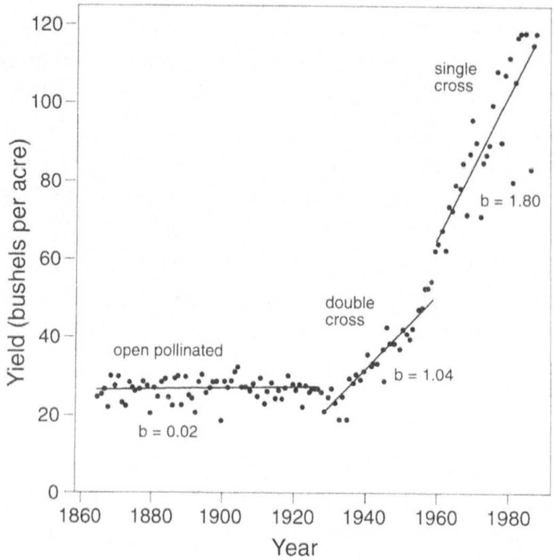

Fig. 7. Mean grain yield of maize in the United States from 1860s to 1990s (1 bu/acre at 15.5% moisture = 53.0 kg dry matter/ha). (Tollenaar et al. 1994c)

has been the result of genetic improvement by maize breeders, changes in cultural management, climate, and the interactions among these factors. Cardwell (1982) concluded that about 40% of the yield increase in Minnesota from 1930 to 1980 could be attributed to changes in cultural practices. He showed yield increases due to higher N application, higher plant density, use of herbicides, narrower row width, earlier planting dates, and fall ploughing, whereas yields decreased due to changes in rotation, decline in soil organic matter, increase in insect pests, and soil erosion. The contribution of genetic improvement through maize breeding to rate of yield improvement in the US has been estimated between 60 and 80% (Russell 1984; Duvick 1992). An estimate of the relative contribution of each of the factors that have been involved in the historic yield improvement is difficult to make because of interactions between genotype by environment and genotype by cultural practices. Specifically, relative differences in grain yield between new and old hybrids are larger under unfavorable conditions than under optimal conditions for maize growth (e.g., high vs. low plant density).

The consistent large rate of increase in maize grain yield across decades cannot be attributed to the improvement in one major physiological trait such as harvest index, leaf photosynthetic rate, LAI, or leaf angle. Yield improvement in various crops has been attributed to an increase in harvest index, i.e., the proportion of total above-ground dry matter allocated to the harvested economic component (Gifford 1986), but yield improvement in maize is not associated with a change in harvest index (Crosbie 1982; Tollenaar 1989b). Harvest index of new hybrids is larger than that of old hybrids under suboptimal conditions for maize growth, but no differences in harvest index are apparent among hybrids when conditions are optimal for maize growth, such as relatively low plant density, no water stress, and high soil N (Tollenaar et al. 1997). Another physiological attribute that would appear to be amenable to improvement is leaf photosynthetic rate. No differences in maximum leaf photosynthetic rate, however, have been reported for a range of old and new US and Canadian maize hybrids (Crosbie 1982; Dwyer and Tollenaar 1989). Leaf area per plant has increased slightly in Canada (Tollenaar 1991) but not in the US (Crosbie 1982), although maximum LAI has increased from old to new hybrids due to the increase in optimum plant density for grain yield. The increase in PPFD interceptance due to increased maximum LAI from, for instance, 3 to 4 or 4 to 5, is relatively small (see Sect. 3.2). Similarly, mean leaf angle has increased from old to new hybrids in the US (Crosbie 1982), but the potential improvement in canopy photosynthetic rate associated with the higher leaf-angle distribution is only approximately 20% (see Sect. 3.2.2), which is small relative to the yield improvement (Fig. 7).

Grain yield improvement in maize has been associated with increased plant density tolerance and "stay-green". The optimum plant density for grain yield of maize hybrids has increased. Duvick (1984) showed that for Corn-Belt hybrids from the 1930s to the 1980s grown at three plant densities (30 000, 47 000, 64 000 plants/ha), the oldest hybrids attained their highest yields at the lowest plant density, hybrids from the 1960s and 1970s yielded highest at the medium plant density, and highest yields of hybrids from the 1980s were attained at the highest plant density (Fig. 8). Grain yield at the higher plant densities is lower in the old hybrids due to a lower harvest index at the high than at the low plant density (Tollenaar et al. 1994c). The other difference between old and new maize hybrids that has been consistently reported in the literature in increased "stay-green" of new maize hybrids (Crosbie 1982; Tollenaar 1991), i.e., leaf senescence occurs later in new than in old hybrids. Consequently, most of the difference in dry matter accumulation between old and new hybrids is attributable to the grain-filling period, in general, and the second half of the grain-filling period, in particular (Tollenaar et al. 1994a,b). The increased tolerance of plant-density stress and increased "stay-green" of modern hybrids appears to be the most obvious manifestation of the increase in stress tolerance of old compared to new hybrids. Differences in stress tolerance between old and new hybrids grown in Ontario have been shown for low soil N, low soil moisture, low night temperature, and herbicide damage (Tollenaar et al. 1994a,b).

The differential response of old vs. new maize hybrids to stresses other than high plant density is often difficult to show. For instance, the reduction in grain yield due to low soil N was similar for new and old hybrids (Duvick 1984). Results of a long-term field study with one old and one new maize hybrid grown under low soil N conditions and weed interference showed that the relative reduction in yield due to weed interference was larger in the old than in the new hybrid (Tollenaar et al. 1997). The reduction in yield relative to maize grown weed-free at high soil N, was 17 and 54% in the weedy treatment at high and low soil N, respectively, for the old hybrid, and 11 and 40% in the weedy

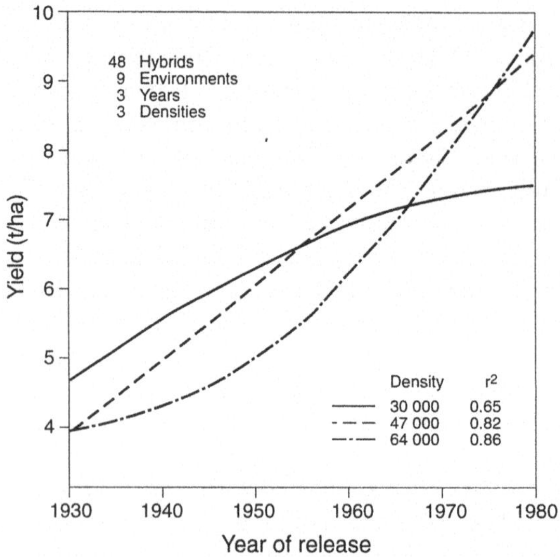

Fig. 8. Regression of maize grain yield at three plant densities vs. year of hybrid release. (Duvick 1984)

treatment at high and low soil N, respectively, for the new hybrid. The impact of moisture stress on canopy photosynthesis of the same hybrids has been investigated in an indoor study (Nissanka et al. 1997). Canopy photosynthesis around silking was not significantly different for the well-watered controls. However, canopy photosynthesis of the old hybrid declined sooner after water stress was applied and the total reduction due to water stress was 62% greater in the old than in the new hybrid. Canopy photosynthesis during the first day after rewatering was 53% higher in the new than in the old hybrid.

In conclusion, the genetic improvement in maize yield during the past 40 to 60 years has been large. Yield improvement has been the result of increased dry matter accumulation, rather than a change in dry matter distribution, but leaf photosynthesis per se does not appear to have changed. Whereas the increase in dry matter accumulation may be in part attributable to "stay-green" and an increased leaf angle, most of the yield improvement appears to be the result of increased stress tolerance. The adoption of low input management practices in maize production will increase the need for more stress-tolerant genotypes and stress tolerance will continue to be the most important trait in maize breeding for grain yield.

Appendix

Steps to calculate canopy photosynthesis for different leaf angles

1. Sunlit leaf area (SLA):

$$SLA = (1 - e^{-kxLAI})/k$$

2. Absorbed PPFD per unit leaf area (I_A):

$$I_A = k \times 1$$

3. Photosynthesis per unit leaf area (P_l):

$$P_l = (P_{max} \times 1_A \times \alpha + P_{max})$$

4. Canopy photosynthesis (P_c):

$$P_c = P_l \times SLA$$

5. Increase in plant density, while keeping photosynthesis per plant equal (ΔPD):

$$\Delta PD = (\text{new plant density})/(\text{old plant density}$$

6. New sunlit leaf area (SLA^1):

$$SLA^1 = (1 - e^{-k \times LAI \times \Delta PD})/k$$

7. New canopy photosynthesis (P_c):
P'_c is equal to canopy photosynthesis at 30° leaf angle [P_c(30 °C)] times ΔPD and P'_c is equal to leaf photosynthesis at 60° leaf angle [P_c(60 °C)] times new sunlit leaf area (SLA′):

$$P_c(30°) \times \Delta PD = P_l(60°) \times SLA'$$
$$P_c(30°)/P_l(60°) = SLA'/\Delta PD = (1 - e^{-k \times LAI \times \Delta PD})/(k \times \Delta PD)$$

References

Acevedo E, Hsiao TC, Henderson DW (1971) Immediate and subsequent growth responses of maize leaves to changes in water status. Plant Physiol 48:631–636

Afuakwa JJ, Crookston RK (1984) Using the kernel milk line to visually monitor grain maturity in maize. Crop Sci 24:687–691

Albrecht DG, Carberry PS (1993) The influence of water deficit prior to tassel initiation on maize growth, development and yield. Field Crops Res 31:55–69

Atkins CA (1993) Nitrogen assimilation and translocation in relation to plant growth In: Buxton DR, Shibles R, Forsberg RA, Blad BL, Asay KH, Paulser GM, Wilson RF (eds) International Crop Science I. Crop Science Society of America, Madison, Wisconsin, pp 807–811

Barlow EWR, Boersma L (1972) Growth response of corn to changes in root temperature and soil water suction measured with LVDT. Crop Sci 12:251–252

Bassetti P, Westgate ME (1994) Floral asynchrony and kernel set in maize quantified by image analysis. Agron J 86:699–703

Begonia GB, Hesketh JD, Frederick JR, Finke RL, Pettigrew WT (1987) Factors affecting leaf duration in soybean and maize. Photosynthetica 21:285–295

Below FE, Christensen LE, Reed AJ, Hageman RH (1981) Availability of reduced N and carbohydrates for ear development of maize. Plant Physiol 68:1186–1190

Blacklow WM (1972) Influence of temperature on germination and elongation of the radicle and shoot of corn (*Zea mays* L.). Crop Sci 12:647–650

Bonhomme R, Derieux M, Duburco JB, Ruget F (1984) Variations in the ovule number at silking in various corn genotypes. Maydica 24:101–107

Bonnett OT (1966) Inflorescences of maize, wheat, rye, barley, and oats: their initiation and development. Univ Ill Coll Agric Exp Stn Bull 721

Boote KJ (1977) Root-shoot relationships. Proc Soil Crop Sci Soc Fl 36:15–23
Boyer JS (1970) Leaf enlargement and metabolic rates in corn soybean, and sunflower at various leaf water potentials. Plant Physiol 46:233–235
Brouwer R (1962) Distribution of dry matter in the plant. Neth J Agric Sci 10:361–376
Brouwer R, Kleinendorst A, Locher J Th (1973) Growth responses of maize plants to temperature. In: Plant response to climatic factors. Proc Uppsala Symp, Uppsala, Sweden. 15–20 Sept, 1970. UNESCO, Paris, pp 169–174
Brown DM, Bootsma A (1993) Crop heat units for ear and other season crops in Ontario. Ont Ministry Agric Food Factsheet, Toronto, Ontario, Agdex 111/31 ISSN 0225-7882
Canada Grains Council (1995) Canada grains industry statistical handbook 95. Canada Grains Council, Winnipeg MB. ISSN 1201-5679
Cardwell VB (1982) Fifty years of Minnesota corn production: sources of yield increase. Agron J 74:984–990
Carter MW, Poneleit CG (1973) Black layer maturity and filling period variation among inbred lines of corn (*Zea mays* L.) Crop Sci 13:436–439
Chase SS, Nanda DK (1967) Number of leaves and maturity classification in *Zea mays* L. Crop Sci 7:431–432
Christensen LE, Below FE, Hageman RH (1981) The effects of ear removal in senescence and metabolism of maize. Plant Physiol 68:1180–1185
Christy AL, Williamson DR, Wideman AS (1986) Maize source development and activity. In: Shannon JC, Knievel DP, Boyer CD (eds) Regulation of carbon and nitrogen reduction and utilization in maize. American Society of Plant Physiologists, Rockville Maryland, pp 11–20
Cliquet J-B, Deleens E, Bousser A, Martin M, Lescure J-C, Prioul J-L, Mariotti A, Morot-Gaudry J-F (1990a) Estimation of carbon and nitrogen allocation during stalk elongation by ^{13}C and ^{15}N tracing in *Zea mays* L. Plant Physiol 92:79–87
Cliquet J-B, Deleens E, Mariotti A (1990b) C and N mobilization from stalk and leaves during kernel filling by ^{13}C and ^{15}N tracing in *Zea mays* L. Plant Physiol 94:1547–1553
Crafts-Brandner SJ, Poneleit CG (1987) Carbon dioxide exchange rates, ribuose biosphate carboxylase/oxygenase and phosphoenolpyruvate carboxylase activities and kernel growth characteristics of maize. Plant Physiol 84:255–260
Crawford TW Jr, Rendig VV, Broadbent FE (1982) Sources, fluxes and sinks of nitrogen during early reproductive growth of maize (*Zea mays* L.). Plant Physiol 70:1654–1660
Crosbie TM (1982) Changes in physiological traits associated with long-term breeding efforts to improve grain yield of maize. In Loden HD, Wilkinson D (eds) Proc 37th Annu Corn and Sorghum Industry Res Conf, Chicago, 5–9 Dec, Am Seed Trade Assoc, Washington, DC, pp 206–233
Daynard TB (1972) Relationships among black layer formation, grain moisture percentage and heat unit accumulation in corn. Agron J 64:716–719
Daynard TB, Duncan WG (1969) The black layer and grain maturity in corn. Crop Sci 9:473–476
Daynard TB, Kannenberg LW (1976) Relationship between length of actual effective grain-filling periods and grain yield in corn. Can J Plant Sci 56:237–242
Daynard TB, Tanner JW, Hume DJ (1969) The contribution of stalk soluble carbohydrates to grain yield in corn. Crop Sci 9:831–834
Deleens E, Gregory N, Bourdu R (1984) Transition between seed reserve use and photosynthetic supply during development of maize seedlings. Plant Sci Lett 37:35–39
Derieux M, Bonhomme R (1982) Heat unit requirements for maize hybrids in Europe. Results from the European FAO subnetwork. II. Period from silking to maturity. Matautu 27:79–96
Dow EW, Daynard TB, Muldoon JF, Major DJ, Thurtell GW (1984) Resistance to drought and density stress in Canadian and European maize (*Zea mays* L.) hybrids. Can J Plant Sci 64:575–585
Duncan WJ (1975) Maize in crop physiology. In: Evans LT (ed) Crop physiology: some case histories, Cambridge University Press, Cambridge, pp 23–50

Duncan WJ, Hatfield AL, Ragland JL (1965) The growth and yield of corn. II. Daily growth of corn kernels. Agron J 57:221–223

Durand R, Bonhomme R, Derieux M (1982) Seuil optimal des sommes de temperatures. Application au mais (*Zea mays* L.). Agronomie 2:589–597

Duvick DN (1984) Genetic contributions to yield gains of US hybrid maize, 1930 to 1980. In: Fehr WR (ed) Genetic contributions to yield grains of five major crop plants. CSSA Spec Publ 7, CSSA Madison, Wisconsin, pp 15–47

Duvick DN (1992) Genetic contributions to advances in yield of US maize. Maydica 37:68–79

Duvick DN (1993) Breeding maize for new industrial uses. In: Bianchi A, Lupotto E, Motto M (eds) Breeding and molecular biology: accomplishments and future promises. Proc 16th maize and sorghum EUCARPIA, Bergamo, Italy, pp 293–302

Dwyer LM, Stewart DW (1986a) Leaf area development in field-grown maize. Agron J 78:334–343

Dwyer LM, Stewart DW (1986b) Effect of leaf age and position on net photosynthetic rates in corn (*Zea mays* L.) Agric For Meteorol 37:29–46

Dwyer LM, Tollenaar M (1989) Genetic improvement in photosynthetic response of hybrid maize cultivars, 1959 to 1988. Can J Plant Sci 69:81–91

Dwyer LM, Stewart DW, Ma BL, Carrigan L, Tollenaar M (1997) Phenological response to temperature. In: Proc 52nd Annu Northeastern Corn Improvement Conf, February 12–13, 1997, Ottawa, Ontario, pp 54–55

Eagles HA, Hardacre AK (1979) Genetic variation in maize (*Zea mays* L.) for germination and emergence at 10 °C. Euphytica 28:287–295

Edmeades GO, Daynard TB (1979) The relationship between final yield and photosynthesis at flowering in individual maize plants. Can J Plant Sci 59:585–601

Edmeades GO, Tollenaar M (1990) Genetic and cultural improvements in maize production. In: Proc Int Congr of Plant Physiology, New Delhi, India, pp 164–180

Ehleringer J, Pearcy RW (1983) Variation in quantum yield for CO_2 uptake among C_3 and C_4 plants. Plant Physiol 73:555–559

Ellis RH, Summerfield RJ, Edmeades GO, Roberts EH (1992) Photoperiod, temperature and the interval from sowing to tassel initiation in diverse cultivars of maize. Crop Sci 32:1225–1232

Evans LT (1975) Crop physiology. Cambridge University Press, Cambridge, pp 327–355

Fairey NA, Daynard TB (1978) Quantitative distribution of assimilates in component organs of maize during reproductive growth. Can J Plant Sci 58:709–717

Farrar JF (1988) Temperature and the partitioning and translocation of carbon. In: Long SP, Woodward FI (eds) Symposia of the Society for Experimental Biology, No XXXXII. Plants and temperature, Essex, UK, 7–9 Sept 1987, pp 203–236

Gallo KP, Daughtry CST (1986) Techniques for measuring intercepted and absorbed photosynthetically active radiation in corn canopies. Agron J 78:752–756

Gay J-P (1980) Development et croissanse chez le mais: aspects pratiques. AGPM-ITCF, Paris, France

Gifford RM (1986) Partitioning of photo-assimilate in the development of crop yield. In: Gronshaw J, Lucas WJ, Giaquinta RT (eds) Phloem transport. Alan B Liss, New York, pp 535–549

Girardin Ph, Tollenaar M (1993) Effects of intraspecific interference on maize leaf azimuth. Crop Sci 34:151–155

Gmelig Meyling HO (1973) Effect of light intensity, temperature and daylength on the rate of leaf appearance of maize. Neth J Agric Sci 21:68–76

Goudriaan J, van Laar HH (1994) Modelling potential crop growth processes. Kluwer, Dordrecht, the Netherlands

Hageman RH (1986) Nitrate metabolism in roots and leaves. In: Shannon JC, Boyer DP, Boyer CD (eds) Regulation of carbon and nitrogen reduction and utilization in maize. American Society of Plant Physiologists, Rockville, Maryland, pp 105–116

Huber SC (1983) Role of sucrose-phosphate synthase in partitioning of carbon in leaves. Plant Physiol 71:818–821

Hume DJ, Campbell DK (1972) Accumulation and translocation of soluble solids in corn stalks. Can J Plant Sci 52:363–368
Jeppson RG, Crookston RK (1986) Effect of elevated growing-point temperature on maize growth and yield. Crop Sci 26:595–598
Kasperbauer MJ, Karlen DL (1994) Plant spacing and reflected far-red light effect on phytochrome-regulated photosynthate allocation in corn seedlings. Crop Sci 34:1564–1569
Kiesselbach TA (1949) The structure and reproduction of corn. Nebraska Agric Exp Stn Res Bull 161:3–96
Kiniry JR, Bonhomme R (1991) Predicting maize phenology. In: Hodges T (ed) Predicting crop phenology. CRC Press, Boca Raton, Florida, pp 115–131
Kiniry JR, Ritchie JT, Musser RL (1983) Dynamic nature of the photoperiod response in maize. Agron J 75:700–703
Kiniry JR, Jones CA, O'Toole JC, Cabelguenne M, Spanel DA (1989) Radiation-use efficiency in biomass accumulation prior to grain-filling for five grain-crop species. Field Crops Res 20:51–64
Lehenbauer PA (1914) Growth of maize seedlings in relation to temperature. Physiol Res 1:247–288
Lejeune P, Bernier G (1996) Effect of environment on the early steps of ear initiation in maize (*Zea mays* L.). Plant Cell Environ 19:217–224
Ma BL, Dwyer LM (1998) Nitrogen uptake and utilization of two maize hybrids differing in leaf senescence. Plant Soil (in press)
Ma, BL, Dwyer LM, Gregorich E (1997) Corn yield response to soil N amendments. pp. 56. Proc 52nd Annu Northeastern Corn Improvement Conf, February 12–13, 1997
Major DJ (1987) Light and photoperiod. In: Christie BR (ed) CRC Handbook of plant science in agriculture, vol 1. CRC Press, Boca Raton, Florida, pp 195–216
Major DJ, Hunter RB, Kannenberg LW, Daynard TB, Tanner JW (1972) Comparison of inbred and hybrid corn grain yield measured at equal leaf area index. Can J Plant Sci 52:315–319
Major DJ, Brown DM, Bootsma A, Dupuis G, Fairey NA, Grant EA, Green DG, Hamilton RI, Langille J, Sonmor LG, Smeltzer GC, White RP (1983) An evaluation of the corn heat unit system for the short-season growing regions across Canada. Can J Plant Sci 63:121–130
McCullough DE, Girardin Ph, Mihajlovic M, Aguilera A, Tollenaar M (1994) Influence of N supply on development and dry matter accumulation of an old and a new maize hybrid. Can J Plant Sci 74:471–477
Mock JJ, McNeill MJ (1979) Cold tolerance of maize inbred lines adapted to various latitudes in North America. Crop Sci 19:239–242
Monsi M, Saeki T (1953) Über den Lichtfaktor in den Pflanzengesellschaften und seine Bedeutung für die Stoffproduktion. Jpn J Bot 14:22–52
Morita K (1980) Release of nitrogen from chloroplasts during leaf senescence in rice (*Oryza sativa* L.). Ann Bot (Lond) 46:297–302
Moutot F, Huet JC, Morot-Gaudry JF, Pernollet JC (1986) Relationship between photosynthesis and protein synthesis in maize. I. Kinetics of translocation of photoassimilated carbon from the ear leaf to the seed. Plant Physiol 80:211–215
Muldoon JF, Daynard TB, Van Duinen B, Tollenaar M (1984) Comparisons among rates of appearance of leaf tips, collars, and leaf area in maize (*Zea mays* L.). Maydica 29:109–120
Nissanka SP, Dixon MA, Tollenaar M (1997) Canopy gas exchange response to moisture stress in old and new maize hybrids. Crop Sci 37:172–181
Palmer AE, Heichel GH, Musgrave RB (1973) Patterns of translocation, respiratory loss and redistribution of ^{14}C in maize labelled after flowering. Crop Sci 13:371–376
Pendleton JW, Smith GE, Winter SR, Johnston TJ (1968) Field investigations of the relationship of leaf angle in corn (*Zea mays* L.) to grain yield and apparent photosynthesis. Agron J 60:422–424
Penning de Vries FWT, van Keulen H (1993) Interaction of yield-determining processes. In: International Crop Science Society of America, Madison, Wisconsin, pp 831–934

Plett S (1992) Comparison of seasonal thermal indices for measurement of corn maturity in a prairie environment. Can J Plant Sci 72:1157–1162
Porter CL (1967) Taxonomy of flowering plants. WH Freeman, San Francisco
Potter JR, Jones JW (1977) Leaf area partitioning as an important factor in growth. Plant Physiol 59:10–14
Prioul JR, Reyss A, Schwebel-Dugué N (1990) Relationships between carbohydrate metabolism in ear and adjacent leaf during grain filling in maize genotypes. Plant Physiol Biochem 28:485–493
Rajcan I (1997) Physiological changes associated with delayed leaf senescence in maize. PhD Thesis, Univ of Guelph, Guelph
Rawson HM, Evans LT (1971) The contribution of stem reserves to grain development in a range of wheat cultivars of different height. Aust J Agric Res 22:851–863
Reid JF, Zur B, Hesketh JD (1990) The dynamics of maize canopy development. 2. Leaf area growth. Biotronics 19:99–107
Ritchie SW, Hanway JJ, Duncan WG (1993) How a corn plant develops. Spec Rep 48. Iowa State Univ Coop Ext Serv, Ames, Iowa, 27 pp
Rocher JP (1988) Comparison of carbohydrate compartmentation in relation to photosynthesis, assimilate export and growth in a range of maize genotypes. Aust J Plant Physiol 15:677–686
Rocher JP, Prioul J-L (1987) Compartmental analysis of assimilate export in a mature maize leaf. Plant Physiol Biochem 25:531–540
Rocher JP, Prioul J-L, Lecharny A, Reyss A, Joussaume A (1989) Genetic variability in carbon fixation, sucrose-P synthase and ADP-glucose pyrophosphorylase in maize plants of differing growth rate. Plant Physiol 89:416–420
Russell WA (1984) Agronomic performance of maize cultivars representing different eras of breeding. Maydica 29:375–390
Russell WA, Stuber CW (1984) Effects of photoperiod and temperatures on the duration of vegetative growth in maize. Crop Sci 23:847–850
Shaykewich CF (1995) An appraisal of cereal crop phenology modelling. Can J Plant Sci 75:329–341
Sprague GF, Dudley JW (1988) Corn and corn improvement. American Society of Agronomy, Madison, Wisconsin
Stewart DW, Dwyer LM (1993) Mathematical characterization of maize canopies. Agric For Meteorol 66:247–265
Stewart DW, Dwyer LM (1994a) Appearance time, expansion rate and expansion duration for leaves of field-grown maize (*Zea mays* L.). Can J Plant Sci 74:31–36
Stewart DW, Dwyer LM (1994b) A model of expansion and senescence of individual leaves of field-grown maize (*Zea mays* L.). Can J Plant Sci 74:37–42
Stewart DW, Dwyer LM, Andrews CJ, Dugas JA (1997) Modeling carbohydrate production, storage and export in leafy and normal maize (*Zea mays* L.). Crop Sci 37:1228–1236
Swank JC, Below FE, Lambert RJ, Hageman RH (1982) Interaction of carbon and nitrogen metabolism in the productivity of maize. Plant Physiol 70:1185–1190
Ta CT, Weiland RT (1992) Nitrogen partitioning in maize during ear development. Crop Sci 32:443–451
Thiagarajah MR, Hunt LA (1982) Effects of temperature on leaf growth in corn (*Zea mays* L.). Can J Bot 60:1647–1652
Tollenaar M (1983) Potential vegetative productivity in Canada. Can J Plant Sci 63:1–10
Tollenaar M (1989a) Response of dry matter accumulation in maize to temperature: I. Dry matter partitioning. Crop Sci 29:1239–1246
Tollenaar M (1989b) Genetic improvement in grain yield of commercial maize hybrids grown in Ontario from 1959 to 1988. Crop Sci 29:1365–1371
Tollenaar M (1991) Physiological basis of the genetic improvement of maize hybrids in Ontario from 1959 to 1988. Crop Sci 31:119–124
Tollenaar M (1992) Is low plant density a stress in maize? Maydica 57:305–311

Tollenaar M (1999) Effects of photoperiod and irradiance from planting to the 15-leaf stage on rate of leaf appearance and duration of the grain-filling period in maize. Field Crops Res (submitted)
Tollenaar M, Aguilera A (1992) Radiation use efficiency of an old and a new maize hybrid. Agron J: 84:536–541
Tollenaar M, Bruulsema TW (1988) Effects of temperature and duration of kernel dry matter accumulation of maize. Can J Plant Sci 68:935–940
Tollenaar M, Hunter RB (1983) A photoperiod- and temperature-sensitive period for leaf number of maize. Crop Sci 23:457–460
Tollenaar M, Daynard TB, Hunter RB (1979) The effect of temperature on rate of leaf appearance and flowering date in maize. Crop Sci 19:363–366
Tollenaar M, Dwyer LM, Stewart DW (1992) Ear and kernel formation in maize hybrids representing three decades of grain yield improvement in Ontario. Crop Sci 32:432–438
Tollenaar M, Dibo AA, Aguilera A, Weise SF, Swanton CJ (1994a) Effect of crop density on weed interference in maize. Agron J 86:591–595
Tollenaar M, Nissanka SP, Aguilera A, Weise SF, Swanton CJ (1994b) Effect of weed interference and soil N on four maize hybrids. Agron J 86:596–601
Tollenaar M, McCullough DE, Dwyer LM (1994c) Physiological basis of the genetic improvement of corn. In: Slafer GA (ed) Genetic improvement of field crops. Marcel Dekker, New York, pp 183–236
Tollenaar M, Aguilera A, Nissanka SP (1997) Grain yield is reduced more by weed interference in an old than in a new maize hybrid. Agron J 89:239–246
Tolley-Henry L, Raper CD Jr (1986) Nitrogen and dry matter partitioning in soybean plants during onset of and recovery from nitrogen stress. Bot Gaz 147:392–399
Tolley-Henry L, Raper CD Jr, Granato TC (1988) Cyclic variations in nitrogen uptake rate of soybean plants: effects of external nitrate concentration. J Exp Bot 39:613–622
Trenbath BR, Angus JF (1975) Leaf inclination and crop production. Field Crop Abstr 28:231–244
van Doren SK (1995) Effects of enhanced UV-B radiation on Ontario soybean cultivars and maize hybrids. MSc Thesis, Univ of Guelph, Guelph
Wardlaw IF (1968) The control and pattern of movement of carbohydrates in plants. Bot Rev 34:79–105
Wardlaw IF (1987) The physiology and develoment of temperate cereals. In: Lazenby A, Matheson EM (eds) Australian field crops. 1. Wheat and other temperature cereals. Angus and Robertson, Sydney, Australia, pp 58–98
Warrington IJ, Kanemasu ET (1983a) Corn growth response to temperature and photoperiod. II. Leaf-initiation and leaf appearance rates. Agron J 75:755–761
Warrington IJ, Kanemasu ET (1983b) Corn growth response to temperature and photoperiod. III. Leaf number. Agron J 75:762–766
Warsaw H (1985) High yield corn production. In: Munson RD (ed) Proc Physiol, Biochem and Chem Associated with Maximum Yield Corn, St Louis, MO, 11–12 Nov 1985. Found Agron Res Potash Phosphate Inst, St Louis, USA, pp 193–199
Watson SA (1987) Structure and composition. In: Watson SA, Ramstad PE (eds) Corn: chemistry and technology. Amer Association Cereal Chemists, St Paul, Minnesota pp 53–82.
Watson SA, Ramstad PE (1987) Corn: chemistry and technology. Amer Association Cereal Chemists, St Paul, Minnesota
Wolfe DW, Henderson DW, Hsiao TC, Alvins A (1988) Interactive water and nitrogen effects on senescence of maize. II. Photosynthetic decline and longevity of individual leaves. Agron J 80:865–870

Leaf Expansion and Phenological Development: Key Determinants of Sunflower Plasticity, Growth and Yield

V.O. Sadras[1] and N. Trápani[2]

1 Introduction

The physiology of sunflower (*Helianthus annuus* L.) in relation to field performance, management and improvement of the crop has been reviewed recently (Connor and Sadras 1992; Connor and Hall 1997; see also previous reviews by Knowles 1978; Robinson 1978). A number of aspects of sunflower breeding including achievements, methods, future directions, and the physiological changes associated with cultivar improvement have also received attention in recent publications (Skoric 1992; Friedt 1992; Sadras and Villalobos 1994; cf. also Fick 1978; Whelan 1978; Sneep et al. 1979). The present chapter does not, in consequence, attempt to analyze the physiological control of growth and yield of sunflower. We concentrate, instead, on the analysis of leaf expansion and phenological development. These two topics have been identified as key aspects of sunflower physiology in relation to yield and yield stability (Connor and Sadras 1992; Sadras and Villalobos 1994). Leaf expansion and phenological development have also been regarded as critical determinants of yield for other crop species (Passioura et al. 1993).

The ability of the crop to produce commercially acceptable yield in diverse environments is related to its phenotypic plasticity, i.e., the property by which genotypes are able to express a range of phenotypes in response to the physical environment (Bradshaw 1965). Phenotypic plasticity is a general characteristic of the species of the genus *Helianthus* (Heiser et al. 1969) that is also present in commercial sunflower (Connor and Sadras 1992). The exploitation of this characteristic is a challenge to the development of new cultivars and cultural techniques (Connor and Sadras 1992). Such developments require, in turn, a better understanding of the control of both leaf expansion and phenological development as two major determinants of phenotypic plasticity.

In this chapter, the responses of crop yield to sowing date are outlined to illustrate the plasticity of sunflower. The roles of leaf expansion and the pattern of phenological development are discussed. Sunflower yield responses to plant population are compared with the responses of other crops to highlight the

[1] CSIRO Plant Industry, Cotton Research Unit, Narrabri 2390, NSW, Australia
[2] I.F.E.V.A., Facultad de Agronomía, Universidad de Buenos Aires, Avda. San Martín 4453, 1417 Buenos Aires, Argentina

Crop Yield, Physiology and Processes
D. L. Smith, C. Hamel (Eds.)
© Springer-Verlag Berlin Heidelberg 1999

importance of leaf expansion as a determinant of sunflower plasticity. The physiological controls of leaf expansion and phenological development are then examined and the notion is advanced that a better insight into these processes would contribute greatly to the understanding of sunflower responses to variable agronomic environments. The responses of leaf area to the limiting factors of crop yield in most agronomic situations, viz. water and nitrogen availability, are emphasized. The current knowledge of the physiology of sunflower phenological development is summarized and compared with the skill of the models developed to predict the duration of the main phenophases. Further research needs related to the physiology of leaf expansion and phenological development are indicated.

2 Phenotypic Plasticity

Two examples that illustrate sunflower plasticity are introduced. First, the experiment by Gimeno et al. (1989) is presented to show how the patterns of leaf area and phenological development contribute to the successful growth and yield of winter-sown crops. Second, the responses of sunflower yield to plant density are compared with those of other crops including maize (*Zea mays*), wheat (*Triticum aestivum*) and soybean (*Glycine max*). The responses of leaf area to plant density are underlined in this comparison.

2.1 Yield Responses to Winter Sowing

Sunflower is a typical "summer" crop (Robinson 1971), i.e., it is usually sown in spring, grows during spring and summer and is harvested between late summer and autumn. In many growing areas, however, the high evaporative demand of the spring-summer season coincides with the absence of rainfall. Severe water deficits associated with this climatic pattern impose serious restrictions to crop yield. To analyze the feasibility of planting sunflower in winter as a drought-escape strategy in this kind of environment, a 4-year experiment was carried out by Gimeno et al. (1989) in southern Spain. Three hybrids of early, medium or late maturity were sown each year. Seven to ten sowing dates at 15–30 day intervals spanned the period from 15 December to 15 April. Normal sowing dates in the area are around mid-March. Rainfall in this Mediterranean environment is concentrated between late autumn and spring and average minimum temperature during the winter months is about 4 °C.

Grain- and oil yield increased as sowing dates were advanced from April to December (Fig. 1). Averaged over years and hybrids, sowing on December 15 increased grain yield by 55% and oil yield by 65% over the previously accepted normal 15 March sowing date. Grain yield can be considered as the product of shoot dry weight and harvest index (harvest index = grain dry weight/shoot dry weight; Donald and Hamblin 1976). Increases in both shoot dry weight and

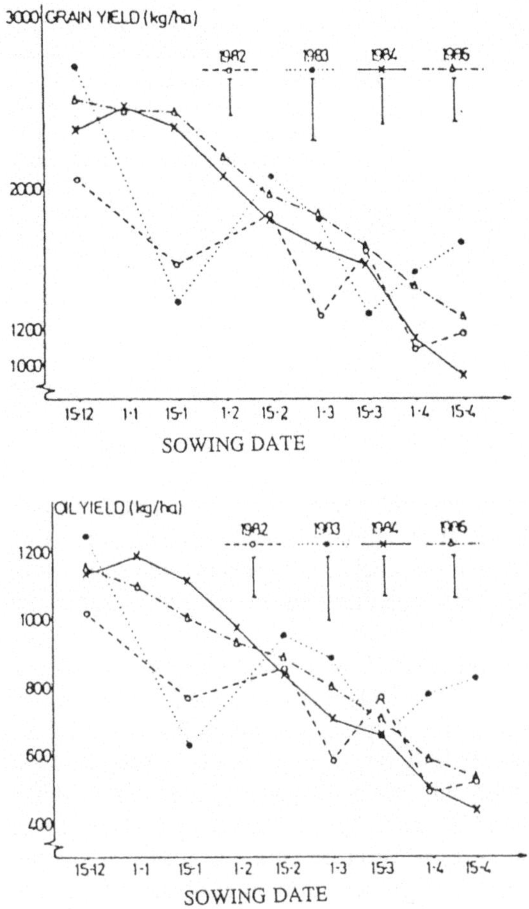

Fig. 1. Grain- and oil-yield responses to sowing date in southern Spain. Data are average of three hybrids differing in length of cycle (cultivar × date-of-sowing interaction were not significant, $P = 0.05$). The years and least significant differences ($P = 0.05$) for each year are shown. (Gimeno et al. 1989)

harvest index contributed to the greater yield of winter-sown crops. The increasing harvest index with advanced sowing date was probably related to a greater availability of water for the postanthesis period (cf. Sadras and Connor 1991). Temperature effects on the duration of grain filling and on grain number are also to be considered. Differences in biomass production were associated with differences in leaf area index (cf. Figs. 5 and 6 of Gimeno et al. 1989). Figure 2 shows the dynamics of leaf area index in relation to the time of flowering for these crops. The greater temperatures experienced for the spring-sown crop determined a much faster development than that for the winter-sown crop, i.e., there was a ca. 2-week difference in the time of anthesis in relation to a 3-month difference in sowing time (Fig. 2). The comparatively longer photoperiod during the early phases of the spring-sown crop may also have contributed to its faster development (cf. Sect. 4.2). The longer duration of the sowing to flowering phase in the winter-sown crop contributed to its

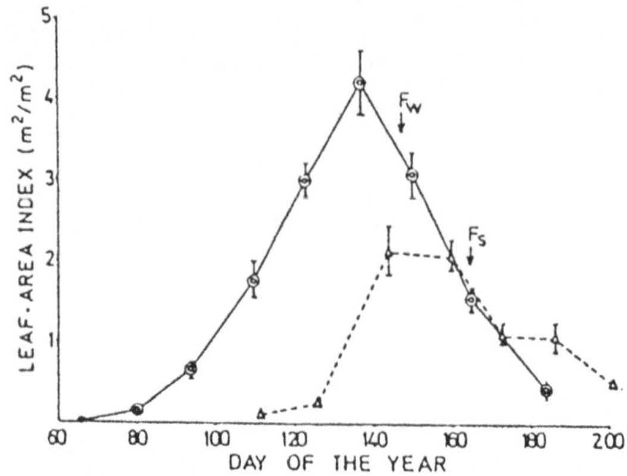

Fig. 2. Comparison of leaf area index of crops sown on 15 December (*circles*) and 15 March (*triangles*). Flowering times for the winter (F_W) and spring (F_S) crops are indicated. (Gimeno et al. 1989)

larger leaf area index in comparison with the spring-sown crop. In addition to this developmental effect, a direct effect of temperature on both leaf expansion rate and duration of expansion was most likely an important determinant of the differences depicted in Fig. 2 (cf. Rawson and Hindmarsh 1982; Sadras and Hall 1988).

Two conclusions relevant to the purpose of the present study can be drawn from the experiment of Gimeno et al. (1989). First, the yield advantage of winter sowing dates over the conventional spring sowing is the result of the great sunflower plasticity. This is a striking response indeed for a crop that has been included in a "subtropical-tropical" group of species (Angus et al. 1981). It is noteworthy that this experiment included hybrids normally used in spring sowing. The performance of winter crops, therefore, could be improved further by extending sunflowers' plasticity through selection of cultivars adapted to low temperatures (Downes 1985).

The second conclusion refers to the role of leaf area and the pattern of development in the observed responses. The slower development of the winter-sown crop contributed to its larger leaf area. The advancement of the date of sowing contributed to a better match between the pattern of crop development and the pattern of water availability (Gimeno et al. 1989). Thus, early-sown crops had a greater fraction of water available for the filling period, that was reflected in improved harvest index. The greater leaf area duration in winter sowing contributed to the greater yield by increasing accumulation of biomass. The relationships between leaf area, light interception and biomass production are discussed further in Section 3.1.

2.2 Yield Responses to Plant Population

Yield responses to plant population need to be known for practical purposes, as planting density is a major management variable used in matching crop requirements to the environmental offer of resources. These responses are strongly dependent on both plant plasticity and resource availability. Figure 3 compares the yield responses to plant population of sunflower, maize, wheat and soybean. Although plasticity may be active in processes important in determining vegetative and/or reproductive phases, the following discussion concentrates on the former. Soybean (Fig. 3d) and wheat (Fig. 3c) show a close-to-maximum yield at the lowest plant populations and little or no yield decline at greater plant populations. Branching (soybean) and tillering (wheat) are important characteristics by which plants may adapt their size to the availability of resources. Conversely, single-stemmed maize plants are unable to fully exploit the resources available at low plant populations and show substantial yield reductions with respect to the optimum range of plant population (Fig. 3a). Decline of maize yield at plant density greater than the optimum is related to sterility caused by population stress (Tollenar 1977). A maize-like response to plant population could be expected from the morphological characteristics of commercial sunflower (i.e., no branching, determinate growth habit). Data in Fig. 3b and other studies (e.g., Alessi et al. 1977; Jones 1984; Cardinali et al. 1985; Villalobos et al. 1994) show, however, a marked yield stability over ranges of plant population that include those of most commercial crops. Both compensation between yield components and leaf area plasticity contribute to this response in sunflower (Alessi et al. 1977; Jones 1984; Sadras and Hall 1988; Sadras et al. 1989). For instance, leaf area of plants grown at $2\,m^{-2}$ was about twice that of plants grown at $5\,m^{-2}$ so that, on a ground-area basis, leaf area was similar for both crops (Fig. 4). This example shows, again, the relevant role of leaf area in determining sunflower plasticity. The importance of leaf expansion rate in the observed response of plant leaf area to plant population (Fig. 4) is discussed in Section 3.2.

3 Leaf Area

Two kinds of crop responses to factors such as plant population, water and nitrogen availability can be distinguished according to the main processes and mechanisms involved, viz. responses that are dependent on or independent of assimilate availability. This classification is useful for the present analysis in spite of the fact that, in some cases, the differentiation between these categories is not clear. The early changes in dry matter partitioning of plants grown at high populations that are mediated by changes in the spectral composition of light (Ballaré et al. 1990) are an example of plant responses that are unrelated to assimilate shortage. Another example in this category is the reduction in floret number of sunflower exposed to water stress before anthesis (Hall et al. 1985). This reduction does not seem to be related to assimilate shortage but to some direct effect of water stress on floret survival (Sadras et al. 1993a). On the other hand, the small size of plants grown under nitrogen or water stress is the result, at least in part, of reduced carbon assimilation. In most cases, sunflower

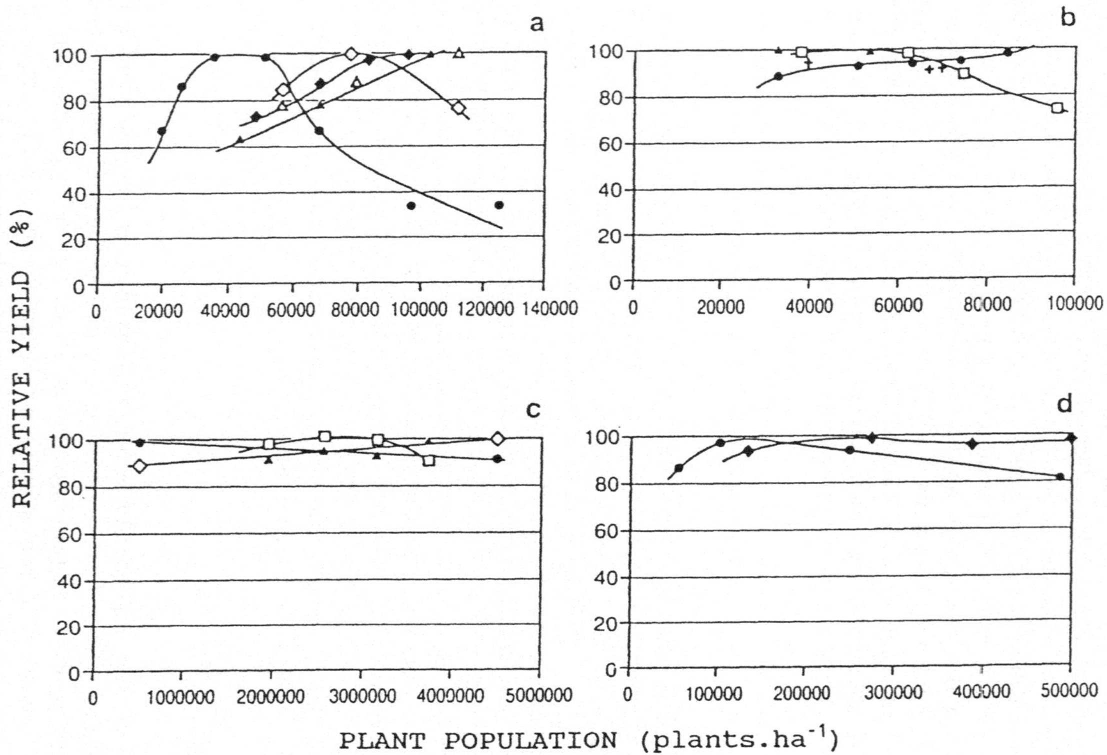

Fig. 3a–d. Yield responses to plant population for **a** maize, **b** sunflower, **c** wheat and **d** soybean. Yield is expressed as percent of maximum. *Different symbols and curves* correspond to cultivars and/or availability of water and nitrogen. (Andrade et al. 1992)

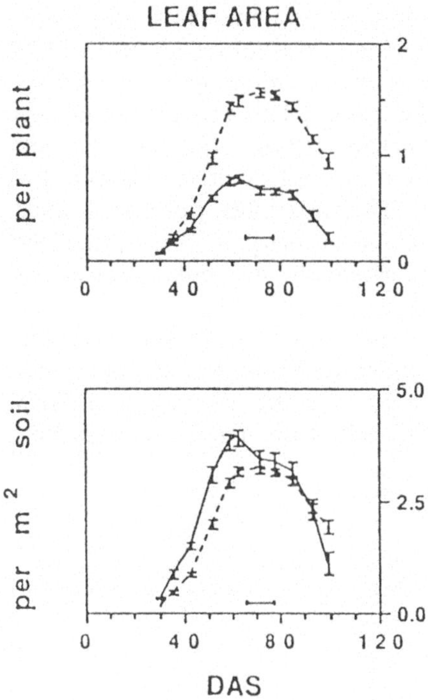

Fig. 4. Effect of plant population on leaf area on a plant or ground area basis. *Solid and dashed lines* are for plant populations of 5 and 2 plants m^{-2}. *Horizontal bars* show the first to last anthesis period. (Sadras et al. 1989)

assimilation per unit leaf area is little affected by environmental stresses, whereas reduction in leaf area accounts for most of the loss in crop ability to assimilate CO_2 (Connor and Sadras 1992). The knowledge of leaf area responses to these factors is thus important in order to understand crop responses. In this section, the relationships between leaf area, light interception and crop growth are discussed. Leaf area components are considered and the sensitivity of leaf expansion to the main environmental stresses is emphasized. Finally, the physiological mechanisms that regulate leaf expansion under water and nitrogen deficits are analyzed.

3.1 Leaf Area, Light Interception and Crop Growth

Crop biomass accumulation can be described as a function of the quantity of photosynthetically active radiation (PAR) absorbed by the canopy and the conversion efficiency of PAR into biomass (RUE) (Warren Wilson 1967). Absorbed PAR can be equated, for the present purposes, to intercepted PAR (IPAR) and crop biomass can thus be expressed as RUE × IPAR. Sunflower RUE varies with ontogeny (Trápani et al. 1992) and it may also vary with leaf nitrogen content (Connor et al. 1993; Hall et al. 1995), cultivar (Sadras et al. 1991a), water availability (Guiducci 1988) and plant population (Villalobos et al. 1994). In spite of this, most of the variation in biomass production among seasons, environments and cultivars can be ascribed to variations in IPAR (cf. Monteith 1977). Intercepted PAR depends on the amount of PAR incident upon the crop and the proportion of PAR intercepted by the canopy. This proportion, in turn, depends on leaf area index and the coefficient of PAR extinction. The latter coefficient varies with ontogeny (Rawson et al. 1984); cultivar (Sadras et al. 1991b) and water stress (Guiducci 1988). This variability, however, is small compared with the usual variation in leaf area so that light interception (Fig. 5a) and biomass production (Fig. 5b) are essentially dependent on leaf area index (cf. also Connor and Sadras 1992).

The relationship between leaf area, light interception and biomass accumulation (Fig. 5), among other factors, determines the strong correlation between leaf area and yield reported for sunflower in a number of studies (CETIOM 1992; Rawson and Turner 1983; Hall et al. 1985; Giménez and Fereres 1986; Connor et al. 1993; Fig. 6).

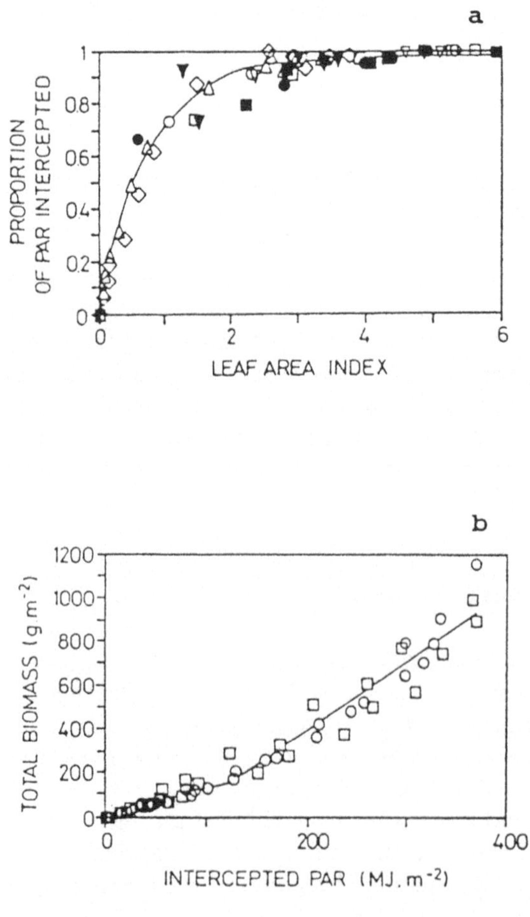

Fig. 5a,b. Relationship between leaf area index, light interception and growth in sunflower. **a** Proportion of intercepted PAR as a function of leaf area index. **b** Total biomass (i.e., shoot + root) as a function of intercepted PAR. (Trápani et al. 1992)

3.2
Leaf Area Components

Leaf area index is the product of plant leaf area and plant population. As previously discussed, however, these are not independent variables in sunflower (Sect. 2.2; Fig. 4).

Plant leaf area results from the balance between leaf growth and senescence. Before anthesis, growth normally dominates over senescence and leaf area increases, reaching a maximum value around anthesis. Leaf senescence, although important in determining crop performance, has been a widely neglected topic in sunflower studies and shall not be analyzed in this chapter.

Leaf growth depends on leaf number and area of individual leaves which can be considered, in turn:

$$\text{leaf number} = \text{rate of leaf primordia initiation} \times \text{duration of leaf primordia production},$$

$$\text{individual leaf area} = \text{rate of expansion} \times \text{duration of expansion}.$$

The effects of environmental factors upon plant leaf area can be interpreted in terms of the effects on the rates of leaf initiation and expansion and the duration of these processes.

Steer et al. (1986) found leaf numbers of 33 and 32 for plants grown at 5 and $15\,m^{-2}$, respectively. Similarly, Villalobos et al. (1992) found little effect of plant population on leaf number of four hybrids grown in a range from 0.5 to $10\,m^{-2}$. Only extremely high plant populations, well outside the range of com-

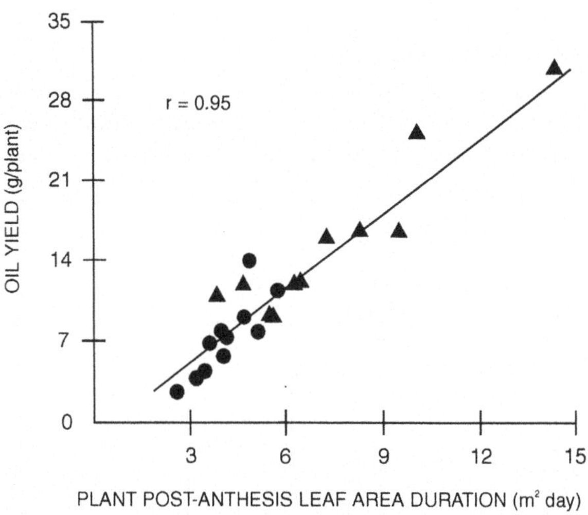

Fig. 6. Relationship between oil yield and leaf area duration from anthesis to maturity for sunflower lines B14 (*circles*) and R16 (*triangles*) grown under different water regimes. (Hall et al. 1985)

mercial crops (ca. 50 m^{-2}), significantly reduced sunflower leaf number (López-Pereira and Sadras 1993). The duration of leaf expansion was also unaffected by plant population in the experiment of Steer et al. (1986). The substantial effect of plant population on plant leaf area (e.g., Fig. 4) is, therefore, related to changes in the rate of leaf expansion (Table 1). Table 1 shows not only the large response of leaf expansion rate to plant population but also the plasticity in the response as a threefold change in plant population was reflected in an approximately threefold change in the rate of leaf expansion (cf. Fig. 4). This response of leaf expansion rate is a major determinant of the plastic response of sunflower to plant population discussed in Section 2.2.

Similar to plant population, water and nitrogen availability have little effect on components of leaf area other than the rate of leaf expansion. Leaf number is hardly affected by water and nitrogen supply under field conditions (e.g., Connor and Jones 1985; Sadras et al. 1993b; Connor et al. 1993). The duration of leaf expansion may be unchanged or extended by water and nitrogen deficits (Connor and Jones 1985; Steer and Hocking 1983; Steer et al. 1986; Trápani et al. 1993; Trápani and Hall 1996). Reduction in leaf expansion rate is, thus, the main determinant of the reduction in the leaf area of plants grown under nitrogen or water deficit. The effects of water and nitrogen on leaf area components and other aspects of growth and yield including stomatal conductance and photosynthesis are discussed in the next sections.

Summarizing, crop growth depends on light interception and, hence, on leaf area. The rate of leaf expansion is the most important variable in regulating plant leaf area before anthesis. The mechanisms that regulate leaf expansion under water and nitrogen deficits are analyzed in the next sections.

3.3 Responses of Leaf Expansion to Water Availability

The great sensitivity of leaf expansion to water deficits has been widely recognized (Hsiao et al. 1976). In sunflower, the effects of water deficits on leaf

Table 1. Effects of plant population and nitrogen supply on the rate of sunflower leaf expansion. (Data from Steer et al. 1986)

Treatment[a]	Leaf expansion rate[b] (cm^2 day^{-1})
Np	12–13
np	6–8
NP	4–5
nP	1–2

[a] Code for treatment: Nitrogen: N = 300 kg N/ha, n = nil. Plant population: P = 15 m^{-2}, p = 5 m^{-2}.
[b] Ranges for mid-position leaves (from nodes 14 to 20).

expansion have been investigated under both controlled and field conditions (e.g., Connor and Jones 1985; Takami et al. 1981). Under field conditions, the marked sensitivity of leaf expansion to water deficits contrasts with the limited responsiveness of stomata (Connor and Jones 1985; Cox and Jolliff 1987; Sadras et al. 1993c). This supports the view that reduced rate of leaf expansion, and hence reduced light interception, is the main mechanism by which drought-stressed sunflower crops regulate transpiration before anthesis (Connor and Sadras 1992).

Leaf expansion under water deficits may be regulated by hydraulic and/or nonhydraulic mechanisms. Biophysical considerations in relation to the hydraulic control of leaf expansion have been presented by Hsiao et al. (1985) and Dale (1988). As turgor pressure is the driving force of cell expansion, turgor decline may explain, in some cases, the reduction in leaf expansion rate (LER) found in water-stressed plants. Other hydraulic mechanisms that regulate leaf expansion irrespective of changes in turgor pressure have been proposed (Steudle 1985; Boyer and Nonami 1990).

Under certain conditions, however, soil drying may induce reductions in LER irrespective of changes in tissue-water status (Davies and Zhang 1991). Such observations have led to the hypothesis that roots are able to sense soil drying. As soil dries, roots produce nonhydraulic signals that are transported to the shoot, where they reduce leaf growth and stomatal conductance (Davies and Zhang 1991).

The notion that hormonal root signals are important in transducing environmental conditions into growth responses is generally accepted (Davies and Zhang 1991) yet "there is surprisingly little definitive evidence for the role of any hormone in regulating either shoot or root growth in soils of low water potential" (Munns and Sharp 1993). Moreover, with few exceptions (e.g., Tardieu et al. 1993), the evidence supporting the role of chemically based root signals in the regulation of shoot responses derives from experiments with young plants growing in pots in controlled environments.

3.3.1
Growth and Water Relations of Plants in Controlled Environments

Responses to water deficits of laboratory-grown plants do not necessarily reflect the responses of plants grown in the field for several reasons:

Restricted soil volume may affect shoot growth irrespective of water and nutrient supply (McConnaughay et al. 1993). Chemical root signals are primarily responsible for shoot responses normally observed in plants subjected to high soil strength (Masle and Passioura 1987; Andrade et al. 1993). Thus, mechanical stress at the root/pot-wall interface will generate "background" root signals that probably impose restrictions to most studies with pot-grown plants. Additionally, growth of secondary roots is enhanced in pot-grown plants (Ben-Porath and Baker 1990) while secondary and tertiary roots have been indicated as having a particular controlling influence on shoot growth via root signals (Sharp and Davies 1989). Furthermore, shoot growth is responsive not only to pot size but also to its shape (McConnaughay et al. 1993).

Acclimation mechanisms are important in the control of sunflower leaf expansion (Matthews et al. 1984) and photosynthesis (Wise et al. 1990) under water deficits. The rate of onset of stress (sensu Ritchie 1981) strongly influences osmotic adjustment and overall plan-water relations; its implications for studies with pot-grown plants have been recognized for a long time (cf. Ritchie 1981) yet they are often ignored in many laboratory studies. The large intraspecific variability of osmotic adjustment in sunflower (Chimenti and Hall 1993) compounds the problems of laboratory experiments in which stress develops at unrealistically high rates. The large daily fluctuations in leaf water potential that induce physiological acclimation in the field have also been suggested as a serious limitation to comparisons between plants grown in the field and controlled environments (Wise et al. 1990).

Rooting substrates have dramatic effects on key plant properties. Hydraulic conductivity of sand-grown roots may be up to ten times lower than that for roots grown hydroponically (Brar et al. 1990). Many controlled-environment studies of water relations in sunflower were conducted in highly artificial hydroponic systems where this sort of effect has been ignored (e.g., Neals et al. 1989; Masia et al. 1994; Ternesi et al. 1994).

Root density of annual species in the field normally declines exponentially with soil depth (Gerwitz and Page 1974; Sadras et al. 1989). This pattern of root distribution – which is strongly influenced by water and nutrient distribution (e.g., Robinson 1994) – is rarely found in pot-grown plants (e.g., Henson et al. 1989).

Reverse water flow, i.e., the transfer of water from zones of wet soil to dry soil through the root system (e.g., Xu and Bland 1993 and literature cited therein), may be important in some field conditions but is probably absent in most experiments with pot-grown plants.

Soil disturbance during potting may generate unrealistic soil environments in terms of structure, bulk density, nutrient amount and distribution (e.g., Johnson et al. 1995).

Artificial light environments may also generate abnormal plant responses with regard to water relations. For example, Lu et al. (1993) showed contrasting responses of adaxial and abaxial stomata depending on whether plants were grown in growth chambers or under sunlight.

In summary, there are many important sources of divergence between responses of plants grown in the field and controlled environments. Studies on the physiology of sunflower stomata illustrate the risk of extrapolating results from controlled to field environments. In the field, stomata of preflowering plants are markedly insensitive to soil-water deficits (Connor and Jones 1985; Sadras et al. 1993c). Thus, stomatal responses to soil drying attributed to root signals in many laboratory studies (Gollan et al. 1986; Neales et al. 1989; Zhang and Davies 1989) are probably artifacts derived from one or more of those artificial growing conditions.

3.3.2
Leaf Expansion in Field-Grown Sunflower

The relevance of hydraulic and nonhydraulic mechanisms in the control of LER in field-grown sunflower was evaluated in experiments carried out on a deep sandy-loam soil in a mediterranean environment under spring and summer conditions (Sadras et al. 1993d). Soil-water content, LER and Ψ_l were measured in "wet", "dry" and "split-root" treatments. The study was designed to test the following criteria:

1. Leaf responses are better described by soil-water content than by leaf water potential (Ψ_l) (e.g., Turner et al. 1985).
2. Ψ_l response to soil drying lags behind LER response (e.g., Zhang and Davies 1989).
3. Plants with roots split between two containers so that part of the root system grows in dry soil while the other part is well watered respond as plants grown in completely dry soil (Davies and Zhang 1991).

Two approaches were used to estimate the onset of Ψ_l and LER responses to soil drying. First, the onset of the response for each variable was taken as the *time* when the first significant differences between "dry" and "wet" treatments are detected by analysis of variance. This is the usual procedure to evaluate the timing of Ψ_l and LER responses. The accuracy of measurements, however, may be insufficient to detect early, small changes in Ψ_l (Henson et al. 1989; Matsuda and Rayan 1990). For this reason, a delayed response of Ψ_l with respect to the response of LER may be either indicative of true non-hydraulic effects or an artifact related to the fact that, with current techniques, LER can be measured with greater accuracy than Ψ_l. Importantly, the ability to detect treatment effects on LER and Ψ_l depends not only on the intrinsic accuracy of the techniques used to measure these variables but also on sample size (Sadras et al. 1993d). To overcome some of these restrictions, a second approach that estimates the onset of LER and Ψ_l responses was developed. In this approach, based on the model used to describe leaf responses to plant available water (PAW) (Ritchie 1981; Rosenthal et al. 1987), the onset of the response for each variable is taken as the *PAW threshold* for which Ψ_l and LER of "dry" plants begins to decline with respect to those of "wet" plants. PAW thresholds were estimated using regression techniques that allow for statistical comparisons.

The results of these experiments (summarized in Table 2) show that: (1) responses of LER to soil drying were better described by Ψ_l than by soil-water content; (2) estimates of PAW thresholds indicate that LER responses did not precede Ψ_l response to soil drying. Analysis of variance for the timing of first

Table 2. Hydraulic and non-hydraulic control of leaf expansion rate in field-grown sunflower. (Sadras et al. 1993d)

Criterion[a]	Observation
1. Leaf responses are better described by soil-water content than by Ψ_l	No
2. Ψ_l response to soil drying lags behind LER response[b]	
Time	Yes
Soil-water content (PAW)	No
3. Split-root plants respond as plants grown in completely dry soil	No

[a] Affirmative answers to these criteria are in favour of the hypothesis of nonhydraulic control of leaf expansion.
[b] The relative timing of Ψ_l and LER responses was evaluated in two ways: (1) on a time basis, (2) on a soil-water content basis. This means that the "day" and the "PAW" when changes in Ψ_l and LER of "dry" with respect to "wet" plants were observed have been taken for comparisons (see text).

observed difference, conversely, showed that changes in LER preceded changes in Ψ_l in the "dry" treatment; and (3) split-rooted plants did not respond as "dry" controls, as expected in the case of a significant root signal effect, but maintained LER similar to "wet" controls. Altogether, these results indicate that hydraulic effects were of more importance than non-hydraulic root signals in the control of sunflower leaf expansion. In conclusion, there is good evidence to support the idea that leaf expansion of plants grown in controlled environments may be affected by "root signals". Limited evidence in the field, however, suggests a more important role of hydraulic signals. Factors that could be involved in the discrepancy between results found in laboratory and field experiments have been discussed before (Sect. 3.3.1).

Detailed studies dealing with gradients of water potential (Matyssek et al. 1991a,b) and the role of the epidermis in the control of leaf expansion (Dale 1988) suggest that subtle hydraulic mechanisms may operate in the control of leaf expansion in addition to turgor-mediated processes. The limited information on the effects of abscisic acid on wall biochemistry and leaf growth suggests that a possible mechanism exists whereby water stress could inhibit tissue expansion by enhancing abscisic acid concentration (Dale 1988). Importantly, interactions between hydraulic and non-hydraulic signals have been shown to play a role in the control of stomatal conductance in some species (Tardieu and Davies 1993; Tardieu et al. 1993), indicating that these mechanisms are not mutually exclusive. Further work is necessary to evaluate the role of such interactions in the control of sunflower leaf expansion in the field.

3.4
Effects of Nitrogen on Growth and Development

Crop growth and yield respond dramatically to nitrogen supply. Investigation on a number of species, including sunflower, document this statement (Novoa and Loomis 1981; Steer and Hocking 1983; Angus and Moncur 1985; Milford et al. 1985; Trewavas 1985; Steer et al. 1986; Hirose 1988; van Keulen et al. 1989; Connor et al. 1993).

Phenological development of sunflower is little affected by nitrogen availability. Experiments under both field (Connor et al. 1993; Trápani and Hall 1996) and controlled conditions (Steer and Hocking 1983) showed that only very low nitrogen supply extended the time to anthesis by a few days while time to physiological maturity was not affected (Connor et al. 1993).

Nitrogen effects on crop growth may be mediated by effects on leaf area (Sects. 3.1 and 3.2) and/or photosynthesis per unit leaf area. In this section we discuss the effects of nitrogen availability on leaf area components. The limited information existing on the mechanisms that underlie the response of leaf expansion to nitrogen is analyzed. Comparison between the effects of nitrogen on leaf expansion and carbon fixation is also considered.

3.4.1
Effects of Nitrogen on Leaf Area

The effects of nitrogen supply on sunflower leaf area have been investigated in a number of studies (Radin and Boyer 1982; Steer and Hocking 1983; Steer et al. 1986; Connor et al. 1993; Trápani and Hall 1996). Nitrogen affects the components of plant leaf area (i.e., leaf number and individual leaf area, Sect. 3.2) differentially. Steer and Hocking (1983) investigated these effects in controlled-environment experiments. They found that both the duration and the rate of leaf primordia initiation were affected only when nitrogen shortage was extreme. In other species nitrogen supply affected leaf cell production and expansion (Morton and Watson 1948; Trevawas 1985; MacAdam et al. 1989); in sunflower information on this aspect is lacking. Nitrogen supply had little or no effect on final leaf number of plants grown in the field (Connor et al. 1993; Sadras et al. 1993b; Trápani and Hall 1996).

Duration of leaf expansion in pot plants under controlled conditions was not affected by nitrogen (Steer and Hocking 1983) while only severe shortage of nitrogen did so in the field (Steer et al. 1986; Trápani and Hall 1996). Leaf expansion rate was markedly responsive to nitrogen supply both under controlled and field conditions (cf. Table 1; Steer and Hocking 1983; Steer et al. 1986; Connor et al. 1993; Trápani and Hall 1996). Thus, leaf expansion rate is the main process determining the response of plant and crop leaf area to nitrogen supply.

Mechanisms of LER Responses to Nitrogen. The mechanisms involved in the response of sunflower leaf expansion to nitrogen have been explored by Radin and Boyer (1982). They studied plant-water relations as they play a key role in the process of cell growth and thus of leaf expansion (Taiz 1984; Cosgrove 1986, 1987; Boyer 1988; Walker 1992). Radin and Boyer (1982) analyzed the responses of plant hydraulic conductivity and leaf turgor to nitrogen levels in detailed experiments conducted with pot plants under controlled conditions. Lower hydraulic conductivity between root and the expanding leaf and reduced turgor – even in plants well supplied with water – were measured under restricted nitrogen availability. Reduced nitrogen and NO_3^--nitrogen compound levels were lower in the expanding leaves of plants growing under nitrogen stress while neither cell-wall extensibility nor the threshold turgor for growth (as estimated by the rate of growth/turgor relationship) was affected by nitrogen supply.

Chapin (1991) proposed a centralized stress-response system in plants: the various types of environmental stresses – including water and nitrogen shortages – would activate a common mechanism controlling plant growth rate and allocation. A change in plant hormone levels (i.e., declines in cytokinins; increases in abscisic acid) would affect leaf expansion through effects on cell-wall extensibility. Alternatively, the altered hormonal levels would reduce root hydraulic conductance, leaf turgor and thereby expansion. Data on several species (Chapin et al. 1988 and literature therein cited; Taylor et al. 1993) are consistent with this hypothesis. Recent experiments by Coleman and Schneider (1996), however, showed that an ABA-deficient mutant of tomato (*flacca*) did not behave as predicted by the hypothesis of Chapin (1991) in response to low soil resource availability.

The effects of nitrogen on turgor of expanding sunflower leaves measured by Radin and Boyer (1982), together with the data of Chapin et al. (1988) and Taylor et al. (1993) on changes of cell-wall extensibility in other species, and the current knowledge on the mechanisms of cell expansion (see Cosgrove 1987 for a review) suggest that changes in leaf cell-wall properties, turgor and osmotic potential, would be involved in LER response to nitrogen. The role of hormone levels as elicitors of the responses deserves further attention. The commonality of effects of water and nitrogen stress on leaf expansion make hypotheses such as those of Chapin (1991) attractive frameworks to improve knowledge.

3.4.2
Leaf Expansion Rate as Affected by Nitrogen Availability: Comparison with Rate of Photosynthesis

Plant and crop growth are functions of leaf area and photosynthesis per unit leaf area. Both components of growth are affected by leaf nitrogen content, although to different extents. While the effects of leaf nitrogen content on photosynthesis have been studied in a series of species (Sinclair and Horie 1989 and literature therein cited) and in sunflower (Just et al. 1989; Connor et al. 1993), information on the relationship between LER and leaf nitrogen content is limited. Nitrogen use in the growing zone of grass leaves was interpreted as sequential rather than competitive between synthetic processes linked to leaf growth and photosynthesis (Gastal and Nelson 1994).

A recent field study on sunflower (Trápani and Hall 1996) has shown that LER in the quasi-linear expansion phase (i.e., LER_{max}) is strongly associated with specific leaf nitrogen (SLN, $g N m^{-2}$) above a high SLN threshold (i.e., $LER_{max} = 0$). The response is curvilinear and dependent on leaf position (Fig. 7). The SLN threshold for LER_{max} ($1.8 g N m^{-2}$) was much higher than the corresponding one found for $P_{max} = 0$ ($0.3 g N m^{-2}$; P_{max} = photosynthesis rate in saturating irradiance; Connor et al. 1993; Fig. 7). These findings indicate that leaf expansion shows a greater sensitivity to leaf nitrogen than photosynthesis. Trápani and Hall (1996) suggested that in sunflower photosynthesis can contribute to plant carbon gain at a lower nitrogen cost than leaf expansion. These results are in accordance with the early findings that nitrogen deficiency in plants restricts growth primarily through a decrease in LER rather than in rate of photosynthesis per unit leaf area (Watson 1947, 1952).

In considering the different processes that result in crop growth and yield, leaf expansion appears as one of the most sensitive to both water (Hsiao et al. 1976; Bradford and Hsiao 1982) and nitrogen stress (Steer and Hocking 1983; Steer et al. 1986; Connor et al. 1993; Trápani and Hall 1996). This fact reinforces the concept that leaf expansion plays a key role in the adaptative mechanisms

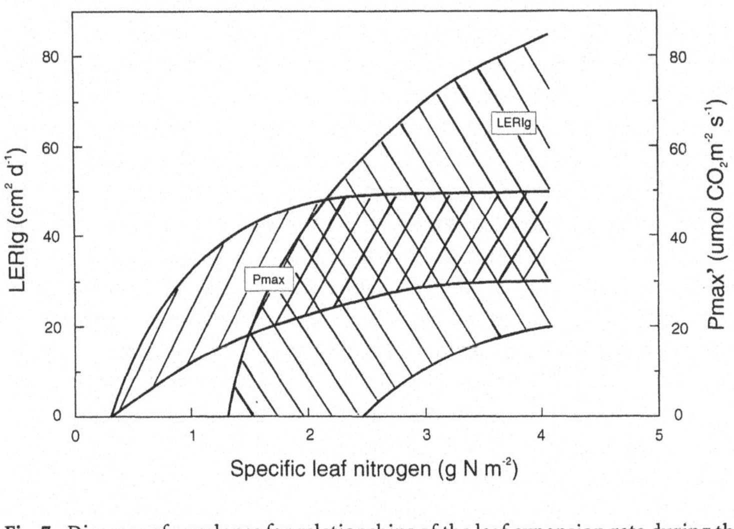

Fig. 7. Diagram of envelopes for relationships of the leaf expansion rate during the quasi-linear phase (LER_{lg}) and photosynthesis at high irradiance (P_{max}') and specific leaf nitrogen of sunflower leaves 8, 12, 16 and 20. Plants were individually grown in the field in 30-l containers. Irrigation and five nitrogen levels were applied (i.e., from 0.25 to 11.25 g N per plant). (Trápani and Hall 1996)

exhibited by sunflower in response to environmental stresses (Connor and Sadras 1992; Sadras et al. 1993c).

In conclusion, the rate of leaf expansion is the main factor in controlling sunflower leaf area and consequently canopy development light interception and growth in response to nitrogen supply. Effects of nitrogen deficits on crop growth that are mediated by changes in photosynthesis are secondary (Connor et al. 1993; Hall et al. 1995). The responses of sunflower growth and yield to nitrogen are extensively documented (e.g., Steer et al. 1986; Connor et al. 1993). Simulation studies with sunflower (Villalobos et al. 1994) and other species (Angus and Moncur 1985; Hirose 1988; van Keulen et al. 1989) have also described these relationships. The physiological mechanisms involved in the response of leaf expansion to nitrogen, however, deserve further attention. Nitrogen stress seems to affect leaf expansion by mechanisms that imply changes in plant-cell water relations and cell-wall properties, as water stress does. The application of new concepts (Chapin 1991) and methodologies in this field (see Cosgrove 1987; Boyer 1988; Taylor et al. 1993) should contribute to extended knowledge on the determinants of leaf expansion.

4 Phenological Development

4.1 Phenological Pattern and Yield

The phenological pattern is a key aspect of crop adaptation to specific environments. Biomass production usually increases with cycle length, as longer cycle is associated with greater leaf number, leaf area duration and, hence, light interception (Sadras and Villalobos 1994; Sect. 3.1). In many cases, the greater production of biomass is reflected in greater grain and oil yield. This is illustrated in Fig. 8, which compares 14 hybrids released in Australia between 1975 and 1986 (White 1988). Sunfola 68-3, an open-pollinated variety, was taken as check cultivar for the comparison of yields in two sites, New South Wales and Central Queensland. In NSW, the more favorable environment, the best-yielding cultivar doubled Sunfola 68-3 yield. In Queensland, a drier site, the best-yielding cultivar outyielded the check by 70%. Irrespective of the site, most of the variation among hybrids was accounted for by cycle length, i.e., time to anthesis accounted for 81 and 65% of yield variance in NSW and Queensland, respectively. Increase in yield per unit increase in time to anthesis, as expressed by the slopes of the regressions in Fig. 8, was greater in the more favorable environment. In South Australia, Potter and McCloud (1985) also found a positive association between yield and time to anthesis in 12 experiments involving 42 genotypes and plant populations between 5 (dryland) and 10 (irrigated) plants m^{-2}.

Simulation models have proved to be useful tools to investigate crop responses to variable environments and management practices. The sunflower

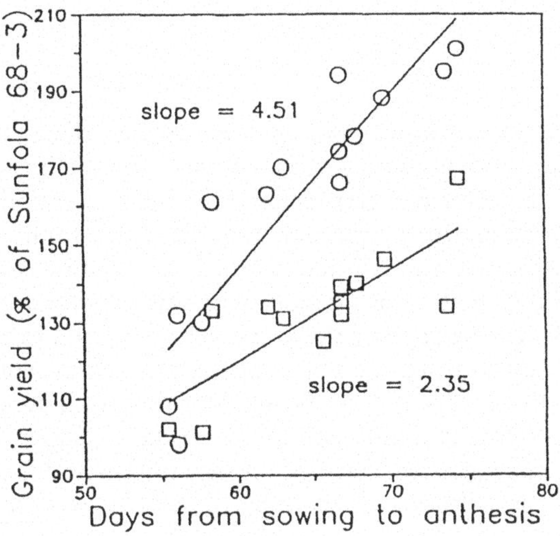

Fig. 8. Grain yield of sunflower hybrids as a function of time to anthesis in two regions of Australia. New South Wales (*circles*) and Queensland (*squares*). (After White 1988)

model of Villalobos et al. (1996) was used to establish the phenological patterns that match various environments in southern Spain (Sadras and Villalobos 1994). Simulation results were in agreement with the experimental data in Fig. 8 for sites with no water limitations to growth, i.e., the greater yields were attained by the longer-cycle cultivars. The greater biomass production of long-cycle cultivars, however, was not reflected in higher yield under severe terminal drought. This was because little water availability during grain filling generated reductions in harvest index that offset the high biomass production. Different phenological patterns, that maximize yield according to the varying patterns of water availability, were thus established. The plasticity of sunflower phenological development in combination with appropriate cropping strategies allows crop cycle to be matched to rainfall patterns, as discussed in Section 2.1.

4.2
Physiological Basis and Modelling of Phenological Development

This section outlines the extent to which the physiological basis of sunflower development has been established in comparison with the accuracy of the models developed to predict the duration of the main phenophases, i.e., sowing to emergence, emergence to anthesis, and anthesis to physiological maturity. It is not our purpose to fully examine sunflower phenological development, as it has been recently reviewed by Connor and Sadras (1992) and Connor and Hall (1997), and summarized in a mathematical model by Villalobos et al. (1996).

4.2.1 Sowing to Emergence

The sowing to emergence phase includes two physiologically distinct processes, i.e., seed germination and growth and emergence of seedlings. Temperature is the environmental factor that has received the most attention in studies of sunflower germination and/or emergence. Typically, these studies attempted to characterize seed responses to temperature in relation to intraspecific variability, tests of germination or methods to predict emergence in the field. Little attention, however, has been paid to the physiological basis of sunflower seed responses to temperature. A similar picture arises from the analysis of reports dealing with the effects of other factors including seed dormancy, seed size, substrate water potential and salinity. For instance, experiments dealing with seed dormancy have concentrated on its duration and the methods to break it, whereas little is known concerning the mechanisms of dormancy in this species.

Angus et al. (1981) used a thermal-time approach to predict the effects of temperature on the duration of the sowing to emergence phase in the field. Using a similar approach, Villalobos et al. (1996) proposed a method that also takes into account sowing depth and soil-water content. These models have proved to produce sound estimates of the duration of this phenophase for observed values of time of emergence between 7 and 32 days (Villalobos et al. 1996). The applicability of these models would be greatly extended by improved knowledge of the effects of soil water content.

4.2.2 Emergence to Anthesis

For the present analysis, this phase can be divided as follows: emergence to end of juvenile phase, end of juvenile phase to floral initiation (floral stage 1.3, Marc and Palmer 1978, 1981), and floral initiation to anthesis.

The existence of a juvenile phase when plants are insensitive to photoperiod has been established for a number of species including sunflower (Vince-Prue 1975; Marc and Palmer 1978). There is no direct evidence of intraspecific variability for the duration of the juvenile phase. However, estimates for three hybrids using a model that assumes the existence of a juvenile phase and photoperiod sensitivity showed a range for this phase from 245 to 295 °C day^{-1} (base temperature 4 °C; Villalobos et al. 1996).

Temperature, radiation and photoperiod are major environmental factors affecting the time of floral initiation. Rawson et al. (1984) found that the accelerating effect of increasing temperature from 15 to 27 °C was smaller and less consistent under high radiation (25 MJ m^{-2} day^{-1}) than under low radiation (9 MJ m^{-2} day^{-1}). The experiments by Rawson and Hindmarsh (1982) indicate that sunflower behaves as a long-day plant for the phenophase end of the juvenile phase to floral initiation. Both experiments (Rawson and Hindmarsh

1982) and optimization techniques (Villalobos et al. 1996) indicate a considerable intraspecific variability in the sensitivity of photoperiodic response that leads to floral initiation. This long-day response is either reversed to a short-day response or lost, depending on the cultivar, for the development after floral initiation. The duration of the period when plants are able to show this short- or neutral-day response is unknown.

The time from emergence to anthesis thus depends on genotype, temperature, radiation and photoperiod. Light quality, air (CO_2), water and nitrogen availability may also affect the duration of this phase but their relevance is comparatively small in most field situations (Connor and Sadras 1992).

A series of modifications in the physiology of determinate crops occurs around anthesis that includes, for instance, dramatic changes in the pattern of biomass partitioning (Trápani et al. 1994). Accurate methods to predict anthesis are thus critical to the performance of crop simulation models. The prediction of the time of anthesis may also be useful for breeding purposes. For these reasons, a considerable number of studies dealing with sunflower phenology have concentrated on the prediction of the time to anthesis. Photoperiod and temperature effects have been integrated in regression models (Goyne et al. 1977). More mechanistic models that are based on detailed studies of plant responses to these factors have been proposed. In this group, the model of Hammer et al. (1982) considers the phases emergence to head visible, and head visible to anthesis. Head visible, the earliest sign of reproductive development that does not require apex dissection, is used in many studies because it can be determined easily and non-destructively. It has a major disadvantage, however, derived from the inclusion of both pre- and postfloral initiation developmental processes (Connor and Sadras 1992). The opposite effect of photoperiod before and after floral initiation discussed above seriously reduces the physiological meaning of head visible as an indicator of reproductive development.

In order to increase the physiological basis of their model, Villalobos et al. (1996) divided the emergence to anthesis phase into three phases, viz. emergence to end of juvenile phase, end of juvenile phase to floral initiation, and floral initiation to anthesis. It is interesting to note that about 5 years elapsed between the early empirical model of Goyne et al. (1977) and the more mechanistic one by Hammer et al. (1982), whereas it took more than 10 years for the publication of a model with a wider physiological basis (Villalobos et al. 1996). It is also worthwhile to note that the physiological basis included in the model of Villalobos et al. (1996) is mostly from work previous to or contemporaneous with Hammer et al.'s model (i.e., Marc and Palmer 1978, 1981; Rawson and Hindmarsh 1982). It seems that the skill of Hammer et al.'s model (1982) to predict the time of anthesis (cf. Hammer et al. 1982; Sadras and Hall 1988) might have made the effort to develop models with better predictive capability unnecessary. In fact, a great step in the ability of models to predict the time to anthesis was achieved with the early mechanistic models in comparison with the earlier regression techniques while comparatively little additional accuracy was gained with the more evolved models (cf. Villalobos et al. 1996). At present there is, in consequence, a great gap between the ability of models to predict the time to anthesis (that is probably close to its maximum) and the poor knowledge on the physiology of sunflower reproductive development.

4.2.3
Anthesis of Physiological Maturity

To date, maximum seed dry weight is the only accurate indicator of physiological maturity (Connor and Sadras 1992). Thus, by definition, growth processes are important in determining the time of physiological maturity together with true developmental processes. This contrasts with other phenostages such as floral initiation that involve clearly differentiated developmental processes. Physiological maturity, then, could be expected to be affected by any factor that influences a plant's ability to fix carbon and/or to translocate current or reserve assimilate to the grain. For instance, terminal drought or nitrogen stress, which are known to accelerate leaf senescence, could shorten the filling period and advance physiological maturity. There is, however, limited empirical evidence to support this proposal. Terminal water stress accelerated maturity in the study of Anderson et al. (1978) but not in the experiments of Whitfield et al. (1989) and Hall et al. (1985). The usually great but variable contribution of reserve assimilate to grain filling in sunflower (Hall et al. 1985, 1989, 1990; Sadras et al. 1993a) might explain these differences.

Rawson et al. (1984) showed a strong effect of the temperature x radiation interaction on the time of physiological maturity. At low temperature (15 °C) the phase anthesis to maturity lasted 60 and 49 days at low and high radiation, respectively. On the other hand, little effect of radiation was found at high temperature (27 °C). Detailed experiments to evaluate the effect of temperature on the rate and duration of grain filling as in other species (e.g., Sofield et al. 1977) are lacking in sunflower. Moreover, no attempt has been made to determine the mechanisms that control sunflower grain filling, and hence physiological maturity.

Few models have been developed to predict the time of physiological maturity and even fewer attempts have been made to test them. Anderson et al. (1978) developed a regression model that includes temperature and soil-water content as independent variables. Duration of the anthesis to physiological maturity phase was assumed to vary among cultivars and to be under thermal control in the model of Villalobos et al. (1996).

5
Conclusions

The performance of winter-sown standard cultivars (i.e., selected for spring sowing) clearly illustrates sunflower's wide phenotypic plasticity. Sunflower plasticity is also evident in its responses to plant population. The flexible responses to varying environments of both leaf expansion and the pattern of phenological development have been shown to play a major role in the determination of the plastic responses of the crop.

Leaf area is critical in determining light interception and crop growth. Before anthesis, leaf expansion rate is the main determinant of leaf area. Leaf expansion rate is very sensitive to water and nitrogen deficits in comparison, for instance, with stomatal conductance or photosynthetic rate. The mechanisms involved in leaf expansion responses to water are better understood than those related to nitrogen deficits. A well-developed theoretical framework on cell and tissue water relations (e.g., Tyree and Jarvis 1982; Cosgrove 1987; Boyer 1988) exists, but field experiments are necessary to further understand the control of leaf expansion in agronomic situations. Particularly, interactions between hydraulic and non-hydraulic mechanisms of control of leaf expansion need to be investigated. A theoretical framework for the physiology of nitrogen effects on leaf expansion is lacking. Limited evidence indicates that nitrogen could affect expansive processes by modifying leaf water relations and cell wall properties. Research at various organization levels – viz. cell, tissue, organ, plant – are necessary to advance present knowledge of the physiology of nitrogen effects on leaf expansion.

The physiological mechanisms underlying sunflower phenological development are poorly understood. Although there is considerable knowledge on the physiology of seed growth and germination in general (e.g., Bewley and Black 1985), specific responses of sunflower seed to the main plant and environmental factors have received little attention. Particularly, the physiology of seed growth has been a largely neglected topic. Experiments such as those carried out with other species to examine the role of temperature (e.g., Sofield et al. 1977) and water (Egli 1990) on the control of seed-filling duration are necessary in sunflower.

Understanding of thermal and photoperiodic control of reproductive development is also quite limited. Experimental work is required to characterize, in principle, the juvenile phase, the period an photoperiod ranges for which both the pre- and postfloral initiation responses are effective and also the genotypic variability for these responses. In spite of the complexity of the mechanisms involved in the transition from the vegetative to the reproductive stage and the narrow understanding of the processes involved, current models are able to accurately estimate the time to anthesis. At an early stage, the necessity to improve model performance stimulated research into the physiology of development. Current model accuracy to predict anthesis can hardly be improved, but this should not mask the need for further research on the little-understood physiology of sunflower development.

References

Alessi M, Power JE, Zimmerman DC (1977) Sunflower yield and water use as influenced by planting date, population, and row spacing. Agron J 69:465–469

Anderson WK, Smith RCG, Mc William JR (1978) A systems approach to the adaptation of sunflower to new environments. I. Phenology and development. Field Crops Res 1:141–152

Andrade FH, Margiotta FA, Martínez RM, Heiland P, Uhart S, Cirilo A, Frugone M (1992) Densidad de plantas en maíz. Boletín Técnico 108, Estación Experimental Agropecuaria Balcarce, Buenos Aires, 19 pp

Andrade AP, Wolfe DW, Fereres E (1993) Leaf expansion, photosynthesis, and water relations of sunflower plants grown on compacted soil. Plant Soil 149:175–184

Angus JF, Moncur MW (1985) Models of growth and development of wheat in relation to plant nitrogen. Aust J Agric Res 36:537–544

Angus JF, Cunningham RB, Moncur MW, Mackenzie DH (1981) Phasic development in field crops. I. Thermal response in the seedling phase. Field Crops Res 3:365–378

Ballaré CL, Scopel AL, Sánchez RA (1990) Far-red reflected from adjacent leaves: an early signal of competition in plant canopies. Science 247:329–332

Ben-Porath AA, Baker DN (1990) Tap-root restriction effects on growth earliness and dry weight partitioning of cotton. Crop Sci 30:809–814

Bewley JD, Black M (1985) Seeds: physiology of development and germination. Plenum Press, New York, 367 pp

Boyer JS (1988) Cell enlargement and growth-induced water potentials. Physiol Plant 73:311–316

Boyer JS, Nonami H (1990) Xylem hydraulics, turgor and wall properties during growth. British Society for Plant Growth Regulation. Monograph 21, Long Ashton Research Station, Long Ashton, Bristol, pp 45–52

Bradford KJ, Hsiao TC (1982) Physiological responses to moderate water stress. In: Lange OL, Nobel PS, Osmond CB, Ziegler H (eds) Physiological plant ecology II vol 12B. Springer, Berlin Heidelberg New York, pp 264–324

Bradshaw AD (1965) Evolutionary significance of phenotypic plasticity in plants. Adv Genet 13:115–155

Brar GS, McMichael BL, Taylor HM (1990) Hydraulic conductivity of cotton roots as influenced by plant age and rooting medium. Agron J 83:264–266

Cardinali FJ, Orioli GA, Pereyra VR (1985) Comportamiento de dos híbridos de girasol a bajas densidades de siembra. Rev Fac Agron, Univ Buenos Aires 6:131–139

CETIOM (1992) Physiologie due tournesol. CETIOM-PROLEA. Les points techniques du CETIOM, Paris, 66 pp

Chapin FS III (1991) Integrated responses of plants to stress. A centralized system of physiological responses. Bioscience 41:29–36

Chapin FS III, Walter CHS, Clarkson DT (1988) Growth response of barley and tomato to nitrogen stress and its control by abscisic acid, water relations and photosynthesis. Planta 173:352–366

Chimenti CA, Hall AJ (1993) Genetic variation and changes with ontogeny of osmotic adjustment in sunfloer (*Helianthus annuus* L.). Euphytica 71:201–210

Coleman JS, Schneider KM (1996) Evidence that abscisic acid does not regulate a centralized whole-plant response to low soil-resource availability. Oecologia 106:277–283

Connor DJ, Hall AJ (1997) Sunflower physiology, Sunflower science and technology. Agronomy Monograph No 35, ASSA, CSSA SSSA, Madison Wisconsin, pp 113–182

Connor DJ, Jones TR (1985) Response of sunflower to strategies of irrigation. II. Morphological and physiological responses to water shortage. Field Crops Res 12:91–103

Connor DJ, Sadras VO (1992) Physiology of yield expression in sunflower. Field Crops Res 30:333–389

Connor DJ, Hall AJ, Sadras VO (1993) Effect of nitrogen content on the photosynthetic characteristics of sunflower leaves. Aust J Plant Physiol 20:251–263

Cosgrove DJ (1986) Biophysical control of plant cell growth. Annu Rev Plant Physiol 37:377–405

Cosgrove DJ (1987) Wall relaxation and the driving forces for cell expansive growth. Plant Physiol 84:561–564

Cox WJ, Jolliff GD (1987) Crop-water relations of sunflower and soybean under irrigated and dryland conditions. Crop Sci 27:553–557

Dale JE (1988) The control of leaf expansion. Annu Rev Plant Physiol Plant Mol Biol 39:267–295

Davies WJ, Zhang J (1991) Root signals and the regulation of growth and development of plants in drying soil. Annu Rev Plant Physiol Plant Mol Biol 42:55–76

Donald GM, Hamblin J (1976) The biological yield and harvest index of cereals as agronomic and plant breeding criteria. Adv Agron 28:361–405

Downes RW (1985) Factors affecting germination of sunflower under low temperature conditions. In: Proc XI Int Sunflower Conf, 10–13 March, Mar del Plata, Argentina. Asociación Argentina de Girasol, Buenos Aires, pp 87–92

Egli DB (1990) Seed water relations and the regulation of the duration of seed growth in soybean. J Exp Bot 41:243–248

Fick GN (1978) Breeding and genetics In: Carter JF (ed) Sunflower science and technology, 19 Agronomy Series, ASSA, CSSA SSSA, Madison, Wisconsin, pp 279–338

Friedt W (1992) Present status and future prospects of biotechnology in sunflower breeding. Field Crops Res 30:425–442

Gastal F, Nelson CJ (1994) Nitrogen use within the growing leaf blade of tall fescue. Plant Physiol 105:191–197

Gerwitz A, Page ER (1974) An empirical mathematical model to describe root systems. J Appl Ecol 11:773–781

Giménez C, Fereres E (1986) Genotypic variability in sunflower cultivars under drought. II Growth and water relations. Aust J Agric Res 37:583–597

Gimeno V, Fernández-Martínez JM, Fereres E (1989) Winter planting as a means of drought escape in sunflower. Field Crops Res 22:307–316

Gollan T, Passioura JB, Munns R (1986) Soil water status affects the stomatal conductance of fully turgid wheat and sunflower leaves. Aust J Plant Physiol 13:459–464

Goyne PJ, Woodruff DR, Churchett JD (1977) Prediction of flowering in sunflowers. Aust J Exp Agric Anim Husb 17:475–481

Guiducci M (1988) Effect of water deficit on leaf area and PAR absorption of a sunflower summer crop. In: Plesnicar M (ed) Proc 12th Int Sunflower Conf, 25–29 July, 1988, Novi Sad, Yugoslavia. Yugoslav Association of Producers of Plant Oil and Fat, Novi Sad, vol 1, pp 89–94

Hall AJ, Chimenti CA, Vilella F, Freier G (1985) Timing of water stress effects on yield components of sunflower. In: Proc XI Int Sunflower Conf, 10–13 March, Mar del Plata, Argentina. Asociación Argentina de Girasol, Buenos Aires, pp 131–136

Hall AJ, Connor DJ, Whitfield DM (1989) Contribution of pre-anthesis assimilates to grain filling in irrigated and water-stressed sunflower crops. I. Estimates using labelled carbon. Field Crops Res 20:95–112

Hall AJ, Whitfield DM, Connor DJ (1990) Contribution of pre-anthesis assimilates to grain filling in irrigated and water-stressed sunflower crops. II. Estimates from a carbon budget. Field Crops Res 24:273–294

Hall AJ, Connor DJ, Sadras VO (1995) Radiation-use efficiency of sunflower crops: effects of specific leaf nitrogen and ontogeny. Field Crops Res 41:65–57

Hammer GL, Goyne PJ, Woodruff DR (1982) Phenology of sunflower cultivars. III. Models for predictions in field environments. Aust J Agric Res 33:263–274

Heiser CB, Smith DM, Clevenger S, Martin WC (1969) The North American sunflower. Mem Torrey Bot Club 22:1–218

Henson IE, Jensen CR, Turner NC (1989) Leaf gas exchange and water relations of lupin and wheat. I Shoot responses to soil water deficits. Aust J Plant Physiol 16:401–413

Hirose T (1988) Modelling the relative growth rate as a function of plan nitrogen concentration. Physiol Plant 72:185–189
Hsiao TC, Acevedo E, Fereres E, Henderson DW (1976) Stress metabolism. Water stress, growth, and osmotic adjustment. Philos Trans R Soc Lond B 273:479–500
Hsiao TC, Silk WK, Jing J (1985) Leaf growth and water deficits. In: Baker NR, Davies WJ, Ong CK (eds) Control of leaf growth. Cambridge University Press, Cambridge, pp 239–266
Johnson DW, Walker RF, Ball JT (1995) Lessons from lysimeters: soil N release from disturbance compromises controlled environment study. Ecol Appl 5:395–400
Jones OR (1984) Yield, water-use efficiency, and oil concentration and quality of dryland sunflower grown in the southern high plains. Agron J 76:229–235
Just D, Saux C, Richaud C, André M (1989) Effect of nitrogen stress on sunflower gas exchange. I. Photorespiration and carbon partitioning. Plant Physiol Biochem 27:669–677
Knowles PF (1978) Morphology and anatomy. In: Carter JF (ed) Sunflower science and technology, 19 Agronomy Series, ASSA, CSSA SSSA, Madison, Wisconsin, pp 55–87
López-Pereira M, Sadras VO (1993) Efecto de densidades extremas en la duración de las etapas fenológicas dl cultivo de girasol *Helianthus annuus*. Actas 20 Reunión Argentina de Fisiología Vegetal, 10–12 November, 1993, Bariloche, Argentina, pp 158–159
Lu Z, Quiñones MA, Zeiger E (1993) Abaxial and adaxial stomata from Pima cotton (*Gossypium barbadense* L.) differ in their pigment content and sensitivity to light quality. Plant Cell Environ 16:851–858
MacAdam JW, Volenec JJ, Nelson CJ (1989) Effects of nitrogen on mesophyll cell division and epidermal cell elongation in tall fescue leaf blades. Plant Physiol 89:549–556
Marc J, Palmer JH (1978) Determination of the length of the vegetative and pre-floral stages in the day-neutral plan *Helianthus annuus* by chilling pulses. J Exp Bot 29:367–373
Marc J, Palmer JH (1981) Photoperiodic sensitivity of inflorescence initiation and development in sunflower. Field Crops Res 4:155–164
Masia A, Pitacco A, Braggio L, Giulivo C (1994) Hormonal responses to partial drying of the root system of *Helianthus annuus*. J Exp Bot 45:69–76
Masle J, Passioura JB (1987) The effect of soil strength on the growth of young wheat plants. Aust J Plant Physiol 14:643–656
Matsuda K, Rayan A (1990) Anatomy: a key factor regulating plant tissue response to water stress. In: Katterman F (ed) Environmental injury to plants. Academic Press, New York, pp 63–88
Matthews MA, van Volkenburgh E, Boyer JS (1984) Acclimation of leaf growth to low water potentials in sunflower. Plant Cell Environ 7:199–206
Matyssek R, Tang AC, Boyer JS (1991a) Plants can grow on internal water. Plant Cell Environ 14:925–930
Matyssek R, Maruyama S, Boyer JS (1991b) Growth-induced water potentials may mobilize internal water. Plant Cell Environ 14:917–923
McConnaughay KDM, Berntson GM, Bazzaz FA (1993) Limitations to CO_2-induced growth enhancement in pot studies. Oecologia 94:550–557
Milford GFJ, Pocock TO, Jaggard KW, Biscoe PV, Armstrong MJ, Last PJ, Goodman PJ (1985) An analysis of leaf growth in sugar beet. IV The expansion of the leaf canopy in relation to temperature and nitrogen. Ann Appl Biol 107:335–347
Monteith JL (1977) Climate and the efficiency of crop production in Britain. Philos Trans R Soc Lond B 281:277–294
Morton AG, Watson DJ (1948) A physiological study of leaf growth. Ann Bot NS 12:281–310
Munns R, Sharp RE (1993) Involvement of abscisic acid in controlling plant growth in soils of low water potential. Aust J Plant Physiol 20:425–437
Neales TF, Masia A, Zhang J, Davies WJ (1989) The effects of partially drying part of the root system of *Helianthus annuus* L. on the abscisic acid content of the roots, xylem sap and leaves. J Exp Bot 40:1113–1120
Nelson CJ, Choi SY, Gastal F, Coutts JH (1992) Nitrogen effects on relationships between leaf growth and leaf photosynthesis. In: Murata N (ed) Research in photosynthesis vol IV. Kluwer, Dordrecht, pp 789–792

Novoa R, Loomis RS (1981) Nitrogen and plant production. Plant Soil 58:177–204
Passioura JB, Condon AG, Richards RA (1993) Water deficits, the development of leaf area and crop productivity. In: Smith JAC, Griffiths H (eds) Water deficits. Plant responses from cell to community. Bios, Oxford, pp 253–264
Potter TD, McCloud PI (1985) Evaluation of sunflower cultivars in South Australia. Aust J Exp Agric 25:178–182
Radin WR, Boyer JS (1982) Control of leaf expansion by nitrogen nutrition in sunflower plants. Role of hydraulic conductivity and turgor. Plant Physiol 69:771–775
Rawson HM, Hindmarsh JH (1982) Effects of temperature on leaf expansion in sunflower. Aust J Plant Physiol 9:209–217
Rawson HM, Turner NC (1983) Irrigation timing and relationships between leaf area and yield in sunflower. Irrig Sci 4:167–175
Rawson HM, Dunstone RL, Long MJ, Begg JE (1984) Canopy development, light interception and seed production in sunflower as influenced by temperature and radiation. Aust J Plant Physiol 11:255–265
Ritchie JT (1981) Water dynamics in the soil-plant-atmosphere system. Plant Soil 58:81–96
Robinson D (1994) The responses of plants to non-uniform supplies of nutrients. New Phytol 127:635–674
Robinson RG (1971) Sunflower phenology – year, variety, and date of planting effects on day and growing degree-day summations. Crop Sci 11:635–638
Robinson RG (1978) Production and culture. In: Carter JF (ed) Sunflower science and technology, 19 Agronomy Series. ASSA, CSSA SSSA, Madison, Wisconsin, pp 89–143
Rosenthal WD, Arkin GF, Shouse PJ, Jordan WR (1987) Water deficit effects on transpiration and leaf growth. Agron J 79:1019–1026
Sadras VO, Connor DJ (1991) Physiological basis of the response of harvest index to the fraction of water transpired after anthesis. A simple model to estimate harvest index for determinate species. Field Crops Res 26:227–239
Sadras VO, Hall AJ (1988) Quantification of temperature, photoperiod and population effects on plant leaf area in sunflower crops. Field Crops Res 18:185–196
Sadras VO, Villalobos FJ (1994) Physiological characteristics related to yield improvement in oilseed sunflower (*Helianthus annuus*). In: Slafer G (ed) Genetic improvement of field crops: current status and development. Marcel Dekker, New York, pp 287–319
Sadras VO, Hall AJ, Trápani N, Vilella F (1989) Dynamics of rooting and root length: leaf area relationships as affected by plant population in sunflower crops. Field Crops Res 22:45–57
Sadras VO, Whitfield DM, Connor DJ (1991a) Transpiration efficiency in crops of semi-dwarf and standard-height sunflower. Irrig Sci 12:87–91
Sadras VO, Whitfield DM, Connor DJ (1991b) Regulation of evapotranspiration and its partitioning between transpiration and soil evaporation by sunflower crops. A comparison between hybrids of different stature. Field Crops Res 28:17–37
Sadras VO, Connor DJ, Whitfield DM (1993a) Yield, yield components and source-sink relationships in water-stressed sunflower. Field Crops Res 31:27–39
Sadras VO, Hall AJ, Connor DJ (1993b) Light-associated nitrogen distribution profiles in flowering canopies of sunflower (*Helianthus annuus*) altered during grain growth. Oecologia 95:488–494
Sadras VO, Villalobos FJ, Fereres E, Wolfe DW (1993c) Leaf responses to soil water deficits: Comparative sensitivity of leaf expansion rate and leaf conductance in field-grown sunflower (*Helianthus annuus* L.). Plant Soil 153:189–194
Sadras VO, Villalobos FJ, Fereres E (1993d) Leaf expansion in field-grown sunflower in response to soil and leaf water status. Agron J:564–570
Sharp RE, Davies WJ (1989) Regulation of growth and development by roots of plants growing with a restricted supply of water. In: Jones HG, Flowers TJ, Jones MP (eds) Plants under stress. Soc Exp Biol Semin Ser 39 Cambridge Univ Press, Cambridge, pp 71–93
Sinclair TR, Horie T (1989) Leaf nitrogen, photosynthesis, and crop radiation-use efficiency: a review. Crop Sci 29:90–98

Skoric D (1992) Achievements and future prospects of sunflower breeding. Field Crops Res 30:231–270

Sneep J, Murrty BR, Utz HF (1979) Current breeding methods. In: Sneep J, Hendriksen AJT (eds) Plant breeding perspectives. Center for Agricultural Publishing and Documentation, Wageningen, pp 104–233

Sofield I, Evans LT, Cook MG, Wardlaw (1977) Factors influencing the rate and duration of grain filling in wheat. Aust J Plant Physiol 4:785–797

Steer BT, Hocking PJ (1983) Leaf and floret production in sunflower (*Helianthus annuus* L.) as affected by nitrogen supply. Ann Bot 52:267–277

Steer BT, Coaldrake PD, Pearson CJ, Canty PJ (1986) Effects of nitrogen supply and population density on plant development and yield components of irrigated sunflower (*Helianthus annuus* L.). Field Crops Res 13:99–115

Steudle E (1985) Water transport as a limiting factor in extension growth. In: Baker NR, Davies WJ, Ong CK (eds) Control of leaf growth. Cambridge University Press, Cambridge, pp 35–55

Taiz L (1984) Plant cell expansion: regulation of cell wall mechanical properties. Annu Rev Plant Physiol 11:333–339

Takami S, Turner NC, Rawson HM (1981) Leaf expansion of four sunflower (*Helianthus annuus* L.) cultivars in relation to water deficits. I. Patterns during plant development. Plant Cell Environ 4:399–407

Tardieu J, Davies WJ (1993) Integration of hydraulic and chemical signalling in the control of stomatal conductance and water status of droughted plants. Plant Cell Environ 16:341–349

Tardieu J, Zhang J, Gowing DJG (1993) Stomatal control by both [ABA] in the xylem sap and leaf water status: a test of a model for droughted or ABA-fed field-grown maize. Plant Cell Environ 16:413–420

Taylor G, McDonald AJS, Standenberg I, Freer-Smith PH (1993) Nitrate supply and the biophysics of leaf growth in *Salix viminalis*. J Exp Bot 44:155–164

Ternesi M, Andrade AP, Jorrin J, Benlloch M (1994) Root-shoot signalling in sunflower plants with confined root systems. Plant Soil 166:31–36

Tollenar M (1977) Sink-source relationship during reproductive development in maize. A review. Maydica 22:49–75

Trápani N, Hall AJ (1996) Effects of level of insertion and nitrogen supply on the expansion of leaves of field-grown sunflower (*Helianthus annuus* L.). Plant Soil 184:331–340

Trápani N, Hall AJ, Sadras VO, Vilella F (1992) Ontogenic changes in radiation use efficiency of sunflower (*Helianthus annuus* L.) crops. Field Crops Res 29:301–316

Trápani N, Fossati M, Hall AJ (1993) Expansión y contenido de nitrógeno foliares en girasol. Actas 20 Reunión Argentina de Fisiología Vegetal, 10–12 November, 1993, Bariloche, Argentina, pp 198–199

Trápani N, Hall AJ, Villalobos FJ (1994) Pre-anthesis partitioning of dry matter in sunflower (*Helianthus annuus* L.) crops. Field Crops Res 37:235–246

Trewavas A (1985) A pivotal role for nitrate and leaf growth in plant development. In: Baker NR, Davies WJ, Ong CK (eds) Control of leaf growth. Cambridge University Press, Cambridge, pp 77–94

Turner NC, Schulze E-D, Gollan T (1985) The responses of stomata and leaf gas exchange to vapour pressure deficits and soil water content. Oecologia 65:348–355

Tyree MT, Jarvis PG (1982) Water in tissues and cells. In: Lange OL, Nobel PS, Osmond CB, Ziegler H (eds) Physiological plant ecology II, vol 12B. Springer, Berlin Heildelberg New York, pp 36–74

van Keulen H, Goudriaan J, Seligman NG (1989) Modelling the effects of nitrogen. In: Russell G, Marshall B, Jarvis PG (ed) Plant canopies: their growth, form and function. Cambridge University Press, Cambridge, pp 83–103

Villalobos FJ, Sadras VO, Soriano A, Fereres E (1992) Responses of sunflower plant leaf area to plant population. In: Scaife A (ed) Proc 2nd Congress of the European Society for Agronomy, 23–28 August 1992, Warwick University, pp 146–147

Villalobos FJ, Sadras VO, Soriano A, Fereres E (1994) Planting density effects on dry matter partitioning and productivity of sunflower genotypes. Field Crops Res 36:1–11
Villalobos FJ, Hall AJ, Ritchie JT, Orgaz F (1996) OILCROP-SUN: a development, growth and yield model of the sunflower crop. Agron J 88:403–415
Vince-Prue D (1975) Photoperiodism in plants. McGraw-Hill, Maidenhead, 444 pp
Walker S (1992) Expansive growth, water stress and osmotic adjustment of crops: a review. S Afr J Sci 88:366–370
Warren Wilson J (1967) Ecological data on dry matter production by plants and plant communities. In: Bradley EF, Denmead DT (eds) The collection and processing of field data. Wiley-Interscience, New York, pp 77–123
Watson DJ (1947) Comparative physiological studies on the growth of field crops. II The effect of varying nutrient supply on net assimilation rate and leaf area. Ann Bot NS 11:375–407
Watson DJ (1952) The physiological basis of variation in yield. Adv Agron 4:101–145
Whelan EDP (1978) Cytology and interspecific hybridization. In: Carter JF (ed) Sunflower science and technology, 19 Agronomy Series. ASSA, CSSA SSSA, Madison, Wisconsin, pp 339–369
White KA (1988) Status and developments in sunflower breeding in Australia. In: Proc 7th Australian Sunflower Workshop, Moama, February 1988
Whitfield DM, Connor DJ, Hall AJ (1989) Carbon dioxide balance of sunflower (*Helianthus annuus* L.) subjected to water stress during grain filling. Field Crops Res 20:65–80
Wise RR, Frederick JR, Alm DM, Kramer DM, Hesketh JD, Crofts AR, Ort DR (1990) Investigation of the limitations to photosynthesis induced by leaf water deficit in field grown sunflower (*Helianthus annuus* L.). Plant Cell Environ 13:923–931
Xu X, Bland WL (1993) Reverse water flow in sorghum roots. Agron J 85:384–388
Zhang J, Davies WJ (1989) Sequential response of whole plant water relations to prolonged soil drying and the involvement of xylem sap ABA in the regulation of stomatal behaviour of sunflower plants. New Phytol 113:167–174

Cotton: Factors Associated with Assimilation Capacity, Flower Production, Boll Set, and Yield

J.J. HEITHOLT[1]

1 Introduction

Nearly all cotton yield physiology research has been conducted on either *Gossypium hirsutum* (upland cotton) or *G. barbadense* (Pima or extra-long staple cotton) since these two species represent more than 90% of the world's cultivated cotton crop. Therefore, this chapter will primarily review the literature dealing with upland and Pima cotton genotypes. The history of introductions that led to upland and Pima cottons was described by Meredith (1991). The role of genetics in improving yield and fiber quality of cotton were presented in reviews by Meredith (1984a, 1991) and Culp (1994).

Previous reviews have discussed much of the unique yield physiology and growth of cotton, ranging from basic biochemistry to effects of agronomic practices (Benedict 1984; Hearn and Constable 1984; Mauney and Stewart 1986; Stewart 1986; Wells 1989; Hearn and Fitt 1992; Munro 1995; Hake et al. 1996a). Subsequently, there have been many advances in cotton yield physiology; thus, not all can be covered here. The five areas that will be discussed are: (1) the characteristics of yield components; (2) canopy development, light interception, and assimilation capacity; (3) flower production and fruit set physiology; (4) environmental, genotype, and management effects on fruit growth; and (5) characteristics of yield physiology and breeding that make cotton research more challenging than research on other crops.

Many physiological variables are associated with yield. However, the relationship between a given physiological variable and yield among genotypes may be different from the relationship between that same variable and yield across environments or treatments (within one genotype). In this chapter, a distinction between genetic and environmental variation will be made when discussing the relationship between a physiological variable and yield. Because the yield physiology of cotton, as well as of most crops, is a function of its behavior in a monoculture field community (as opposed to growth chamber or greenhouse isolated plant culture), this chapter will focus mainly on research conducted under agronomic field conditions. In addition, a glossary of cotton terminology related to yield physiology and a drawing (Fig. 1) depicting the morphology of a cotton plant are provided.

[1] United State Department of Agriculture, Agricultural Research Service, P.O. Box 345, Stoneville MS 38776, USA

Crop Yield, Physiology and Processes
D. L. Smith, C. Hamel (Eds.)
© Springer-Verlag Berlin Heidelberg 1999

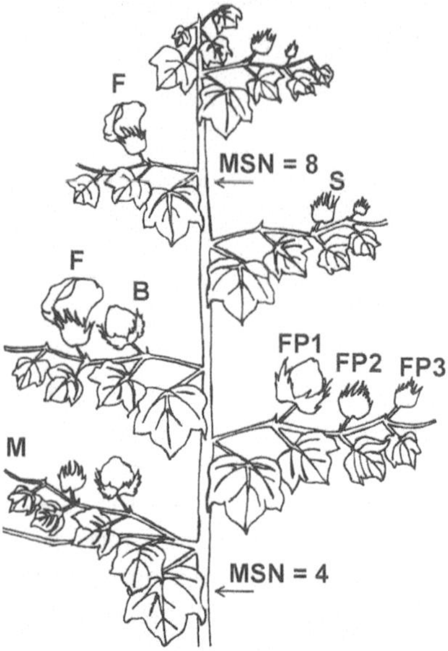

Fig. 1. Morphology of a cotton plant and its fruiting forms. *MSN* Main stem node; *FP* fruiting position; *M* monopodial or vegetative branch; *S* square (floral bud) about 3 days preanthesis; *F* flower (0 DPA); *B* boll (6 DPA). The figure is intended solely for demonstrating the morphological terms used in this chapter. The plant size, number of main stem nodes, number of fruiting forms, and relative scale shown in this figure may not be truly representative of many field-grown plants

1.1 General Characteristics of Yield

Despite being cultured like annual crops, cotton differs from annual crops because it exhibits much of the xerophytic, woody perennial physiology (Hearn 1980; Hearn and Constable 1984) found in some of its ancestors. Fruiting and cottonseed maturation do not necessarily lead to plant death and translocation of leaf and stem assimilates to fruit (Wells 1995). Instead, as the harvestable parts of the crop (lint and seed located in bolls) mature, a considerable portion of the plant's assimilate can be partitioned to nonreproductive plant parts rather than to yield. Assuming the availability of adequate water and fertility, cotton crops can produce excess vegetative biomass that is often associated with reduced yield (Heitholt 1994). To circumvent this possibility, applications of growth retardants (usually gibberellin synthesis inhibitors) are used by commercial growers and researchers in some environments.

Typically, cotton has a crop growth rate during bloom of 17 to 19 $g\,m^{-2}$ day^{-1} (Kerby et al. 1987, 1990b). However, assuming ideal environmental conditions, Baker and Hesketh (1969) reported that cotton could theoretically have a crop growth rate of 42 $g\,m^{-2}\,day^{-1}$ during July and a yield potential of 3650 $kg\,ha^{-1}$. To achieve this yield, nearly 240 bolls m^{-2} (at 1.5 g lint $boll^{-1}$) must mature. Unfortunately, there are many reasons why this yield is not realized. One reason is that the potential (or maximum) number of fruit or seed per unit area that cotton can produce is lower than in many crops. Generally, only one sympodial branch develops at each main stem node with three to five fruiting sites per branch and only one fruiting form develops per fruiting site (Fig. 1). Additional fruiting sites would require formation of additional monopodial branches or more main stem nodes (and their accompanying sympodial branches). Fruit or flower number limitations cannot be overcome by increasing plant density (Guinn et al. 1981; Heitholt 1995). A high-yielding cotton cultivar usually develops only 150 to 200 flowers m^{-2} (Guinn et al. 1981; Guinn 1985; Heitholt 1993) rather than 250 flowers m^{-2} (the theoretical mature boll value above). Consequently, it can be argued that factors causing reduced flower production or increased fruit shed have a rather immediate impact on yield potential.

The time required for young floral buds to develop into mature fruit is longer for cotton than for most crops. Fruiting forms can be susceptible to stress (from floral bud to young boll) for up to 40 days. Early-season fruit loss, regardless of the cause, not only delays maturity (Pettigrew et al. 1992) but can also lead to excess dry matter partitioning to nonreproductive (i.e., vegetative) plant parts and to reduced yield in some normal-leaf genotypes (Kennedy et al. 1986). However, in some environments, cotton can compensate for early-season fruit loss (Sadras 1994; Jones et al. 1996a).

1.2
Relative Importance of Yield Components

Lint rather than seed is the primary harvested product for cotton. Although cotton breeders have selected directly for lint biomass (Meredith 1991), genetic increases in cotton yield, like other row crops, have resulted in increased total seed biomass per unit area. The close association of lint and seed biomass is expected, in part, because the seed coat epidermis physically supports fiber growth. If lint is not determined, researchers report seed cotton biomass rather than seed biomass to indicate yield. Literature that reports seed cotton biomass rather than lint needs to be interpreted cautiously for two reasons: (1) the seed cotton biomass may include foreign matter (plant parts such as leaves, stems, and burs) which overestimates lint yield and (2) lint percentage varies among cultivars, management systems, and environments. Lint, seed, and seed cotton biomass are closely related to the number of bolls per unit area (Wells and Meredith 1984b). This is analogous to the relationship between fruit number and yield in some grain crops. Other components, such as lint mass per seed, number of seed per boll, and lint per boll are generally less likely to be correlated with yield than is boll number (Meredith 1984a).

Several equations are available for relating yield to its components. Lint yield can be defined as the product of total above-ground dry matter per unit area and the percentage of that biomass that is lint (i.e., the harvest index). Like most crops, the genetic gains in cotton lint yield achieved by modern cultivars have occurred because of increases in the partitioning of above-ground biomass to fruit (Wells and Meredith 1984b; Meredith and Wells 1989; Singh et al. 1992).

Second, yield can be defined as the product of the number of bolls produced per unit area and the mass of lint per boll (a common measure of boll size). Of these two components, increases in yield are generally associated with the number of bolls (both among genotypes and environments: Wells and Meredith 1984b; Pettigrew 1994). Boll size has been negatively correlated with genetic variation in yield (Meredith and Bridge 1971). Although higher yields due to favorable environmental conditions (within a genotype) can be associated with larger bolls (Mauney et al. 1978), boll size is often unrelated to yield (Heitholt et al. 1993). Artificial or selective removal of a proportion of fruiting forms can result in fewer but larger bolls than untreated control plants (Jones et al. 1996a; Heitholt 1997). In this case, boll size would be inversely related to yield differences among treatments. Boll size (lint mass per boll) equals the product of number of seed per boll, number of fibers per seed, average length per fiber, and mass per unit fiber length (fineness or micronaire which provides an estimate of weight per unit length). It has recently been suggested that the number of fibers per seed (which has been estimated to be between 10 000 and 15 000) may be an important yield component. However, this hypothesis has not been thoroughly tested.

Third, yield can be defined as the product of seed cotton biomass and the fraction of lint in seed cotton (lint biomass/seed cotton biomass), the lint percentage. Much of the genetic gain in lint yield has resulted from increases in lint percentage (Meredith 1984a, 1991; Fig. 2). Both components (seed cotton biomass and lint percentage) are closely and positively associated with yield but the correlation between yield and seed cotton biomass is often higher (Wells and Meredith 1984b). The strong mathematical relationship between yield (i.e., lint) and seed cotton (lint plus seed) is expected in part because the two variables are not independent. Likewise, lint and lint percentage [lint/(lint + seed)] are not independent variables.

Other yield components often prove useful in explaining yield variation among treatments. Yield is equal to the product of seed number per unit area and lint per seed. The number of seed per unit area was strongly correlated with lint yield when planting dates, genotypes, and row spacings were compared (Heitholt et al. 1993). Lint yield can also be described as the product of bolls per unit area, locules (carpels) per boll, and lint per locule. Whether the number of locules per boll is correlated with yield has not been reported.

Yield can also be expressed as a function of crop growth characteristics during the growing season. Among these characteristics are crop growth rate, duration of growing season, and partitioning to reproductive parts. Using these characteristics, yield can be described as a function of the total PAR (photosynthetically active radiation) per unit area, percentage of PAR captured, radiation use efficiency or net CO_2-exchange rate, and a partitioning factor (harvest index). As will be discussed later, increases in each of these variables have been associated with yield increases.

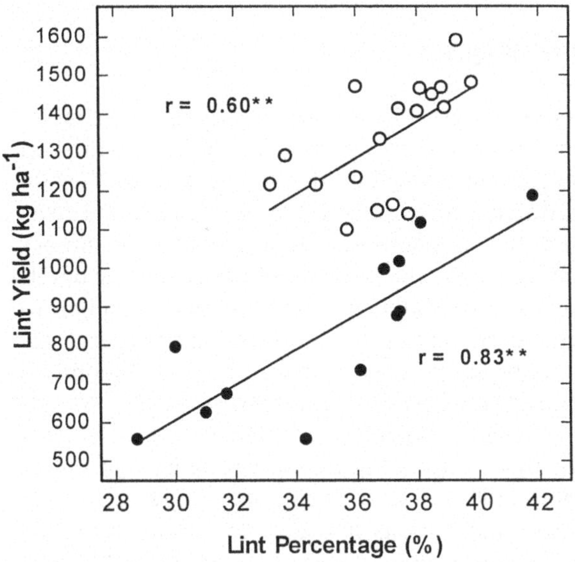

Fig. 2. The relationship between lint yield and lint percentage of cottons. Values are averaged across years for Pettigrew and Meredith (1994, *upper plot*) and across planting dates for Wells and Meredith (1984b, *lower plot*)

2 Root and Canopy Development

Cotton shoot growth is typical of most row crops but it often develops a larger and deeper root system than other crops. Upon germination, cotton seedlings initiate a long taproot. By midseason the taproot has been reported to extend as deep as 3 m (McMichael 1986). This pattern of root growth allows cotton to adapt to dry conditions (Hearn 1980) better than other crops.

Because uniform plant densities and healthy plants are critical for maximum yields, much research has focused on seed germination, seedling growth, and response to stress. Metabolism during seed development and germination was reviewed by Benedict (1984), Galau (1986), and Trelease et al. (1986). Stand establishment and its importance to yield was reviewed by Bird (1986), Christiansen and Rowland (1986), and Wanjura (1986).

Vegetative growth is characterized by successive development of the main stem (primary axis) nodes. Axillary branches differentiate from each main stem node. At lower nodes, monopodial branches (similar to the primary axis) can develop, but from approximately the fifth main stem node (and upward), only sympodial branches develop (Fig. 1). Generally, more fruit are produced on sympodial branches than monopodial branches. Hearn and Constable (1984), Mauney (1986), and McMichael (1986) reviewed cotton shoot and root morphological development.

2.1
Morphological Developmental Stages

The number of main stem nodes and plant height are usually used to characterize above ground vegetative growth. Although the number of main stem nodes and plant height are important (as will be shown later), alone they do not necessarily relate to reproductive development. For example, in one environment a mature crop may be short and have few nodes, whereas in another environment a crop with a similar height and node number may just be developing its first flower. Likewise, cultivars differ in height and node number. To describe progress of growth, the growth phases of (1) seedling emergence, (2) first square, (3) squaring (floral bud stage), (4) first bloom, (5) peak bloom, (6) boll filling, and (7) boll opening, have proved adequately definitive.

Assuming adequate water, nutrients, and solar radiation are available, temperature is the primary factor that controls the rate of vegetative and reproductive development. Constable (1991) described the phenology of fruiting form development of two cotton cultivars as it related to temperature. The effect of temperature on fruit developmental rate was found to differ between earlier developing, lower bolls than for later developing, younger bolls on the same plant. For example, 310 degree-days (DD) were required for a square to reach anthesis at node 6, but 243 DD were required for node 20. Thus, both temperature and morphological position (or maturation stage) affect the rate of square development.

2.2
Solar Radiation Interception

Investigating the physiological explanation for higher yields from F_1 cotton hybrids, Wells et al. (1988) showed that hybrids exhibited greater early-season canopy CER (CO_2-exchange rate) and leaf area than the parents. Single-leaf CER, midseason canopy CER, and midseason leaf areas of the hybrids and parents were similar. Heitholt et al. (1992) found that okra-leaf (a highly cleft leaf shape variant, Fig. 3) cotton grown in narrow rows had greater yield than in wide rows due to increases in early season (floral bud stage) PAR interception. Midseason (flowering) PAR interception and leaf areas were unaffected by row spacing. Wells et al. (1986) provided another example where genetic or management treatments altered the amount of prebloom PAR interception and cotton yield without altering the amount of postbloom PAR interception. Yield was also associated with PAR interception when artificially altered by reflectors or shade cloth (Pettigrew 1994).

These results indicate that increased early-season light capture and growth in cotton, concurrent with the floral bud stage of growth, but slightly before peak bloom and the boll-filling stage, can result in greater yield. However, in soybean (*Glycine max* L. Merr.), prebloom shading reduced the amount of PAR intercepted, but when shade was removed just before bloom, yield was unaf-

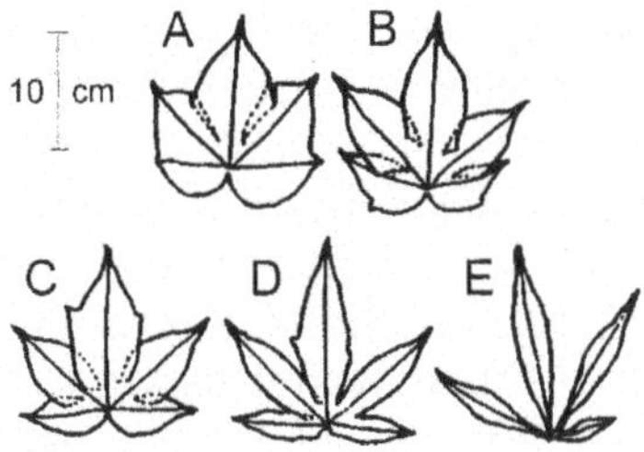

Fig. 3A–E. Examples of cotton leaf morphology variants. **A** Normal-leaf. **B** Sub-okra leaf. **C** F_1 normal-leaf × okra-leaf. **D** Okra-leaf. **E** Super-okra leaf. (Wells et al. 1986)

fected (Jiang and Egli 1995). Together, this suggests that cotton utilizes prebloom radiation more effectively than soybean. The reason for this is unclear but might be related to photosynthate partitioning or fruit biomass at anthesis. Cotton floral buds are large (100 to 200 mg dry weight, Constable and Rawson 1980b; Heitholt and Schmidt 1994) relative to soybean blooms (2 mg dry weight, Heitholt et al. 1986). However, reproductive biomass per plant at first bloom, relative to total biomass, is very small in both cotton (Kerby et al. 1987) and soybean. Other explanations for the difference between cotton and soybean might be related to potential fruiting numbers and the long time required for cotton floral buds to develop into mature fruit (discussed earlier). For cotton, it is likely that cotton floral buds having a chance to develop to maturity need to be initiated by early bloom. Therefore, the amount of PAR intercepted before bloom may be more critical in cotton than soybean in determining final fruit numbers.

Logically, the greater the size of the canopy, the greater the yield potential. This idea was supported by Kerby et al. (1990b), who showed that boll load at cutout was related to vegetative growth rate between the early floral bud and early flower stages. However, in some genotype comparisons (Wells and Meredith 1984b; Lu et al. 1994) or some Mid-South USA environments (J.J. Heitholt, unpubl.), excessive late-season leaf area in cotton is associated with reduced yield. Heitholt (1994) showed that an LAI (leaf area index) of 4 was optimal in a 3-year Mid-South USA study. An LAI of less than 3 or greater than 5 was associated with reduced yield. Because an LAI of 5 or greater is required to support the very high-yielding crops grown in areas of high solar radiation such as Arizona and Australia, the negative relationship between vegetative biomass and yield found in the Mid-South USA suggests that high biomass might be the consequence and not the cause of reduced yield.

2.3 Canopy Architecture

Canopy architecture varies depending upon genotype, management, and weather. Genotypic variation includes leaf morphology variants such as okra-leaf and sub-okra leaf (Fig. 3) and species differences such as *G. hirsutum* and

G. barbadense (Sassenrath-Cole 1995). Variation in maturities and growth types also exist. Although okra-leaf types have been shown to yield more than normal-leaf types under some circumstances (Andries et al. 1969; Wilson 1986; Heitholt et al. 1992) normal-leaf types outyield okra-leaf in most US environments. Okra-leaf types exhibited greater harvest indices or lower leaf area indices than their normal-leaf isolines in three Mid-South studies (Kennedy et al. 1986; Wells and Meredith 1986; Heitholt et al. 1992). In Australia, okra-leaf types comprised nearly 50% of cotton area (for 7 years preceding 1994) and yields were similar to those of normal-leaf (Thomson 1995). The open canopy characteristic of okra-leaf reduces incidence of boll rot (Andries et al. 1969) and may provide greater insect resistance (Wilson 1990; Butter et al. 1992; Brook et al. 1992).

Comparing four different cotton leaf shapes, as well as F_1 (okra × normal) and F_2 hybrids, Wells et al. (1986) showed that super-okra leaf and okra-leaf types had lower LAI and canopy PAR interception than normal-leaf type or the F_1 hybrids. However, the open canopy types (sub-okra, okra-, and super-okra leaf) experienced greater PAR in middle portions (one-third and two-thirds canopy height) of the canopy and potentially greater CER (Wells et al. 1986). Kerby et al. (1980) also reported evidence suggesting a greater CER in the middle portion of okra-leaf canopies. The better distribution of PAR may allow partial compensation for reduced overall interception. The sub-okra leaf trait may offer the best compromise between okra-leaf and normal-leaf. Meredith (1984b) and Meredith and Wells (1987) found that sub-okra leaf populations outyielded their normal-leaf isolines by 5%. Meredith et al. (1996) showed that isolines of a high-yielding upland cotton cultivar (DES 119) expressing combinations of sub-okra, nectariless, and semismooth leaf surface yielded as much as the parent cultivar which exhibited normal-leaf shape, hairy leaf surface, and leaf nectaries on the abaxial midrib.

A second alternative to conventional normal-leaf canopies are the short-statured, short-season genotypes. In short-season environments, such as the Texas High Plains, cultivars with early maturity and cold tolerance are essential for efficient production. These types also form less leaf area than long-season genotypes. Although one might suspect that genotypes with lower LAI and reduced PAR interception would reduce yield in most environments, this is not always the case. In California, Kerby et al. (1990a) reported that two short-statured early-maturing genotypes produced as much lint as two conventional Acala cultivars. Although these two genotypes produced less leaf area and total above-ground dry matter, their yield was equivalent to the conventional types because of a greater harvest index.

Third, canopy architecture is also affected by fertility and other management practices (such as growth regulators and plant spacings). Wullschleger and Oosterhuis (1990b) showed that N deficiency reduced yield by reducing leaf area index, canopy size, and light interception whereas CER on a leaf area basis was unaffected. Potassium deficiencies caused reductions in shoot biomass (Mullins and Burmester 1990; Brouder and Cassman 1994) and also reduced yield. In addition to undersized canopies resulting from nutrient deficiency, oversize canopies can result from excessive N fertilizer. This excessive vegetative growth can lead to delayed maturity (Boquet et al. 1994). Midseason foliar applications of the growth retardant mepiquat chloride are often considered for optimizing canopy architecture (e.g., improving canopy light distribution). In a summary of 35 trials, Kerby (1985) showed that mepiquat chloride consistently reduced the final number of main stem nodes. York (1983a,b) showed that cultivar did not affect the mepiquat chloride response. However, York (1983a,b) found that mepiquat chloride was more effective when high plant densities were used (13.6 and 23.5 plants m^{-2}). Bader and Niles (1986) and Gwathmey et al. (1995) showed that mepiquat chloride increased PAR to middle portions of the canopy. In most cases, mepiquat chloride also reduced plant height (Bader and Niles 1986; McConnell et al. 1992; Williford 1992).

3 CO_2-Assimilation Capacity

In concert with increased boll numbers, past genetic gains in cotton yield have resulted from increases in harvest index and lint percentage and not from greater dry matter production (Meredith 1984a; Wells and Meredith 1984b). Although Meredith and Wells (1989) indicated that increases in harvest index may account for future yield gains, eventually a maximum will be reached and further gains will require increases in radiation-use efficiency or CER. Two possible sources of future gains involve increased early-season PAR interception and increased photosynthetic rate per unit leaf area or per photon flux. Yield increases due to higher early-season PAR interception (from hybrids and narrow rows) have been demonstrated (Wells et al. 1988; Heitholt et al. 1992); however, genetic advances in upland cotton yields from 1905 to 1978 were unrelated to early-season leaf biomass, LAI, or PAR interception (Wells and Meredith 1984a). As will be discussed later, recent genetically advanced lines appear to exhibit higher single-leaf CO_2-exchange rates than older lines (Cornish et al. 1991; Pettigrew and Meredith 1994). In the case of Pima cotton, these increases in single-leaf CO_2-exchange rates have been mostly attributed to increases in stomatal conductance (Cornish et al. 1991; Lu et al. 1994). Whether these advanced genetic lines exhibit an increased internal photosynthetic capacity (i.e., ribulose bisphosphate carboxylase:oxygenase activity, ribulose bisphosphate regeneration capacity) remains a possibility but has not been thoroughly investigated.

3.1 Canopy CO_2-Exchange Rate

Although it is logical to suspect that canopy CER would be related to genotypic variation in yield, only a few researchers have investigated this possibility. Wells et al. (1986) showed that canopy CER was strongly correlated to yield when canopies varying in leaf morphology were tested. In one year, okra-leaf had both lower canopy CER and yield than the normal-leaf. In another year, both variables were similar among the leaf types. Sub-okra leaf often exhibited the highest canopy CER of the six genotypes which could have resulted from its more uniform PAR distribution throughout the canopy.

The average age of leaves in the canopy increases during the season. Because CER decreases as leaves age (Constable and Rawson 1980a; Wullschleger and Oosterhuis 1990a), canopy photosynthesis also declines (Peng and Krieg 1991). Shading of older leaves can further exacerbate the leaf-aging effect (Kerby and Hake 1996) but it has been shown that the effects of leaf age are the most detrimental (Peng and Krieg 1991). It is thought that boll growth may be limited because CER capacity of leaves subtending developing bolls is not well coordinated with growth of those bolls. Yields might be higher if upper leaves did not exhibit such a large decline in CER as a function of age. The decline in CER with age is not as critical for lower leaves if they are shaded by more apical leaves because PAR becomes the limiting factor.

3.2 Single-Leaf CO_2-Exchange Rate

Differences in CER among *Gossypium* species were reported by El-Sharkawy et al. (1965) in Arizona. There was nearly a twofold difference between the lowest and highest species with *G. hirsutum* and *G. barbadense* among the highest. Large differences were also found in area per leaf with *G. hirsutum* and *G. barbadense* exhibiting leaf areas that were two- to tenfold greater than other *Gossypium* species. Elmore et al. (1967) also found significant variation in single-leaf CER rates among cotton species and races.

In studies using only obsolete cotton genotypes, no relationship was found between yield and single-leaf CER. However, positive relationships have been reported in several studies that include modern Pima cultivars and one with upland cultivars. Advanced Pima cotton lines exhibited greater morning CER and stomatal conductance than obsolete lines (Cornish et al. 1991). Genotypic differences in afternoon CER and stomatal conductance were not as distinct. CO_2-response curves indicated that genotypic differences resulted from mesophyll factors as well as stomatal factors. Yield of 18 upland cotton genotypes was shown to correlate to morning single-leaf CER during boll filling (Pettigrew and Meredith 1994). Radin et al. (1994) showed that yield of 12 segregating F_3 and F_4 Pima cotton lines was positively correlated to leaf stomatal conductance. Although conductance was not associated with single-leaf CER or leaf water potential, it was negatively correlated to leaf temperature indicating that high yielding types exhibited more efficient leaf cooling than lower yielding types. Lu et al. (1994) also found that advanced (i.e., high-yielding) Pima lines exhibited greater stomatal conductance and cooler leaf temperatures than obsolete lines. In a related study, Lu and Zeiger (1994) reported that high-yielding Pima genotypes exhibited greater single-leaf CER than obsolete Pima lines. In one study that did not measure CER directly, Rosenthal and Gerik (1991) showed that genotypes differed significantly in radiation-use efficiency. One possible explanation for this finding is that the genotypes also differed in CER per unit leaf area.

Other approaches to identifying genotypes with greater stomatal conductance and yield have recently been used. Gerik et al. (1995a) reported that variation in yield among ten cotton genotypes was positively correlated to carbon isotope discrimination of upper leaves during prebloom and negatively correlated to water-use efficiency. Although values of carbon isotope discrimination varied slightly due to environment, Gerik et al. (1995b) showed that ranking among cultivars was stable across environments and years. McDaniel (1995) also showed that carbon isotope discrimination was negatively correlated to water-use efficiency in segregating interspecific hybrids. Because of the relationship between stomatal conductance and yield, the findings of Cornish et al. (1991), Radin et al. (1994), Lu and Zeiger (1994), Gerik et al. (1995a), and McDaniel (1995) all indicate that high-yielding genotypes exhibit lower water-use efficiency than low-yielding types.

Other sources of genetic variation in CER exist in cotton. Since super-okra leaf and okra-leaf types have clearly exhibited higher yield per LAI, it seems likely that their CER per unit leaf area would be higher. Kerby et al. (1980) showed that okra- and super-okra leaf tended to have higher $^{14}CO_2$ uptake rates per unit leaf area than normal-leaf. Karami et al. (1980) also showed that super-okra leaf tended to have higher midday photosynthetic rates than normal-leaf. Wells et al. (1986) found no difference among leaf types in single-leaf CER when $1000\,\mu mol\,m^{-2}\,s^{-1}$ was used as the minimum PAR for CER determination. Pettigrew et al. (1993) found that super-okra leaf and okra-leaf types had higher CER than the normal-leaf type when $1500\,\mu mol\,m^{-2}\,s^{-1}$ was used as the minimum PAR for CER determination. Pettigrew et al. (1993) also showed that super-okra leaf and okra-leaf types had higher specific leaf weights, higher chlorophyll concentration per unit leaf area, lower conductance, and greater leaf thickness than normal-leaf. The difference in CER between leaf types was shown to be consistent in three genetic backgrounds (Table 1).

When only results from normal-leaf commercial cultivars are considered, the previous studies have generally indicated that genotypic differences in CER were predominately due to differences in conductance (stomatal factors). Whether future high-yielding genotypes will exhibit a greater internal CO_2-fixation capacity (nonstomatal factors) than current genotypes is not known, but may be important to genetic progress.

3.3 Fruiting-Form CO_2-Exchange Rate

Because the bracts and capsule wall on cotton bolls are capable of fixing CO_2, there has been considerable interest in their contribution to yield. Elmore

Table 1. Single-leaf light-saturated CO_2-exchange rates of okra-leaf and normal-leaf isolines from three cotton genetic backgrounds in 1990. (J.J. Heitholt, unpubl.)

Genetic background	July		August	
	Okra-leaf	Normal-leaf	Okra-leaf	Normal-leaf
		$\mu mol\, m^{-2} s^{-1}$		
DES 24-8	36.5	29.0	28.0	24.2
MD 65-11	36.4	28.4	26.2	22.9
STV 213	39.4	30.8	25.5	21.9
LSD (0.05)[a]	3.8		2.9	
Mean	37.4	29.4	26.5	23.0

[a] LSD (0.05) for comparing the six treatments within each sampling date.

(1973) and Wullschleger et al. (1991) showed that bracts were 28 and 14%, respectively, as efficient as leaves in fixing CO_2. Constable and Rawson (1980b) found that net photosynthesis per unit biomass was greater in young squares than bolls. For squares (fruit before anthesis), they also found that bracts would produce up to 50% of their own carbon requirement. However, for bolls (fruit after anthesis) the capsule wall and bracts supplied only 8% of the boll requirement. After injecting $^{14}CO_2$ into bolls of varying ages, Wullschleger et al. (1991) showed that capsule walls could refix a portion of the CO_2 respired from within bolls. It is not known whether genotypic differences in capsule wall CER exist or whether capsule wall CER is correlated to leaf CER differences. Photosynthate contribution by capsule walls may be greater in open canopy architectures because of greater PAR at the capsule surface.

4 Fruiting-Form Formation

Although some accessions and races of *Gossypium* were found to be either short- or long-day types (Mauney and Phillips 1963), modern cultivars are photoperiod-insensitive (day-neutral). Photoperiod insensitivity exhibited by modern cultivars is generally considered a requirement for cotton production in temperate regions (Lee 1984). A more detailed discussion of cotton's photoperiod requirement was published by Lee (1984).

Because of its photoperiod insensitivity, cotton's reproductive growth is controlled primarily by temperature although temperature can interact with photoperiod (Mauney 1966). Since temperature controls the developmental rate of reproductive growth, temperature summations (growing degree-days) are used to predict both individual boll development and stage of whole plant maturity. Once floral buds are initiated, processes that control the number of flowers and boll retention will have a significant impact on yield.

4.1
Floral Bud (Square) and Boll Retention

Because lint yield is strongly correlated with the final number of bolls produced (Guinn et al. 1981; Wells and Meredith 1984b; Heitholt et al. 1992), understanding the factors that control square and boll retention is important. Once visible, a square requires about 20 days (or approximately 400 degree-days using a 12 °C base) to reach anthesis, but this can depend on location of the square on the main stem (Constable 1991).

Although there is agreement that some small square shed occurs, there is disagreement as to its extent and cause. The disagreement regarding the extent of small square shed arose because small squares that shed are not conspicuous and because their numbers appear to be much lower than the number of flowers (bolls) that shed. As for the disagreement regarding the causes of small square shed, two conflicting hypotheses regarding its cause have been proposed. One hypothesis contends that small square shed occurs due to strictly biotic stress (e.g., insect damage) and not due to physiological causes. This hypothesis is supported by the observations that small squares that are shed are often found to have punctures indicative of insect damage. Also, nutritional and assimilate requirements for small squares are small and direct resource limitation is not likely to cause abscission at that stage. The alternative hypothesis proposes that either physiological or biotic stress can cause small square shed. This hypothesis is supported by observations that small squares were shed even under greenhouse conditions where insect pests could be closely monitored. In one greenhouse study, Heitholt (1993) found that 13 to 42% of squares less than 2 mm in width failed to develop into large squares. The percentage of these small squares that were retained varied among genotypes.

In contrast to small squares, there is agreement that large square retention is generally very high (i.e., fully expanded squares are unlikely to shed during the ten days prior to anthesis) unless they encounter insect or severe environmental stress (Mauney and Henneberry 1976; Constable 1991). McMichael et al. (1980) also found small squares susceptible (i.e., subject to shed) when the plants were water stressed whereas large squares were not.

As mentioned earlier, boll shed rates of 50% or more following anthesis are common (Heitholt 1993). Among the events that occur prior to boll shed is the formation of a distinct abscission layer between the peduncle and sympodial branch. It is not know whether the formation of the abscission layer is the first event in the process of a boll's demise or whether abscission layer formation is a secondary result of other primary physiological events. Bolls that are retained, however, usually do not begin significant dry matter accumulation until about 9 days postanthesis (DPA) (Guinn and Brummett 1988; Kerby and Hake 1996). Most boll abscission occurs before 9 DPA, suggesting that factors other than assimilate supply may be limiting boll retention.

4.2
Physiology of Boll Retention/Shed

Because boll loss in cotton can significantly reduce yield, understanding its cause has been the subject of many research projects. Guinn (1982) reviewed the literature pertaining to the causes of square and boll shed in cotton. Both hormonal concentrations and factors affecting assimilate availability were shown to be associated with boll shed. Hearn and Constable (1984) proposed a mechanism whereby both hormones and assimilates interacted to achieve shed. Guinn and Brummett (1987) found that the concentrations of indole acetic acid in 3-day-old bolls and the abscission zone separating the peduncle and sympodial branch were positively correlated with retention. In concert, this same study showed the concentration of abscisic acid in the 3-day-old bolls and the abscission zone to be inversely related to boll retention. Guinn and Brummett (1988) found that the concentration of indole acetic acid was high in floral buds but decreased after anthesis. Conversely, abscisic acid concentration was low at anthesis but increased afterwards. Since bolls are greatly susceptible to abscission during the first few days immediately following anthesis (Guinn and Brummett 1987; Heitholt 1993), the data suggest that abscission may be mediated by the concentrations of these two hormones.

Several reports support an "assimilate hypothesis" for boll retention that suggests that the supply or rate of assimilate supplied to the young boll is critical for its continued development. Using environmental alterations to reduce boll set, Conner et al. (1972) and Saleem and Buxton (1976) provided data that support this hypothesis. Guinn (1974) reported that increasing CO_2 concentration from 350 to 1000 ppm decreased boll shedding and increased glucose and fructose concentration in leaves. It is not clear how leaf assimilate concentration translates to boll assimilate levels.

Direct evidence implicating inadequate assimilate metabolism in bolls as the cause of shedding has been difficult to obtain. Although Heitholt and Schmidt (1994) reported genotypic and sympodial position (FP1 vs. FP2) variation in boll retention and ovary carbohydrates (at −5, 0, and 2 DPA), they showed that concentrations of carbohydrates and nitrogen in the ovaries were unrelated to retention. This latter study should not be interpreted to indicate that assimilate plays no function in the abscission process. Instead, it appears that the physiological connection between assimilate and retention needs to be investigated differently.

Another factor that may regulate the square and boll retention involves the quantity or quality of PAR reaching the fruiting form (Constable 1991; Kasperbauer 1994). Evidence supporting a photomorphogenetic process can be found by observing square set in okra-leaf and Pima cottons (both open canopy types relative to normal-leaf type). Okra-leaf (Heitholt 1993) and Pima (Kerby and Hake 1996) both set more squares than normal-leaf upland cotton. By measuring light penetration into canopies subjected to vegetative bud removal, Sadras (1996a) also provided indirect evidence that square retention may be related to light intensity reaching the square. In soybean, Heindl and Brun (1983) and Myers et al. (1987) found that chances of a soybean flower setting a pod increased as its exposure to red light increased. It is not known what mechanisms explain the increased square set in okra-leaf and Pima cotton although sink activity of their floral buds may be important. Whatever the mechanism is for square set, there is clearly a different mechanism for boll set. Normal-leaf flowers are exposed to less direct PAR but have much greater chances of developing into mature bolls than flowers of super-okra leaf and okra-leaf (Wells and Meredith 1986; Kerby and Buxton 1981; Heitholt 1993).

4.3 Flower Production and Yield

Within a given genotype, seasonal flower production per unit area is more closely related to yield than boll retention percentage (Heitholt 1993, 1995). Because the petals of a cotton flower are white (cream-colored for some upland germplasm and yellow for Pima) on the day of anthesis but turn pink within 24 h, bloom counts can easily be obtained (there is no chance of a flower being counted twice on separate days) and seasonal totals calculated. The importance of seasonal flower production was demonstrated by Ehlig and Donovan (1972) who compared two planting configurations, (1) a single row per bed combined with a low plant density or (2) two rows per bed with a high plant density. The two rows per bed treatment produced 35% more flowers per unit area and 21% greater yield. Boll retention (i.e., percentage) was slightly reduced in two rows per bed treatment. Bruce and Romkens (1965) reported that cotton yield increases from irrigation were related to increased flower production per unit area but not boll retention. Micinski et al. (1992) showed that flower production per unit area was related to yield when delayed plantings reduced yield. Examining flower production, boll retention, and yield differences between two years (with identical treatments), it was shown that in the year with the higher yield, plants produced more flowers but had lower boll retention than in the year with the lower yield (Heitholt 1993). Yield increases due to narrow rows were associated with increased flower production per unit area rather than improved boll retention (Heitholt 1995). These studies support the idea that increases in flower production per unit area are often more important than boll retention in explaining why selected cultural practices increase yield. Increased flower production may be a reflection of increased plant size. Under cultural practices that limit plant size, it is easy to understand why flower production is positively associated with yield.

Others have found that both total flower numbers per unit area and boll retention contribute to yield differences. For example, Guinn and Mauney (1984) found that irrigation increased flower numbers by 9% and boll retention by 10%. Likewise, Cook and El-Zik (1993) reported that irrigation increased total flower numbers by 29% and boll retention by 16%. Radin et al. (1992) reported that yield increases from irrigation were related to increased flower production during peak bloom and a longer flowering period. Cotton plants grown in a greenhouse atmosphere enriched with CO_2 exhibited a 60% increase in total flower numbers and a 37% increase in boll retention (Mauney et al. 1978). The increases in boll retention in these studies, although important, was not as large as the increase in flower numbers. This response by cotton suggests that flower production should be strongly considered as a selection criteria for different management practices and possibly genetic enhancement programs. El-Zik and Thaxton (1989) found that advanced multiadversity resistant germplasm exhibited greater flower production than other germplasm. The data supporting the importance of flower production by itself should not be extrapolated across leaf type morphologies because it is well known that okra-leaf types produce more flowers than normal-leaf types (Wells and Meredith 1986; Heitholt 1993) but generally do not produce more bolls or lint.

Genotypes with greater flower production (especially those with altered leaf morphology such as the okra-leaf type) may have an advantage if rainfall is frequent during peak bloom. This hypothesis is based on the susceptibility of pollen to rainfall. Depending on its intensity, rainfall can disrupt pollen growth in open flowers and hinder fertilization. In a 1995 field study when rain did not occur on the day of anthesis, boll retention decreased when anthers were treated (washed) with 5 ml of water at 15.30 h (mid-afternoon or 3:30 pm) or earlier (Fig. 4). Therefore, flowers exposed to heavy rainfall on the day of anthesis are likely to be poorly pollinated. Unpollinated cotton blooms usually abscise within two weeks of anthesis. The number of flowers susceptible to rainfall-induced pollination failure depends on the genotype and management. A previous report of flower production as a function of time indicated that peak flowering may last approximately 20 days (Heitholt 1995). A cotton crop with a high yield potential (7.5 flowers m^{-2} day^{-1} or a seasonal total of 150 flowers m^{-2}) may be able to withstand more shed than a crop with a low yield

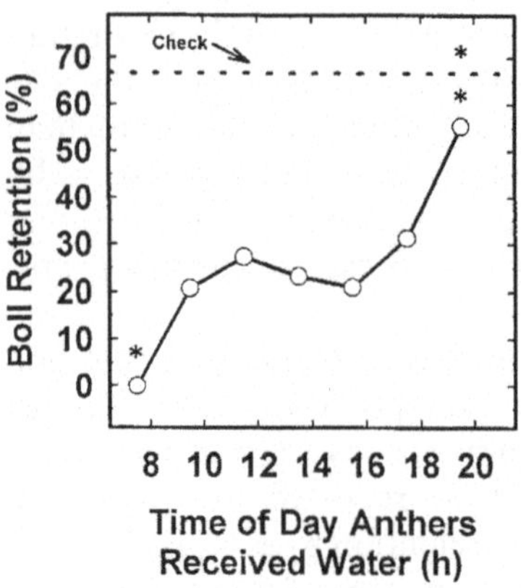

Fig. 4. Effect of spraying 5 ml of H_2O on anthers of Deltapine 50 cotton flowers at selected times on the day of anthesis. *Asterisks* for the check and 19$^1/_2$ h (actually 7:30 pm or 19.30 h) points indicate that those retention values were significantly ($P < 0.05$) greater than all other points. Likewise, the *asterisk* for 7$^1/_2$ h (7:30 am or 07.30 h) indicates that retention for that time was significantly lower than all other times. *Points* represent the mean of three replicates with ten flowers per time per replicate

potential (5 flowers m^{-2} day^{-1} or a seasonal total of only 100 flowers m^{-2}). If, for example, rainfall occurs on 6 days during peak bloom, nearly 45 (high potential) and 30 flowers m^{-2} (low potential) could be shed due to pollination failure. This would leave 105 flowers m^{-2} for the high-yield potential crop but only 70 flowers m^{-2} for the low potential crop. The consequence would be delayed maturity for both crops. However, the number of available flowers (and therefore bolls) might limit yield in the low potential crop. Thus, genotypes or cultural practices that produce a greater number of flowers may yield more when subjected to frequent rainfall during peak bloom than genotypes or cultural practices producing a limited number of flowers because the former may be able to withstand (compensate) the rainfall-induced flower loss.

An example of how increased flower numbers may have increased yield potential was observed in 1994 at Stoneville, MS, where rain fell for 9 days from 9 to 26 July (the peak bloom period). Yield of okra-leaf types was greater than normal-leaf types in three experiments that year (Table 2) whereas in a typical year, yield of these two leaf types is similar. One possible reason for the okra-leaf yield increase over normal-leaf could be due to flower numbers. Assuming the rain prevented pollination of flowers at anthesis, the normal-leaf plants produced only 100 flowers m^{-2} capable of maturing compared to nearly 150 flowers m^{-2} for okra-leaf. Although flower production was most likely involved in the yield differences among genotypes, other explanations for the okra-leaf advantage need to be mentioned here. First, normal-leaf has a greater tendency

Table 2. The effect of leaf type (okra vs. normal) on seasonal flower production, estimated flowers lost to rain, flowers not exposed to rain at anthesis, boll production, and lint yield in 1994 in Stoneville, Mississippi. Values in Experiment 1 and 2 were averaged across two row spacings. Values were averaged across genetic backgrounds in Exp. 2 and Exp. 3. Rain fell for 9 days during the flowering period in July 1994

Experiment	Genetic background	Leaf type	Total seasonal flowers	Flowers lost to rain[e]	Flowers not exposed to rain	Mature bolls produced	Boll set[f] percentage (%)	Yield (kg ha^{-1})
				no. m^{-2}				
1	DES 24-8 ne	Okra	206[a]	54	152	71[a]	47	1040[a]
1	DES 24-8 ne	Normal	121	31	90	56	62	860
2	Four[c]	Okra	202[a]	54	148	65[a]	44	965[a]
2	Four[c]	Normal	129	31	98	56	57	834
3	Two[d]	Okra	-	-	-	63[b]	-	962[b]
3	Two[d]	Normal	-	-	-	56	-	868

[a,b] The difference between leaf types was significant at the 1 or 5% level.
[c] The four genetic backgrounds for the okra-normal isolines in Exp. 2 were: MD51ne, Tamcot HQ95, DES 119, and Stoneville 6413.
[d] The two genetic backgrounds for the okra-normal isolines in Exp. 3 were: MD51ne and DES 24-8ne.
[e] In both Exps. 1 and 2, the number of open flowers per day for the 9 days' rainfall averaged 6 m^{-2} day^{-1} for okra-leaf and 3.5 m^{-2} day^{-1} for normal-leaf. As stated in the text, it was assumed that flowers that opened on one of the 9 days of rain failed to pollinate successfully and subsequently abscised.
[f] Boll Set Percentage = (the number of mature bolls)/(flowers not exposed to rain).

to produce excess vegetative growth (at the expense of fruit) under wet conditions than does okra-leaf. Second, because cloudy weather accompanied the rainy period, decreased carbohydrate assimilation may also have contributed to some of the genotype differences.

4.4 Fruit Loss and Yield Compensation

Despite studies that have shown the importance of early-season fruit set to yield, other studies have shown that cotton has the ability occasionally not only to compensate for fruit loss but also to produce a higher yield as a result. In this section, reproductive compensation is defined as the production of "replacement" fruit for fruit lost earlier so that final yield is similar to control (unaffected) plots. I will not discuss vegetative compensation (replacement of leaves lost prematurely). As mentioned earlier, cotton normally produces only one floral bud per fruiting site. Therefore, to achieve compensation, cotton must set a higher percentage of existing flowers, produce additional flowers on more distal positions of sympodial branches, produce more flowers on monopodial branches, or produce a greater number of main stem nodes with additional sympodial branches (and thus more fruiting sites). In a short growing season, compensation from lost fruit must come from already existing floral buds. Floral buds initiated during midseason in response to older fruit loss may not have enough time to produce mature fruit.

The literature concerning compensatory growth was recently reviewed by Sadras (1994). One hypothesis advanced in that review suggested that yield compensation was more likely to be successful when yield potential of the crop was limited but not extremely low. When yield potential was high, yield compensation to fruit loss was less likely. The rationale behind this idea is based on the possibility that low yield potential environments result in inadequate LAI, PAR capture, number of fruiting sites, and root development (Sadras 1994; Kerby and Hake 1996). Kerby and Hake (1996) proposed that a first fruiting node at main stem 5 or less indicated inadequate vegetative growth and a strong possibility of premature cutout. This idea agrees with the potential beneficial effect of early-season fruit loss in poor environments advanced by Sadras (1994). Early-season fruit loss is associated with increased LAI and increased biomass partitioning to leaves (Kennedy et al. 1986; Sadras 1994; Jones et al. 1996b). Early-season fruit loss can also increase biomass partitioning to roots. The "larger" plants that result from early-season fruit loss may have a yield potential that is equal to or greater than control plants if the growing season is not shortened by adverse weather (Sadras 1994; Jones et al. 1996a). However, under very stressful conditions (high plant density and low soil N), early season square loss can increase dry matter allocation to roots and prevent yield compensation (Sadras 1996b). In contrast to early-season fruit loss, late-season fruit loss is likely to decrease yield. Jones et al. (1996a) characterized compensation in cotton subjected to flower removal during the 4th week of bloom and later and found yield reductions of 13 to 33%.

4.5
Importance of Fruiting Positions

First fruiting positions (FP1, Fig. 1) produce the largest fruit (Heitholt and Schmidt 1994), the most fruit, and the greatest proportion of yield (Jenkins et al. 1990a,b). In addition, more bolls and higher retention are found at FP1 than FP2 (Kerby et al. 1986, 1987). Fruiting position affects not only yield components but also fiber properties. In a study of eight genotypes, greater fiber length (Heitholt and Schmidt 1994), and greater fiber wall thickness (Heitholt 1997) were found for FP1 lint than FP2 lint. Also, it is generally accepted that high-yielding canopies are associated with a high ratio between the number of FP1 bolls to the number of FP1 sites. As a consequence of this emphasis on FP1 and the likelihood that FP1 bolls may reduce retention of adjacent FP2 bolls (Kerby and Buxton 1981), physiologists and breeders may need to increase their emphasis on producing plants that exhibit a high retention of not only FP1 but also FP2. In support of this idea, Heitholt (1997) showed that removal of FP2 squares for 6 weeks significantly reduced yield, indicating that FP2 fruit can be important contributors to yield. Future yield increases are unlikely to come from increases in plant density or the number of main stem nodes per unit area. Therefore, future yield increases may require genotypes or environments that result in a high retention of both FP1 and FP2 sites, assuming that assimilate supply does not prevent simultaneous high FP1 and FP2 retention.

5
Carbohydrate Metabolism and Utilization

Efficient production of carbohydrate in leaves and its transport to bolls is essential for a high-yielding cotton crop. Because of the perennial nature of cotton physiology, photosynthate is not necessarily partitioned efficiently for maximum yield. For lint to grow, photoassimilates need to be translocated from leaves to elongating and thickening fiber cells located on the seed coat (epidermis). High-yielding cotton genotypes need to efficiently synthesize cellulose in fiber cells as well as synthesize storage carbohydrates, proteins, and lipids in seeds.

5.1
Leaf Assimilate Physiology

It has been proposed that photosynthate partitioning within cotton leaves may be related to CER. Leaf starch was negatively correlated with CER and the correlation was higher in cotton than soybean (Krieg 1986). Hendrix and Huber (1986) reported diurnal fluctuations of sucrose phosphate synthase and sucrose synthase activities in cotton leaves. They found that sucrose export rate was generally related to sucrose concentration and sucrose phosphate synthase activities during the day. Hendrix and Grange (1991) also found that the rate of carbon export from leaves at night was related to leaf starch concentration at the end of the photoperiod. Although leaf starch can sometimes be negatively associated with chlorophyll fluorescence (Warner and Burke 1993) and CER (Warner et al. 1995), high leaf starch is found in CO_2-enriched cotton (Hendrix 1992; Hendrix et al. 1994) while high CER is maintained (Hileman et al. 1992). Leaf morphology may also play a role in partitioning of photoassimilates. Kerby et al. (1980) reported that leaves from a super-okra leaf type exported ^{14}C-assimilate more rapidly than normal-leaf types. In other work, leaves of the okra-leaf type were found to contain less sucrose and starch than the normal-leaf type (J.J. Heitholt, unpubl.).

5.2 Fiber Assimilate Physiology

Since the review by DeLanghe (1986), characterization of assimilate utilization by ovules has recently been reported (Wafler and Meier 1994). Although several studies have characterized the conversion of sucrose to cellulose in cotton bolls, it is not known whether the metabolic processes involved could help explain yield differences among genotypes. Basra et al. (1990) showed that both acid invertase and sucrose synthase were important enzymes for sucrose utilization during the elongation phase. By 25 DPA, when most incoming sucrose is expected to be converted to cellulose, sucrose synthase activity predominated. The difference in sucrose synthase was correlated with the difference in fiber length between *G. hirsutum* and *G. arboreum*. Also, glucose and fructose concentrations in fibers were greater than sucrose concentration by 10 DPA and later. In contrast, Heitholt and Schmidt (1994) found that sucrose concentration was greater than glucose or fructose in young cotton ovaries. Kloth (1992) showed that specific activity of malate dehydrogenase of 12 DPA fiber was correlated with micronaire. Although yield of these genotypes was not reported, micronaire has been correlated with yield (Wells and Meredith 1984b; Heitholt 1993) raising the possibility that future studies may find that enzyme activities of fiber may be correlated with yield.

6 Environmental Effects

It is well known that cotton yield is dependent upon solar radiation, water, temperature, soil fertility, CO_2 concentration, and relative humidity. Reports on the physiological effects of solar radiation (Heitholt et al. 1992; Pettigrew 1994) and water (Hearn 1980) are numerous and will not be included here. Recent insights into the physiological effects of temperature, fertility, and CO_2 concentration are also numerous but a few are briefly discussed here.

6.1 Temperature

Optimal temperature for cotton often depends on other factors such as water, relative humidity, and solar radiation. In general, however, cool temperatures (<15 °C) are detrimental to cotton growth and yield, as indicated by its failure to grow well at high latitudes. High temperatures (>35 °C) are also detrimental because of reduced CER, reduced pollen viability, and increased night respiration. The importance of maintaining leaf temperatures in the optimum range is illustrated by the Pima cotton leaf conductance studies discussed earlier (Lu et al. 1994). Genotypes that exhibited greater stomatal conductance and thereby cooler leaf temperatures produced greater yield. Reddy et al. (1991) reported an optimal day/night temperature for cotton of 30 °C during the day and 20 °C at night for both vegetative and reproductive growth. Burke et al. (1988) reported that the temperature optimum for cotton glyoxylate reductase was between 23.5 and 35.0 °C. Burke et al. (1990) suggested using 27.5 °C as the best average temperature for optimizing cotton growth. Warner and Burke (1993) demonstrated that night temperatures of 20 °C increased predawn starch concentration by three times and decreased PS II chlorophyll fluorescence by approximately 50%.

Night temperatures have been the focus of several studies. Gipson and Joham (1968) showed that 21 °C was an optimal night temperature but also reported genotypic variation in response to cool nights. Brown et al. (1995) showed that artificially increasing night temperatures (2 to 5 °C above ambient) decreased yield in a desert environment. Temperature stress effects on yield are not restricted to effects on vegetative parts since Haigler et al. (1991) showed that low night temperatures decreased elongation and dry matter growth rate of fibers grown on ovules in vitro.

6.2 Nutrition

The importance of various mineral nutrients to cotton yield has been recently reviewed (Kerby and Adams 1985; Radin and Mauney 1986; Hodges 1992; Cassman 1993); therefore, only selected aspects of nutrition related to yield physiology will be discussed. Excess N fertility can reduce yield and delay maturity (Boman and Westerman 1994). Reduced yield under K or N deficiency are due to reductions in vegetative growth, canopy CER (Wullschleger and Oosterhuis 1990b), or above-ground biomass (Cassman et al. 1989). Constable et al. (1992) also summarized the effects of N and tillage systems on clay soils and found that the N requirement was lower under minimum tillage than under complete tillage.

6.3
CO_2-Enrichment and Yield

Numerous reports of CO_2-enriched yield increases in cotton have been reported. Idso et al. (1988) and Kimball and Mauney (1993) showed CO_2-enrichment did not affect root:shoot ratio. In contrast, Mauney et al. (1994) found CO_2-enrichment increased yield and root:shoot ratio. Kimball and Mauney (1993) reported yield increases of 60% when CO_2 concentration was increased from 350 to 650 μmol mol^{-1}. Harvest index was unchanged, indicating that biomass of all plant parts responded to the CO_2.

7
Monitoring and Management

Monitoring can be defined as collecting nutritional, developmental, morphological, and entomological data (e.g., petiole N, plant height, plant "maps" of flower and boll locations, and insect counts). There are two objectives for monitoring. First, and more difficult, is to provide data that might lead to immediate management decisions. Prior to flowering, data on square set, main stem node counts, and height to node ratio (especially the upper five nodes) are useful (Bourland et al. 1992; Kerby and Hake 1996). In the early season, poor square set generally indicates insect damage, but other possibilities cannot be ruled out. At the early square stage (20 days prior to first bloom), height to node ratios of 3 cm node^{-1} or less indicate inadequate vegetative growth. A ratio of 3 to 5 cm node^{-1} is desired but a ratio above 5 cm node^{-1} indicates a need to reduce vegetative vigor (Kerby and Hake 1996). Height to node ratios at first flower should be around 5 cm node^{-1}. Values of 7 cm node^{-1} indicate too much vegetative growth. Management practices that adjust vegetative growth when vegetative vigor (e.g., height to node ratios) is outside the desired ranges may be useful. Remedial management practices must be applied relatively quickly if yield potential is to be maintained. Inadequate growth (low height to node ratios) indicates a need for more N, K, or water. Excessive vegetative growth (height to node ratio above 5 cm) indicates insect damage of squares, poor soil conditions, insect damage, diseases, or a need for less N or water. Since N and water cannot be easily removed, growth retardants are an option. As emphasized by Hearn and Constable (1984), improving retention would not improve yield unless assimilates are available to fill bolls.

The second objective for obtaining mid- and late-season morphological data, is to provide a morphological/physiological explanation for the season's performance (i.e., yield). Kerby and Hake (1996) provided an example where monitored variables explained 43% of the variation in yield in California cotton from 1982 to 1991 as shown with the equation:

$$\text{Yield (kg/ha)} = 746 - 106\ (\text{FFB}) + 12.7\ (\text{R}) + 58.6\ (\text{NFN}),$$

where FFB = number of nodes between cotyledon to first fruiting branch, R = percent retention of the bottom five FP1 sites, and NFN = number of main stem nodes in 95% fruiting zone.

In another example of important research data obtained from monitoring, Hake and Stair (1994) showed that yield was strongly related to the number of main stem nodes present above the uppermost white flower at first open flower. The number of main stem nodes above white flower (NAWF) at first bloom is often equal to 7 or 8 but when cutout is reached, NAWF is often 4 (McConnell et al. 1993) or 5 (Bourland et al. 1992). Therefore, NAWF is often a rapid method for monitoring crop maturity.

7.1 Growth Regulators

One of the longest controversies regarding cotton management is the use of the growth retardant mepiquat chloride (1,1-dimethylpiperidinium chloride). The use and efficacy of mepiquat chloride in California was recently reviewed by Kerby et al. (1996). In order to alleviate the reduced fruit set and excess vegetative growth that occur from insect damage or excess N and water, mepiquat chloride is often sprayed over-the-top between the preanthesis stage and the midbloom stage. Despite dozens of studies having tested the effects of mepiquat chloride, results on yield are still inconclusive. In a 3-year Mid-South USA study, mepiquat chloride was found to be ineffective when planting occurred in late April or early May but it increased yield when cotton was planted late (mid-May) (Cathey and Meredith 1988). Kerby et al. (1986) also suggested that mepiquat chloride would be more likely to increase yield when cotton was planted late. Walter et al. (1980) found that mepiquat chloride reduced plant height, internode length, and leaf area but increased leaf thickness. Fernandez et al. (1992) showed that mepiquat chloride inhibited leaf area expansion, daily transpiration, and net daily C uptake. Although mepiquat chloride did not affect leaf water relations, it did reduce water-use efficiency under water stress. Environmental factors obviously have a major effect on the efficacy of mepiquat chloride. In general, the effects of mepiquat chloride on canopy architecture seem to be more consistent than its effects on yield. Unfortunately, whether the decision to apply mepiquat chloride is correct or not often depends upon the weather that follows the application.

7.2 Crop Termination

Foliar chemicals are often applied to remove cotton leaves and to stimulate boll opening prior to mechanical harvesting. Although several factors determine whether defoliation is needed, the main goal of defoliation is to maintain quality fiber and to improve harvest efficiency. Succulent leaves contain cell sap that can stain lint during harvest. Senescent leaves that remain in the canopy can be broken up and entangled with lint during harvest. Both succulent and senescent leaves reduce harvest efficiency. Reasons for stimulating boll opening include enhanced fiber drying and improved harvest efficiency. For further details, the reader is referred to reviews of cotton defoliation provided by Cathey (1986) and by Roberts et al. (1996).

Timely termination of cotton growth prior to harvest can greatly affect harvestable yield. Snipes and Baskin (1994) showed defoliation at 60% open bolls did not reduce yield or fiber quality when compared to plots defoliated later. Defoliation at 40% open bolls or lower reduced yield but increased length and strength of remaining lint. Presumably, the early defoliation caused the shed of late bolls that normally contribute shorter and weaker fiber. Kerby et al. (1992) and Supak et al. (1993) reported that the number of main stem nodes above the uppermost open ("cracked") FP1 harvestable boll (typically two or three nodes) was a useful indicator for terminating growth of the crop. Delaying crop termination or harvesting of open cotton can reduce yield and fiber quality.

8 Breeding and Yield Physiology Research Challenges

Changes in available cotton genotypes (e.g., transgenics with herbicide and insect resistance), changes in environmental regulations, requirements for better quality fiber, and demand for early maturity have changed the cotton industry and affected applied cotton yield physiology research objectives. Instead of using the same management for all genotypes, selected combinations of genotypes and management are now used. This makes yield physiology research more challenging. Regulatory changes, especially rules that prohibit certain insecticide applications, can subsequently alter insect ecology and cultivar performance. Altered insect ecology, increased damage by secondary pests, and increased costs for developing new pesticide chemistry place a premium on preserving beneficial insects. Early maturity is desired because it can drastically reduce input costs (e.g., pesticides), especially in irrigated areas. Finally, mill-processing technology is now more efficient but requires stronger fiber to maintain that efficiency (Deussen 1992). Considered together, these recent changes have altered the objectives of many cotton research programs.

8.1 Treatment and Genotype Comparisons

Unfortunately for cotton, it is more costly to compare management treatments or experimental lines than it is for crops like soybean or maize (*Zea mays* L.). Difficulty in mechanically harvesting small plots (cleaning harvesting equipment between samples is slow but necessary to maintain sample or genetic purity). Ginning and delinting seed are two additional processes required of cotton breeders but not required of grain crop researchers. Furthermore, cotton researchers must not only evaluate seed chemistry as do researchers of other crops but also must determine fiber properties. For further details on procedures required to sample bolls and to measure yield, yield components, and other physiological factors of cotton the reader is referred to Meredith et al. (1975), Wells (1989), and Keisling and Smith (1994). On the positive side, producing F_1 hybrids, which yield more than parents (Meredith 1990), for subsequent germplasm enhancement might be considered easier for cotton than many other self-pollinated crops because cotton flowers are large and 30 seed can be produced per cross (pers. observ.). Emasculation to prevent self-pollination is easy because the day prior to anthesis is easily identified (Holt et al. 1994).

8.2 Management Intensity

Economical cotton production requires more intensive management and monitoring than other field crops (Kerby and Hake 1996). Both morphological development and insect pests are closely monitored in cotton. Morphological development data obtained from monitoring, if evaluated rapidly may help determine fertility and water needs, as well as the rate and timing of growth-retardant and defoliant applications. Equal in importance to water, cotton insects are another critical factor controlling yield. Although most crop species experience yield-reducing insect pest infestations, more insecticides were applied to control cotton insects (1.95 $kg\,ha^{-1}\,yr^{-1}$ in the US in 1994, mostly in the mid-South USA) than in other crops such as maize (0.24 $kg\,ha^{-1}\,yr^{-1}$; USDA 1995). Cotton's greater economic value per unit area is strongly linked to these intensive applications. Physiological reasons for cotton's greater susceptibility to pests is unclear. Three factors that contribute to cotton's greater sensitivity to insect pests are: (1) many cotton insects attack fruit, rather than leaves, (2) many of these pests are small and difficult to detect (Hearn and Fitt 1992), and (3) cotton's extended fruiting cycle that can last nearly 70 days from visible floral bud (pinhead square) to open boll.

9 Summary

Cotton yield is the result of a complex integration of competing vegetative and reproductive growth. Establishing a canopy with an LAI of 3.5 to 4 is needed to produce an optimal yield in many environments whereas a greater LAI may be necessary for top yield in high production environments. Numerous studies supported the importance of producing many flowers during early and mid-bloom. One advantage of producing many fruiting sites and flowers is that it facilitates greater compensatory growth if an excessive period of shed occurs early. Obtaining good fruit set and flower production is essential for obtaining high yield. Combinations of new genotypes and management practices that increase utilization of PAR and increase flowering may increase cotton yield in the near future.

10
Glossary of Cotton Yield Physiology Terms

Because cotton vegetative and reproductive growth differ substantially from grain crops, cotton agronomists and physiologists have established terms or expressions that have helped describe various aspects of cotton's growth and development. A brief list of these terms and their definitions are provided below:

anthesis -	the time when the flower opens.
bloom -	flower. The term bloom is usually used to indicate that the flower is at the day of anthesis.
boll -	the developing fruit (ovary) of cotton. The boll contains the capsule and carpels (which produce the seed and lint).
boll retention -	the number of mature bolls formed divided by the number of blooms. In some cases the denominator refers to the seasonal total number of blooms in other cases it may refer to a specific group of tagged blooms.
bur -	mature capsule tissue of the boll that has dried and "opened."
bract -	a leaf-like structure. Typically in cotton, three bracts subtend the flower and are attached near the receptacle. Prior to anthesis, bracts envelop the flower and are approximately 8 cm^2 each. After anthesis, bracts remain attached to the boll throughout boll development. There are usually three bracts per fruiting form.
carpel -	one cavity of a compound pistil or a cavity of the pistil (boll) that develops around the ovules.
capsule -	the outer boll wall structure, also see carpel.
cutout -	stage of growth corresponding to anthesis of the last flower likely to produce a mature boll. Based on this definition, an exact single date of cutout cannot be easily determined. However, cotton is usually considered to be within 1 week of cutout when NAWF = 5 (Bourland et al. 1992).
fruiting position (FP) -	relative location of fruit or sympodial leaf on the sympodia. First position leaves and fruit are proximal to the main stem whereas second and third

	position fruit are located distal to position one (Fig. 1). FP1 refers to a first position fruiting site, FP2 to a second position fruiting site, etc.
lint -	single cell trichomes (fibers) that initiate on the seed coat epidermis between 0 and 3 DPA. At maturity, most fibers are between 22 and 28 mm long and are composed primarily of cellulose. Short trichomes that initiate at 6 to 10 DPA are referred to as fuzz fibers and remain attached to the seed after ginning.
lint percentage -	lint weight (×100) divided by seed cotton (lint + seed) weight. Values for commercial cultivars and production systems often range from 30 to 42%.
locule -	see carpel.
micronaire -	an indirect measure of fiber thickness (linear density, roughly microgram per inch) that is strongly related to the fiber's perimeter (circumference) and maturity. Typical values range from 3.0 to 5.5.
monopodial branch (monopodia) -	indeterminate shoot growth similar to a main stem often emanating from lower main stem nodes.
nodes above cracked boll (NACB) -	the number of main stem nodes that are present apical to the uppermost harvestable FP1 boll with broken sutures and visible lint. This number is often used as an indicator for timing of defoliation.
nodes above white flower (NAWF) -	the number of main stem nodes that are present apical to the uppermost position one white bloom. NAWF is often used as indicator for remaining flowering potential and date of cutout.
Pima cotton -	*Gossypium barbadense* L., extra-long staple or Egyptian cotton
plant map -	a diagram indicating morphological location of all fruits as a function of main stem node and fruiting position (see Fig. 1).
position -	see fruiting position.
seed cotton -	unginned combination of seed plus lint.
square -	a developing flower and its subtending bracts (square refers only to prebloom or preflowering stage). The first square of the season is often visible around 40 days after planting. Anthesis occurs approximately 20 to 25 days after the square becomes visible. In cotton literature, a square is sometimes referred to as a floral bud.

square retention -	the number of floral buds at selected fruiting sites that produce flowers divided by the number of fruiting sites. To simplify collection of monitoring data, square retention data is sometimes restricted to the first fruiting position.
sympodial fruiting branch (sympodia) -	a branch originating directly from a main stem leaf axil (usually from main stem node number four or greater) that produces fruit and leaves.
upland cotton -	*Gossypium hirsutum* L.
yield -	(for most reports including this chapter) lint biomass per unit area. Others may use yield to refer to seed cotton biomass per unit area.

Acknowledgments. The author thanks Drs. C.T. Bryson, G.A. Constable, J.T. Cothren, D.B. Egli, T.A. Kerby, L. Libous-Bailey, W.R. Meredith, Jr., W.T. Pettigrew, V.O. Sadras, G.F. Sassenrath-Cole, and the anonymous reviewers for helpful comments. The author also thanks E. Keene and T. Houston for help with the figures.

References

Andries JA, Jones JE, Sloane LW, Marshall JG (1969) Effects of okra leaf shape on boll rot, yield, and other important characters of upland cotton, *Gossypium hirsutum* L. Crop Sci 9:705–710

Bader RF, Niles GA (1986) Response of short- and full-season cotton cultivars to mepiquat chloride. I. Morphological and phenological variables. In: Brown JM (ed) Proc Beltwide Cotton Conf 4–9 Jan, Las Vegas, Nevada, Natl Cotton Council of Am, Memphis, TN, pp 513–517

Baker DN, Hesketh JD (1969) Respiration and the carbon balance in cotton (*Gossypium hirsutum* L). In: Brown JM (ed) Proc Beltwide Cotton Production Research Conf 7–8 Jan, New Orleans, Louisiana, Natl Cotton Council of Am, Memphis, Tennessee, pp 60–64

Basra AS, Sarlach RS, Nayyar H, Malik CP (1990) Sucrose hydrolysis in relation to development of cotton (*Gossypium* spp) fibres. Ind J Exp Biol 28:985–988

Benedict CR (1984) Physiology. In: Kohel RJ, Lewis CF (eds) Cotton. Agron Monogr 24 ASA, CSSA, SSSA, Madison, Wisconsin, pp 151–200

Bird LS (1986) Seed quality and stand establishment. In: Mauney JR, Stewart JMcD (eds) Cotton physiology. I. The Cotton Foundation, Memphis, Tennessee, pp 543–550

Boman RK, Westerman RL (1994) Nitrogen and mepiquat chloride effects on the production of nonrank, irrigated, short-season cotton. J Prod Agric 7:70–75

Boquet DJ, Moser EB, Breitenbeck GA (1994) Boll weight and within-plant yield distribution in field-grown cotton given different levels of nitrogen. Agron J 86:20–26

Bourland FM, Oosterhuis DM, Tugwell NP (1992) Concept for monitoring the growth and development of cotton plants using main-stem node counts. J Prod Agric 5:532–538

Brook KD, Hearn AB, Kelly CF (1992) Response of cotton to damage by insect pests in Australia: pest management trials. J Econ Entomol 85:1356–1367

Brouder SM, Cassman KG (1994) Evaluation of a mechanistic model of potassium uptake by cotton in vermiculitic soil. Soil Sci Soc Am J 58:1174–1183

Brown PW, Zeiher CA, Silvertooth JC (1995) Response of upland cotton to elevated night temperatures. I. Results of field studies. In: Richter DA, Armour J (eds) Proc Beltwide Cotton Conf 4–7 Jan, San Antonio, Texas. Natl Cotton Council of Am, Memphis, Tennessee, p 1129

Bruce RR, Romkens MJM (1965) Fruiting and growth characteristics of cotton in relation to soil moisture tension. Agron J 57:135–140

Burke JJ, Mahan JR, Hatfield JL (1988) Crop-specific thermal kinetic windows in relation to wheat and cotton biomass production. Agron J 80:553–556

Burke JJ, Hatfield JL, Wanjura DF (1990) A thermal stress index for cotton. Agron J 82:526–530
Butter NS, Vir BK, Kaur G, Singh TH, Raheja RK (1992) Biochemical basis of resistance to whitefly *Bemisia tabaci* Genn. (Aleyrodidae: Hemiptera) in cotton. Trop Agric Trinidad 69: 119–122
Cassman KG (1993) Cotton. In: Bennett WF (ed) Nutrient deficiencies in crop plants. Amer Phytopathol Soc Press, St Paul, Minnesota, pp 111–119
Cassman KG, Kerby TA, Roberts BA, Bryant DC, Brouder SM (1989) Differential response of two cotton cultivars to fertilizer and soil potassium. Agron J 81:870–876
Cathey GW (1986) Physiology of defoliation in cotton production. In: Mauney JR, Stewart JMcD (eds) Cotton physiology I. The Cotton Foundation, Memphis, Tennessee, pp 143–154
Cathey G, Meredith WR Jr (1988) Cotton response to planting date and mepiquat chloride. Agron J 80:463–466
Christiansen MN, Rowland R (1986) Germination and stand establishment. In: Mauney JR, Stewart JMcD (eds) Cotton physiology I. The Cotton Foundation, Memphis, Tennessee, pp 535–541
Conner JW, Krieg DR, Gipson JR (1972) Accumulation of simple sugars in developing cotton bolls as influenced by night temperatures. Crop Sci 12:752–754
Constable GA (1991) Mapping the production and survival of fruit on field-grown cotton. Agron J 83:374–378
Constable GA, Rawson HM (1980a) Effect of leaf position, expansion and age of photosynthesis, transpiration and water use efficiency of cotton. Aust J Plant Physiol 7:89–100
Constable GA, Rawson HM (1980b) Photosynthesis, respiration and transpiration of cotton fruit. Photosynthetica 14:557–563
Constable GA, Rochester IJ, Daniells IG (1992) Cotton yield and nitrogen requirement is modified by crop rotation and tillage method. Soil Tillage Res 23:41–59
Cook CG, El-Zik KM (1993) Fruiting and lint yield of cotton cultivars under irrigated and non-irrigated conditions. Field Crops Res 33:411–421
Cornish K, Radin JW, Turcotte EL, Lu Z, Zeiger E (1991) Enhanced photosynthesis and stomatal conductance of Pima cotton (*Gossypium barbadense* L.) bred for increased yield. Plant Physiol 97:484–489
Culp TW (1994) Genetic contributions to yield in cotton. In: Slafer GA (ed) Genetic improvement of field crops. Marcel Dekker, New York, pp 321–360
DeLanghe EAL (1986) Lint development. In: Mauney JR, Stewart JMcD (eds) Cotton physiology I. The Cotton Foundation, Memphis, Tennessee, pp 325–349
Deussen H (1992) Improved cotton fiber properties – the textile industry's key to success in global competition. In: Proc Cotton Fiber Cellulose: Structure, Function, and Utilization Conf Natl Cotton Council of Am, 28–31 Oct. Savannah, Georgia, pp 43–63
Ehlig CF, Donovan TJ (1972) Rates of flower production and boll retention by cotton on single and double row beds. In: Proc Beltwide Cotton Prod Res Conf, Memphis, Tennessee, 10–12 Jan, Natl Cotton Council of Am, Memphis Tennessee, pp 40–42
Elmore CD (1973) Contributions of the capsule wall and bracts to the developing cotton fruit. Crop Sci 13:751–752
Elmore CD, Hesketh JD, Muramoto H (1967) A survey of rates of leaf growth, leaf aging and leaf photosynthetic rates. J Ariz Acad Sci 4:215–219
El-Sharkaway M, Hesketh J, Muramoto H (1965) Leaf photosynthetic rates and other growth characteristics among 26 species of *Gossypium*. Crop Sci 5:173–175
El-Zik KM, Thaxton PM (1989) Genetic improvement for resistance to pests and stresses in cotton. In: Frisbie RE, El-Zik KM, Wilson LT (eds) Integrated pest management systems and cotton production. John Wiley, New York, pp 192–224
Fernandez CJ, Cothren JT, McInnes KJ (1992) Carbon and water economies of well-watered and water-deficient cotton plants treated with mepiquat chloride. Crop Sci 32:175–180
Galau GA (1986) Differential gene activity in cotton embryogenesis. In: Mauney JR, Stewart JMcD (eds) Cotton physiology I. The Cotton Foundation, Memphis, Tennessee, pp 425–439

Gerik TJ, Faver KL, El-Zik K, Gannaway JW, Thaxton PM (1995a) Relationship between carbon isotope discrimination and cotton growth and leaf gas exchange. In: Richter DA, Armour J (eds) Proc Beltwide Cotton Conf, 4–7 Jan, San Antonio, Texas, Natl Cotton Council of Am, Memphis, Tennessee, p 1127

Gerik TJ, Faver KL, El-Zik K, Gannaway JW, Thaxton PM (1995b) Stability of carbon isotope discrimination over diverse cultivars and environments. In: Richter DA, Armour J (eds) Proc Beltwide Cotton Conf, 4–7 Jan, San Antonio, Texas, Natl Cotton Council of Am, p 1127

Gipson JR, Joham HE (1968) Influence of night temperature on growth and development of cotton (*Gossypium hirsutum* L.) I. Fruiting and boll development. Agron J 60:292–295

Guinn G (1974) Abscission of cotton floral buds and bolls as influenced by factors affecting photosynthesis and respiration. Crop Sci 14:291–293

Guinn G (1982) Causes of square and boll shedding in cotton. US Department of Agriculture Tech Bull 1672, Washington, D.C., 21 pp

Guinn G (1985) Fruiting of cotton. III. Nutritional stress and cutout. Crop Sci 25:981–985

Guinn G, Brummett DL (1987) Concentrations of abscisic acid and indoleacetic acid in cotton fruits and their abscission zones in relation to fruit retention. Plant Physiol 83:199–202

Guinn G, Brummett, DL (1988) Changes in abscisic acid and indoleacetic acid before and after anthesis relative to changes in abscission rates of cotton fruiting forms. Plant Physiol 87:629–631

Guinn G, Mauney JR (1984) Fruiting of cotton. II. Effects plant moisture status and active boll load on boll retention. Agron J 76:94–98

Guinn G, Mauney JR, Fry KE (1981) Irrigation scheduling and plant population effects on growth, bloom rates, boll abscission, and yield of cotton. Agron J 73:529–534

Gwathmey CO, Wassel OM, Michaud CE (1995) Pix effects on canopy light interception by contrasting cotton varieties. In: Richter DA, Armour J (eds) Proc Beltwide Cotton Conf, 4–7 Jan, San Antonio, Texas, Natl Cotton Council of Am, Memphis, Tennessee, p 1153

Haigler CH, Rao NR, Roberts EM, Huang J-Y, Upchurch DR, Trolinder NL (1991) Cultured ovules as models for cotton fiber development under low temperatures. Plant Physiol 95:88–95

Hake K, Stair K (1994) Yield probability forecasts using plant monitoring. In: Herber DJ, Richter DA (eds) Proc Beltwide Cotton Conf 5–8 Jan, San Diego, California, Natl Cotton Council of Am, Memphis, Tennessee, p 1279

Hake SJ, Kerby TA, Hake KD (1996a) Cotton production. ANR Publications, University of California, Oakland, California, 417 pp

Hearn AB (1980) Water relationships in cotton. Outlook Agric 10:159–166

Hearn AB, Constable GA (1984) Cotton. In: Goldsworthy PR, Fisher NM (eds) The physiology of tropical food crops. John Wiley, Bath Avon, pp 495–527

Hearn AB, Fitt GP (1992) Cotton cropping systems. In: Pearson CJ (ed) Ecosystems of the world vol 18. Field crop ecosystems. Elsevier, Amsterdam, pp 85–142

Heindl JC, Brun WA (1983) Light and shade effects on abscission and ^{14}C-photoassimilate partitioning among reproductive structures in soybean. Plant Physiol 73:434–439

Heitholt JJ (1993) Cotton boll retention and its relationship to lint yield. Crop Sci 33:486–490

Heitholt JJ (1994) Canopy characteristics associated with deficient and excessive cotton plant densities. Crop Sci 34:1291–1297

Heitholt JJ (1995) Cotton flowering and boll retention in different planting configurations and leaf shapes. Agron J 87:994–998

Heitholt JJ (1997) Floral bud removal from specific fruiting positions in cotton: Effect on yield and fiber quality. Crop Sci 37:826–832

Heitholt JJ, Schmidt JH (1994) Receptacle and ovary assimilate concentrations and subsequent boll retention in cotton. Crop Sci 34:125–131

Heitholt JJ, Egli DB, Leggett JE (1986) Characteristics of reproductive abortion in soybean. Crop Sci 26:589–595

Heitholt JJ, Pettigrew WT, Meredith WR Jr (1992) Light interception and lint yield of narrow row cotton. Crop Sci 32:728–733

Heitholt JJ, Pettigrew WT, Meredith WR Jr (1993) Growth, boll opening rate, and fiber properties of narrow-row cotton. Agron J 85:590–594
Hendrix DL (1992) Influence of elevated CO_2 on leaf starch of field grown. In: Hendrey GR (ed) FACE: free-air CO_2 enrichment for plant research in the field. Crit Rev Plant Sci 11:223–226
Hendrix DL, Grange RI (1991) Carbon partitioning and export from mature cotton leaves. Plant Physiol 95:228–233
Hendrix DL, Huber SC (1986) Diurnal fluctuations in cotton leaf carbon export, carbohydrate content, and sucrose synthesizing enzymes. Plant Physiol 81:584–586
Hendrix DL, Mauney JR, Kimball BA, Lewin K, Nagy J, Hendrey GR (1994) Influence of elevated CO_2 and mild water stress on nonstructural carbohydrates in field-grown cotton tissues. Agric For Meteorol 70:153–162
Hileman DR, Bhattacharya NC, Ghosh PP, Biswas PK Lewin KF, Hendrey GR (1992) Responses of photosynthesis and stomatal conductance to elevated carbon dioxide in field grown cotton. In: Hendrey GR (ed) FACE: free-air CO_2 enrichment for plant research in the field. Crit Rev Plant Sci 11:227–231
Hodges SC (1992) Nutrient deficiency disorders. In: Hillocks RS (ed) Cotton diseases. CAB International, Wallingford, England, pp 355–403
Holt SJ, Stewart JMcD, McNew RW (1994) Flower bud development in greenhouse-grown cotton. Crop Sci 34:973–976
Idso SB, Kimball BA, Mauney JR (1988) Effects of atmospheric CO_2-enrichment on root:shoot ratio of carrot, radish, cotton, and soybean. Agric Ecosyst Environ 21:293–299
Jenkins JN, McCarty JC Jr, Parrott WL (1990a) Effectiveness of fruiting sites in cotton: yield. Crop Sci 30:365–369
Jenkins JN, McCarty JC Jr, Parrott WL (1990b) Effectiveness of fruiting sites in cotton: boll size and boll set percentage. Crop Sci 30:857–860
Jiang H, Egli DB (1995) Soybean seed number and crop growth rate during flowering. Agron J 87:264–267
Jones MA, Wells R, Guthrie DS (1996a) Cotton response to seasonal patterns of flower removal: I. Yield and fiber quality. Crop Sci 36:633–638
Jones MA, Wells R, Guthrie DS (1996b) Cotton response to seasonal patterns of flower removal: II. Growth and dry matter allocation. Crop Sci 36:639–645
Karami E, Krieg DR, Quisenberry JE (1980) Water relations and C-14 assimilation of cotton with different leaf morphology. Crop Sci 20:421–426
Kasperbauer MJ (1994) Cotton plant size and fiber development responses to FR/R ratio reflected from soil surface. Physiol Plant 91:317–321
Keisling TC, Smith CW (1994) Research plot lengths for spindle harvested cotton. Commun Soil Sci Plant Anal 25:2981–2988
Kennedy CW, Smith WC Jr, Jones JE (1986) Effect of early season square removal on three leaf types of cotton. Crop Sci 26:139–145
Kerby TA (1985) Cotton response to mepiquat chloride. Agron J 77:515–518
Kerby TA, Adams F (1985) Potassium nutrition of cotton. In: Potassium in agriculture. American Society of Agronomy, Madison, Wisconsin
Kerby TA, Buxton DR (1981) Competition between adjacent fruiting forms in cotton. Agron J 73:867–871
Kerby T, Hake K (1996) Monitoring cotton's growth. In: Hake SJ, Kerby TA, Hake K (eds) Cotton Production Manual. ANR Publications, University of California, Oakland, California, pp 335–355
Kerby TA, Buxton DR, Matsuda K (1980) Carbon source-sink relationship within narrow-row cotton canopies. Crop Sci 20:208–213
Kerby TA, Hake K, Keeley M (1986) Cotton fruiting modification with mepiquat chloride. Agron J 78:907–912
Kerby TA, Keeley M, Johnson S (1987) Growth and development of acala cotton. Univ of Calif, Agric Exp Stn, Div Natur Resour, Bull 1921, 13 pp

Kerby TA, Cassman KG, Keeley M (1990a) Genotypes and plant densities for narrow-row cotton systems. I. Height, nodes, earliness, and location of yield. Crop Sci 30:644–649

Kerby TA, Cassman KG, Keeley M (1990b) Genotypes and plant densities for narrow-row cotton systems. II. Leaf area and dry-matter partitioning. Crop Sci 30:649–653

Kerby TA, Supak J, Banks JC, Snipes CE (1992) Timing defoliations using nodes above cracked boll. In: Herber DJ, Richter DA (eds) Proc Beltwide Cotton Conf, Nashville, Tennessee, Natl Cotton Council of Am, pp 155–156

Kerby TA, Weir B, Keeley M (1996) The uses of pix. In: Hake SJ, Kerby TA, Hake KD (eds) Cotton production manual. ANR Publications, University of California, Oakland, California, pp 294–304

Kimball BA, Mauney JR (1993) Response of cotton to varying CO_2, irrigation, and nitrogen on yield and growth. Agron J 85:706–712

Kloth RH (1992) Variability of malate dehydrogenase among cotton cultivars with differing fiber traits. Crop Sci 32:617–621

Krieg DR (1986) Feedback control and stress effects on photosynthesis. In: Mauney JR, Stewart JMcD (eds) Cotton physiology I. The Cotton Foundation, Memphis, Tennessee, pp 227–243

Lee JA (1984) Cotton as a world crop. In: Kohel RJ, Lewis CF (eds) Cotton. Agronomy Monograph 24. ASA-CSSA-SSSA, Madison, Wisconsin, pp 6–25

Lu Z, Zeiger E (1994) Selection for higher yields and heat resistance in Pima cotton has caused genetically determined changes in stomatal conductances. Physiol Plant 92:273–278

Lu Z, Radin JW, Turcotte EL, Percy R, Zeiger E (1994) High yields in advanced lines of Pima cotton are associated with higher stomatal conductance, reduced leaf area and lower leaf temperature. Physiol Plant 92:266–272

Mauney JR (1966) Floral initiation in upland cotton *Gossypium hirsutum* L. in response to temperatures. J Exp Bot 17:452–459

Mauney JR (1986) Carbohydrate production and distribution in cotton canopies. In: Mauney JR, Stewart JMcD (eds) Cotton physiology I. The Cotton Foundation, Memphis, Tennessee, pp 183–191

Mauney JR, Henneberry TJ (1976) Plant bug damage and shed of immature cotton squares in Arizona. In: Brown JM (ed) Proc Beltwide Cotton Conf, 9–11 Jan, Dallas Texas, Natl Cotton Council of Am, pp 41–42

Mauney JR, Phillips LL (1963) Influence of daylength and night temperature on flowering of *Gossypium*. Bot Gaz 124:278–283

Mauney JR, Stewart JMcD (eds) (1986) Cotton physiology I. The Cotton Foundation, Memphis, Tennessee, 786 pp

Mauney JR, Fry KE, Guinn G (1978) Relationship of photosynthetic rate to growth and fruiting of cotton, soybean, sorghum, and sunflower. Crop Sci 18:259–263

Mauney JR, Kimball BA, Pinter PJ Jr, LaMorte RL, Lewin KF, Nagy J, Hendrey GR (1994) Growth and yield of cotton in response to a free-air carbon dioxide enrichment (FACE) environment. Agric For Meteorol 70:49–67

McConnell JS, Baker WH, Frizzell BS, Varvil JJ (1992) Response of cotton to nitrogen fertilization and early multiple applications of mepiquat chloride. J Plant Nutr 15:457–468

McConnell JS, Baker WH, Miller DM, Frizzell S, Varvil JJ (1993) Nitrogen fertilization of cotton cultivars of differing maturity. Agron J 85:1151–1156

McDaniel RG (1995) Carbon isotopes and cotton breeding. In: Richter DA, Armour J (eds) Proc Beltwide Cotton Conf, 4–7 Jan, San Antonio, Texas, Natl Cotton Council of Am, p 1127

McMichael BL (1986) Growth of roots. In: Mauney JR, Stewart JMcD (ed) Cotton physiology I. The Cotton Foundation, Memphis, Tennessee, pp 29–38

McMichael BL, Guinn G, Fry KE (1980) The effects of moisture deficits on square shedding. In: Brown JM (ed) Proc Beltwide Cotton Conf 6–10 Jan. St Louis, Missouri, Natl Cotton Council of Am, Memphis, Tennessee, p 38

Meredith WR Jr (1984a) Quantitative genetics. In: Kohel RJ, Lewis CF (eds) Cotton. Agronomy Monograph 24, ASA-CSSA-SSSA, Madison, Wisconsin, pp 131–150

Meredith WR Jr (1984b) Influence of leaf morphology on lint yield of cotton - enhancement by the sub-okra trait. Crop Sci 24:855–857

Meredith WR Jr (1990) Yield and fiber quality potential for second-generation cotton hybrids. Crop Sci 30:1045–1048

Meredith WR Jr (1991) Contributions of introductions to cotton improvements. In: Use of plant introductions in cultivar development, part 1. CSSA Spec Publ 17, Crop Science Society of Am, Madison, Wisconsin, pp 127–146

Meredith WR Jr, Bridge RR (1971) Breakup of linkage blocks in cotton, *Gossypium hirsutum* L. Crop Sci 11:695–698

Meredith WR Jr, Wells R (1987) Sub-okra leaf influence on cotton yield. Crop Sci 27:47–48

Meredith WR Jr, Wells R (1989) Potential for increasing cotton yields through enhanced partitioning to reproductive structures. Crop Sci 29:636–639

Meredith WR Jr, Ramey HH, Bridge RR (1975) Determination of yield, yield components, and fiber properties as influenced by selective and nonselective samplings. Crop Sci 15:432–435

Meredith WR Jr, Pettigrew WT, Heitholt JJ (1996) Sub-okra, semi-smoothness, and nectariless effect on cotton performance. Crop Sci 36:22–25

Micinski S, Colyer PD, Nguyen KT, Koonce KL (1992) Cotton white flower counts and yield with and without early-season pest control. J Prod Agric 5:126–130

Mullins GL, Burmester CH (1990) Dry matter, nitrogen, phosphorus, and potassium accumulation by four cotton cultivars. Agron J 82:729–736

Munro JM (1995) Cotton and its production. In: Matthews GA, Tunstall JP (eds) Insect pests of cotton. CAB International, Wallingford, UK, pp 3–26

Myers RL, Brun WA, Brenner ML (1987) Effect of raceme-localized supplemental light on soybean reproductive abscission. Crop Sci 27:273–277

Peng S, Krieg DR (1991) Single leaf and canopy photosynthesis response to plant age in cotton. Agron J 83:704–708

Pettigrew WT (1994) Source-to-sink manipulations effects on cotton lint yield and yield components. Agron J 86:731–735

Pettigrew WT, Meredith WR Jr (1994) Leaf gas exchange parameters vary among cotton genotypes. Crop Sci 34:700–705

Pettigrew WT, Heitholt JJ, Meredith WR Jr (1992) Early season floral bud removal and cotton growth, yield, and fiber quality. Agron J 84:209–214

Pettigrew WT, Heitholt JJ, Vaughn KC (1993) Gas exchange differences and comparative anatomy among cotton leaf-type isolines. Crop Sci 33:1295–1299

Radin JW, Mauney JR (1986) The nitrogen stress syndrome. In: Mauney JR, Stewart JMcD (eds) Cotton physiology I. The Cotton Foundation, Memphis, Tennessee, pp 91–105

Radin JW, Reaves LL, Mauney JR, French OF (1992) Yield enhancement in cotton by frequent irrigations during fruiting. Agron J 84:551–557

Radin JW, Lu A, Percy RG, Zeiger E (1994) Genetic variability for stomatal conductance in Pima cotton and its relation to improvements of heat adaptation. Proc Natl Acad Sci 91:7217–7221

Reddy VR, Reddy KR, Baker DN (1991) Temperature effect on growth and development of cotton during the fruiting period. Agron J 83:211–217

Roberts BA, Curley RG, Kerby TA, Wright SD, Mayfield WD (1996) Defoliation, harvest, and ginning. In: Hake SJ, Kerby TA, Hake KD (eds) Cotton production manual. ANR Publications, University of California, Oakland, California, pp 305–323

Rosenthal WD, Gerik TJ (1991) Radiation use efficiency among cotton cultivars. Agron J 83:655–658

Sadras VO (1994) Compensatory growth in cotton after loss of reproductive organs. Field Crops Res 40:1–18

Sadras VO (1996a) Population-level compensation after loss of vegetative buds: interactions among damaged and undamaged cotton neighbours. Oecologia 106:417–423

Sadras VO (1996b) Cotton compensatory growth after loss of reproductive organs as affected by availability of resources and duration of recovery period. Oecologia 106:432–439

Saleem MB, Buxton DR (1976) Carbohydrate status of narrow row cotton as related to vegetative and fruit development. Crop Sci 16:523–526

Sassenrath-Cole GF (1995) Dependence of canopy light distribution on leaf and canopy structure for two cotton (*Gossypium*) species. Agric For Meteorol 77:55–72

Singh M, Singh VP, Bhardwaj SN (1992) Breeding suitable plant type for high productivity in upland cotton (*Gossypium hirsutum*). Ind J Agric Sci 62:319–324

Snipes CE, Baskin CC (1994) Influence of early defoliation on cotton yield, seed quality, and fiber properties. Field Crops Res 37:137–143

Stewart JMcD (1986) Integrated events in the flower and fruit. In: Mauney JR, Stewart JMcD (ed) Cotton physiology I. The Cotton Foundation, Memphis, Tennessee, pp 261–300

Supak JR, Kerby TA, Banks JC, Snipes CE (1993) Use of plant monitoring to schedule chemical crop termination. In: Herber DJ, Richter DA (eds) Proc Beltwide Cotton Conf, Jan, New Orleans, Louisiana, Natl Cotton Council of Am, pp 1194–1196

Thomson NJ (1995) Commercial utilisation of the okra leaf mutant of cotton - the Australian experience. In: Constable GA, Forrester NW (eds) Challenging the future. Proc World Cotton Research Conf I. Brisbane, Queensland, 24–28 Feb 1994, pp 393–401

Trelease RN, Miernyk JA, Choinski JS Jr, Bortman SJ (1986) Synthesis and compartmentation of enzymes during cottonseed maturation. In: Mauney JR, Stewart JMcD (eds) Cotton physiology I. The Cotton Foundation, Memphis, Tennessee, pp 441–462

USDA (1995) Agricultural chemical usage. 1994 Field crops summary. USDA, NASS, ERS, Washington DC, March, 106 pp

Wafler U, Meier H (1994) Enzyme activities in developing cotton fibres. Plant Physiol Biochem 32:697–702

Walter H, Gausman HW, Rittig FR, Namken LN, Escobar DE, Rodriguez RR (1980) Effect of mepiquat chloride on cotton leaf and canopy structure and dry weights of its components. In: Brown JM (ed) Proc Beltwide Cotton Conf, Natl Cotton Council of Am, Memphis, Tennessee, pp 32–35

Wanjura DF (1986) Field environment and stand establishment. In: Mauney JR, Stewart JMcD (eds) Cotton physiology I. The Cotton Foundation, Memphis, Tennessee, pp 551–554

Warner DA, Burke JJ (1993) Cool night temperatures alter leaf starch and photosystem II chlorophyll fluorescence in cotton. Agron J 85:836–840

Warner DA, Holaday AS, Burke JJ (1995) Response of carbon metabolism to night temperature in cotton. Agron J 87:1193–1197

Wells R (1989) Measurements of lint production in cotton and factors affecting yield. Mod Methods Plant Anal 10:278–294

Wells R (1995) Seasonal patterns of carbohydrate storage in cotton. In: Richter DA, Armour J (eds) Proc Beltwide Cotton Conf, 4–7 Jan, San Antonio, Texas, Natl Cotton Council of Am, p 1137

Wells R, Meredith WR Jr (1984a) Comparative growth of obsolete and modern cotton cultivars. I. Vegetative dry matter partitioning. Crop Sci 24:858–862

Wells R, Meredith WR Jr (1984b) Comparative growth of obsolete and modern cotton cultivars. III. Relationship of yield to observed growth characteristics. Crop Sci 24:868–872

Wells R, Meredith WR Jr (1986) Normal vs. okra-leaf yield interactions in cotton. II. Analysis of vegetative and reproductive growth. Crop Sci 26:223–228

Wells R, Meredith WR Jr, Williford JR (1986) Canopy photosynthesis and its relationship to plant productivity in near-isogenic cotton lines differing in leaf morphology. Plant Physiol 82:635–640

Wells R, Meredith WR Jr, Williford JR (1988) Heterosis in upland cotton. II. Relationship to leaf area to plant photosynthesis. Crop Sci 28:522–525

Williford JR (1992) Production of cotton on narrow row spacing. Trans ASAE 35:1109–1112

Wilson FD (1986) Pink bollworm resistance, lint yield, and lint yield components of okra-leaf cotton in different genetic backgrounds. Crop Sci 26:1164–1167

Wilson FD (1990) Relative resistance of cotton lines to pink bollworm. Crop Sci 30:500–504
Wullschleger SD, Oosterhuis DM (1990a) Photosynthesis of individual field-grown cotton leaves during ontogeny, Photosynth Res 23:163–170
Wullschleger SD, Oosterhuis DM (1990b) Canopy development and photosynthesis of cotton as influenced by nitrogen nutrition. J Plant Nutr 13:1141–1154
Wullschleger SD, Oosterhuis DM, Hurren RG, Hanson PJ (1991) Evidence for light-dependent recycling of respired carbon dioxide by the cotton fruit. Plant Physiol 97:574–579
York AC (1983a) Cotton cultivar response to mepiquat chloride. Agron J 75:663–667
York AC (1983b) Response of cotton to mepiquat chloride with varying N rates and plant populations. Agron J 75:667–672

Jute

P. PALIT[1]

1 Introduction

Jute is second only to cotton in importance among vegetable fibres. The bulk of the world's supplies of jute is grown in India, Bangladesh, China, Thailand and Nepal (Anonymous 1995). The two cultivated species *Corchorus capsularis* L. and *C. olitorius* L. of the family Tiliaceae yield jute, which is a secondary phloem fibre obtained from the bark of the stem. Properly extracted fibres are fairly lustrous, creamy white to brownish in colour with moderate tenacity (20–25 $g\,tex^{-1}$) but rough to touch. Unlike cotton, jute is a multicellular and multiconstitutent fibre. Apart from α cellulose ($\approx$60%) the other two major constituents are hemicellulose ($\approx$23%) and lignin ($\approx$14%). Due to its biodegradibility and annual renewability, jute fibre has diversified commercial and household applications (Table 1).

Historical records show that jute fibre (*Patta* in Sanskrit) was in use in ancient India (Watt 1889). Towards the end of the 18th century, Europeans, in their search for flax, started using Indian jute for packaging purposes (Kundu et al. 1959). There is considerable disagreement on the origin and natural distribution of the two cultivated species. *Corchorus capsularis* is found wild in many parts of India and China. Kundu (1951) concluded that it is native to the Indo-Burma region, including South China. It is not found in Africa and Australia. In contrast, *C. olitorius* has been recorded to occur in North Australia, Asia and Africa. Kundu (1951, 1956) proposed Africa as its primary centre of origin with India or Indo-Burma as a secondary centre. This species has, however, become naturalized in most tropical regions (Edmonds 1990). Both the cultivated species are diploid ($2n = 14$) and exhibit regular meiosis (Sharma and Roy 1958; Datta 1971). They are essentially self-pollinated with average natural crossing of 2–5% in *C. capsularis* and 5–10% in *C. olitorius* (Dutt and Ghosh 1962). This implies that jute varities are mostly pure lines, and chances of having large genetic variability within the selected varieties are low. Jute crop improvement through intervarietal hybridization involving distant geogrpahic types resulted in a number of cultivars with many agronomically and commercially desirable traits. Induction of polyploidy, mutation

[1] Central Research Institute for Jute and Allied Fibre, ICAR, Barrackpore, 24 Parganas, West Bengal 743 101, India

Crop Yield, Physiology and Processes
D. L. Smith, C. Hamel (Eds.)
© Springer-Verlag Berlin Heidelberg 1999

Table 1. Use of jute in commerce and industry. (IJO, Dhaka, Bangladesh)

Traditional uses	Environmental management applications	Timber substitutions	Valuable byproducts
Packaging	*Geotextiles*		*Jute sticks*
Hessian, sacks and bags	Soil erosion	Paper pulp	Fences
Rugs	Agricultural mulching	Cardboard	Walls
Fabrics	Silt fences	Particle board	Cooking fuel
Carpet backings	Drainage	Wall papers/coverings	Tender leaves as vegetables
Felts	Highway construction	Plastic reinforcement	
Upholstery		Jute composites	
Handicrfats			

breeding or interspecific hybridization has not yet produced any distinctly high-yielding types (Ghosh 1963).

The aspects of jute physiology have been comprehensively discussed earlier by Sen Gupta (1953a,b) and Kar (1959). Dealing with the physiological basis of jute growth and yield necessitates a primary consideration of some intrinsic aspects of its cultivation and yield determinants. They are:

1. The time of crop harvest for economic yield (stem fibre) and that for the means of propagation (seeds) is spaced at an interval of 6 weeks or more due to specific photoperiod requirements of vegetative growth favoured by long-day conditions and reproductive growth by short-day. Therefore for commercial purposes separate crops are raised for fibre (March–September) and for seed (June–December) in the jute-growing regions of the Northern Hemisphere.
2. The secondary phloem tissue which determines the yield capacity of jute arises from cambial activity that starts within 3–4 days after seed germination (Ghosh et al. 1943). Though cambial activity continues till fruiting, the commencement of flowering which coincides with cessation of growth in length of the vegetative stem greatly reduces the quality of the fibre (Kundu 1953; Johansen et al. 1985a). Selection programmes are thus biased against early flowering types.
3. To obtain a good harvest of slender, unbranched vegetative stems, the plants are grown at high density (400–500 thousand plants ha^{-1}) in a rectangle of nearly 1:4 (Ghosh 1983).
4. After harvesting, the stems are immersed in water for retting – a process of separating the embedded fibre from the stem bark through partial rotting brought about by the natural microbes present in the retting water. This process has a strong influence on the quality of jute fibre (Bhattacharya 1974).

2
Seed: Viability and Germination

Jute seeds are albuminous and very small in size. The seeds of the two cultivated species differ in size; while 1 g of properly dried *C. capsularis* seeds contains 289 ± 7 seeds, the same contains 555 ± 20 seeds of *C. olitorius* (Palit and Bhattacharyya 1981). Although endosperm and seed coat occupy the major portion of the seed, it is the size of the embryo which shows more positive correlation with the seedling weight and yield within a species (Saha and Basak 1982). An initial capital gain by larger embryos may help to establish larger individuals and hence greater yield.

Good-quality jute seeds are raised in areas where there is no rainfall during their maturity. As the seeds of cultivated species have no dormancy (Kundu and Sarma 1955; Basak and Pal 1975), rain during their maturation often initiates germination in situ, bursting the pod open and facilitating the spread of the seed-borne fungus *Macrophomina phaseolina* (Ghosh 1983). At harvest the seeds usually contain nearly 21% moisture which, if not reduced to 7–8% by repeated sunning, leads to non-viability within a short time (Ghosh and Basak 1951). Non-viability, however, is not inflicted by oxygen tension during storage (Baso and Bhattacharyya 1974). Long-stored seeds lose viability, presumably due to the accumulation of some toxic metabolites (Jain and Saha 1971). Leaching out of autotoxic metabolites from stored seeds during soaking-drying treatment seems to be conducive to minimizing loss of viability during subsequent storage (Basu et al. 1978). The presence of various concentrations of polyamines like spermidine, putrescine and their metabolic inhibitors in germinating seeds has been reported (Pandit and Ghosh 1990) but their physiological role in maintaining seed vigour still remains unexplored. Germination and vigour of 1–3-year-old seeds can be restored by treating them with 100 ppm GA (Fahmy 1974) or thiourea (Chakraverty 1975).

The minimal water requirement for germination initiation by radicle emergence is about 150–170% of the seed dry weight. The delayed germination of *C. olitorius* seeds in the petri dish, as well as in soil, may be due to their higher water potential compared to that of *C. capsularis*, which inhibits water intake and lengthens the time needed to reach the minimal hydration level (Palit and Bhattacharyya 1981). The higher amount of soluble sugar present in the *C. capsularis* seeds (Datta 1971) perhaps regulates their osmolarity. Although Verma and Arora (1978) indicate that testing for germination is independent of light and darkness, other laboratory experiments suggest that light is necessary for maximum germination (Okusanya 1979). The seeds could germinate in up to 1% NaCl solutions, but the time elapsing before germination is increased with increasing salinity. Field experiments in saline soil areas of West Bengal with salinity ranging between 15–40 m mohs cm^{-1} showed a better germination performance by *C. capsularis* varieties than *C. olitorius* (Kar and De Sarkar 1957). The critical water potential beyond which jute seeds do not germinate is -0.3 M Pa for *C. olitorius* and -0.5 M Pa for *C. capsularis* (Palit and Bhattacharyya

1981). Decreasing water potential of the imbibing media up to the seeds' critical level does not hamper their protein metabolism but progressively decreases the ATP content and energy charge (EC). With further decrease in water potential, protein synthesis is affected, restricting the radicle growth even though the seeds remain otherwise metabolically active with EC value little above 0.5 (Palit 1987a). During water stress, seeds also accumulate free proline, which gradually disappears on relieving the stress by rewatering (Palit and Singh 1991).

3 Seedling Growth and Initial Field Operations

Jute seeds are generally oversown with four to five times the number of seeds required for the desired plant density (Ghosh 1983). This is considered necessary for the following reasons:

1. A single germinating seedling is too delicate to break through the soil crust that may form if there is rain soon after sowing, and therefore requires the help of other seedlings for emergence.
2. Since jute is directly sown (broadcast or by seed drill) in well-pulverized moderately dry soil, the tender seedlings are often subjected to long periods of high soil moisture tension which increases seedling mortality. Therefore a high seed rate ensures an adequate population.
3. The thick stand in the initial stage helps to check weed growth in the field.

The overcrowded seedlings are thinned two or three times within a span of 30 to 40 days after sowing (DAS) to obtain a uniform crop of desired density. As the rectangularity in line-sown crop is high there is a tremendous competition between plants within rows and a wide gap prevails between rows. The radiation-use efficiency (RUE) at this stage is extremely low due to irregular leaf area distribution over the field (Palit 1993). Most of the radiation falling on the gaps accelerates weed growth. A closer row planting may improve RUE but, with age, excessive plant density will increase variability in size of individuals and influence self-thinning, thus reducing the productivity (Palit et al. 1988). The frequency distribution of plant weight is approximately normal shortly after seedling emergence but as time passes the distribution tends to become increasingly positively skewed which results in a mixture of small and large individuals in the final population (Johansen et al. 1985a; Palit et al. 1988). The situation becomes more critical if the manual thinning of seedlings is not done as early as possible (Ghosh et al. 1983). Although the realitive growth rate ($g\,g^{-1}\,day^{-1}$) of both the species increases exponentially upto 40–50 DAS (Johansen et al. 1985a) the initial dry matter production is only about 0.011–0.064 $g\,plant^{-1}\,day^{-1}$ during this stage (Palit 1993). The young tender seedlings are tolerant to repeated treading of the operator for weeding and thinning but are very sensitive to moisture stress. If they are subjected to low humidity at or below 19% and available soil moisture is below 20%, the crop tends to wilt and premature flowering is induced in many of the varieties (Rahman et al. 1980; Ghosh 1983; Patel and Mandal 1983). This premature flowering due to water stress has been discussed by Sen Gupta (1953a) and Johansen et al. (1985b). They consider light quality, intensity and day length to be more dominant factors than any other environmental stresses. Similarly, water stagnation in the field also shows lasting effects on plant growth and yield (Chaudhuri and Bask 1969). It is interesting that such water tension or logging does not cause any significant growth and yield reduction if applied to older plants (P. Palit and A.C. Bhattacharyya, unpubl. data) presumably due to more rapid development of xylem elements and maintenance of a high leaf water potential (Sen Gupta 1947; Basak et al. 1973) and due to the chances of developing adventitious roots along the stem on flooding (Ghosh 1983).

4 Root System

The root system is of less importance than the stem in a dense population of unbranched jute plants. Evaluated by the ^{14}C assimilate distribution, the root growth amounts to only 10–18% of the shoot growth at various stages (Palit 1985). The roots of the two species are distinct in their lengths and extent of spreading. The tap root system in *C. capsularis* is relatively shorter with numerous branches and sub-branches which spread profusely in the upper soil layer. In *C. olitorius* the tap root is relatively longer with few but well-developed and deeply penetrated branches (Kar and De Sarkar 1961; Choudhury and Bask 1969). By using ^{32}P as tracer, roots of both the species have been shown to penetrate up to 60 cm or more into the soil and the lateral absorbing roots are confined to a radius of 10 cm (Goswami and Saha 1969). Plants of *C. olitorius*, with a deeper tap root system, show a better water-use efficiency than those of *C. capsularis* under conditions of low soil moisture (Patel and Mandal 1983). It has been shown that *C. capsularis* varieties with higher cation exchange capacity can absorb more Ca and have greater N requirements than the *C. olitorius* varieties which can absorb more K from a K-deficient soil and have lower N requirements (Gopalakrishnan et al. 1970).

Jute plants readily produce adventitious roots from any place on the stem, but particularly at the basal region under the influence of stimuli like stagnation of water, extremely humid conditions in the field or application of 2,4-D. Though these adventitious roots on the stem help to cope with the water-logged conditions, they greatly reduce the quality of stem fibre due to their endogenous origin from the secondary cambium (Ghosh 1983).

5 Leaf Area, Light Interception and Canopy Photosynthesis

Jute plants exhibit a kind of dimorphism in leaf size and shape, being larger in vegetative phase but smaller and narrower in the reproductive phase. The first true leaf is initiated at 6–8 DAS. The initial leaf area development in both the species is slow, being 3–5 $cm^2 plant^{-1} day^{-1}$ during the first 20 DAS, but once the thinning and weeding operations are completed, leaf area develops very quickly, and at around 40 DAS the LAI exceeds 4. By 60 DAS, LAI exceeds 6, and 100% canopy closure is achieved. LAI further increases to about 8 before being levelled off at a value around 4 during 100–120 DAS (Fig. 1). The relative rate of new leaf emergence at the apical region is proportional to the relative rate of plant height increase. After emergence, each leaf continues to grow in size for 10–15 days. The fully expanded lamina of *C. olitorius* is 18–20 cm length and 7.5–8 cm in maximum breadth. The corresponding size of *C. capsularis* is 16–17 cm by 6–7 cm. The petiole length varies from 4–8 cm in both the species (Palit and Bhattacharyya 1982). From top to bottom the leaf inclination ranges from 31° to −22°, so the jute resembles a broadleaved and almost planophile canopy with droopy lower leaves. In general, the leaves retain a considerable amount of their photosynthates and translocate only about 30% from older leaves, to about 55% from the younger ones (Palit 1985). This perhaps helps in their longer life span, 40–50 days in *C. capsularis* and 30–40 days in *C. olitorius*. Formation of a separation layer at the abscission zone is abrupt and strictly acropetalous (Sen 1968). In *C. olitorius* the rate of leaf fall from the bottom is faster than the emergence rate at the top, especially towards the later

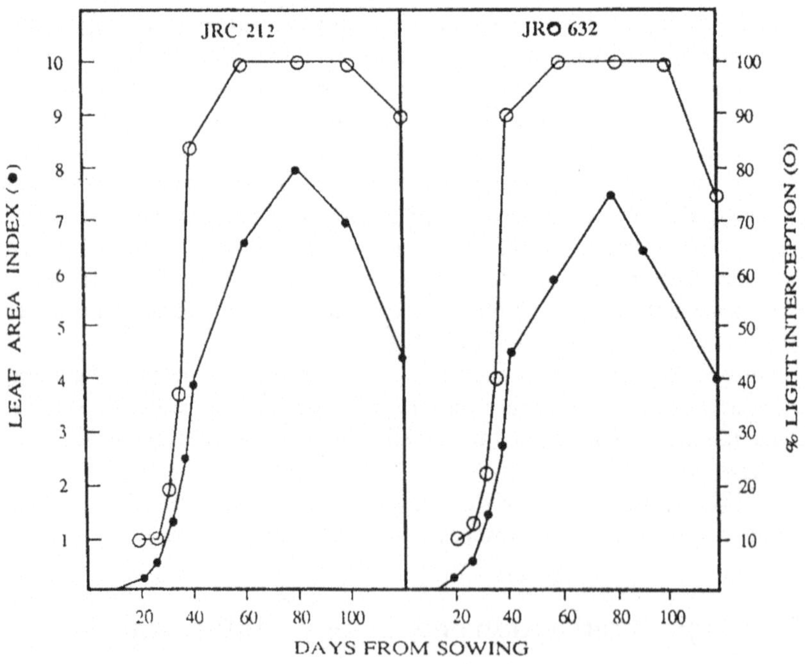

Fig. 1. Leaf area index and percent light interception by line sown crops of two jute species (*C. capsularis* cv. JRC 212 and *C. olitorius* cv. JRO 632) at different growth stages

phase of plant growth. In *C. capsularis* leaf fall at this stage is sluggish and thus it retains more leaves per plant and produces a relatively higher LAI (Palit and Bhattacharyya 1982). Presence of a greater proportion of younger leaves and the reduction of poorly photosynthesizing older leaves in *C. olitorius* is one of the reasons for its rejuvinated net assimilation rate (NAR) even in ageing plants (Gopalakrishnan and Goswami 1970; Palit and Bhattacharyya 1982).

With the increase in canopy height the leaf-bearing portion of the plants progressively reduces from about 80% at 40 DAS to only about 30% at around 100 DAS, when the larger portion of the lower canopy becomes leafless (Fig. 2). Irrespective of growth stage and crop height the leaf areas remain highest in the topmost 10–20-cm layers and lowest for the basal section. The leaf area distribution in the middle layers remains more or less uniform and together they make up the main bulk of the crop LAI. The PhAR extinction coefficient (K) recorded below the top layer at any growth stage beyond 40 DAS is close to 0.9, indicating that most of the incident light is absorbed at the upper layer. K at the bottom of the canopy, however, attenuates with plant age (Palit and Bhattacharyya 1987) and does not always agree with the usual radiation foliage relationship ($K = \cos\alpha$) (Kuroiwa and Monsi 1963). This is perhaps due to widely variable leaf inclinations below the top layer which scatter the incident light into different proportions of direct and diffused PhAR and do not allow proper estimations of K at the base of the canopy (Palit and Bhattacharyya 1987). Such irregular radiation distribution may affect the functioning of the leaves down the canopy in a complicated way (Boardman 1977). Moreover, with

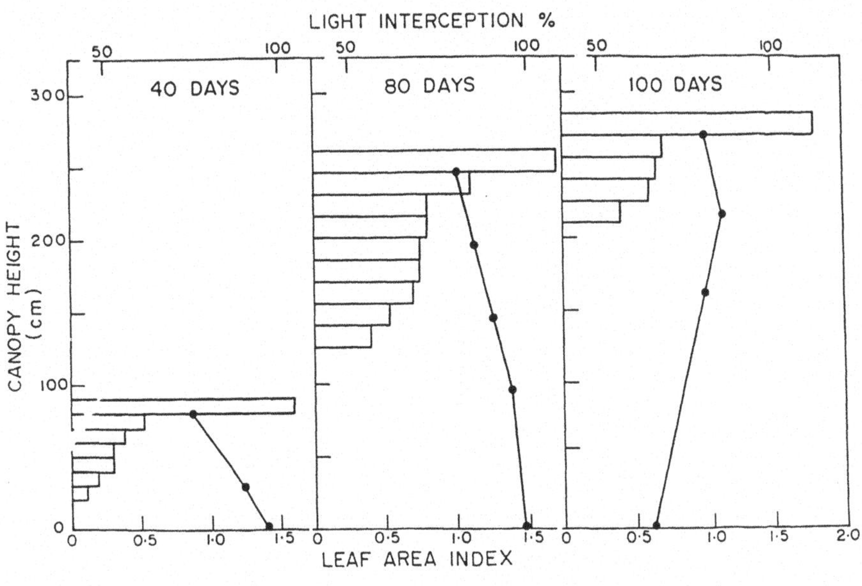

Fig. 2. Vertical distribution of leaf area (*blocks*) and PhAR interception profiles (*lines*) at different growth stages of jute canopy

high LAI (Palit 1993) during the early stages, the mean PhAR received by the lower leaves remains constantly low. The reduced net photosynthetic rate (P_N) of the lower leaves under field conditions compared to the similar leaves of isolated plants may be the result of such radiation effects. These lower leaves, however, retain considerable chlorophyll and do not shed readily, thus indicating their light limitation of photosynthetic efficiency (Palit and Bhattacharyya 1987).

Poor photsynthesis by the lower leaves results in a decreased canopy photosynthetic rate (P_C) although the maximum number of leaves are available. It shows that P_C is related more to the average efficiency of the leaves than to leaf area (Palit and Bhattacharyya 1982). An exponential decline in RGR, apparent until final harvest (Johansen et al. 1985a), is also proportional to the decline in hours of bright sunshine with the onset of monsoon conditions (Fig. 3). All these findings explain why jute growth and hence yield is limited by the poor PhAR availability within the canopy. Jute cultivars should then be selected with more uniformly inclined shorter leaves for greater light penetration down the canopy. Plants with such a leaf structure would also tolerate the high population pressure efficiently to give better yield (Palit and Bhattacharyya 1988).

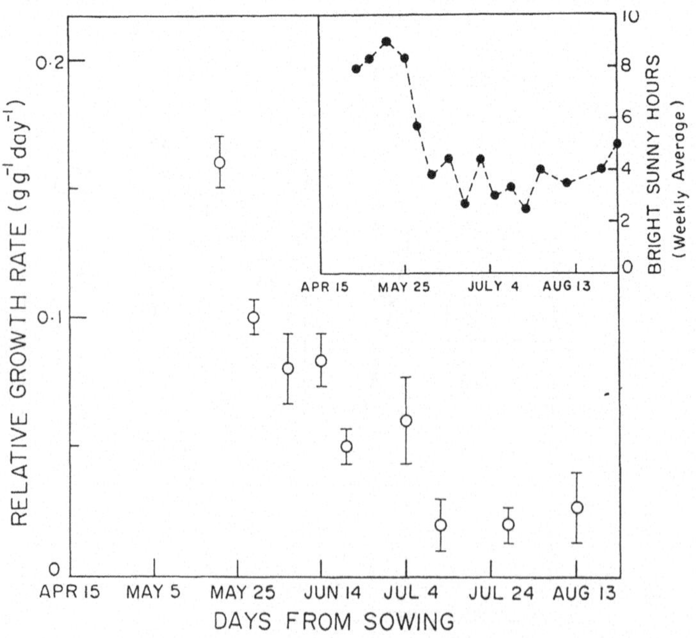

Fig. 3. Relative growth rate of jute canopy during the vegetative season in comparison with daily amounts of bright sunny hours (plants sown on 15 April)

6 CO_2 Metabolism and Carbon-Use Efficiency

Characterization of photosynthetic CO_2 metabolism has shown jute to be a C_3 plant (Palit and Bhattacharyya 1984a). The activity of RuBP carboxylase, the key enzyme for C_3 CO_2 fixation, in jute is about five times higher than its PEP carboxylase activity under the same assay condition (Table 2). These activities are comparable to that in sunflower, a C_3 dicotyledonous plant, but are quite different from maize, a C_4 plant. Anatomy of a jute leaf shows a distinct layer of bundle-sheath cells surrounding the vascular bundle (Gopalakrishnan and Saha 1977), but these cells contain very few chloroplasts (Kundu and Sen 1958) and the early and major CO_2 fixation product is 3-PGA rather than malate or aspartate (Palit and Bhattacharyya 1984a). Glycolic acid, another early product of C_3 photosynthesis and a substrate for photorespiration is rapidly oxidized in jute, which shows nearly four times more glycolate oxidase activity than maize (Table 2). The light-to-dark ratio of CO_2 evolution in CO_2 free air following photosynthesis in $^{14}CO_2$ is 2.77 in jute compared to only 0.11 in maize. All these indicate that jute loses a considerable amount of its recently fixed CO_2 through photorespiration (Palit and Bhattacharyya 1984a).

The rate of CO_2 fixation in a commercial variety (JRO 632) is equivalent to 1.5 μmol mg^{-1} chlorophyll min^{-1}. However, in an isogenic chlorophyll mutant even a one third reduction in leaf chlorophyll did not reduce this photosynthetic efficiency significantly (P. Palit, unpubl. data). The photosynthetic rates of some other chlorophyll-deficient mutants have been shown to have lower light saturation than the chlorophyll-sufficient varieties (Sen et al. 1983; Choudhury and Saha 1988). Therefore, there is scope for increasing jute photosynthesis per unit chlorophyll. Such a plant type may be ideal since the leaves will be thinner and will allow more light transmission for higher overall canopy photosynthesis.

The total respiratory CO_2 efflux from a field-grown plant, measured as percent ^{14}C loss from the whole plant for 15 days against the original amount of ^{14}C assimilated on day 0, shows a gradual decline with age (Palit 1985, 1993). This decline is proportional to the decline in protein to biomass ratio (Palit 1993), indicating a reduction in maintenance respiration (Stahl and McCree 1988). An increased proportion of non-respiring tissue (wood) to respiring

Table 2. Comparison of the activity of the photosynthetic enzymes in leaf extracts of jute, maize and sunflower. (Palit and Bhattacharyya 1984a)

Plant	PEP carboxylase	RuBP carboxylase	Glycolate oxidase
	[μmoles mg^{-1} chl min^{-1}]		
Jute	0.64	3.33	1.10
Maize	6.50	1.08	0.30
Sunflower	0.80	4.08	Not determined

tissue (green leaves and bark) with age indicates a change to a more woody life form (Schulze 1983; Palit 1993). Thus the carbon-use efficiency in jute increases from about 24% during 20–40 DAS to about 80% during 100–120 DAS. Such increased growth effiency (dry matter produced per unit respiratory loss) in a ageing plant, however, does not contribute to the yield capacity of jute.

7 Assimilate Translocation and Partitioning

The translocation of the photosynthetic product from a leaf is largely controlled by the demand in the developing sink which may be a structural or a storage organ. The storage organs ultimately constitute the yield of most crops, and improvement of yield has come largely from increases in the partitioning of the assimilates into the harvested storage organs (Gifford and Evans 1981). In contrast, only the "structural growth" of the vegetative plant is the determinant of economic yield in jute and the development of an exclusive storage organ is not apparent until flowering. Palit and Bhattacharyya (1982) have shown that assimilate production before flowering (source strength) is almost equal to its utilization for structural growth (sink strength). Of the structural materials, the major sinks are developing leaves, wood and stem bark. Since jute fibre is obtained only from the bark, the relative allocation of photosynthates to it is most important. The carbon assimilated by the plant at early stages is mostly distributed to the leaves as the development of photosynthetic surface is then most important. At around 60 DAS the development of wood and bark seems equally active as the assimilated carbon during this stage is almost equally distributed to leaves, wood and bark. At later stages most of the photosynthates accumulate in the wood, indicating the preponderance of wood formation (Table 3). The migration coefficient or the ratio of fibre weight to total plant dry weight during growth increases gradually up to 80 DAS and then becomes constant until harvest at 120 DAS in both the species. This fractional allocation of assimilates to the economic yield results in a low harvest index of 20–25% for jute (Palit and Bhattacharyya 1982; Johansen et al. 1985a). As the cambial activity normally results in the formation of several times more xylem cells than the phloem (Fahn 1982), it would be rather difficult to increase the harvest index of jute significantly. The precise control mechanism of cambial differentiation is also not clear (Aloni 1988) although Mitra (1984) demonstrated that inheritence of bast fibre in jute is monogenic and recessive. The significance of photosynthetic distribution during the second month of plant growth should be better recognized for its role in increasing fibre development as the bark-to-fibre ratio is constant (Palit and Bhattacharyya 1982). The foliar application of growth regulators (Stant 1961; Mukherjee and Datta 1964; Chakraverty 1968; Kar 1968) and N_2-fixing bacteria (Pal and Sen 1988) may also increase cambial activity and fibre yield of jute.

Table 3. Distribution of ^{14}C assimilated by jute plants at different growth stages. (Palit 1985)

Plant part	10^{-5} × radioactivity (dpm) per plant		^{14}C content (as % of total originally assimilated)	
	t = 0	t = 15	t = 0	t = 15
	Fed $^{14}CO_2$ 40 days after sowing			
Leaf	382	37	98	9.6
Wood	0	22	0	5.6
Bark	8	11	2	2.8
Root	0	8	0	2.0
Total	390	78	100	20.0
	Fed $^{14}CO_2$ 62 days after sowing			
Leaf	430	69	99	16.0
Wood	0	63	0	14.5
Bark	4	59	11	3.6
Root	0	19	0	4.4
Total	434	210	100	48.5
	Fed $^{14}CO_2$ 105 days after sowing			
Leaf	409	76	96	17.9
Wood	0	135	0	31.6
Bark	17	43	4	10.0
Root	0	55	0	12.9
Total	426	309	100	72.4

The amount of ^{14}C in each plant was calculated by multiplying dpm/mg of a part by its dry weight. The percentage of ^{14}C in each part was computed against the original amount of ^{14}C assimilated. t = 0, immediately after $^{14}CO_2$ assimilation. t = 15, 15 days after $^{14}CO_2$ assimilation.

The direction of assimilate movement from different source leaves follows the general trend found in most plants (Wardlaw 1990), that is, from upper leaves to the apical region, from the lower leaves to the lower portions including roots and from the leaves in the intermediate position to both directions (Palit 1985). In the apical region the apical meristem appears to be the strongest sink for the translocated assimilates and only the crown leaves can sustain its demand for stem growth in length. The leaves in the middle and lower portion are the essential contributors to the cambial meristem since deblading these leaves restricts the stem growth in diameter (Palit 1985). An induced mutant of *C. olitorius* with a faster rate of leaf abscission also showed similar growth restriction (Sen 1968). Therefore, jute leaves at different positions on the stem seem to supply assimilates specifically to different parts for the overall growth of the stem and the selection for better growth should be biased towards smaller but persistent leaves.

8 Photoperiod and Flowering

Both the cultivated species of jute are typical short-day-requiring plants for flowering (Sen Gupta and Sen 1946; Bose et al. 1974). The transition from the vegetative to reproductive phase is marked by forking of the apex into two or more branches. In the Northern Hemisphere, known cultivars start flowering in late August or early September irrespective of the sowing time. In general, *C. olitorius* is more responsive to short light periods than *C. capsularis*, though genotypes which flower early or late also exist in both species (Hossain and Sen 1978; Joshua and Thakare 1980; Ghosh 1983). Cold methanol extracts of the stem tips from early flowering lines have been found to induce earliness in a late-flowering variety (Islam et al. 1980). Photoperiod stimulus can be induced soon after germination (Sen Gupta and Sen 1949). The critical light period is 12.5h (Sen Gupta 1953a; Fawusi and Ormrod 1981). In *C. olitorius* four short-day cycles of 10h natural day light followed by 14h darkness have been found sufficient to induce flowering. The number of flowers and fruits produced on a plant increases as the number of short-day cycles is increased from 3 to 30. Plants given 3 to 7 short-day cycles produce flowers on the main stem only, but when they are maintained in short days for longer periods, flowers are also produced on the branches (Nwoke 1980). The inflorescence is extraaxillary and flowering occurs in periodic flushes within a span of 3–6 weeks in *C. capsularis* and about 10 weeks in *C. olitorius*. The seeds formed during the first half of the flowering range show higher germination percentage than those formed in the last week of the flowering period. An optimum seed setting with optimum germination percentage is found during the period of maximum flush of flowering (Kar 1959).

9 Improvement of Jute Productivity: Problems and Prospects

It is evident from Table 1 that each part of the jute plant has commercial value. Therefore improvement of jute productivity should not always be considered in terms of fibre only. High biomass production is equally important for paper and other industries. Considering an average photon flux density (PFD) during April–August in a typical jute-growing locality (22°50′N, 88°22′E) equal to 39.87 $mol\,m^{-2}day^{-1}$, photosynthetic efficiency of 1.25 $g\,mol^{-1}$ and 53% respiratory loss (cf. Table 3) a daily potential biomass production of 49.77 $g\,m^{-2}day^{-1}$ has been predicted (Palit 1993). As this daily potential productivity (DPP) tends to be proportional to the amount of solar radiation intercepted by the crop's leaf surface at different growth stages (cf. Fig. 1) the total potential primary productivity (TPP) of jute may be calculated to be 40.81 $t\,ha^{-1}$ using the equation

$$TPP = DPP \sum_{f,d} fd,$$

where f is the fraction of intercepted radiation and d is the duration of each f (Palit 1993). With a 20% harvest index the potential fibre productivity would be 8.16 $t\,ha^{-1}$ provided other yield-reducing environmental stresses are absent. Under similar conditions, commercial cultivars have produced fibre yield of 3.3–3.9 $t\,ha^{-1}$ amounting to 40–48% of the potential (Ghosh 1983). In a situation where the RUE of a closed canopy exceeds 3 g (dry matter) MJ^{-1} (Palit 1993), further increase in primary production in a field-grown crop would be difficult (Monteith 1972; Gosse et al. 1986). Only the genotypes with high RUE before canopy closure may improve biomass production. In a sense, such genotypes will be faster-growing, have shorter field life and fit in location-specific two- or three-crop rotation. Gopalakrishnan and Goswami (1970) provided evidence for growth-stage-dependent genotypic differences in the rate of dry matter production in *C. olitorius*. Such differences in average specific leaf weight per plant are, however, absent in both species (Palit and Bhattacharyya 1984b). Significant differences in final shoot dry weight among parents and crosses were reported by Basak et al. (1973). Such inbred lines are practically homozygous and any variability they show may be the result of environmental factors. The key to the success of genetic variability in the hybrids mainly lies in the genetic diversity of the parents, and heritability of the component characters. In a world collection of jute germplasm, Palit et al. (1996) showed high diversity in characters such as plant height, harvest index, cambial activity and fibre strength within each origin country and mutants. Irrespective of specific origins and traits, cambial activity (measured as fibre : wood ratio) showed lowest (42–72%) and plant height showed highest (71–93%) diversity. Existence of such large polymorphic variation within each origin is an indication of the possibility of finding rare alleles in each origin of the abovementioned traits. The higher order of genetic coefficients of variation, and higher estimates of genetic advance indicate that improvements in these traits through selection would be more effective in the case of *C. olitorius* than in *C. capsularis* genotypes (Palit et al. 1996). The programme to assemble a worldwide collection of jute germplasm is comparatively recent (Arangzeb 1988) and the availability of more and more new collections every year should provide ample opportunities for crop physiologists to show the way to further advances in jute productivity.

References

Aloni R (1988) Vascular differentiation in the plant. In: Roberts LW, Graham PB, Aloni R (eds) Vascular differentiation and plant growth regulators. Springer, Berlin Heidelberg New York, pp 39–62

Anonymous (1995) Jute, Kenaf and allied fibres statistics, June, 1995, FAO, Rome

Arangzeb S (1988) Genetic resources of wild and weedy *Corchorus* and *Hibiscus* species and their value in breeding research. In: Denton IR (ed) Proc IJO/BJRI Training course on collection, conversation and characterisation of jute and kenaf germplasm. Oct 1988, IJO, Dhaka, Bangladesh

Basak SL, Pal AK (1975) Genetics of seed dormancy in cultivated jute. Seed Res 3:11–15

Basak SL, Choudhuri BB, Jana MK (1973) Effect of fertilizer on combining ability of jute. SABRAO News Lett 5:15–21
Basu RN, Chattopadhay K, Bandopadhay PK, Basak SL (1978) Maintenance of vigour and viability of stored jute seeds. Seed Res 6:1–13
Bhaduri SK, Haribabu M, Sen SK, Dasgupta PC, Majumder P (1985) Constituents of jute seed. Indian J Agric Sci 55:750–753
Bhattacharya SK (1974) Retting of jute – a key process that needs more attention. Jute Bull 36:194–198
Boardman NK (1977) Comparative photosynthesis of sun and shade plants. Annu Rev Plant Physiol 28:355–377
Bose RG, Bhattacharyya JP (1974) Jute seed storage and oxygen requirements. Curr Sci 43:756–757
Bose TK, Bose T, Basu RN (1974) Photoperiod responses in jute. Indian J Plant Physiol 17:78–84
Chakraverty RK (1968) The effects of growth regulating substances on the growth, yield and fibre quality of jute (*Corchorus olitorius* L.). In: Sircar SM (ed) Proc Int Symp Plant Growth Substances. Dept of Botany, Univ of Calcutta, Calcutta, pp 337–352
Chakraverty RK (1975) Germination, viability and seedling growth in two species of *Corchorus*. Sci Cult 41:395–396
Chaudhuri BB, Bask SL (1969) Moisture stress and excess on cultivated jute. Indian J Agric Sci 39:666–667
Choudhury PK, Saha SP (1988) In search of photosynthetic parameters of significance to fibre yield in jute. In: Int Congr of Plant Physiology, 12–20 Feb 1988, New Delhi
Datta PC (1971) Chromosomal biotypes of jute.Bull Bot Soc Bengal 25:89–95
Dutt N, Ghosh SK (1962) Measurement of natural crossing effected by insects in *olitorius* and *capsularis* jute. Indian J Agric Sci 32:242–250
Edmonds JM (1990) Harbarium survey of African *Corchorus* L. species. Systematic and ecogeographic studies on crop genopods 4. IBPGR/IJO. Rome, Italy
Fahmy R (1974) Soaking jute seeds in different concentration of gibberellic acid, 2,4-D and colchicine and its effect on germination and organic compounds formed in the seedlings. Biochem Physiol Pflanz 165:141–148
Fahn A (1982) Plant anatomy, 3rd edn. Pergamon, Oxford, 544 pp
Fawusi MOA, Ormrod DP (1981) Photoperiod responses of *Corchorus olitorius* L. in controlled environments. Ann Bot 48:635–638
Ghosh SS, Rao KR, Patil JS (1943) Anatomical studies on jute (*Corchorus*) with special reference to the formation of fibre. In: Agric Res Mem 1 Indian Central Jute Commitee Calcutta, pp 24–38
Ghosh T (1983) Handbook on jute. FAO Plant Production and Protection Paper 51, FAO, Rome
Ghosh T, Basak M (1951) Effect of insolation and chemical treatment in relation to storing of jute seed. Indian Phytopathol 4:38–44
Gifford RM, Evans LT (1981) Photosynthesis, carbon partitioning and yield. Annu Rev Plant Physiol 32:485–509
Gopalakrishnan S, Goswami NN (1970) Note on varietal differences in leaf area development and net assimilation rate in tossa jute (*C. olitorius* L.) Indian J Agric Sci 40:552–555
Gopalakrishnan S, Saha S (1977) Studies on the foliar characteristics of white and tossa jute in relation to photosynthesis. Indian J Agric Sci 47:41–47
Gopalakrishnan S, Goswami NN, Bagchi D (1970) Cation exchange capacites of the roots of different varieties of jute in relation to uptake of Ca and K. In: Proc Dept Atomic Energy Symp on Radiation and Radioisotopes in soil studies and plant nutrition. 21–23 Dec 1970, Bangalore
Gosse G, Varlet-Grancher C, Bonhomme R, Chartier M, Allirand JM, LeMaire G (1986) Production maximale de matière sèche et rayonment solaire intercepté par un couvert végétal. Agronomie 6:47–56
Goswami NN, Saha JR (1969) Radiophosphorus tracing of the root pattern of jute. J Indian Soc Soil Sci 17:507–513

Hossain M, Sen S (1978) Early and photoinsensetive mutants in jute. Indian J Genet Plant Breed 38:179–181

Islam MS, Ilias M, Biswas AK, Naznin S (1980) Biochemical investigation of earliness in radiation-induced early-flowering mutant lines of jute *Corchorus capsularis* L. Indian J Expl Biol 18:1478–1486

Jain NK, Saha JR (1971) Effect of storage length on seed germination in jute. Agron J 63:636–638

Johansen C, Wasque M, Ahmed MM, Begum S (1985a) Plant growth curves and fibre quality changes of jute (*Corchorus* spp.) grown in Bangladesh. Field Crops Res 12:387–395

Johansen C, Wasque M, Begum S (1985b) Effect and interaction of photoperiod, temperature, water stress and nitrogen on flowering and growth in jute. Field Crops Res 12:397–406

Joshua DC, Thakare RG (1980) Short day-insensitive jute mutant. Mutat Breed News Lett 16:6–7

Kar BK (1959) Physiology of the golden fibre – its nature and production. Presidential address, Section of Agric Sci. In: Proc 46th Indian Sci Congr 21–28 Jan 1959, University of Delhi, Delhi, pp 166–184

Kar BK (1968) Physiological effects of hormones on growth and development of jute plants. In: Sircar SM (ed) Proc Int Symp Plant Growth Substances, Dept of Botany, Univ of Calcutta, Calcutta, pp 317–322

Kar BK, De Sarkar BK (1957) Investigations on the salt tolerance in jute (*Corchorus*). In: Proc 44th Indian Sci Congr Part III, 14–20 Jan 1957, University of Calcutta, Calcutta, p 386

Kar BK, De Sarkar BK (1961) Investigations on the physiology of jute. V. growth and development of root system in *C. capsularis* and *C. olitorius*. Indian J Plant Physiol 4:1–20

Kundu BC (1944) Anatomy of jute stem with special reference to cambial activity and distribution of fibres in relation to leaf-trace system. JR Asiat Soc Beng (Science) 10:27–52

Kundu BC (1951) Origin of jute. Indian J Genet Plant Breed 11:95–99

Kundu BC (1953) Quality of Indian jute. Jute Bull 15:450–462

Kundu BC (1956) Jute – world's foremost bast fibre. Econ Bot 10:203–240

Kundu BC, Sarma MS (1955) Jute seed II. Production and distribution of improved varieties. Jute Bull 13:361–363

Kundu BC, Sen S (1958) Development and structure of the foliage leaf of some of the varieties of *Corchorus capsularis* L. and *C. olitorius* L. Bull Bot Soc Bengal 12:91–106

Kundu BC, Basak KC, Sarkar PB (1959) Jute in India. Indian Central Jute Committee, Calcutta

Kuroiwa S, Monsi M (1963) Theoritical analysis of light factor and photosynthesis in plant communities. (1) Relationships between foliage structure and direct, diffuse and total solar radiations. J Agric Meteorol 18:143–151

Mitra GC (1984) Genetic control of secondary phloic fibres in jute (*Corchorus capsularis* L.) Genetica 63:9–11

Monteith JL (1972) Solar radiation and productivity in tropical ecosystems. J Appl Ecol 9:747–766

Mukherjee RK, Datta C (1964) Effect of gibberellic acid on growth of jute. Indian J Plant Physiol 7:102–115

Nwoke FIO (1980) Effects of numbers of photoperiodic cycles on induction and development of flowers and fruits in *Corchorus olitorius* L. (var. Oniyaya Epen) Ann Bot 45:569–576

Okusanya OT (1979) Quantitative analysis of the effects of photoperiod, temperature, soil salinity and soil types on the germination and growth of *Corchorus olitorius*. Oikos 33:444–450

Pal D, Sen SP (1988) Comparative effects of associative symbiosis with the root and leaf surfaces of maize and jute plants. In: Sen SP, Palit P (eds) Biofertilizers – potentialities and problems. Plant Physiology Forum, Naya Prokash, Calcutta, pp 211–218

Palit P (1985) Translocation and distribution of ^{14}C-labelled assimilate associated with growth of jute (*Corchorus olitorius* L.). Aust J Plant Physiol 12:527–534

Palit P (1987a) Adenine nucleotide, energy charge and protein synthesis in germinating jute (*Corchorus capsularis* L.) seeds under water stress. Indian J Exp Biol 25:135–138

Palit P (1993) Radiation and carbon use efficiency in field grown jute (*Corchorus* spp.) in relation to potential primary production. Photosynthetica 28:369–375

Palit P. Bhattacharyya AC (1981) Germination and water uptake of jute seeds under water stress. Indian J Exp Biol 19:848–852

Palit P, Bhattacharyya AC (1982) Source-sink control of jute productivity. Indian J Plant Physiol 25:187–200

Palit P, Bhattacharyya AC (1984a) Characterization of the type of photosynthetic carbon dioxide fixation in jute (*Corchorus olitorius* L.) J Exp Bot 35:169–173

Palit P, Bhattacharyya AC (1984b) Measurement of leaf area per plant of white jute (*Corchorus capsularis* L.) and tossa jute (*C. olitorius* L.) using the average specific leaf weight value. Trop Agric (Trinidad) 61:59–62

Palit P, Bhattacharyya AC (1987) Interception of radiant energy, canopy photosynthesis and growth of cultivated jute (*Corchorus olitorius* L) Photosynthetica 21:453–461

Palit P, Bhattacharyya AC (1988) Photosynthesis and yield of jute (*Corchorus olitorius*) in relation to foliar arrangements. In: Sinha SK, Sane PV, Bhargava SC, Agrawal PK (eds) Proc Int Congr of Plant Physiology, vol I. Soc Plant Physiol Biochem IARI, New Delhi, pp 487–490

Palit P, Singh VK (1991) External water potential as the effector for proline accumulation in imbibing jute (*Corchorus capsularis* L.) seeds. Indian J Exp Biol 29:793–794

Palit P, Bhattacharyya AC, Samanta BK (1988) Reassessment of the effects of broadcast and line sowing on the fibre yield of jute. Exp Agric 24:471–475

Palit P, Sasmal BC, Bhattacharyya AC (1996) Germplasm diversity and estimate of genetic advance of four morpho-physiological traits in a world collection of jute. Euphytica 90:49–58

Pandit M, Ghosh B (1990) Control of ornithine decarboxylase activity in jute seeds by antizyme. J Biosci 15:82–91

Patel CS, Mandal AK (1983) Effect of moisture regimes and level of fertilizer application on yield and water requirement of jute (*Corchorus olitorius* L. and *Corchorus capsularis* L.) J Agric Sci Camb 101:311–316

Rahman MM, Choudhury SH, Prodhan SA (1980) Characters of jute at young stage under drought condition. Bangladesh J Agric Sci 5:1–4

Saha A, Basak SL (1982) Inheritance of embryo size and importance of embryo and endosperm size in white jute. Seed Res 10:72–78

Schulze ED (1983) Root-shoot interactions and plant life forms. Neth J Agric Sci 31:291–303

Sen S (1968) An induced mutant of *Corchorus olitorius* L. with enhanced abscission rate. Ann Bot 32:863–866

Sen SP, Saha S, Choudhuri P, Pal D (1983) Scope of augmenting potentiality in jute than at present by physiological means. In: FAO Consultation of Experts on Jute & Kenaf improvement. 12–15 Sept 1983, Dhaka

Sen Gupta JC (1947) Variation in osmotic values in jute plants. J Indian Bot Soc 26:157–175

Sen Gupta JC (1953a) Studies on the physiology of the jute plant. I. Effect of the time of sowing and vernalization on growth and development of jute plants. Agricultural Research Mem 6.I, Indian Jute Committee, Calcutta

Sen Gupta JC (1953b) Studies on the physiology of the jute plant. II Mineral Nutrition of jute plants. Agricultural Research Mem 6.II, Indian Jute Committee, Calcutta

Sen Gupta JC, Sen NK (1944) On photoperiodic effect of jute plants. Indian J Agric Sci 14:196–202

Sen Gupta JC, Sen NK (1946) Photoperiodism in jute. Nature 153:655–656

Sharma AK, Roy M (1958) Cytological studies on jute and its allies. Agro Lusit 20:6–15

Stahl RS, McCree KJ (1988) Ontogenetic changes in the respiration coefficients of grain sorghum. Crop Sci 28:111–113

Stant MY (1961) The effect of gibberellic acid on fibre cell length. Ann Bot 25:453–462

Verma MM, Arora N (1978) Further studies on seed-testing procedures for jute (*Corchorus capsularis* and *C. olitorius*) seeds. Seed Res 6:151–157

Wardlaw IF (1990) The control of carbon partitioning in plants. New Phytol 116:341–381

Watt G (1889) Dictionary of the economic products of India, vol II. Dept Revenue and Agril. Govt. of Inida, Calcutta, pp 534–562

Sugarcane

S.E. Lingle[1]

1 Introduction

1.1 Economic Importance

Sugarcane (*Saccharum* spp.) is the plant kingdom's version of the pig: it grows fast, reproduces abundantly, and everything is used except the rustling of the leaves. A native of the tropics, sugarcane is grown extensively in tropical and subtropical areas throughout the world. This grass crop is produced predominantly for its juice which is extracted for sucrose, but can also be used for fermentation products such as ethanol and acetic acid. The by-product molasses is used as an animal feed additive and to produce furfural and ethanolic liquors (e.g., rum). Bagasse, the fibrous residue remaining after juice extraction, is often burned to produce the steam required to power the sugarcane mill. Excess steam can be used for cogeneration of electricity. Bagasse can also be used to fabricate fiber products, such as paper and fiberboard (Heinz et al. 1994; Polack 1995; Purchase 1995). Worldwide, approximately 12 million ha of sugarcane were harvested for sugar production in 1994–1995, yielding 758 milliont of stalks from which were extracted 78 million t of sucrose. In the United States in 1993–1994, sugarcane was grown on 357 kha, producing 27 million t of stalks yielding 3 million t of sucrose, and 1 million t of black strap molasses (Buzzanell and Lockley-Jolly 1994).

The above-ground biomass of the sugarcane crop consists of the photosynthetic canopy (leaves and leaf sheaths), and juicy, fibrous stalks. The green and dead leaves and leaf sheaths contribute little sucrose and many nonsucrose impurities, so they are considered as "trash". To remove this trash, sugarcane may be burned before or after harvest or stripped mechanically or by hand. The total stalks taken to the sugar mill constitute cane yield. Sucrose yield averages about 10% of the cane yield.

Crop management is somewhat dependent on the desired end-product. For instance, for sucrose production the crop is managed to maximize juice quality (sucrose concentration and purity) at harvest. Low sucrose concentration and high concentrations of nonsucrose sugars (e.g., glucose and fructose) and non-

[1]United States Department of Agriculture, Agricultural Research Service, 1100 E. Lee Boulevard, P.O. Box 19687, New Orleans, Louisiana 70179–0687, USA

Crop Yield, Physiology and Processes
D. L. Smith, C. Hamel (Eds.)
© Springer-Verlag Berlin Heidelberg 1999

sugar impurities (e.g., minerals and organic acids) decrease the efficiency of sucrose extraction and crystallization (Chen 1985). In addition, excess fibrous material increases milling and transport costs. However, for fermentation products juice quality is of less concern since all sugars are fermentable. Most of this discussion of the physiology of growth and yield in sugarcane presumes sucrose to be the economic product.

1.2 Botanical Description

There are six species of the genus *Saccharum* native to Southeast Asia and India: *S. officinarum* L.; *S. spontaneum* L.; *S. robustum* Brandes and Jeswiet ex Grassl; *S. sinense* Roxb.; *S. barberi* Jesw.; and *S. edule* Hassk (Daniels and Roach 1987). Commercially produced cultivars are usually interspecific hybrids, primarily of *S. officinarum* and *S. spontaneum*, but also of all other *Saccharum* species except *S. edule*, since it is infertile (Daniels and Roach 1987; Sreenivasan et al. 1987).

All of the known *Saccharum* species are polyploids. Each has varying basic chromosome numbers and ploidy levels. Total chromosome numbers range from $2n = 36$ to $2n = 170$. Estimates of the basic number range from 6 to 10, with a ploidy level from 6× to 12× (Sreenivasan et al. 1987). Recent studies using molecular markers to establish linkage groups suggest that the most common basic chromosome numbers for this group are 8 (*S. spontaneum*) and 10 (*S. officinarum*) (Glaszmann et al. 1989; D'Hont et al. 1994). Since varying degrees of introgression and chromosome sorting occur during interspecific hybridization and backcrossing, the genetics of sugarcane cultivars are extremely complex.

1.3 Anatomy and Morphology

The morphology of the sugarcane plant is similar to that of maize (*Zea mays* L.) and sorghum [*Sorghum bicolor* (L.) Moench]. However, sugarcane is vegetatively propagated. Whole stems or stem pieces (seed pieces or setts) are planted, and new shoots arise from buds that occur at each node on the sett. The first roots (sett roots) arise from root initials on the node of the sett, and these support the young shoots until they develop their own shoot roots (Fig. 1) (Artschwager 1925; van Dillewijn 1952). Most cultivars tiller strongly at the crown, although the number of tillers is also dependent on plant spacing within and between rows (van Dillewijn 1952; Humbert 1968).

Once established, the root system of sugarcane is mostly fibrous and concentrated in the top 30 to 60 cm of soil. The roots can extend outward from the crown for several meters. Buttress roots are thicker than fibrous roots and may reach a depth of 1.2 m, while extremely long "rope" roots penetrate 4.5 m or deeper (van Dillewijn 1952). The root system of sugarcane is the least-studied

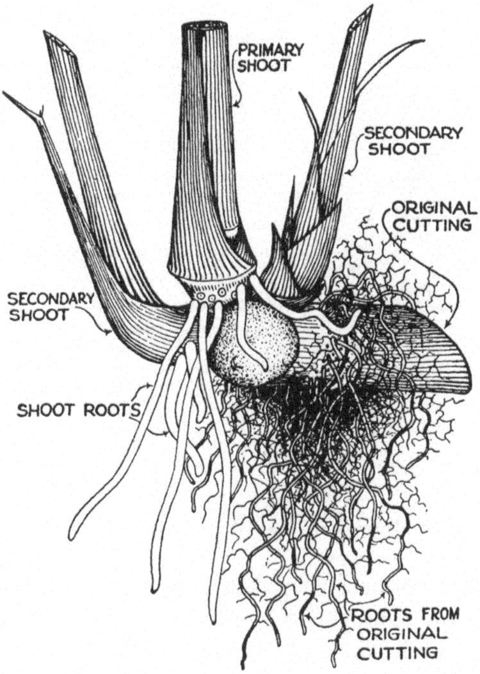

Fig. 1. Diagram of initial growth from a single-bud cutting of sugarcane. (Martin 1938 Sugarcane Diseases in Hawaii. Experiment Station, Hawaiian Sugar Planters' Association, Honolulu)

part of the sugarcane plant, so relatively little is known about it. Most of the water and mineral absorption occur in the fibrous roots, while the buttress roots anchor the large stalks in the soil so that stalks remain upright during high winds. The function of the rope roots, and whether they occur in most or only some clones is unknown (van Dillewijn 1952).

The apical meristem of the stalk produces the new cells that become the stalk body with its attached leaves (Artschwager 1925). The apical meristem is surrounded by a cylinder of immature leaves (spindle) with the most immature leaves to the inside. As leaf blades and sheaths elongate, the leaves unfurl from the spindle, forming a whorl. Leaf blades reach full elongation approximately coincidently with emergence of the collar (dewlap) between blade and sheath from the leaf whorl, but the leaf sheath continues to elongate after the collar is visible (van Dillewijn 1952). Below the collar the leaf sheath encircles the stem, extending downward about 30 cm to its point of attachment at the base of a node.

Leaf blades are alternate, long and lanceolate, with a prominent midrib and parallel veination (Artschwager 1940). The leaves have a specialized anatomy. The vascular bundles are surrounded by a bundle sheath, which, in turn, is surrounded by several layers of mesophyll cells (Fig. 2; Artschwager 1925). The chloroplasts in the bundle sheath are different from those in the mesophyll, and the biochemical processes of carbon fixation are differentially distributed among tissue types (Furbank and Taylor 1995). The anatomical and biochem-

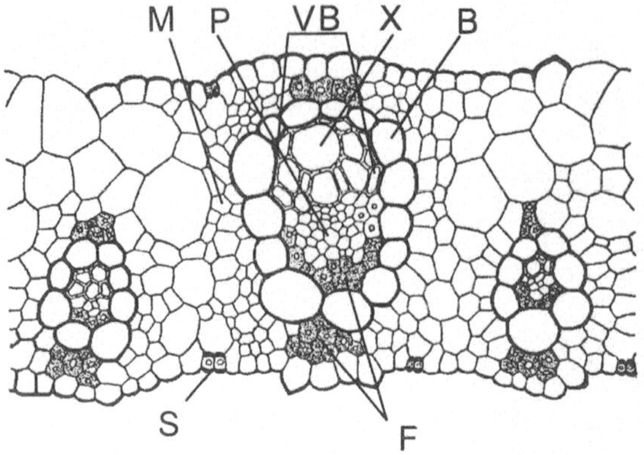

Fig. 2. Diagram of a cross-section of the lamina of a sugarcane leaf. The vascular bundles (*VB*) contain xylem (*X*), phloem (*P*), and fibers (*F*). Bundles are surrounded by a single layer of bundle-sheath cells (*B*) and then by several layers of mesophyl cells (*M*). Stomata (*S*) are located on both leaf surfaces, but are more common on the lower side. (Artschwager 1940)

ical specialization of cells allows for efficient carbon fixation via the C_4 pathway of photosynthesis (see below).

Sugarcane stalks consist of a series of nodes, where leaf sheaths attach, and internodes, which form the bulk of the sucrose storage tissue (Fig. 3). Both nodes and internodes are usually solid. Most of the internode tissue is parenchyma (unspecialized cells with primary and secondary walls), the primary sink for sucrose storage in the stalk. Vascular bundles are scattered throughout the parenchyma cylinder (Fig. 4), oriented vertically, with no vascular connections between bundles in the internode (Artschwager 1925). Internodes lengthen due to elongation of cells. After elongation, parenchyma cells are 200 to 300 μm long and 80 to 150 μm in diameter (Oworu et al. 1977a; Moore and Cosgrove 1991). The volume of fully elongated storage tissue is approximately 20% cell walls and intercellular space (apoplast), 72% intracellular space (symplast), and 8% vascular bundle (Hawker 1965). The apoplast is largely fluid-filled (Hawker 1965; Dong et al. 1994). About 95% of the parenchyma cell volume is a large vacuole that accumulates sucrose. Sucrose also accumulates in the cytoplasm and cell-wall space (apoplast). In some internodes, rapid expansion of the cells cause the interior parenchyma to stretch apart, forming air-filled voids. These broken cells and the void are called pith. Pithiness decreases the amount of juice in a stalk, and is therefore undesirable. Pithiness is a genotype characteristic, but it tends to increase following inflorescence initiation and development (Humbert 1968), and also during periods of very rapid stalk growth (Clements 1980).

The vascular bundles in the internodes are surrounded by a bundle sheath of cells with thick walls (Fig. 5), which lignify and suberize with internode age (Jacobsen et al. 1992). Vascular bundles are smaller and closer together at the periphery of the stalk than in the center. About 50% of the bundles are located within 3 mm of the rind (Walsh et al. 1996). There are 1200 to 1500 vascular bundles in a typical cross-section of stem (Artschwager 1925).

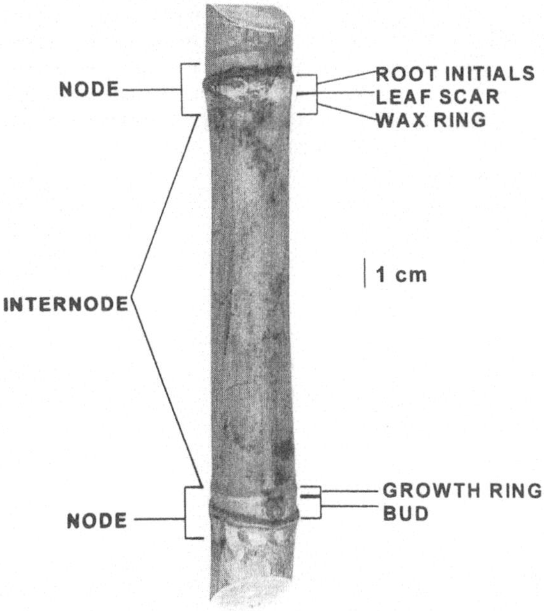

Fig. 3. Sugarcane stem section. The bulk storage tissue is the internode. Each internode is subtended by a growth ring and then a node, where the leaf sheath attaches. A vegetative bud and root initials are located at the node

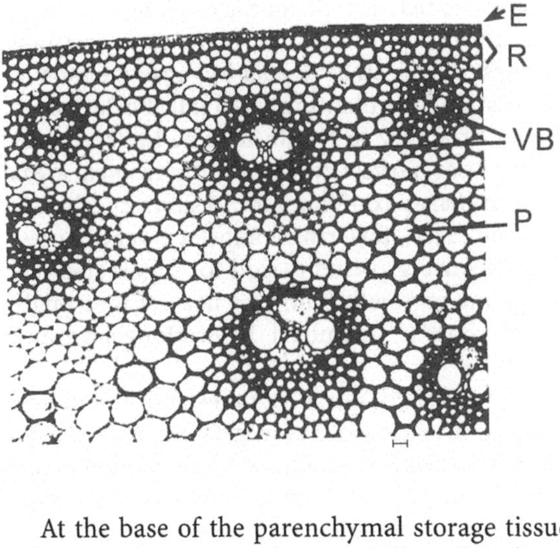

Fig. 4. Cross-section of sugarcane internode. The internode has a single-cell layer epidermis (*E*), several layers of thick-walled cells that form the rind (*R*), and vascular bundles (*VB*) scattered throughout the storage parenchyma (*P*). (Artschwager E 1940)

At the base of the parenchymal storage tissue is a short cylinder of intercalary meristem (growth ring). Cells within the growth ring will elongate if the stalk lodges, thus returning the stalk apex to a vertical orientation (Artschwager 1940). Just below the growth ring is the node where the leaf sheath attaches. On the node are root initials and a vegetative bud. Like the leaves, the buds on successive nodes alternate sides. Sugarcane stalks usually show strong apical dominance, and the growth of the lateral buds is suppressed unless the terminal meristem is damaged or flowers (Humbert 1968).

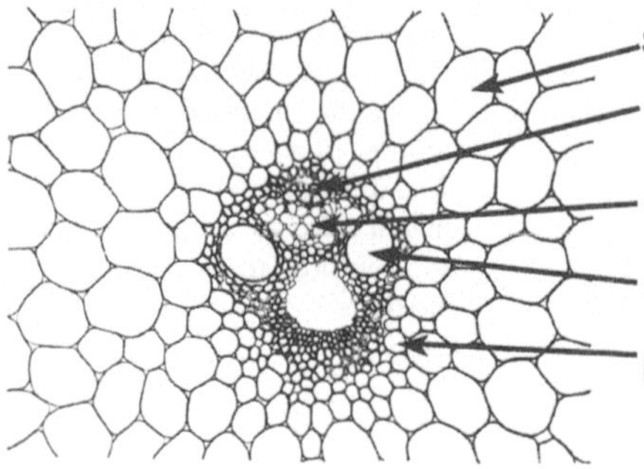

Fig. 5. Vascular bundle and adjacent parenchyma (*P*) tissue from a sugarcane internode. The bundle contains xylem (*X*), phloem (*P*), fibers (*F*), and is surrounded by a bundle sheath (*B*). (Clements 1980)

2 Growth and Yield Dynamics

2.1 Stand Establishment

Sugarcane is a perennial crop, regrowing from vegetative buds on the crown after the crop has been harvested. Regrowth of the sugarcane field is called ratooning. Distinctions are usually made between the initial or plant cane crop, and subsequent or ratoon crops. In most cultivars the plant cane and first ratoon crops tend to have the highest yields. Stalk population and size tend to decrease with each successive harvest, and the planting will finally be destroyed and replanted. Worldwide there is variability in length of time between harvests from 1 to 2 or more years. Two-year harvest cycles occur in tropical areas where frost does not damage the terminal apex of the stalk. In subtropical areas cold winter temperatures or, in some cases, flowering compel a 1-year harvest cycle. In some of these areas the effective growing period for sugarcane may be less than 9 months, since sugarcane regrowth following harvest is strongly inhibited at soil temperatures below 18 °C (Humbert 1968; Mongelard and Mimura 1971; Bacchi and Sousa 1977).

Since sugarcane is vegetatively propagated, initial stand establishment and therefore yield are dependent on germination of the vegetative buds on the stalks. Healthy buds on nodes of stalk sections as young as the elongation phase are capable of germinating, although bud growth in intact stalks is usually suppressed by apical dominance. Percent germination and time to germination after cutting tend to increase with bud age (Clements 1980; Singh and Kanwar 1986). Water content of the node may influence germination; the bud does not sprout until the water content increases to 70% (Singh 1975). Humbert (1968) states that warm, moist soil is desirable for rapid germination in sugarcane. Reduction of soil water potential below field capacity increasingly reduces germination percentage of buds, although sett roots can sprout even at −3.03 MPa (Singh and Srivastava 1974).

Very little is known about bud and root germination and growth between planting and shoot emergence. Clements (1940) concluded that sett size determined the size of the young plant; that is, the greater the size of the bud and the more reserves in adjacent stalk available for initial growth, the greater the size and the faster the growth of the new shoot. Bud position also influences early growth (Clements 1980). Placing the sett so that the bud is on top slightly enhances shoot emergence but delays shoot-root development. Buds oriented to the side or beneath the sett are delayed in emergence but have enhanced shoot-root development. However, Clements (1940) concluded that there is not enough difference in development with bud position for it to be economically useful to specifically orient the setts at planting.

The number of buds per sett does not appear to significantly affect stand establishment. It might be expected that when whole stalks are planted, the earliest sprouting buds would suppress the growth of the others. Alternatively, shorter setts are more subject to invasion by soil organisms through the cut surfaces of the stalk. In practice, there is little difference in stand establishment between whole stalk and cut setts as long as the buds are not damaged during cutting (Benda et al. 1981). Setts 45–60 cm long with several buds are easier to handle than several-meter long, possibly curved stalks, but planting whole stalks is still practiced in many areas.

The optimum temperature for germination is 30 °C (Gascho et al. 1973). Temperatures below 16 °C inhibit bud germination, and both high (50 °C) and low (10 °C) temperatures during germination suppress tillering, growth, and yield in the plant and ratoon crops (Gascho et al. 1973). Therefore, planting is scheduled to avoid high or low soil temperatures during germination.

Like many grasses, sugarcane produces new tillers from the base of each stalk until light, temperature, or other environmental factor becomes limiting (van Dillewijn 1952). Unlike many grasses, which appear to have distinct tillering and elongation stages, sugarcane stalks begin to grow and elongate as soon as they are formed, even while new tillers are being formed. Thus, a single plant will have tillers of many different heights and maturities. MacColl (1976) showed that cultivars that produced many tillers after the earliest tillers were elongating had reduced sugar concentrations in the early growth. He speculated that the late tillers were drawing sugar from the elongating stalks.

In field situations, tillering effectively ceases at the time of canopy closure, mostly due to reduced light at the soil surface (van Dillewijn 1952). Canopy closure is the time of maximum stalk numbers, since the most immature stalks die after this time. It is not known if this is due to insufficient carbohydrate assimilation caused by shading, or if the plant triggers tiller death in some manner (van Dillewijn 1952). Tillering ability of the plant is not permanently inhibited by canopy closure, however. If stalk lodging creates holes in the crop canopy, new tillers will emerge. Since immature stalks have many green leaves and poor quality juice, new tillers appearing near harvest reduce the juice quality of the entire crop.

It is generally accepted that wide intra- and inter-row plant spacing increases tiller numbers per plant over those in narrower spacings, and stalks in wide spacings are thicker and shorter than stalks in narrow spacings. However, narrow plant spacing usually increases the millable stalk population per area and cane yield (Irvine and Benda 1980; Richard et al. 1991; Singh and Singh 1993), especially in areas with a short cropping season (Irvine et al. 1980a). Yadav (1993) concluded that high planting density increased the density of primary stalks, which had a positive correlation with yield. A disadvantage of high planting density is an increase in lodging, which makes mechanical harvesting more difficult (Irvine et al. 1980b; Richard et al. 1991).

2.2
Growth and Sugar Accumulation

A logistic growth curve describes biomass accumulation over time in many crops (Charles-Edwards 1982), including sugarcane (Fig. 6; McMartin 1979; Coale et al. 1993). Dry matter initially increases exponentially from near zero, then linearly, and finally plateaus. The initial exponential phase following planting or ratooning corresponds to the establishment of new tillers. The linear phase in sugarcane is called grand growth; during this phase tillers elongate. Eventually, environmental conditions begin to restrict growth. Shaded tillers and old leaves die and are shed and biomass plateaus because new

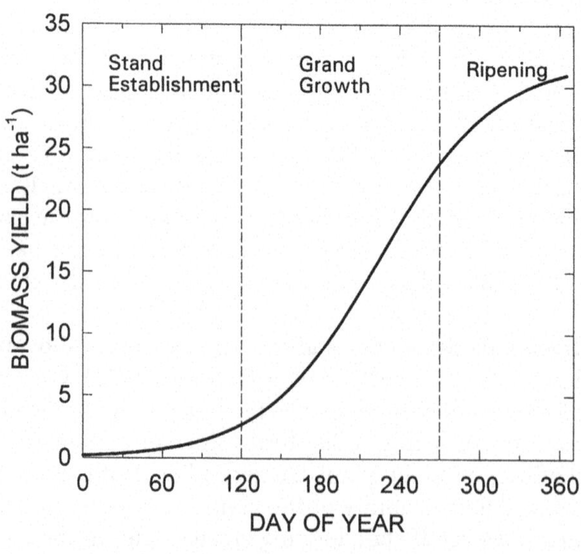

Fig. 6. Idealized sugarcane biomass accumulation curve. During stand establishment new tillers are formed. Most stalk elongation occurs during grand growth. During ripening, growth slows, and the sucrose concentration in the juice increases as nonsucrose impurities decrease

growth just compensates for the lost dry matter. Duration of each of the phases and the rate of growth during grand growth vary with cultivar. These are strongly influenced by seasonally variable temperature, solar radiation, and moisture (Clements et al. 1952). In tropical areas where sugarcane is grown for 2 years, sugarcane may have a bimodal growth curve, with two periods of grand growth associated with the optimum growth environment in each year (Clements and Kubota 1943).

Sugarcane is a warm-climate crop, and its growth is very much influenced by temperature. Inman-Bamber (1994) calculated that the base temperature for leaf appearance is 10 °C, while the base temperature for tiller appearance is 16 °C. Plant elongation rate is linearly related to temperature (Glasziou et al. 1965; Inman-Bamber 1994). Bacchi and Sousa (1977) determined 18 °C to be the minimum temperature for stalk elongation. Increasing minimum temperature up to 30 °C increases stalk length, stalk weight, and number of internodes per stalk (Clements et al. 1952; Glasziou et al. 1965). Temperatures above 30 °C tend to decrease the components of biomass (Glasziou et al. 1965). Cool root temperatures limit growth more than cool air temperatures (Mongelard and Mimura 1971, 1972).

The differential response of the various plant organs to temperature may affect crop architecture. For instance, temperatures near 16 °C would allow tiller appearance but not stem elongation. Prolonged periods at or below 16 °C would prolong the tillering period. Once temperatures increased to the stem elongation threshold, these tillers would start elongation together, close the canopy, and inhibit further tillering. This would produce a crop of uniform maturity. High early tiller number is positively correlated with high cane and sucrose yield (MacColl 1976). This same effect could be achieved by selecting cultivars with a lower threshold temperature for tillering.

Muchow et al. (1994) showed that during the linear growth phase, aboveground biomass accumulation per hectare is linearly related to intercepted solar radiation. This confirmed observations of the relationship between sugarcane growth or yield and solar radiation made previously by others (Legendre 1975; Clements 1980). Intercepted radiation in crops is a function of leaf area index and solar radiation (Sinclair and Horie 1989). Since the quantity of radiation available to the plant is dependent on uncontrollable factors such as day length and cloud cover, radiation interception can only be maximized by encouraging quick canopy closure during crop development. This can be accomplished by increasing the planting rate or decreasing distance between rows.

The elongation of sugarcane stalks is adversely affected at soil water potential lower than −0.05 MPa (Clements et al. 1952). At water potentials greater than this, elongation rate is somewhat independent of soil water potential. Since elongation of a particular internode occurs only at a specific developmental stage, water stress during that stage limits the ultimate length of the internode. If a brief water stress is subsequently relieved, succeeding internodes will be unaffected by the period of restricted moisture. When a sugarcane stalk is stripped of older leaves, periods of low soil water potential are clearly indicated by short internodes.

Plant nutrition is also the key to high growth rates. Nitrogen deficiency especially decreases total dry matter yield (Samuels and Landrau 1952). Accumulation of nitrogen by sugarcane follows the same curve as dry matter accumulation (Coale et al. 1993). Leaf nitrogen content is positively correlated with elongation rate (Clements 1980), although the relation is not a simple one. Thomas and Oerther (1976) indicated that cane yield was maximized at a leaf nitrogen content of 1.9%. In sugarcane, elongation rate of the stalk is very well modeled by the elongation rate of the expanding sheaths. The composition of the leaf sheath is used in Hawaii to monitor the nutritional status of the plant (Clements 1980).

From a study conducted over several years at several locations, Thompson (1978) calculated maximum biomass accumulation rates to range from 5.4 to 47.9 $g\,m^{-2}\,day^{-1}$. The highest rates of biomass accumulation occurred during short periods most conducive to growth. When averaged over the entire growing period, biomass accumulation was about 9 $g\,m^{-2}\,day^{-1}$ in nonirrigated plots, and about 11.5 $g\,m^{-2}\,day^{-1}$ in irrigated plots.

Sugar begins to accumulate in sugarcane internodes as they start elongating. Accumulation proceeds in a logistic fashion (Fig. 7) until a maximum concentration is reached. This maximum is influenced by cultivar and environment. Sucrose, glucose, and fructose are the major sugars found in sugarcane. Glucose and fructose concentrations are highest during the elongation phase of the internode. Concentrations of these sugars decrease as the elongated internodes mature. In mature, healthy internodes, sucrose is essentially the only sugar found (Lingle and Smith 1991).

Sugar accumulation is not uniform across the diameter or down the length of an internode. Sugar concentrations tend to be lower both near the rind and in the central core than in the central cylinder (Fernandes and Benda 1985). The vascular bundles near the rind are the first to lignify and suberize (Jacobsen et al. 1992), and this may reduce the efficiency of sucrose transport from the phloem to the storage cells. The cells in the central core are those most likely to stretch apart during elongation, forming voids that do not contain juice. Also, vascular bundles are farther apart in the central core. Oworu et al. (1977a) showed that sucrose concentration is higher in cultivars with more closely spaced vascular bundles. Along the length of the internode, sugar concentration is greatest in the middle and lower towards both nodes (Cornelison 1944; Fernandes and Benda 1985). In elongating sections, sugar concentrations are usually higher in nodes than internodes, while in fully elongated sections the sugar concentration of the internode exceeds that of the adjacent nodes (Fernandes and Benda 1985; Celestine-Myrtil and Parfait 1987).

Internodes continue to accumulate sugar until the amount of assimilate translocated into the storage tissue is in equilibrium with that used for respiration or remobilized to other plant parts. Das (1934) and Clements (1959) con-

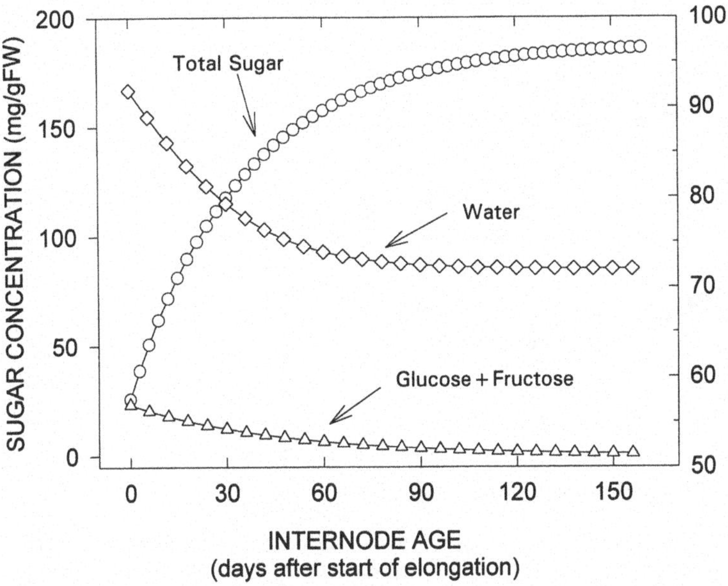

Fig. 7. Sugar accumulation in a sugarcane internode. Total sugar accumulates in a logistic fashion. Glucose and fructose concentrations and water content decrease as the internode matures

cluded that sugar content per internode is maximum coincident with the death of the subtending leaf. However, sucrose concentration in the juice continues to increase because the water content of the internode decreases (Fig. 7). Sucrose continues to be imported into all internodes throughout the life of the stalk (Hartt et al. 1963).

Because the stalk is a continuum of nodes and internodes in all stages of development, there is a gradient of sugar concentration along the stem (Fig. 8; Das 1934; Fernandes and Benda 1985; Celestine-Myrtil and Parfait 1987). Early in crop development the gradient is steep. As grand growth ceases and the crop matures, two things happen: (1) sugar concentrations in the oldest internodes increase to a maximum; and (2) the youngest internodes have higher concentrations of sugar than internodes of the same age during grand growth. This last change makes the sugar gradient along the stem less steep, and results in an increase in quality of whole-stalk juice.

The immature portion of the stalk still supported by green leaves is the most metabolically active part of the stalk. Concentrations of nonsucrose impurities in the juice are highest in internodes subtended by green leaves. Concentrations of the reducing sugars glucose and fructose are highest in elongating internodes and decrease with internode age (Fig. 7; Celestine-Myrtil and Parfait 1987; Lingle and Smith 1991). Other soluble nonsugars such as amino and organic acids are also highest in elongating internodes

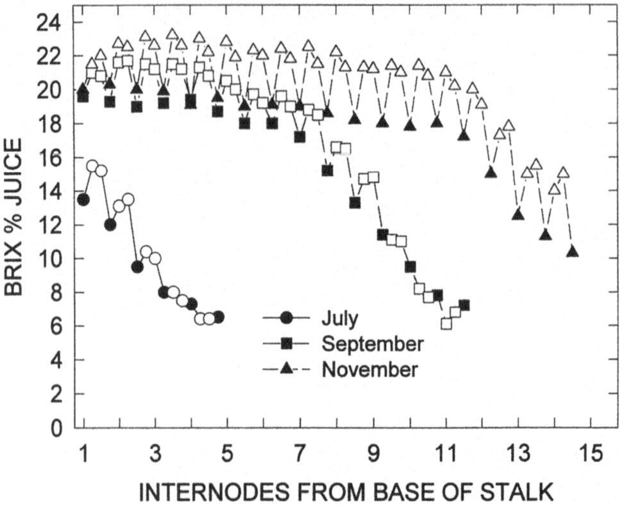

Fig. 8. Total soluble solids (Brix) in juice along the stem of sugarcane during growth and ripening. *Solid symbols* represent nodes, and *open circles* represent the range of concentrations within the internode. As growth continues, progressively more of the stalk has reached a high soluble solids concentration. (After Fernandes and Benda 1985)

(Celestine-Myrtil-Marlin 1990), as is starch (Wood 1962). Because new tiller production ceases at canopy closure, the immature portion of the crop decreases with time so that the concentrations of nonsucrose impurities in whole stalks also decrease.

Ripening in sugarcane is an increase in sucrose yield per ton of cane. This increase in sucrose yield is often the result of decreases in nonsucrose juice impurities. Since most impurities are associated with the growing portion of the stalk, conditions unfavorable for growth, such as cool temperatures, restricted soil moisture, and, especially, low available soil nitrogen are usually associated with ripening (Humbert 1968). Sunny warm days, cool nights, and low plant nitrogen at the end of the growing season are considered ideal for achieving maximum sucrose concentration. In most production areas, the crop is managed to restrict nitrogen and water availability at the end of the growth period to promote ripening before harvest. Within the constraints of the local climate, these practices help maximize sucrose concentration and minimize impurities without adversely limiting cane yield.

Since photoassimilate in sugarcane both sustains growth and is stored, the growth and storage processes are inherently competitive. Environmental or cultural conditions which promote growth, such as warm temperature, sufficient water, and available soil nitrogen, tend to decrease sucrose concentration in juice. However, because these conditions usually increase cane yield more than they reduce sucrose concentration, sucrose yield per hectare is usually maximized by the same conditions which maximize cane yield. In some studies comparing cultivars, sugarcane yield and sucrose concentration are negatively correlated (e.g., $r = -0.24$, Milligan et al. 1990). However, others have found no significant relationship between sugarcane yield and sucrose concentration (e.g., $r = 0.06$, Tai et al. 1982). Therefore, it is possible to select for increased sucrose concentration without sacrificing sugarcane yield.

It is clear that cane yield and sucrose concentration have strong genetic components. Biochemical and physiological processes, controlled by gene expression, determine how much carbohydrate is produced by the leaf, how and where it is transported, and how it is allocated within growing or storing organs. Thus, the synthesis and partitioning of sucrose within the sugarcane plant is fundamental to the achievement of high cane yield and juice quality.

3 Sucrose Synthesis and Partitioning

3.1 Photosynthesis

Like maize, sugarcane uses the C_4 pathway of photosynthesis. The four-carbon organic acid, oxaloacetic acid, is the first product of CO_2 fixation in the mesophyll cells (Fig. 2). Oxaloacetic acid is converted into malic acid by malate dehydrogenase. The malate is then transported to the bundle sheath where it is decarboxylated, and the released CO_2 is fixed into sucrose via the C_3 ribulose-bisphosphate carboxylase reaction (Hatch and Slack 1966; Hatch 1987; Furbank and Taylor 1995). Chloroplasts from the mesophyll are anatomically distinct from those of the bundle sheath. The enzymes involved in the different stages of C_4 metabolism are also distributed differently between the cell types. Phosphoenolpyruvate carboxylase, which initially fixes CO_2 into oxaloacetate, is primarily expressed in the mesophyll, while NADP-dependent malic enzyme, which decarboxylates malate, and ribulose bisphosphate carboxylase, which refixes the released CO_2, are primarily expressed in the bundle sheath. Most of the sucrose synthesis also occurs in the bundle sheath (Hatch 1987; Furbank and Taylor 1995), and it is loaded from the bundle sheath to the phloem.

Carbon assimilation rate is influenced by many factors, especially light intensity, age of the plant, and soil and leaf water content. This is especially true under field conditions. Estimates of maximum carbon assimilation in leaves at full sunlight range from about $18\,\mu mol\,m^{-2}\,s^{-1}$ in Hawaii during winter (Grantz 1989) to $52\,\mu mol\,m^{-2}\,s^{-1}$ in Australia during summer (Bull and Tovey 1974). Carbon assimilation rate apparently reaches a maximum at a photon flux density of about $1100\,\mu mol\,m^{-2}\,s^{-1}$ in field-grown sugarcane (Grantz 1989). Hartt and Burr (1965) showed that older leaves had greater photosynthetic rates than younger leaves, but leaves on young stalks had greater rates of photosynthesis than leaves of the same physiological age on older stalks.

Stomatal conductance of leaves determines the availability of CO_2 for photosynthesis. In sugarcane, carbon assimilation rate is linearly correlated to stomatal conductance (Grantz et al. 1987). Grantz et al. (1987) showed that at photon flux densities greater than $40\,\mu mol\,m^{-2}\,s^{-1}$, stomatal conductivity declines with increasing vapor pressure difference. However, this increased stomatal conductance did not decrease intercellular CO_2 concentrations below the CO_2 saturation concentration of $100\,cm^3\,m^{-3}$ (Hartt and Burr 1965; Grantz et al. 1987). Therefore, assimilation rate was only slightly decreased by the vapor pressure difference. In addition, Meinzer and Grantz (1990) found that while stomatal conductance per leaf area increased with leaf area up to $0.2\,m^2$ $plant^{-1}$, stomatal conductance per leaf area then decreased. This stabilized transpiration rates per plant, and maintained leaf water content over a range of growing conditions. This adjustment may help explain the observation of Hartt and Burr (1965) that leaves on young stalks had greater photosynthetic rates than leaves on older stalks.

Nutrient deficiencies also decrease photosynthetic rate. Plants experiencing severe N deficiency had only about 40% of the carbon fixation rate of those with adequate N (Hartt and Burr 1965; Hartt 1970). There is a high correlation between photosynthetic rate and N content of the leaf. Maximum rates were observed in leaves with about 2% total N (Hartt and Burr 1965). Potassium deficiency reduced photosynthetic rates slightly (Hartt and Burr 1965; Hartt 1969), while phosphorus-deficient leaves had about 60% of the photosynthetic rate of nondeficient leaves (Hartt and Burr 1965). However, in all of these studies, the nutrient deficiencies were extreme, induced by elimination of the nutrient from the nutrient solution in solution-grown plants. It is not known what effect mild deficiencies may have on photosynthetic rate.

3.2 Translocation and Partitioning

Sucrose is transported very rapidly from leaves to the storage tissue in sugarcane, moving at about $84\,cm\,h^{-1}$ (Hartt et al. 1963). According to Hartt et al. (1963), photosynthate moves basipetally down the leaf and sheath, enters the node, and travels down vascular bundles in the center of the internode. Downward movement may continue for as many as eight node-internode sections before lateral movement takes any ^{14}C-photosynthate to vascular bundles leading upward. Transport of photosynthate between vascular bundles probably occurs in the nodes, since vascular bundles are closer together in the nodes, and there are lateral vascular traces that may allow for photosynthate exchange between bundles (Artschwager 1925).

Environmental factors influence both the movement of ^{14}C-photosynthate from the leaf and its partitioning among the different sinks. Cold air temperature (15 °C) relative to the roots (22 °C) reduced the rate of ^{14}C-photosynthate export from leaf exposed to $^{14}CO_2$ (Hartt 1965). This treatment reduced the partitioning of assimilated carbon to the stalk, and increased partitioning to the roots. Cold root temperature (17 °C) relative to the air (22 °C) reduced the rate of export from the leaf and reduced partitioning to the roots (Hartt 1965). Low soil moisture reduced rate of export from the leaf and increased partitioning to sucrose-storing internodes (Hartt 1967). Generally, translocation rate, partitioning, and growth are more sensitive to environmental changes than is photosynthesis. Conditions which reduce growth can lead to an increase in sugar concentration in the leaf sheaths (Alexander 1967; Alexander and Montalvo-Zapata 1972; Clements 1980). This increase suggests that rates of translocation and partitioning, not photosynthesis, determine sugar concentration in sugarcane storage tissue. Hartt (1965, 1967) concluded that the environmental conditions associated with ripening in sugarcane restrict photosynthesis less than growth. Thus, ripening is the result of altered photoassimilate translocation and partitioning, not photosynthesis.

3.3 Stem Anatomy and Sucrose Storage

There is great variation in sugar accumulation among genotypes within the *Saccharum* complex. Concentration of total soluble solids in juice range from 0.2% in a *S. robustum* clone to 24.5% in an *S. officinarum* clone (Ramana Rao et al. 1985; Sreenivasan and Nair 1991). Commercial cultivars are capable of accumulating sucrose concentrations to 800 mM (27% of juice) in the storage tissues (Welbaum and Meinzer 1990). Bull and Glasziou (1963), assuming 70% water content and a structural minimum of 5% fiber in mature stalks, estimated a theoretical maximum sucrose concentration of 25% of total fresh weight (33% of juice).

Storage of sucrose requires the movement of sucrose from the sieve tubes of the phloem (phloem unloading), and sequestering that sucrose in compartments where it cannot be utilized or remobilized to other plant parts. There are two major types of phloem unloading, apoplastic and symplastic (Patrick 1990). In apoplastic unloading, sucrose is unloaded from the phloem into the cell-wall space. In this instance, cleavage by acid invertase in the apoplast may promote sucrose unloading by decreasing the concentration of sucrose in the apoplast adjacent to the phloem. In symplastic unloading, sucrose is transported from the phloem directly into the symplast of adjacent cells.

There are still questions about the mechanism for phloem unloading and sucrose transport in sugarcane stalks (Moore 1995). In one model, sucrose is unloaded from the phloem into the apoplast of the storage parenchyma. There, sucrose is cleaved by acid invertase to form glucose and fructose, which are actively loaded into the cytoplasm of the storage parenchyma. Sucrose is then resynthesized by the combined action of sucrose-phosphate synthase and sucrose phosphatase, and transported into the vacuole (Glasziou and Gayler 1972a). Until recently, this model was widely accepted. Acid invertase is very active in the cell-wall space of elongating internodes (Glasziou 1961; Glasziou and Bull 1965), so much of the sucrose in the apoplast is likely to be cleaved. When ^{14}C-fructosyl-sucrose (sucrose labeled with ^{14}C only in its fructose moiety) was supplied to tissue disks, some of the ^{14}C-sucrose recovered from the tissue contained ^{14}C-glucose, indicating sucrose cleavage and resynthesis (Hawker and Hatch 1965). Also, cells from sugarcane suspension cultures are unable to transport sucrose, but can transport glucose (Maretzki and Thom 1972).

While apoplastic cleavage of sucrose may occur, it does not appear to be necessary for sucrose transport into storage parenchyma. In stalk slices incubated in ^{14}C-fructosyl-sucrose, most of the ^{14}C in a long-term storage compartment was still associated with fructose (Lingle 1989). Also, fluorosucrose, which cannot be cleaved by invertase but can be transported by sucrose carriers (Hitz et al. 1985), was taken up by sugarcane stalk slices at rates similar or higher than those of sucrose (Thom and Maretzki 1992), showing that invertase action was not necessary for sucrose transport. In elongating internodes, which have abundant acid invertase activity in the cell-wall space, some sucrose cleavage probably occurs prior to transport into the cell. However, sucrose cleavage is not obligatory for phloem unloading and transport into storage parenchyma.

As the internode elongates and matures, the vascular bundles become increasingly lignified and suberized, as do the adjacent storage parenchyma cells (Jacobsen et al. 1992). Jacobson found that this lignification and suberization prevents the movement of apoplastic tracer dyes from the vascular bundles into the apoplast of the storage parenchyma. Thus, in fully elongated internodes, sucrose transport from symplast to apoplast must occur outside the bundle sheath, and is not a direct step in unloading sucrose from the phloem. This was recently confirmed by Walsh et al. (1996).

As the internode matures, membranes of the parenchyma become increasingly leaky (Glasziou 1960; Hawker 1965; Glasziou and Gayler 1972b). This increases the solute concentration in the cell-wall apoplast, until it approximately equals that in the vacuoles (Hawker 1965; Welbaum and Meinzer 1990). Membrane leakiness may keep the turgor of the parenchyma cells below that of the phloem, allowing continued phloem unloading even when sucrose concentrations in the storage compartment become high (Patrick 1990; Moore and Cosgrove 1991). Since the apoplast contains similar concentrations of sugars to the symplast, the entire volume of the storage parenchyma except the vascular tissue is available for sucrose storage.

Oworu et al. (1977b) showed that storage tissue disks from genotypes with high final sucrose concentrations were able to absorb more sucrose in vitro than those from genotypes with lower sucrose concentrations. The differences were attributed to higher in vitro sucrose uptake rates in high-sucrose clones than in low-sucrose clones. These authors (Oworu et al. 1977a) also showed that higher sucrose uptake rates were associated with smaller storage parenchyma cells with thicker cell walls and more vascular bundles per area.

3.4 Sucrose Metabolism

Whether sucrose which reaches the internode is utilized or stored depends on the activity of several enzymes involved in sucrose metabolism. These are: acid and neutral invertases, which cleave sucrose into glucose and fructose; sucrose synthase (SS), which uses uridine-diphosphate (UDP) to cleave sucrose to fructose and UDP-glucose (it can also synthesize sucrose from fructose and UDP-glucose); and sucrose-phosphate synthase (SPS), which synthesizes sucrose-phosphate from fructose-6-phosphate and UDP-glucose (the sucrose-phosphate is immediately cleaved to sucrose and phosphate by sucrose phosphatase, and sucrose-phosphate does not accumulate in the cell) (ap Rees 1984). All of these enzymes are present in sugarcane internodes, although their relative activities change as the internode develops (Moore 1995).

In elongating internodes, there is abundant acid invertase activity from two different soluble acid invertases in the apoplast and the vacuole (Glasziou and Waldron 1964; Glasziou and Bull 1965). Elongation rate is clearly associated with total soluble acid invertase activity (Hatch and Glasziou 1963; Glasziou and Bull 1965), induced by auxin (Sacher et al. 1963). Soluble acid invertase activity is responsible for the relatively high amounts of glucose and fructose in elongating internodes (Celestine-Myrtil and Parfait 1987; Lingle and Smith 1991). A decrease in soluble acid invertase activity is related to the decrease in growth rate and the reduction in glucose and fructose that occurs with natural or induced ripening (Su et al. 1992).

Surprisingly, three enzymes that cleave sucrose are active in sucrose-accumulating internodes. Activities of both cell wall-bound acid invertase and neutral invertase activity are higher in mature than in elongating internodes (Hatch and Glasziou 1963; Matsui and Kitagawa 1985; Batta and Singh 1986). Neutral invertase is a cytoplasmic enzyme that may be responsible for cleaving sucrose in support of glycolysis and respiration (ap Rees 1984).

The sucrose-cleavage activity of SS is also high in elongating internodes (Lingle and Smith 1991). Unlike soluble acid invertase, which is extremely low in mature internodes, sucrose synthase activity persists in mature internodes, and may even increase in mature internodes during natural ripening (Lingle and Smith 1991; Lingle and Irvine 1994). The two forms of SS in plants (Chourey 1981), designated SS1 and SS2, are expressed in immature internodes, but only SS2 occurs in fully elongated internodes (Buczynksi et al. 1993). The activity of SS2 is localized in or near the phloem in many plant tissues (Tomlinson et al. 1991; Nolte and Koch 1993) including sugarcane (Buczynksi et al. 1993). Thus, this enzyme may play a key role in sucrose unloading from phloem to storage tissue.

The role of SPS activity in sugar metabolism in sugarcane internodes is not well known. Hatch (1964) demonstrated the synthesis of sucrose-phosphate by stem tissue slices. SPS activity peaks during sucrose accumulation in sugarcane suspension cultures (Wendler et al. 1990; Goldner et al. 1991). However, the relationship of SPS activity to sucrose accumulation in stalks is unclear. Komor et al. (1996) observed no significant differences in SPS activity among internodes along the stem. However, other data suggests that SPS activity may be higher in mature, sucrose-storing internodes than in elongating internodes (S.E. Lingle, unpubl.). In a comparison of high-sucrose and low-sucrose *Saccharum* genotypes, Zhu et al. (1996) found that the difference between SPS and acid invertase activities was highly correlated ($r^2 = 0.86$) with sucrose concentration in stalks.

Clearly, sugarcane is not the simple sink it was originally considered, but a very dynamic system (Fig. 9). In elongating internodes, entering sucrose may be unloaded from the phloem to the apoplast, where it can be cleaved by acid invertase. The glucose and fructose can be transported into the cell, where the combined action of several enzymes including sucrose-phosphate synthase and sucrose-phosphate phosphatase resynthesize sucrose. It can then be transported into the vacuole, where another acid invertase cleaves some of it into glucose and fructose. Alternately, sucrose may be transported intact from apoplast into the cytoplasm, where neutral invertase or sucrose synthase may cleave it. In a fully elongated internode, sucrose may be unloaded through symplastic connections into the cytoplasm of adjacent cells, and then leak into the apoplast. At this stage, sucrose synthase activity is probably associated with the phloem, so that neutral invertase is responsible for cleaving sucrose for gly-

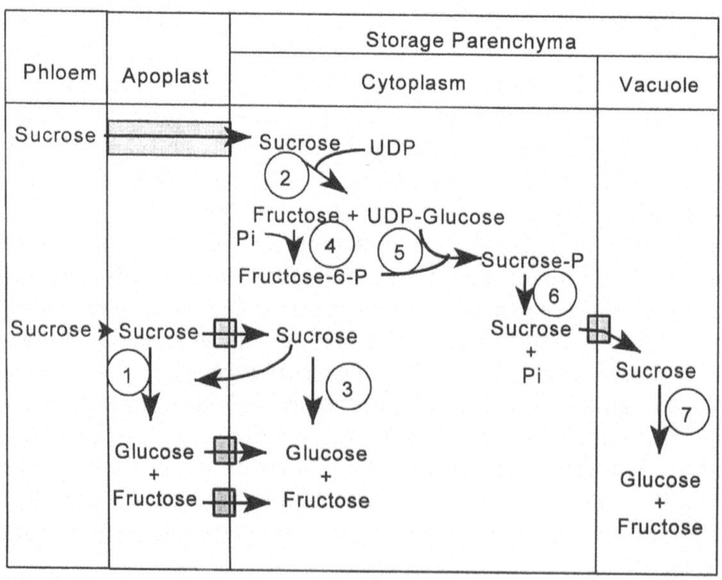

Fig. 9. Schematic representation of pathways for sucrose transport and metabolism in storage parenchyma. Enzymes indicated are: *1* cell-wall acid invertase; *2* sucrose synthase; *3* neutral invertase; *4* hexokinase; *5* sucrose-phosphate synthase; *6* sucrose-phosphate phosphatase; and *7* vacuolar acid invertase. Activities of all enzymes vary with internode development

colytic energy production. Because vacuolar acid invertase is low, glucose and fructose concentrations are low. The presence of a cell wall-bound invertase indicates that some of the sucrose in the apoplast is cleaved. The products of that cleavage may be transported back into the cell where sucrose is resynthesized.

All of these interlinked processes are clearly under some sort of control. Although it appears wasteful to both cleave and synthesize sucrose at the same time in the same cell, the simultaneous presence of the sucrose catabolic and synthetic pathways allows the plant to respond quickly to changes in its environment (Moore 1995).

4 Conclusion: Prospects for Increasing Sucrose Yield

When sucrose is the desired product of sugarcane production, sucrose yield per hectare is of ultimate concern. However, sucrose yield per hectare is a function of both cane yield per hectare and sucrose concentration in the cane. The same sucrose yield can be achieved by a field with high cane yield but lower sucrose concentration as with a field with lower cane yield but higher sucrose concentration. Therefore, greater sucrose yield can be achieved by increasing total biomass produced by the crop, increasing sucrose concentration in the crop, or both.

Increasing cane yield has been a successful means in increasing sucrose yield (Milligan et al. 1990; Heinz et al. 1994). Increasing cane yield might be accomplished by increasing the efficiency of photosynthesis in sugarcane. Photosynthetic rates vary among *Saccharum* species and hybrid cultivars (Irvine 1967; Nose et al. 1994). However, under nonstress conditions photosynthetic rate per leaf area does not appear to limit either cane yield or sucrose accumulation in sugarcane (Irvine 1975). Muchow et al. (1994) calculated maximum radiation-use efficiency in sugarcane during the linear growth phase to be $1.75\,g\,MJ^{-1}$, the theoretical maximum for C_4 crops (Sinclair and Horie 1989). Thus, maximum photosynthetic rates may not be improvable. However, there is likely to be variation in the degree to which various limiting factors, such as temperature, water availability, and nitrogen fertilization, suppress photosynthesis.

Fields yielding the highest biomass achieve high leaf area index quickly after planting or ratooning, and retain that leaf area until harvest. This can be accomplished by increasing plant population, decreasing row spacing, and maintaining high soil nitrogen contents to encourage rapid growth. Selection of plants that begin growth when soil temperatures are below the current optimum may increase biomass accumulation by increasing the duration of maximum leaf area index.

Increased cane yield increases harvest, transport, and milling costs. Where extra bagasse has additional economic value such as electricity or fiber board production, the increased costs of higher cane yield may be offset. If excess bagasse has no value, or if milling capacity cannot be economically increased, increasing sucrose concentration may be a more economic approach to increasing sucrose yield.

The capacity of sugarcane stalks to accumulate sucrose is limited by the translocation, partitioning, and metabolism of sucrose. Selection for high stalk sucrose by traditional plant breeding has resulted in increases in sucrose yield (Irvine and Richard 1983; Breaux 1984; Gravois et al. 1991). However, although there appear to be differences in sucrose biochemistry between *Saccharum* species with extremely high or low sucrose concentrations, clear differences in sucrose transport and biochemistry among improved genotypes are not observed. Genetic modification through biotechnology offers hope for insight and improvement in this area. Many of the key genes of sucrose metabolism have been isolated and cloned. This makes it possible both to examine gene expression related to sucrose accumulation and to manipulate enzyme activity by transforming plants. In time, this may lead to sugarcane genotypes with enhanced capability to store sucrose.

References

Alexander AG (1967) High sucrose levels and abnormal enzyme activity as a function of nutritional stress in sugarcane. J Agric Univ P R 51:325–333

Alexander AG, Montalvo-Zapata R (1972) Ripening activity of C.P. 41845 in sugarcane having nitrate- and gibberellic acid-stimulated growth regimes. Crop Sci 12:654–657

ap Rees T (1984) Sucrose metabolism. In: Lewis DH (ed) Storage carbohydrates in vascular plants. Cambridge Univ Press, Cambridge, pp 53–73

Artschwager E (1925) Anatomy of the vegetative organs of sugar cane. J Agric Res 30:197–241

Artschwager E (1940) Morphology of the vegetative organs of sugarcane. J Agric Res 60:503–549

Bacchi OOS, Sousa JAGC (1977) Minimum threshold temperature for sugarcane growth. Proc Int Soc Sugar Cane Technol 16:1733–1742

Batta SK, Singh R (1986) Sucrose metabolism in sugar cane grown under varying climatic conditions: synthesis and storage of sucrose in relation to the activities of sucrose synthase, sucrose-phosphate synthase and invertase. Phytochemistry 25:2431–2437

Benda GTA, Fanguy HP, Irvine JE, Breaux RD (1981) Field planting of short sugarcane seed pieces in Louisiana. Proc Am Soc Sugar Cane Technol 8:40–48

Breaux RD (1984) Breeding to enhance sucrose content of sugarcane varieties in Louisiana. Field Crops Res 9:59–67

Buczynksi SR, Thom M, Chourey P, Maretzki A (1993) Tissue distribution and characterization of sucrose synthase isozymes in sugarcane. J Plant Physiol 142:641–646

Bull TA, Glasziou KT (1963) The evolutionary significance of sugar accumulation in *Saccharum*. Aust J Biol Sci 16:737–742

Bull TA, Tovey DA (1974) Aspects of modelling sugarcane growth by computer simulation. Proc Int Soc Sugar Cane Technol 15:1021–1032

Buzzanell PJ, Lockley-Jolly F (1994) Sugar and sweetener yearbook. computer file #89019. Economic Research Service, US Dept of Agriculture, Washington, DC

Celestine-Myrtil-Marlin AD (1990) Influence of cane age on sugars and organic acids distribution in sugarcane stalks. Sugar Azúcar (August):17–24

Celestine-Myrtil DA, Parfait A (1987) HPLC analysis of sugars in sugar cane stalks. Int Sugar J 89:217–220

Charles-Edwards DA (1982) Physiological determinants of crop growth. Academic Press, Sydney

Chen JCP (1985) Meade-Chen cane sugar handbook, 11th edn. John Wiley, New York

Chourey PS (1981) Genetic control of sucrose synthetase in maize endosperm. Mol Gen Genet 184:372–376
Clements HF (1940) Factors affecting the germination of sugar cane. Hawaii Plant Rec 44:117–146
Clements HF (1959) Quality in sugar production – field aspects. Rept Hawaii Sugar Technol Assoc (18th Annu Meet):17–24
Clements HF (1980) Sugarcane crop logging and crop control. The University Press of Hawaii, Honolulu
Clements HF, Kubota T (1943) The primary index, its meaning and application to crop management with special reference to sugar cane. Hawaii Plant Rec 47(4):257–297
Clements HF, Shigeura H, Akamine EK (1952) Factors affecting the growth of sugar cane. University of Hawaii Agricultural Experiment Station Technical Bull 18, Honolulu
Coale FJ, Sanchez CA, Izuno ET, Bottcher AB (1993) Nutrient accumulation and removal by sugarcane grown on Everglades histosols. Agron J 85:310–315
Cornelison AH (1944) Vegetative differences influence the composition of sugar cane. Hawaii Plant Rec 48:125–164
Daniels J, Roach BT (1987) Taxonomy and evolution. In: Heinz DJ (ed) Sugarcane improvement through breeding. Elsevier, Amsterdam, pp 7–84
Das UK (1934) The sugar cane plant. A study of millable cane and sucrose formation. Hawaii Sugar Plant Rec 38:33–94
D'Hont A, Lu YH, Gonzales de Leon D, Grivet L, Feldmann P, Lanaud C, Glaszmann JC (1994) A molecular approach to unraveling the genetics of sugarcane, a complex polyploid of the Andropogoneae tribe. Genome 37:222–230
Dong ZM, Canny MJ, McCully ME, Roboredo MR, Cabadilla CF, Ortega E, Rodes R (1994) A nitrogen-fixing endophyte of sugarcane stems. A new role for the apoplast. Plant Physiol 105:1139–1147
Fernandes AC, Benda GTA (1985) Distribution patterns of Brix and fibre in the primary stalk of sugarcane. Sugar Cane (5):8–13
Furbank RT, Taylor WC (1995) Regulation of photosynthesis in C_3 and C_4 plants: a molecular approach. Plant Cell 7:797–807
Gascho GJ, Ruelke OC, West SH (1973) Residual effects of germination temperature in sugarcane. Crop Sci 13:274–276
Glasziou KT (1960) Accumulation and transformation of sugars in sugar cane stalks. Plant Physiol 35:895–901
Glasziou KT (1961) Accumulation and transformation of sugars in stalks of sugar cane. Origin of glucose and fructose in the inner space. Plant Physiol 36:175–179
Glasziou KT, Bull TA (1965) The relation between total invertase activity and internode expansion in sugarcane stalks. Proc Int Soc Sugar Cane Technol 12:575–581
Glasziou KT, Gayler KR (1972a) Storage of sugars in stalks of sugar cane. Bot Rev 38:471–490
Glasziou KT, Gayler KR (1972b) Sugar accumulation in sugarcane. Role of cell walls in sucrose transport. Plant Physiol 49:912–913
Glasziou KT, Waldron JC (1964) Regulation of acid invertase levels in sugarcane stalks by auxin- and metabolite-mediated control systems. Nature 203:541–542
Glasziou KT, Bull TA, Hatch MD, Whiteman PC (1965) Physiology of sugar-cane. VII. Effects of temperature, photoperiod duration, and diurnal and seasonal temperature changes on growth and ripening. Aust J Biol Sci 18:53–66
Glaszmann JC, Noyer JL, Fautret A, Lanaud C, Feldmann P (1989) Molecular genetic markers in sugarcane. Proc Int Soc Sugar Cane Technol 20:872–882
Goldner W, Thom M, Maretzki A (1991) Sucrose metabolism in sugarcane cell suspension cultures. Plant Sci 73:143–147
Grantz DA (1989) Effect of cool temperatures on photosynthesis and stomatal conductance in field-grown sugarcane in Hawaii. Field Crops Res 22:143–155
Grantz DA, Moore PH, Zeiger E (1987) Stomatal responses to light and humidity in sugarcane: prediction of daily time courses and identification of potential selection criteria. Plant Cell Environ 10:197–204

Gravois KA, Milligan SB, Martin FA (1991) Indirect selection for increased sucrose yield in early sugarcane testing stages. Field Crops Res 26:67–73
Hartt CE (1965) The effect of temperature upon translocation of ^{14}C in sugarcane. Plant Physiol 40:74–81
Hartt CE (1967) Effect of moisture supply upon translocation and storage of ^{14}C in sugarcane. Plant Physiol 42:338–346
Hartt CE (1969) Effect of potassium deficiency upon translocation of ^{14}C in attached blades and entire plants of sugarcane. Plant Physiol 44:1461–1469
Hartt CE (1970) Effect of nitrogen deficiency upon translocation of ^{14}C in sugarcane. Plant Physiol 46:419–422
Hartt CE, Burr GO (1965) Factors affecting photosynthesis in sugar cane. Proc Int Soc Sugar Cane Technol 12:590–609
Hartt CE, Kortschak HP, Forbes AJ, Burr GO (1963) Translocation of ^{14}C in sugarcane. Plant Physiol 38:305–319
Hatch MD (1964) Sugar accumulation by sugar-cane storage tissue: the role of sucrose phosphate. Biochem J 93:521–526
Hatch MD (1987) C_4 photosynthesis: a unique blend of modified biochemistry, anatomy and ultrastructure. Biochim Biophys Acta 895:81–106
Hatch MD, Glasziou KT (1963) Sugar accumulation cycle in sugar cane. II. Relationship of invertase activity to sugar content and growth rate in storage tissue of plants grown in controlled environments. Plant Physiol 38:344–348
Hatch MD, Slack CR (1966) Photosynthesis by sugarcane leaves. A new carboxylation reaction and the pathway of sugar formation. Biochem J 101:103–111
Hawker JS (1965) The sugar content of cell walls and intercellular spaces in sugar-cane stems and its relation to sugar transport. Aust J Biol Sci 18:959–969
Hawker JS, Hatch MD (1965) Mechanism of sugar storage by mature stem tissue of sugarcane. Physiol Plant 18:444–453
Heinz DJ, Osgood RV, Moore PH (1994) Sugarcane. In: Encyclopedia of agricultural science, vol 4. Academic Press, San Diego, pp 225–237
Hitz WD, Schmitt MR, Card PJ, Giaquinta RT (1985) Transport and metabolism of 1′-fluorosucrose, a sucrose analog not subject to invertase hydrolysis. Plant Physiol 77:291–295
Humbert RP (1968) The growing of sugar cane, revised edn. Elsevier, Amsterdam
Inman-Bamber NG (1994) Temperature and seasonal effects on canopy development and light interception of sugarcane. Field Crops Res 36:41–51
Irvine JE (1967) Photosynthesis in sugarcane under field conditions. Crop Sci 7:297–300
Irvine JE (1975) Relations of photosynthetic rates and leaf and canopy characters to sugarcane yield. Crop Sci 15:671–676
Irvine JE, Benda GTA (1980) Sugarcane spacing. II. Effects of spacing on the plant. Proc Int Soc Sugar Cane Technol 17:357–367
Irvine JE, Richard CA (1983) Production statistics as evidence of technological improvement: the Louisiana sugar cane industry as a case study. Sugar Cane 1983(1):13–18
Irvine JE, Benda GTA, Richard CA (1980a) Sugarcane spacing. I. Historical and theoretical aspects. Proc Int Soc Sugar Cane Technol 17:350–356
Irvine JE, Richard CA, Garrison DD, Jackson WR, Matherne RJ, Camp C, Carter C (1980b) Sugarcane spacing. III. Development of production techniques for narrow rows. Proc Int Soc Sugar Cane Technol 17:368–376
Jacobsen KR, Fisher DG, Maretzki A, Moore PH (1992) Developmental changes in the anatomy of the sugarcane stem in relation to phloem unloading and sucrose storage. Bot Acta 105:70–80
Komor E, Zingsheim O, Sprügel H (1996) Cycles of sugar transport and sucrose metabolism in sugarcane tissue: Quantitative determination. In: Wilson JR, Campbell JA, Garside AL (eds) Sugarcane: research towards efficient and sustainable production. CSIRO Division of Tropical Crops and Pastures, Brisbane, pp 92–94

Legendre BL (1975) Ripening of sugarcane: effects of sunlight, temperature, and rainfall. Crop Sci 15:349–352
Lingle SE (1989) Evidence for the uptake of sucrose intact into sugarcane internodes. Plant Physiol 90:6–8
Lingle SE, Irvine JE (1994) Sucrose synthase and natural ripening in sugarcane. Crop Sci 34:1279–1283
Lingle SE, Smith RC (1991) Sucrose metabolism related to growth and ripening in sugarcane internodes. Crop Sci 31:172–177
MacColl D (1976) Growth and sugar accumulation of sugarcane. I. Percentage of sugar in relation to pattern of growth. Exp Agric 12:369–377
Maretzki A, Thom M (1972) Membrane transport of sugars in cell suspensions of sugarcane. I. Evidence for sites and specificity. Plant Physiol 49:177–182
Martin JP (1938) Sugarcane diseases in Hawaii. Experiment Station, Hawaiian Sugar Planters' Association, Honolulu
Matsui T, Kitagawa H (1985) Seasonal changes of activities of invertases in relation to sugar accumulation in sugarcane. Nippon Shokuhin Kogyo Gakkaishi 32:655–660
McMartin A (1979) The logistic curve of plant growth and its application to sugarcane. Proc S Afr Sugar Technol Assoc 53:189–193
Meinzer FC, Grantz DA (1990) Stomatal and hydraulic conductance in growing sugarcane: stomatal adjustment to water transport capacity. Plant Cell Environ 13:383–388
Milligan SB, Gravois KA, Bischoff KP, Martin FP (1990) Crop effects on genetic relationship among sugarcane traits. Crop Sci 30:927–931
Mongelard JC, Mimura L (1971) Growth studies on the sugarcane plant. I. Effects of temperature. Crop Sci 11:795–800
Mongelard JC, Mimura L (1972) Growth studies of the sugarcane plant. II. Some effects of root temperature and gibberellic acid and their interactions on growth. Crop Sci 12:52–58
Moore PH (1995) Temporal and spatial regulation of sucrose accumulation in the sugarcane stem. Aust J Plant Physiol 22(4):661–679
Moore PH, Cosgrove DJ (1991) Developmental changes in cell and tissue water relations parameters in storage parenchyma of sugarcane. Plant Physiol 96:794–801
Muchow RC, Spillman MF, Wood AW, Thomas MR (1994) Radiation interception and biomass accumulation in a sugarcane crop grown under irrigated tropical conditions. Aust J Agric Res 45:37–49
Nolte KD, Koch KE (1993) Companion-cell specific localization of sucrose synthase in zones of phloem loading and unloading. Plant Physiol 101:899–905
Nose A, Uehara MY, Kawamitsu Y, Kobamoto N, Nakama M (1994) Variations in leaf gas exchange traits of *Saccharum* including feral sugarcane, *Saccharum spontaneum* L. Jpn J Crop Sci 63:489–495
Oworu OO, McDavid CR, MacColl D (1977a) The anatomy of the storage tissue of sugar-cane in relation to sugar uptake. Ann Bot 41:401–404
Oworu OO, McDavid CR, MacColl D (1977b) A comparison of rates of storage of sucrose in eight clones of sugar-cane as measured by sucrose uptake in vitro. Ann Bot 41:393–399
Patrick JW (1990) Sieve element unloading: cellular pathway, mechanism and control. Physiol Plant 78:298–308
Polack J (1995) Louisiana raw cane sugar in the 20th century. Sugar J 57(9):23–28
Purchase BS (1995) Products from sugarcane. Int Sugar J 97:70–72, 77–80
Ramana Rao TC, Sreenivasan TV, Palanichami K (eds) (1985) Catalogue on sugarcane genetic resources. II. *Saccharum barberi*. Jeswiet. *Saccharum sinsense*. Roxb. Amend Jeswiet. *Saccharum robustum*. Brandes et Jeswiet ex Grassl. *Saccharum edule*. Hassk. Sugarcane Breeding Institute, Coimbatore, India
Richard EPJ, Dunckelman JW, Carter CE (1991) Productivity of sugarcane on narrow rows, as affected by mechanical harvesting. Field Crops Res 26:375–386

Sacher JA, Hatch MD, Glasziou KT (1963) Sugar accumulation cycle in sugar cane. III. Physical and metabolic aspects of cycle in immature storage tissues. Plant Physiol 38:348–354
Samuels G, Landrau PJ (1952) The response of sugarcane to fertilizers. I. The Arecibo cycle, 1944–1950. J Agric Univ Puerto Rico 36:203–229
Sinclair TR, Horie T (1989) Leaf nitrogen, photosynthesis, and crop radiation use efficiency: a review. Crop Sci 29:90–98
Singh G, Singh OP (1993) Performance of sugarcane (*Saccharum officinarum*) varieties at various row spacings when grown under flood-prone conditions. Indian J Agric Sci 63:818–820
Singh O, Kanwar RS (1986) Association of some cane sett assimilate with germination. Sugar Cane 1986(2):7–10
Singh S (1975) Significance of critical nodal water content of sugarcane sett in germination of bud. Sci Cult (India) 41:475–477
Singh S, Srivastava KK (1974) Effects of soil-water potential on germination of sugarcane setts. Indian J Agric Sci 44:184–187
Sreenivasan TV, Nair NV (eds) (1991) Catalogue on sugarcane genetic resources. III. *Saccharum officinarum* L. Sugarcane Breeding Institute, Coimbatore, India
Sreenivasan TV, Ahloowalia BS, Heinz DJ (1987) Cytogenetics. In: Heinz DJ (ed) Sugarcane improvement through breeding. Elsevier, Amsterdam, pp 211–253
Su LY, Cruz AD, Moore PH, Maretzki A (1992) The relationship of glyphosate treatment to sugar metabolism in sugarcane: new physiological insights. J Plant Physiol 140:168–173
Tai PYP, Rice ER, Chew V, Miller JD (1982) Phenotypic stability analyses of sugarcane cultivar performance tests. Crop Sci 22:1179–1184
Thom M, Maretzki A (1992) Evidence for direct uptake of sucrose by sugarcane stalk tissue. J Plant Physiol 139:555–559
Thomas JR, Oerther GFJ (1976) Growth, production, and leaf N content of sugarcane in Texas. Proc Am Soc Sugar Cane Technol 5:28–36
Thompson GD (1978) The production of biomass by sugarcane. Proc S Afr Sugar Technol Assoc 52:180–187
Tomlinson PT, Duke ER, Nolte KD, Koch KE (1991) Sucrose synthase and invertase in isolated vascular bundles. Plant Physiol 97:1249–1252
van Dillewijn C (1952) Botany of sugarcane. The Chronica Botanica Co, Waltham, Massachusetts
Walsh KB, Sky RC, Brown SM (1996) Pathway of sucrose unloading from the phloem in sugarcane stalks. In: Wilson JR, Campbell JA, Garside AL (eds) Sugarcane: research towards efficient and sustainable production. CSIRO Division of Tropical Crops and Pastures, Brisbane, pp 105–107
Welbaum GE, Meinzer FC (1990) Compartmentation of solutes and water in developing sugarcane stalk tissue. Plant Physiol 93:1147–1153
Wendler R, Veith R, Dancer J, Stitt M, Komor E (1990) Sucrose storage in cell suspension cultures of *Saccharum* sp. (sugarcane) is regulated by a cycle of synthesis and degradation. Planta 183:31–39
Wood GH (1962) Some factors influencing starch in sugarcane. Proc S Afr Sugar Technol Assoc 36:123–135
Yadav RL (1993) Cane yield in relation to shoot-population density of sugarcane (*Saccharum* species). Indian J Agron 38:337–338
Zhu YJ, Komor E, Moore PH (1996) Relationship of sucrose metabolism enzymes with sucrose storage in sugarcane. In: Wilson JR, Campbell JA, Garside AL (eds) Sugarcane: research towards efficient and sustainable production. CSIRO Division of Tropical Crops and Pastures, Brisbane, pp 89–91

Sugar Beet

T.H. Thomas[1]

1 Introduction

Sugar beet (*Beta vulgaris* L.), a member of the Chenopodiaceae, is a biennial plant grown almost exclusively as a source of sucrose. However, following sucrose extraction, the pulp, often enriched with molasses and then dried and pelleted, provides a marketable high-energy animal feed. Historically, the crop was developed in central Europe and was grown mainly throughout that continent, but it has now been introduced onto all of the populated continents except Australia. Nevertheless, it is essentially a temperate crop, the majority grown between latitudes of 30° and 60°N in Europe. At present, it is only grown to any extent in the Southern Hemisphere in Chile.

The success of sugar beet as a crop depends on predictable seed germination, early seedling establishment and the rapid development of a leaf canopy which is able to utilize the available solar radiation efficiently. Indeed, a criterion for success used by most farmers and processors is the date by which the leaves of plants in adjacent rows begin to overlap (see Sect. 2).

Following germination, the plant produces a rosette of glabrous, dark green, glossy leaves with prominent midribs and strong petioles. Leaf production continues throughout the season, with few leaves senescing, while the tap root enlarges and accumulates sucrose as its main storage product. In reality, only 90% of the storage organ is root-derived, the crown being mainly hypocotyl tissue. Generally, sucrose concentrations are lower and α-amino nitrogen compounds higher in the crown region than in the remainder of the root, so that some of this tissue is usually removed before processing in the factory. The fresh weight concentration of sucrose in the roots of well-grown, modern varieties is about 18%.

Flowering of sugar beet (Longden et al. 1995), which usually occurs in the 2nd year, is induced by low temperature, but the development of the flowering stem (bolting) and subsequent floral organs (flowering) is dependent on lengthening days. Both seeds and seedlings can be vernalized so that early sowings of the crop may contain bolters if early season temperatures consistently fail to exceed the vernalization threshold of about 12 °C. Bolters are to

[1] IACR, Long Ashton Research Station Department of Agricultural Sciences, University of Bristol, Long Ashton, Bristol, BS18 9Af, UK

Crop Yield, Physiology and Processes
D. L. Smith, C. Hamel (Eds.)
© Springer-Verlag Berlin Heidelberg 1999

be avoided since they may produce seeds which can give rise to weed beet. Additionally, they contain lower sucrose percentages, create harvesting problems, are difficult to slice and, in the pulp driers, make dust which can ignite spontaneously.

Extensive literature dealing with sugar beet physiology and biochemistry details the many attempts to find ways of increasing sucrose yield without increasing juice impurity. This chapter discusses some of the physiological limitations to yield and sucrose production and describes some approaches for improving productivity of the crop.

2 Crop Establishment and Early Development

2.1 Seedling Establishment

There is considerable evidence that crop yield and sugar production is directly related to the amount of radiation intercepted by sugar beet foliage between sowing and harvest (Scott et al. 1973; Fig. 1). Consequently, the time of sowing is a critical element contributing to final crop yield. However, in most sugar beet regions, with the exception of Mediterranean and North African countries, there is little point in sowing before March because the average temperatures near the soil surface are close to or below 3 °C, which is the base temperature for germination. As already mentioned, premature sowings also introduce the risk of a substantial percentage of the crop bolting. In the UK, sowing on 20 March has been shown to incur an acceptable risk of bolting one year in ten. This coincides with the best time to sow in order to achieve a leaf canopy that closes early. In dry, hot climates such as those around the Mediterranean Sea, sowing an overwintered crop in autumn is a standard practice. However, irrigation is necessary under these conditions. A similar approach could be adopted in more northerly latitudes, if a non-bolting sugar beet variety could be developed, possibly through genetic engineering (Longden and Thomas 1989). The potential benefits could be as much as 15% extra sucrose or the possibility of earlier harvesting and processing.

Rapid establishment of uniform and vigorous sugar beet plants can only be achieved with good quality seed, which germinates quickly and synchronously. Most crops are now grown from monogerm, high-processed seeds, drilled to a stand using precision drills. In the UK, the farmer aims to achieve at least 75 000 plants/ha from around 100 000 seeds sown, but higher plant densities are common in other countries. In some countries, the crop is not drilled to a stand. In the UK, plant establishment improved greatly throughout the 1980s, with average field establishment increasing from about 60% in 1981 to around the present level of 83–85% (Kerr and Leaman 1995). The laboratory germination of modern varieties is now nearly 100%. These good germination and

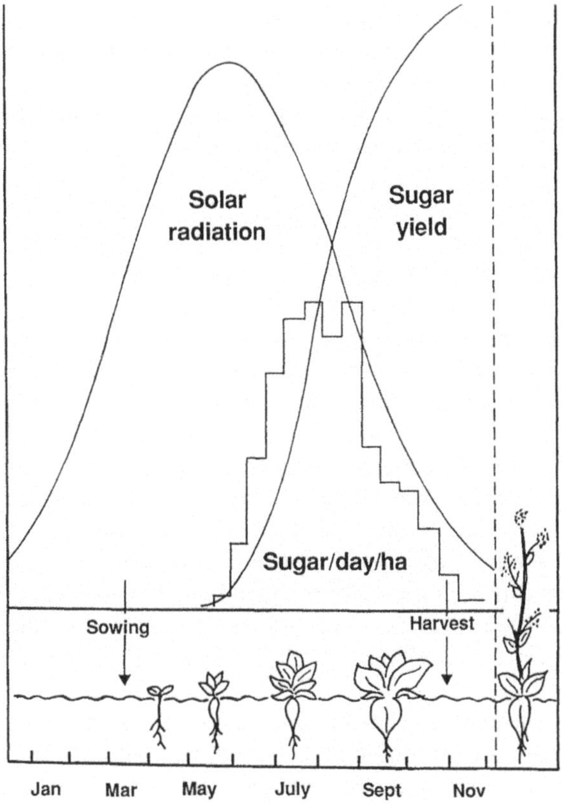

Fig. 1. Relationship between solar radiation, sugarbeet development and sugar accumulation and yield (northwestern Europe). (After Thomas et al. 1993)

emergence values are a consequence of successful breeding and innovations in seed production, processing and pelleting.

2.2 Identifying Seed Vigour

Variations in seedling growth rate can be due to several interacting factors, including initial seed weight (Scott et al. 1974), seed placement and seedbed structure (Durr et al. 1992), and various environmental influences including temperature and water availability. The time taken from germination to emergence of the seedling from the soil can have persistent effects on crop development. Inherent factors in the seed contribute to what is loosely described as seed vigour; its expression is determined by both genetic and environmental factors. Several laboratory vigour tests have been developed; generally, these subject the germinating seedlings to stress conditions using different temperature and moisture regimes. These tests can be reasonable indicators of the difference between laboratory and field performance. However, because the tests are time-consuming and sometimes inconsistent, a rapid biochemical test for seed vigour has been the goal of seed physiologists for many years. Parameters such as solute leakage, energy charge and lipid composition have been examined with limited success, but a recent test based on the quantity of extractable high molecular weight nucleic acids is showing considerable promise (Thomas et al. 1995).

2.3 Improving Seed Performance

Numerous ways of improving the germination of sugar beet seeds have been explored. Treatments involving leaching out inhibitors with water (Scott et al. 1972), steeping seeds in aqueous solutions of plant hormones (Scott et al. 1972; Nelson and Sharples 1980; Nelson et al. 1984), dilute acids (Akeson et al. 1981) or fungicides (Durrant et al. 1988) have met with limited success and have not been exploited commercially. Longden et al. (1979) examined the establishment and growth of sugar beet from either fluid-drilled (chitted and sown in a carrier gel) or advanced seeds (controlled imbibition followed by drying before sowing). The fluid-drilling technique did not give large benefits but the advancement treatment showed promise and has since been modified (Durrant and Jaggard 1988; Durrant and Mash 1990; Thomas et al. 1994b) and developed commercially.

The basic advancement treatment consisted of steeping seeds in an agitated 0.2% (w/v) aqueous suspension of thiram at 25 °C for 8 h, partially drying to 124% of the original weight, incubating at 25 °C for 88 h and then air-drying. The commercial treatment is a modification of this process. The seeds are physiologically active during the hot treatment and are devernalized but they do not germinate because their water content is close to the base water potential for germination (Gummerson 1986).

There are two potential advantages to be gained from sugar beet seed advancement (Thomas et al. 1994b). Firstly, allowing the seeds to complete part of the germination process before sowing reduces the time for seedlings to emerge as compared with non-advanced seeds. This effect could increase sugar yield in the UK by ca. 0.19 $t\,ha^{-1}$ (0.048 $t\,ha^{-1}\,day^{-1}$). Secondly, devernalization of the seeds following controlled imbibition at high temperatures allows earlier sowing. For every growing day gained during March, yield can be increased by about 0.035 $t\,ha^{-1}\,day^{-1}$. The number of days likely to be gained by advancing the earliest sowing date from 20 March to 11 March will depend on whether or not the soil is friable and dry enough to be worked by machinery. It has been calculated that, because of the earlier emergence and crop development, the treatment could be worth an extra 0.35 $t\,ha^{-1}$ of sugar to the UK grower and processor, which is about 5% of the average national yield.

Other approaches for improving seed performance include osmotic priming and matriconditioning. Osburn and Schroth (1989) found that seedling fresh weight averaged 57 and 39% greater with NaCl and PEG osmoprimed seed respectively, compared with non-treated seeds. Priming treatments have also reduced the incidence of preemergence damping-off in soils infested with *Pythium ultimum* (Rush 1991, 1992) and some of the benefits to seedling development have been attributed to this effect. Seed matriconditioning involves presowing treatments with moist solid carriers that have matric properties (Khan et al. 1995). Experiments in the USA involved mixing the seed with water and Micro-Cel E, a synthetic calcium silicate with high water-holding capacity, thus obtaining a water potential of −1.4 MPa. Field emergence at suboptimal temperatures was increased by ca. 10% with reductions in the time to 50% emergence (T_{50}) of up to 2 days. It is suggested that the treatment influences physical, biochemical and physiological events associated with improved performance at low temperatures. However, the commercial potential of seed priming and matriconditioning technology has not yet been determined in extensive field tests.

2.4 Manipulating Seed Production

A further possibility for improving seed performance is by chemical or biotechnological modification of the seed crop. Sugar beet has an indeterminate inflorescence on which the earliest, most mature seeds are largest (Alcaraz 1981); these give rise to the most vigorous seedlings which, in turn, may give rise to plants with the greatest sucrose yield (Thomas et al. 1996). As a result, up to 75% of harvested seed is discarded during grading on a size basis. A shorter more determinate inflorescence could produce seeds of more uniform size. An additional benefit of modifying the inflorescence in this way would be that seed could be harvested early to avoid its being vernalized on the straw, thus reducing the tendency to bolt in the root crop. However, attempts to do this with chemical treatments, particularly growth retardants, have been unsuccessful in sugar beet; the treatments either have little effect or decrease seed yield and germination (Longden 1974).

3 Leaf Growth and Development

During the first stages of seedling development there is little tap root growth, energy being used mainly for leaf and fibrous root development (see Fig. 2). Detailed descriptions of the patterns of leaf development under UK conditions, from initiation to senescence, have been recorded by Milford and coworkers (1985a–d) and for central European conditions by Hodanova (1981). Beet leaves were produced continuously throughout the season, with two to three leaves appearing each week during the summer months, reaching an approximate total of 40 leaves by early October. The leaf area index (LAI) reaches its

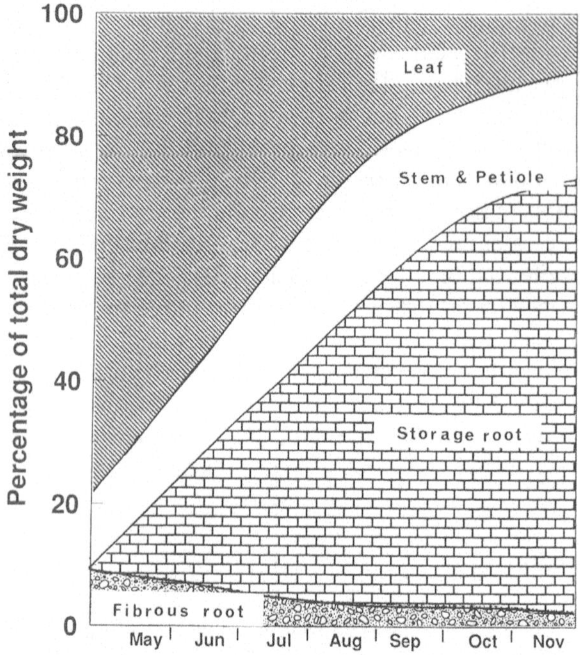

Fig. 2. Sugarbeet dry matter distribution throughout the season. (After Scott et al. 1974)

maximum around the time of full expansion of the largest leaf (ca. leaf no. 12); in order to achieve the maximum rate of growth, the maximum LAI should exceed 3 (Goodman 1966).

Leaf production and expansion early in the season are limited by low temperature. Light quality and duration affect leaf growth independently of temperature (Milford and Lenton 1978). Day-length extensions with light rich in energy from red to far-red wavelengths are particularly effective in increasing leaf growth, partly by causing changes in the production of endogenous hormones, particularly gibberellins (Lenton and Milford 1977). Although it is possible to increase leaf area and petiole growth by treatment with gibberellic acid, unfortunately these effects are not translated into increased accumulation of sugar in the storage root.

The first two leaves usually senesce and die in June, with about 10–12 leaves dying in sequence by the end of September. Continued leaf growth in the autumn may prevent maximum potential sucrose accumulation by the storage root. In countries where the crop matures in warm conditions, late season application of growth retardants such as maleic hydrazide has been shown to prevent the formation of new leaves, resulting in increased dry matter of the harvested root (Wittwer and Hansen 1952), but this is not used commercially.

4 Fibrous Root Development

4.1 Root Growth in the Soil

Extensive field studies by Brown and Dunham (1986) showed that, from the two-leaf seedling stage onward, the fibrous root system developed laterally at about 0.4 cm/day until approximately 80 days after sowing. During this period the roots penetrated to a depth of over 50 cm, by which time the foliage covered over 30% of the ground. When a full leaf canopy had been achieved, the root system had extended down to 1 m, thereafter penetrating to a depth of more than 1.5 m. The average length of root per plant in the top 30 cm of soil exceeded 600 cm by the end of the summer (Fig. 3). In terms of allometric relationships, the fibrous root system accounts for a progressively decreasing proportion of the plant weight, from about 10% of the total in early June to about 3% near harvest (Brown and Biscoe 1985). The roots in the top soil are more active in taking up water and nutrients per unit root length than the deeper roots (Armstrong et al. 1986). Water is removed preferentially from the surface layers but as the root system dries out the surface layers, water uptake is from progressively deeper in the soil profile (Brown et al. 1987).

4.2 Root-to-Shoot Ratios

In comparing seedlings grown either in vermiculite or in nutrient solution, Fick et al. (1972) concluded that many properties of the fibrous root were fairly constant. For instance, the frequency distributions of diameters, and of lengths in different diameter classes, were the same for sugar beet plants of different ages grown in the different media. The ratio of root to shoot growth varies widely among plant species and is strongly modified by external factors. However, there is a general tendency between and within species to maintain a characteristic relationship between root and shoot dry weight. This was clearly demonstrated by Fick et al. (1971) for sugar beet, in leaf and root pruning experiments using seedlings at the 6- or 12-leaf stage. Leaf removal resulted in a rapid decrease in the amount of new growth going to the fibrous roots and an increase in the relative growth rate of leaf blades; the opposite effect was produced by root pruning. Thus, the partitioning of photosynthate between leaves and fibrous roots appears to be regulated by a functional equilibrium in sugar beet.

Recent data obtained from liquid nutrient culture experiments (Thomas 1996) indicate a similarity in the distribution of fibrous root to shoot ratios in seedlings at the eight-leaf stage across three varieties of sugar beet. Most plants had a root-to-shoot ratio within the range 0.27 to 0.32. However,

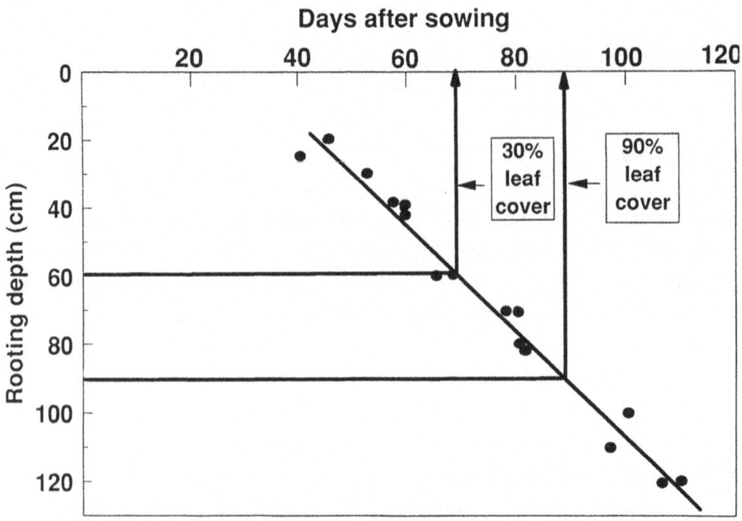

Fig. 3. Increase in effective rooting depth with time for sugarbeet crops averaged over 5 years. (After Brown and Dunham 1986)

the wide spread of values (CV between 55–72%) suggests an opportunity for manipulating allometric relationships in sugar beet by selection and breeding.

Fibrous root growth and function are both dynamic phenomena, with compensatory growth occurring in one part of the system when other regions are suffering from damage or stress (Fick et al. 1975). Similarly, shifts in the root to shoot ratio can be induced by various environmental changes such as drought. In sugar beet, the form and function of fibrous roots have been very much underresearched. With more information it might be possible to introduce more desirable root characteristics into new varieties or to manipulate root system distribution in the soil profile to optimize water and nutrient uptake.

5 Growth and Development of the Storage Root

5.1 Anatomical Development of the Storage Root

The *Beta vulgaris* species is diverse, including quite dissimilar crops such as leafy chard, table beet, fodder beet and sugar beet. Fodder beet, table beet and sugar beet are tap-rooted and secondary vascular meristematic activity in the upper part of the root and in the hypocotyl and stem leads to the formation of the storage root. The percentage of hypocotyl contributing to the storage root is greater in fodder beet; the aim of sugar beet breeders has been to minimize this tissue since it contains less sugar and more impurities than does the root. Depending on the environment, the storage root contains 4 to 15 cambial rings which are initiated in quick succession (Artschwager 1926). Cambial activity declines acropetally, producing a tapered storage root. Studies by Milford (1973) and Rapoport and Loomis (1986) showed that primary cambium initiation was complete 2 weeks after emergence, and the first two secondary cambia were formed during the next week. Six weeks after emergence, when the plants had produced 12 leaves and the storage root was 1.0–1.5 cm in diameter, 6 rings had been formed.

Cambial activity continues throughout the season and does not limit the potential for storage. Under field conditions, plant spacings are such that each plant is unable to achieve its potential root size. Elliott et al. (1984) have pointed out that growth occurs concurrently in all the rings and not consecutively, but that expansion of the storage root involves significant contributions from only about half of the cambia. Thus rings 1 to 6 make up approximately 75% of the mature storage root and those beyond ring 9 contribute very little to final yield.

5.2 Sucrose Accumulation

The way in which sucrose is synthesized and transported in sugar beet has been well documented (Giaquinta 1979; Wyse 1979; Thomas 1986). Sucrose from the phloem is stored in the parenchyma cell vacuoles both in the vascular zones and in the parenchymatous zone itself (Giaquinta 1979). The cells of the vascular zone have the highest sucrose concentration (Zamski and Azenhot 1981) and sucrose levels are greatest in the centre of the root section, which has the largest diameter. It is generally assumed that the major limitations to sucrose accumulation lie within the storage root, i.e., it is controlled by sink strength. For example, when roots of a low-sucrose fodder beet genotype were grafted on to the tops of a high sucrose sugar beet genotype, the sucrose content of the fodder beet increased by only 13%. Reciprocal grafts reduced the sucrose content of the sugar beet root by only 5% (Rapoport and Loomis 1985). However, Milford et al. (1988) suggested that environmental and husbandry effects on yield could be more easily explained by shoot rather than root events determining the allocation of dry matter between growth and sucrose storage in the root. These conclusions were based on the growth of a number of sugar beet crops over several seasons. Using a modelling approach, no evidence of discontinuity was found in carbohydrate partitioning between shoot and root at any stage of crop development. This indicates that sugar beet has no obvious separate growth phases and, unlike sugar cane, does not exhibit a ripening phenomenon.

6 Photosynthesis and Growth

6.1 Carbon Fixation in the Leaf

As a C_3 plant, sugar beet fixes CO_2 by the Calvin cycle. An overall description of photosynthesis and its control in beet has been provided by Elliott and Weston (1993). Sucrose biosynthesis in the leaf is controlled by a mixture of feedforward and feedback mechanisms involving several enzymes including fructose bisphosphate phosphatase (FBPase), sucrose phosphate synthetase (SPSase) and sucrose phosphate phosphatase (SPPase). The partitioning of carbon among various carbon pools in the leaf appears to be determined in part by the relative capacities of some enzymes for starch and sucrose metabolism (Rao et al. 1990). Under low light with low rates of photosynthesis, starch formation is favoured, but in bright light the cytosolic resistance to sucrose synthesis is overcome. The maximum rate of photosynthesis appears to be restricted by the rate of end-product synthesis (Stitt and Quick 1989).

Although plants tend to export sucrose continuously through each 24-h period, in sugar beet export is much less during the night than in the day time (Terry and Mortimer 1972). During the day, carbohydrate is stored temporarily as starch in readiness for export at night when photosynthesis is not occurring. Despite this, 40% of sugar beet leaf sucrose is also in a storage pool and is not being transported (Fondy and Geiger 1982). This provides a further indication that sucrose transport and accumulation is sink- rather than source-limited, though phloem loading may also be a major rate-limiting process. Removal of excess sucrose at the source should decrease the feedback inhibition in the system and so increase the maximum rate of photosynthesis.

6.2 Light and Photosynthesis

The relationship between radiant energy absorbtion and carbon assimilation in sugar beet has been examined by Glauert (1983). There was a close correlation between CO_2 uptake and irradiance, with both increasing progressively through the morning to decline later in the day. Assimilation decreased immediately whenever the sun was obscured by cloud cover. From CO_2 uptake patterns, it was evident that the photosynthesis/light response was similar from mid-June until September and, only in October was there evidence of reduced response due to canopy ageing (Fig. 4). This reflects the continued initiation of new, and slow senescence of old, leaves by sugar beet plants. However, the overall age of the leaf surface increases because new leaves expand little and make only a minor contribution to the light-intercepting area (Hodanova 1981). The diminishing response of the canopy late in the season appears to be

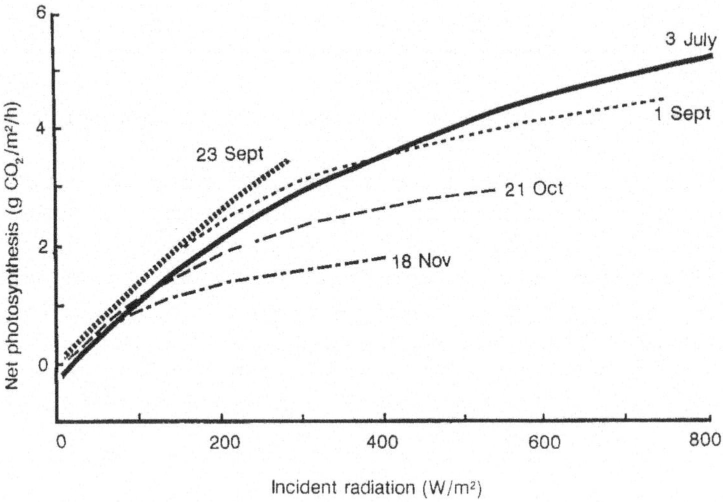

Fig. 4. Relationship between net photosynthesis and incident radiation of a sugarbeet crop during summer/autumn 1980. (After Glauert 1983)

mainly under internal control, since photosynthesis-light response curves are little affected by temperature before November (Glauert 1983).

Glauert's gas-exchange experiments identified that throughout the season sugar beet maintains a conversion coefficient close to 1.9 g/MJ of total radiation intercepted, which is relatively independent of temperature, plant size and age of the crop. Hence, it is suggested that biomass production during the season is directly related to the amount of radiation intercepted by the foliage between sowing and harvest. This proposal was confirmed by taking weekly increments of plant weight and weekly measurements of light interception from field-grown crops of the same variety as used in the gas-exchange work. Light energy intercepted throughout the season is related not only to biomass, but also to root and sucrose yields (Scott and Jaggard 1993). Increases in season-long radiation interception increase root yield proportionally, with a constant proportion of 70–72% being accounted for by sucrose within the dry weight of the storage root. A coefficient of 0.97 g/MJ represents the conversion of intercepted light to sucrose.

6.3 Modification of Photosynthetic Efficiency

Chemical and genetic approaches towards improving photosynthetic efficiency in sugar beet are of obvious interest to the industry. Although there is some indication that photosynthesis may be promoted by certain plant hormones, particularly under rate-limiting conditions, no chemical treatments have been developed which affect photosynthesis directly. Moreover, until the specific roles of hormones in controlling photosynthesis are elucidated, a rational approach to designing photosynthesis-promoting chemicals is unlikely. Likewise, a clearer understanding of the molecular biology of photosynthesis is necessary before biotechnological approaches can be expected to be fruitful.

Although plant growth regulators seem to have little effect on photosynthesis per se, it has been claimed that some growth retardants may increase photosynthetic efficiency by increasing chlorophyll production. Thus, sugar beet plants treated with paclobutrazol demonstrated apparent increases in net photosynthesis per unit area in glasshouse experiments (Jaggard et al. 1982). Unfortunately, treated field-grown plants, although darker green, did not convert radiant energy to dry matter more effectively than control plants. Despite an increased rate of photosynthesis, both root yield and sucrose content were decreased (Glauert and Biscoe 1985).

7
Water Relations and Stress

7.1
Control of Water Uptake

A relationship similar to that described in Section 6 exists between growth and water use. This is because the amount of intercepted solar radiation controls the potential for both photosynthesis and transpiration. Exploitation of this potential by the crop is related to the amount of CO_2 moving into, and water vapour out of, the leaves. Both are regulated by the stomata and boundary layer resistance.

Plants have several mechanisms to keep cells turgid and prevent wilting. Loss of water to the atmosphere can rapidly be reduced by closure of stomata. While this reduces water loss, it also inhibits nutrient flow from the roots and CO_2 uptake into the leaves reducing carbohydrate production and plant growth. At the individual cell level, water uptake and photosynthesis can continue as long as enough soluble compounds are present to keep cells turgid. Thus Na^+, K^+, calcium ions, chloride anions and sucrose are all readily accumulated in sugar beet cell vacuoles. Organic solutes accumulating in the cytoplasm include α-amino N compounds (amino acids, mainly glutamine) and glycine betaine, which increase considerably in response to water stress. Unfortunately, such compounds are melassigenic and so adversely affect sugar extraction in the factory.

The concentration gradient of CO_2 across stomata is relatively stable during the day but the water vapour potential gradient varies greatly with atmospheric humidity. In temperate regions, less water should be used than in arid regions to produce similar biomass. Dunham et al. (1993) determined a dry matter/water use ratio of 5–7 g/kg in the UK compared with 2.3 g/kg in California.

7.2
Stress Tolerance

Results from recent investigations into sugar beet drought stress tolerance indicate that within a ploidy, the most stress-tolerant sugar beet varieties have fewer stomata on the fully expanded leaves (Thomas and Clarke 1995); they presumably lose less water due to transpiration. On the other hand, varieties with more stomata tended to perform better in unstressed conditions. This can be explained by the fact that more CO_2 passes into the leaves, so more sucrose is produced. This indicates that the most productive varieties are not necessarily the most stress-tolerant. When chlorophyll fluorescence is used as an indicator of photosynthetic efficiency following recovery from stress, up to threefold differences have been detected between the stress tolerance of leaf tissue in a range of 25 varieties and breeding lines. Such physiological and biochemical information, coupled with rapid developments in biotechnology, could provide new opportunities to create varieties with improved stress tolerance.

8
Assimilate Partitioning

8.1
Assimilate Movement in the Plant

A detailed description of the fundamental aspects of assimilate transport in sugar beet has been published by Elliott and Weston (1993). It has been suggested that sucrose-metabolizing enzymes in leaves are hormonally controlled, indicating that the hormones ABA, GA and IAA may play crucial roles in carbon partitioning (Daie 1987). This is likely to be a dual action, involving modification of the rate of sucrose synthesis and the rate of loading into the phloem, thus relating to the amount of sucrose available for export. An apoplastic sucrose-loading pathway from the leaf to the phloem has been proposed (Delrot and Bonnemain 1989; Wardlaw 1990). This probably involves a protein-sucrose symport mechanism; a sucrose-binding protein found on the plasmalemma of sieve-tube elements of minor veins in sugar beet may well be the protein-sucrose symport (Warmbrodt et al. 1989). GA promotes and ABA inhibits this loading process (Daie and Wyse 1983). Acropetal transport of assimilate along the phloem pathway is increased by direct auxin effects on the transport process (Patrick and Wareing 1980). Unloading of sucrose from the sieve element is likely to be apoplastic (Wyse 1979; Lemoine et al. 1988).

8.2
Assimilate Storage in the Root

Storage of sucrose in the vacuoles of tap root cells involves a proton-sucrose antiport system (Willenbrink et al. 1984; Briskin et al. 1985). Sucrose is cotransported with potassium and countertransported with protons across the tonoplast of the sink cells (Saftner and Wyse 1981). This system is stimulated by IAA and ABA (Wyse et al. 1980). Since several aspects of storage root growth, sucrose metabolism and sucrose storage seem to be hormone-mediated, it has been suggested that these processes could be modified either by chemical treatment or bioengineering (Thomas et al. 1993; Fig. 5). However, there has been little progress with either of these approaches to date.

Potential increases in sucrose storage volume in the tap root depend on the initiation of new cells and their subsequent expansion and vacuolation. Throughout the season sugar-beet root cells gradually accumulate sucrose to a maximum concentration of 900 mmol dm^{-3}. The storage cells appear always to be in equilibrium and able to generate sufficient osmotic potential to expand. Wyse et al. (1986) demonstrated that sucrose absorption by discs of sugar beet tap root is inhibited by high turgor. This is consistent with inhibition of the tonoplast proton-dependent ATPase, with consequent decreases in vacuolar H^+ concentration causing the sucrose reduction.

Consistent turgor values of between 5 and 8 bar have been demonstrated through pressure probe measurements, yet the osmotic concentrations that occur should correspond to a cell turgor of 12–13 bar. Therefore, it is clear that the turgor of sugar-beet storage cells is closely regulated in some way. Although the regulatory mechanism has not been elucidated, the increased cell turgor

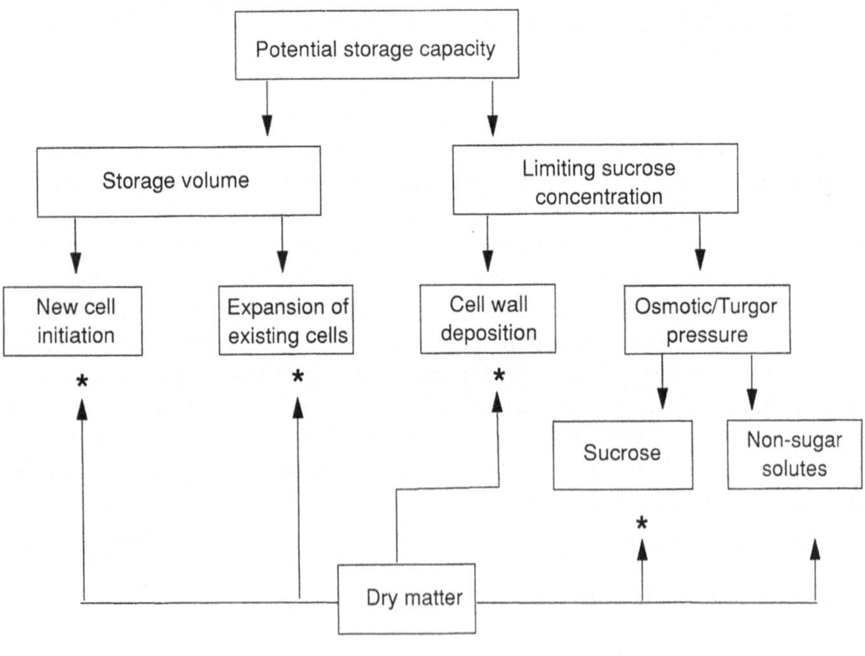

Fig. 5. Some limitations to dry matter partitioning in storage roots. *Asterisks* represent processes known to be under hormonal influence. (After Thomas et al. 1993)

induces leakage of sucrose, amino-N compounds and potassium from the vacuole into the extracellular space. Subsequently, both sucrose and amino-N compounds are preferentially retrieved back into the cell, but potassium remains outside to maintain the osmotic balance (Shewry 1994).

Root yield and sucrose concentrations are not directly related to each other because the ratio is influenced by the way in which incoming photosynthate is partitioned by the root for growth and sugar storage. As mentioned previously, the young, small cells close to the concentric cambia are more efficient at accumulating sucrose per unit volume and weight than the mature, larger cells. Milford (1973) found that as cell size increased to $10–15 \times 10^{-8} cm^3$ the sucrose content per cell increased nearly proportionally with cell volume. However, cells with a higher volume showed a less than proportional sucrose increase, even though water and non-sucrose dry mass continued to increase proportionally. Thus, sucrose concentration is likely to be a function of the relative proportions of large and small cells, and a storage root of many small cells would be more efficient at accumulating sucrose than one composed of fewer large cells.

8.3 Modifying Sucrose Storage Capacity

It has been suggested that productivity could be increased by modifying cambial initiation and activity at an early stage of growth, either chemically (Elliott et al. 1986) or by a molecular biology approach aimed at altering endogenous hormone balance. A possible approach to high sucrose concentrations and yield could be to design large storage roots with more vascular rings than are found in present varieties and with short sucrose transport paths from the phloem to the storage cells of the parenchymatous zone.

Another approach to increasing sink size would be to accelerate loading by increasing the rate of sucrose uptake, leading to earlier maturity and harvest. Morris (1983) presented evidence that the synthesis and activity of invertases may be regulated by endogenous hormones. It would therefore seem possible to use plant-growth regulators to modify the fluxes and patterns of assimilate transport by affecting the invertase levels in competing sinks. However, because the rate-limiting step in sugar transfer in sugar beet is likely to be an energy-dependent transport process located at one of the membrane systems in the phloem/sink complex, this approach may not be feasible. At present we do not know enough about these transport systems to justify speculation on the role of plant hormones in their regulation, though both ABA and IAA have been shown to enhance sink loading in sugar beet (Wyse et al. 1980).

In sugar beet, faster sucrose uptake has been achieved in laboratory experiments that used root tissue discs and cell cultures treated with the growth retardants PIX and BAS 106W (Daie 1987), but there is little indication in the literature that this can be done with field-grown plants. The problems of targeting chemicals to sink regions at critical growth stages, and thus of elucidating their exact mode of action, remain unresolved. In general, there have been few investigations into the direct effects of applied regulatory chemicals on sugar-enzyme activity or energy-dependent sucrose transport.

9 Crop Growth and Sucrose Accumulation

The importance of good plant establishment, correct plant spacings and adequate nutrition as determinants of yield have been described by Scott and Jaggard (1993). Harvest index is affected by season and variety, with top growth generally continuing to increase in weight late in the season when harvest indices are small. Patterns of top growth are considerably modified by nitrogen availability, with a continuous uptake of nitrogen increasing the growth of late-produced leaves (Houba 1973). In commercial crops, nitrogen uptake is usually arrested when top weights reach their maximum (Armstrong et al. 1986), though some situations can occur when nitrogen is continuously available, so promoting top growth with detrimental effects on root yield, quality and sugar yield (Milford et al. 1988). This can be the case when crops are grown on organic soils, or on mineral soils to which organic manures have been applied.

The distribution of sucrose to dry matter in the storage root starts at about 25% and rises to a stable level of about 75% of the dry matter (Scott and Jaggard 1993). The rate at which this change in distribution occurs is highly dependent on seasonal factors (Milford 1973). In general, the concentration on a fresh weight basis tends to increase from less than 5% in early June to reach 15–20% in October and November but the precise level achieved fluctuates with soil water status and rainfall.

A question that has not been addressed is the relevance of plant uniformity to yield and quality. Experiments at Broom's Barn in the mid-1970s (see Thomas et al. 1995) showed a clear correlation between root size and sucrose concentration, i.e., the small roots in a population tended to have higher sucrose concentrations than the large roots. It can be envisaged that crops that have a considerable population of small roots could have a higher sucrose content than those in which all the roots are of medium size. It would be interesting to establish the relationship between sucrose and dry mass production in such populations.

10 Sugar Beet and Climate Change

With the predicted increase in atmospheric CO_2 and temperature, it seems possible that by 2050 sugar beet yields in temperate regions could be increased by up to 30%. This is assuming that water will not be a limiting factor. Such increases could be achieved through a combination of increased radiation interception caused by earlier sowing, hence faster leaf growth, earlier leaf canopy closure, and improved photosynthetic capacity (Thomas et al. 1994a).

Most current bolting-resistant varieties need to receive more than about 35 vernalizing days with maximum temperatures below 12 °C, before any significant bolting occurs. From this information, it can be calculated that, in the UK, a 3 °C rise in average temperature would move the safe sowing date from 20 March to the end of January. It has been calculated that every day gained during March under present climatic conditions increases sucrose yield by about 35 kg/ha. A combination of increased temperature and elevated atmospheric CO_2 may increase this advantage considerably.

The responses of sugar beet to extra CO_2 were demonstrated by Ford and Thorne (1967), who showed that effects on dry weight depended both on effects on leaf area and on net assimilation rate. The latter effects were more important at higher CO_2 concentrations. Doubling the light intensity affected dry weight and leaf area similarly to a tenfold increase in CO_2 concentration, but the root yield was increased relatively more than leaf area.

In the future this interaction between light, CO_2 and temperature could be significant. However, there could be less benefit than anticipated from increased temperatures and CO_2 levels if cloud layers are dense enough to reduce the levels of radiation reaching the crop canopy. Nevertheless, Ford and Thorne's results show that even at low light intensities sugar beet growth can increase with CO_2 concentration up to $1000\,\mu l\,l^{-1}$ (equivalent to parts per million).

The expected effects of higher temperatures and CO_2 would be to increase dry matter production, but this may not necessarily increase sugar yield. Much will depend on how the extra dry matter is partitioned in the plant and what this means in terms of sugar and N-impurity content of the root. In general, plants respond to elevated CO_2 levels by increased carbohydrate synthesis and reduced nitrogen uptake. However, other environmental changes induced by global warming, which might induce plant stress, could modify this relationship.

More recent studies indicate that increased temperature tends to decrease total dry mass and sucrose concentration in sugar beet whereas elevated CO_2 increases total dry mass, with benefits to root dry mass and sucrose content (Demmers-Derks et al. 1996). The consequences of predicted climate change on the competition between sugar beet and *Chenopodium album* (fat hen), a related weed species, have been studied by Houghton (1995), who concluded that increases in temperature are likely to have more significant effects on sugar beet productivity than changes in CO_2. This is because the leaf area of fat hen is affected more than that of sugar beet by temperature, and early leaf growth is a critical determinant of the competitive interaction. Responses to higher CO_2 concentrations are likely to be similar for both species.

11 Sugarbeet in the Future

Research is still needed on ways to enhance yield, particularly investigations on earlier establishment to improve utilization of early season solar radiation, through seed advancement treatments and possibly autumn sowings. This may be very relevant if new lines are developed for alternative products or if bioethanol becomes an economically viable product because it would allow processing outside of the normal campaign period. Beet, being the most productive plant in temperate climates, is a perfect target to use as a carrier for high-value products. Hence work is in progress to introduce genes for protein, special starches, oils, flavours and possibly pharmaceuticals.

A further strategy may be selection and development of low-temperature tolerant lines. It is also important to discover the qualitative and quantitative effects of water stress on crop productivity and to identify the underlying mechanisms as a strategy towards developing stress-resistant lines. Use of the relationship between soil-moisture deficits and crop growth has been invaluable in improving advice on water application. Above all, there is also a need to know more about the genetic, physiological and agronomic factors that affect the relationships between leaf development and accumulation of sucrose and amino-nitrogen compounds in the storage roots of sugar beet. These relationships are important to growers, processors and breeders since they affect parameters such as beet quality, juice purity and stress tolerance.

Significant biotechnological contributions are being made towards the future of sugar-beet growing, aimed mainly at increasing the cost efficiency of production and developing environmentally sympathetic agronomy. The first practical outlets, i.e., herbicide-resistant varieties, should be on the market around the year 2000. Biotechnology and contribute significantly to sugar beet growing in the future, mainly by increasing cost-effectiveness and improving environmental acceptability. However, it is clear that a deeper understanding of sugar beet biochemistry and physiology is necessary to clearly identify future biotechnological targets.

References

Akeson WR, Freytag AH, Henson MA (1981) Improvement of sugar beet seed emergence with dilute acid and growth regulator treatments. Crop Sci 21:307–312

Alcaraz G (1981) Etude de certains aspects de l'hétérogénéite de la semence de betterave sucrière: influence sur le comportement en germination. Thesis, Diplome de Docteur de 3ème cycle, L'Université, Paris

Armstrong MJ, Milford GFJ, Pocock TO, Last PJ, Day W (1986) The dynamics of nitrogen uptake and its remobilization during the growth of sugar beet. J Agric Sci Camb 107:145–154

Artschwager E (1926) Anatomy of the vegetative organs of the sugar beet. J Agric Res 33:143–176

Briskin DP, Thornley WR, Wyse R (1985) Membrane transport in isolated vesicles from sugar beet taproot. II. Evidence for a sucrose/H^+-antiport. Plant Physiol 78:871–875

Brown KF, Biscoe PV (1985) Fibrous root growth and water use of sugar beet. J Agric Sci Camb 105:679–691

Brown KF, Dunham RJ (1986) The fibrous root system: the forgotten roots of the sugar beet crop. Br Sugar Beet Rev 54(3):22–24

Brown KF, Messem AB, Dunham RJ, Biscoe PV (1987) Effect of drought on growth and water use of sugar beet. J Agric Sci Camb 109:421–435

Daie J (1987) Bioregulator enhancement of sink activity in sugar beet. Plant Growth Regul 5:219–228

Daie J, Wyse RE (1983) Regulation of phloem loading in *Phaseolus vulgaris* by plant growth regulators. Proc Plant Growth Regul Soc Am, Meeting, East Lansing, Michigan, pp 139–144
Delrot S, Bonnemain JL (1989) Phloem loading and unloading. Ann Sci Forestry, 47 supplement 786s-786s In: Dreyer E, et al. (eds) Forest tree physiology. Elsevier/INRA, Amsterdam, Paris
Demmers-Derks H, Mitchell RAC, Mitchell VJ, Driscoll SP, Gibbard C, Lawlor DW (1996) Sugar beet under climatic change: photosynthesis and production. Aspects Appl Biol 45:163–170
Dunham RJ, Draycott AP, Messem AB (1993) Irrigation of sugar beet in the UK. Proc 56th Congr Int Inst Sugar Beet Res, Feb 1993, Brussels, pp 25–40
Durr C, Boiffin J, Fleury A, Coulomb J (1992) Analysis of the variability of the sugar beet growth during the early stages. II Factors influencing the seeding size in field conditions. Agronomie 12:527–535
Durrant MJ, Jaggard KW (1988) Sugar beet seed advancement to increase etablishment and decrease bolting. J Agric Sci Camb 110:367–374
Durrant MJ, Mash SJ (1990) Sugar beet seed treatments and early sowing. Seed Sci Technol 18:839–850
Durrant MJ, Payne PA, Prince JWF, Fletcher R (1988) Thiram steep seed treatment to control *Phoma betae* and improve the establishment of the sugar beet plant stand. Crop Protect 7:319–326
Elliott MC, Weston GD (1993) Biology and phsyiology of the sugar beet plant. In: Cooke DA, Scott RK (eds) The sugar beet crop: science into practice. Chapman and Hall, London, pp 37–66
Elliott MC, Hosford DJ, Lenton JR, Milford GFJ, Pocock TO, Smith JE, Lawrence DK, Firby DJ (1984) Hormonal control of storage root growth. In: Jackson MB, Stead T (eds) Growth regulators in root development. Br Plant Growth Regul Gp, Wantage, Monogr 10, pp 25–35
Elliott MC, Hosford DJ, Smith JI, Lawrence DK (1986) Opportunities for regulation of sugar beet storage root growth. Biol Plant 28:1–8
Fick GW, Williams WA, Loomis RS (1971) Recovery from partial defoliation and root pruning in sugar beet. Crop Sci 11:718–721
Fick GW, Williams WA, Ulrich A (1972) Parameters of the fibrous root system of sugar beet (*Beta vulgaris* L.). Crop Sci 12:108–112
Fick GW, Loomis RS, Williams WA (1975) Sugar beet. In: Evans LT (ed) Crop physiology. Cambridge University Press, Cambridge, pp 260–295
Fondy BR, Geiger J (1982) Diurnal pattern of translocation and carbohydrate metabolism in source leaves of *Beta vulgaris*. Plant Physiol 70:671–676
Ford MA, Thorne GN (1967) The effect of CO_2 concentrations on the growth of sugar beet, barley, kale and maize. Ann Bot 31:630–644
Giaquinta RT (1979) Phloem loading of sucrose: involvement of membrane ATPase and proton transport. Plant Physiol 63:44–48
Glauert W (1983) Carbon exchange of a sugar beet crop through a season. Thesis, University of Nottingham, Nottingham
Glauert AW, Biscoe PV (1985) Seasonal variations in the photosynthesis of sugar beet crops and the influence of a chemical regulator. Br Plant Growth Regul Gp Monogr 12, Proc, York, UK, pp 99–110
Goodman PJ (1966) Effect of varying plant populations on growth and yield of sugar beet. Agric Prog 41:82–100
Gummerson RJ (1986) The effect of constant temperatures and osmotic potentials on the germination of sugar beet. J Exp Bot 37:729–741
Hodanova D (1981) Photosynthetic capacity, irradiance and sequential senescence of sugar beet leaves. Biol Plant 23:58–67
Houba VJG (1973) Effect of nitrogen dressings on growth and development of sugar beet. Agric Res Rep 791, Pudoc, Wageningen
Houghton SK (1995) The effect of the weed fat hen (*Chenopodium album* L.) on sugar-beet production under present and future UK climates. Proc 58th Congr Int Inst Sugar Beet Res, June 1995, Beaune, pp 403–407

Jaggard KW, Lawrence DK, Biscoe PV (1982) An understanding of crop physiology in assessing a plant growth regulator on sugar beet. In: McLaren JS (ed) Chemical manipulation of crop growth and development. Butterworths, London, pp 139–150
Kerr S, Leaman M (1995) Varieties for 1996. Br Sugar Beet Rev 63(2):9–12
Khan AA, Ilyas S, Ptasznik W (1995) Integrating low water seed hydration with other treatments to improve cold tolerance. Ann Bot 75:13–19
Lemoine R, Daie J, Wyse R (1988) Evidence for the presence of a sucrose carrier in immature sugar beet tap roots. Plant Physiol 86:575–580
Lenton JR, Milford GFJ (1997) Plant growth regulators and the physiological limitations to yield in sugar beet. Pestic Sci 8:224–229
Longden PC (1974) Harvesting sugar beet seed. J Agric Sci Camb 83:435–442
Longden PC, Thomas TH (1989) Why not autumn-sown sugar beet? Br Sugar Beet Rev 57(3):7–9
Longden PC, Johnson MG, Darby RJ, Salter PJ (1979) Establishment and growth of sugar beet as affected by seed treatment and fluid drilling. J Agric Sci Camb 93:541–552
Longden PC, Clarke NA, Thomas TH (1995) Control of bolting and flowering in sugar beet (*Beta vulgaris*). Proc 58th Congr Int Inst Sugar Beet Res, June 1995, Beaune, pp 71–81
Milford GFJ (1973) The growth and development of the storage root of sugar beet. Ann Appl Biol 75:427–438
Milford GFJ, Lenton JR (1978) Development parameters regulating sugar yield in beet. In: George EF (ed) Opportunities for chemical plant growth regulation. BCPC, Croydon, pp 135–142
Milford GFJ, Pocock TO, Riley J (1985a) An analysis of leaf growth in sugar beet. I: Leaf appearance and expansion in relation to temperature under controlled conditions. Ann Appl Biol 106:163–172
Milford GFJ, Pocock TO, Riley J (1985b) An analysis of leaf growth in sugar beet. II: Leaf appearance in field crops. Ann Appl Biol 106:173–185
Milford GFJ, Pocock TO, Riley J (1985c) An analysis of leaf growth in sugar beet. III: Leaf expansion in field crops. Ann Appl Biol 106:187–203
Milford GFJ, Pocock TO, Jaggard KW, Biscoe PV, Armstrong MJ, Last PJ, Goodman PJ (1985d) An analysis of leaf growth in sugar beet. IV: The expansion of the leaf canopy in relation to temperature and nitrogen. Ann Appl Biol 107:335–347
Milford GFJ, Travis KZ, Pocock TO, Jaggard KW, Day W (1988) Growth and dry-matter partitioning in sugar beet. J Agric Sci Camb 110:301–308
Morris DA (1983) Hormonal regulation of assimilate partitioning: possible mediation by invertase. Br Plant Growth Regul Group News Bull 6(2):23–24
Nelson JM, Sharples GC (1980) Stimulation of tomato, pepper and sugar beet seed germination at low temperatures by growth regulators. J Seed Technol 5:62–68
Nelson JM, Jenkins A, Sharples GC (1984) Soaking and other seed pretreatment effects seed on germination and emergence of sugarbeets at high temperature. J Seed Technol 9:79–86
Osburn RM, Schroth MN (1989) Effect of osmopriming in sugar beet seed on germination rate and incidence of *Pythium ultimum* damping-off. Plant Dis 73:21–24
Patrick J, Wareing PF (1980) Hormonal control of assimilate movement and distribution. In: Jeffcoat B (ed) Aspects and prospects of plant growth regulators. Wessex Press, Wantage, pp 65–84
Rao IM, Fredeen AL, Terry N (1990) Leaf Phosphate status, photosynthesis and carbon partitioning in sugar beet. III. Diurnal changes in carbon partitioning and carbon export. Plant Physiol 92:29–36
Rapoport HF, Loomis RS (1985) Interaction of storage root and shoot in grafted sugarbeet and chard. Crop Sci 25:1079–1084
Rapoport HF, Loomis RS (1986) Structural aspects of root thickening in *Beta vulgaris* L.: comparative thickening in sugarbeet and chard. Bot Gaz 147:270–277
Rush CM (1991) Comparison of seed priming techniques with regard to seeding emergence and *Pythium* damping-off in sugar beet. Phytopathology 81:878–882
Rush CM (1992) Stand establishment of sugar beet seedlings in pathogen-infested soils as influenced by cultivar and seed-priming technique. Plant Dis 76:800–805

Saftner RA, Wyse RE (1981) Hormonal effects on photosynthate transport in the root sink of sugar beet. Plant Physiol 67S:156
Scott RK, Jaggard KW (1993) Crop physiology and agronomy. In: Cooke DA, Scott RK (eds) The sugar beet crop: science into practice. Chapman and Hall, London, pp 179–237
Scott RK, Wood DW, Harper F (1972) Plant growth regulators as a pretreatment for sugarbeet seeds. Proc 11th Weed Control Conf, Brighton, UK, Nov 1992, pp 752–759
Scott RK, English SD, Wood DW, Unsworth MH (1973) The yield of sugar beet in relation to weather and length of growing season. J Agric Sci Camb 81:339–347
Scott RK, Harper F, Wood DW, Jaggard KW (1974) Effects of seed size on growth, development and yield of monogerm sugarbeet. J Agric Sci Camb 82:517–530
Shewry PR (1994) Inherent plant productivity. IACR report for 1994, Rothamsted, UK, pp 12–20
Stitt M, Quick WP (1989) Photosynthetic carbon partitioning: its regulation and possibilities for manipulation. Physiol Plant 77:633–641
Terry N, Mortimer DC (1972) Estimation of the rates of mass carbon transfer by leaves of sugar beet. Can J Bot 50:1049–1054
Thomas TH (1986) Hormonal control of assimilate movement and compartmentation. In: Bopp M (ed) Plant growth substances 1985. Springer Berlin Heidelberg New York
Thomas TH (1996) Form and function of the sugar beet fibrous root system: its importance to crop productivity. Proc 59th Congr Int Inst Sugar Beet Res, Feb 1996, Brussels, pp 143–152
Thomas TH, Clarke NA (1995) Are sugar beet drought stress problems being watered down? Br Sugar Beet Rev 63(3), pp 8–11
Thomas TH, Gartland KMA, Slater A, Elliott MC (1993) Opportunities for manipulation of growth and development. In: Cooke DA, Scott RK (eds) The sugar beet crop: science into practice. Chapman and Hall, London, pp 521–549
Thomas TH, Houghton SK, Clarke NA (1994a) The possible impact of global environment change on sugar beet production. In: World sugar and sweetener yearbook 1994. FO Licht GmbH, Ratzeburg, Germany F5–F18
Thomas TH, Jaggard KW, Durrant MJ, Mash SJ (1994b) The physiological advancement of sugar beet seeds. BCPC Monogr 57. Seed treatment: progress and prospects, BCPC Publications, Brackwell, UK, pp 391–396
Thomas TH, Asher MJC, Dewar AM (1995) Recent progress in sugar-beet seed research. In: World sugar and sweetener yearbook 1995. FO Licht GmbH, Ratzeburg, Germany F5–F22
Thomas TH, Yallop SJ, Redfearn M (1996) Seed size and vigour. IACR Broom's Barn Ann Rep for 1995, pp 7–8
Wardlaw IF (1990) The control partitioning in plants. New Phytol 116:341–381
Warmbrodt RD, Buckhout TJ, Hitz WD (1989) Localisation of a protein, immunologically similar to a sucrose-binding protein from developing soybean cotyledons, on the plasma membrane of sieve-tube members of spinach leavels. Planta 180:105–115
Willenbrink J, Doll S, Getz P, Meyer S (1984) Zuckeraufnahme in isolierten Vakuolen und Protoplasten aus dem Speichergewerbe von Bete-Rüben. Ber Dtsch Bot Ges 97:27–39
Wittwer SH, Hansen CM (1952) Some effects of preharvest foliage sprays of maleic hydrazide on the sugar content in certain field experiments. Proc 7th Gen Meet Am Soc of Sugar Beet Technol, Salt Lake City, Utah, 1952, p 90
Wyse R (1979) Sucrose uptake by sugar beet tap root tissue. Plant Physiol 64:837–841
Wyse RE, Daie J, Saftner RA (1980) Hormonal control of sink activity in sugar beet. Plant Physiol 65:S-662
Wyse R, Zamski E, Tomas AD (1986) Turgor regulation of sucrose transport in sugar beet taproot tissue. Plant Physiol 81:478–481
Zamski E, Azenhot A (1981) Sugarbeet vasculature. I: Cambial development and the three-dimensional structure of the vascular system. Bot Gaz 142:334–343

Potato[1]

J. Vos[2]

1 Introduction

The botanical genus *Solanum* comprises about 2000 species. Less than 10% of these are tuber-bearing. Gene centers are in Central and South America. Potato was used as a food crop by the Incas, probably many centuries before Columbus and his companions landed in South America. It is believed that the potato was introduced into Spain in about 1580 and from there the its cultivation spread through Europe and to other continents. The Latin name for the tetraploid potato genotypes that are grown today is *Solanum tuberosum* L. Large genetic variation has been present among the cultivars in use. This diversity explains the widespread cultivation of potato, while individual cultivars appear best suited to specific environments (niches). Genetic diversity was increased during this century by successful efforts to incorporate resistance genes from other species such as *S. demissum* and *S. vernei.*

The distribution of potato production is influenced by socioeconomic as well as biophysical factors. Potato production requires a minimum period of about 90 to 150 days of suitable temperatures: a low risk of frost and, at the other extreme, not exceeding ca. 23 to 25 °C for prolonged periods of time if acceptable levels of production are to be achieved.

The established way of propagating potatoes is clonal, using "seed" tubers. However, potatoes can also be grown from true botanical seeds. In some areas it is difficult to either produce healthy seed potatoes or store them. Under such conditions, the use of true potato seed, abbreviated to TPS, is a practical alternative, both to multiply seed stocks and to grow consumable potatoes (Pallais 1991). In this chapter we shall assume that potatoes are grown from seed tubers.

[1] In memory of Dr. Date Engbert van der Zaag, influential potato specialist, who died during the preparation of this manuscript, in appreciation of the many discussions, both work-related and private, over the past 25 years

[2] Department of Agronomy, Wageningen Agricultural University, Haarweg 333, 6709 RZ Wageningen, The Netherlands

Crop Yield, Physiology and Processes
D. L. Smith, C. Hamel (Eds.)
© Springer-Verlag Berlin Heidelberg 1999

2
The Basic Plan of the Potato Plant

To understand the physiology of yield formation it is helpful to start with an outline of the basic plan of the potato plant. A potato tuber is a swollen underground stem part. The tuber has an apical bud and scale leaves which are arranged in a 5/13 phyllotactic spiral. Each scale leaf subtends a bud. The top bud and lateral buds are morphologically from different orders (primary axis versus secondary axes), but there are no fundamental differences in the pattern of development. Generally they are referred to as the eyes of a potato tuber, an eye being the leaf scar and the subtended bud. The number of eyes on a tuber depends on tuber weight and tuber shape, 15 to 20 being a representative range.

Near harvest of a potato crop, the number of leaves and leaf primordia on a bud axis of a tuber varies between 4 and 13 (Krijthe 1962; Goodwin 1967) and differs with distance to the top bud (Goodwin 1967). Primordia of successive orders can be found on tuber buds: according to Krijthe (1962), the initiation of secondary primordia follows the development of primary primordia on an apex with a lag of seven positions, i.e., when ten leaf primordia are counted on an apex, secondary initials are present in the axils of the oldest three of them.

In general, the buds are dormant at harvest of the crop. At some point during the period of storage, between harvest and the next time of planting, dormancy is broken and sprouts can emerge on the seed tuber. Upon planting, sprouts grow and become stems with leaves and stolons on which new tubers develop. Figure 1 shows a schematic plan of a potato plant. Stems emerging directly from the mother tuber are called main stems. Only one main stem is drawn in Fig. 1, but depending on pretreatment and size of the seed tuber three to seven main stems per seed tuber can emergence.

In an early stage of growth the apex of the main stem changes and terminates in the initials of an inflorescence (cluster of flowers). Flower initiation generally occurs before tuber initiation and takes place a few weeks after emergence and sometimes even before planting. In this sense potato is a determinate plant.

The total number of nodes between the point of emergence from the mother tuber and the inflorescence is often about 25 (Vos 1995a) although numbers up to 40 have been reported, depending on cultivar, conditions during storage of the seed tuber and conditions upon planting (Firman et al. 1991). The number of leaves on the main stem (i.e., above ground nodes) is not fixed, but about 17 is a typical number. The first one to three leaves are generally small and not subdivided into leaflets; at higher nodes compound leaves are formed. The upper, adaxial leaf surface differs from the lower, abaxial surface. Stomatal frequency is much higher (and stomatal conductance much higher) on the lower than on the upper leaf surface, i.e., about 200 versus 40 stomata mm^{-2} (Vos and Groenwold 1989).

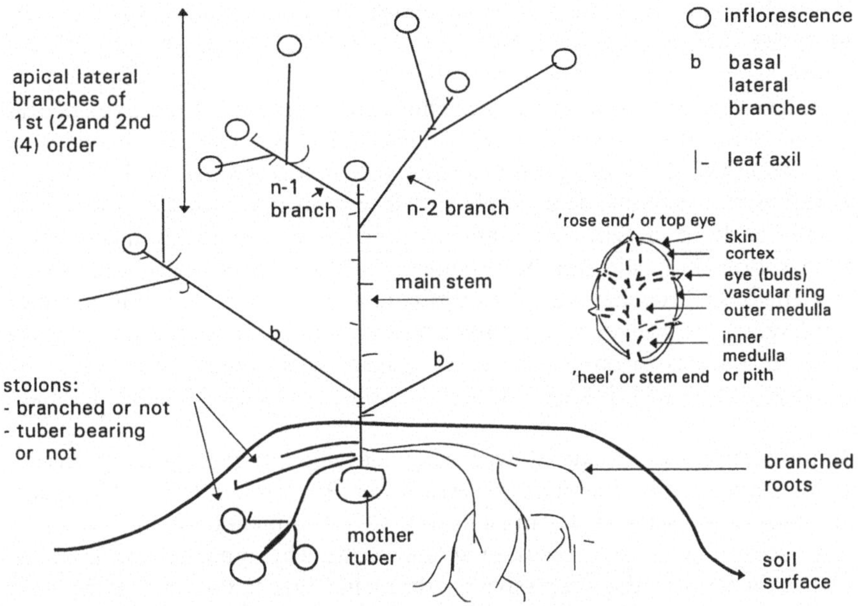

Fig. 1. Schematic representation of a potato plant and its component parts

The number of underground nodes can be influenced by planting depth and is in the range of seven to ten. Roots and stolons emerge from the axils of the scale leaves at these nodes. Stolons are underground branches, which grow more or less horizontally, resembling stems as regards structure and arrangement of tissues. On well-developed sprouts, stolons can be seen before planting. The number of stolons per underground main stem node generally ranges between zero and twelve (Lovell and Booth 1969). Stolons, like the underground part of the main stem, have nodes with rudimentary leaves and buds which can give rise to stolon branches. The end of a stolon or stolon branch can start to swell and become a tuber.

On the main stem, primordia in each axil can produce lateral, leaf-bearing branches. However, branches usually emerge only from certain basal nodes and from particular nodes below the inflorescence. The middle nodes on the stem usually show only rudimentary, transient growth of branches, except under specific conditions such as following decapitation of the plant and under high temperatures. Basal lateral branches are distinguished from apical lateral branches. Basal lateral branches can emerge from underground axils and from axils up to about five to seven nodes above the soil surface. Underground basal lateral branches emerging from nodes close to the mother tuber generally develop stolons and tubers, whereas branches emerging from nodes close to the soil surface do not. Lateral branches connected to the main stem on an underground node tend to form only a small fraction of the total number of stems emerging from the soil (discussed in more detail by Vos 1995a).

Apical branches generally emerge from specific nodes. Calling the leaf subtending the inflorescence leaf n, then apical branches emerge from the axils of leaf number n-1 and n-2. Apical branches also terminate in an inflorescence. A next level (or order) of apical branches can emerge from the axils of leaf positions n-1 and n-2 on each level of these apical branches. In this process several orders or levels of apical branches can be formed with 2, 4, 8, 16, etc. different stem segments. The number of leaves on n-1 and n-2 branches is generally fairly fixed, with n-2 branches showing systematically more leaves than n-1 branches, fair numbers being 5 to 10 (Almekinders and Struik 1996). The basal branches can also terminate in an inflorescence and produce apical laterals themselves. The number of leaves on basal lateral branches till an inflorescence is formed is 23 to 17 (Almekinders and Struik 1996), the number declining with increase in the node number of insertion on the main stem.

Inflorescences can produce flowers which develop into seed-bearing berries, but the sequence of generative processes may be interrupted at any stage; sometimes inflorescences are present in a very rudimentary state only. This varies between cultivars and growing conditions. The apparent continuation of the main stem in the successive n-1 branches makes the term sympodial branching applicable (Danert 1957; Almekinders and Struik 1994).

The potato plant is indeterminate in the sense that there is no fixed end point to the process of branching. The virtually unlimited possibilities to develop new levels of growth from primordia in axils is an important feature of the potato plant, explaining much of its plasticity. This phenomenon is to some extent comparable to tillering in Gramineae species. Suppression or release of suppression of growth of axillary buds is central in many of the potato's reactions to environmental conditions.

3 Physiological Age and Growth Vigour

It has been known for a long time that the condition of the seed tuber can affect the performance of the crop grown from the seed. Kawakami (1952) was the first to draw attention to these phenomena. From the moment a tuber is formed on the mother plant, it undergoes a sequence of physiological changes which are reflected in its sprouting behaviour. The term physiological age is used to refer to this typical sequence of changes. Physiological age implies that the chronological age is not a reliable indicator of the age and the associated performance of a tuber.

Initially, the tuber is in a dormant state: sprouts are not formed even under conditions favourable for growth. The length of the dormant period shows large differences between cultivars. In the literature it is claimed that the dormancy period is reduced when the mother crop experiences higher temperature during tuber growth. However, recent research did not corroborate such a relation. On the contrary, conditions during growth appeared to have little effect on the length of the dormant period (van Ittersum 1992a,b; van Ittersum and Scholte 1992).

Release of dormancy is a gradual process. Often the apical eye develops sprout(s) first. Once this has happened sprouting of the other eyes is suppressed. This is referred to as apical dominance. This dominance can be broken by removing sprouts from the top eye. Apical dominance, resulting in a low number of stems per plant, can be prevented by a rapid change from storage temperature to temperatures favourable for sprout growth prior to planting (chitting treatments). As the seed ages, more sprouts will develop on the mother tuber and the growth vigour of the sprouts increases. The growth vigour is defined as the potential of a seed tuber to produce sprouts and plants rapidly under (standardized) conditions favourable for growth. Growth vigour was shown to correlate well with the early performance of the crop in the field. As the seed ages further, thinner and fewer sprouts are formed, whereas very old seed no longer produces emerging sprouts; instead, small tubers of a next generation are formed directly on the surface of the old mother tuber.

At planting, the seed tuber should be in a physiological state ensuring rapid growth of a sufficient number of stems. When seed potatoes are harvested in the summer in temperate zones and planted in the fall in mediterranean climates the seed may be too young for optimal performance. Dormancy can be shortened and growth vigour can be advanced by treating the tubers with heat shocks (3 weeks at 28 °C) or cold shocks (3 weeks at 2 °C; van Ittersum 1993).

The need has been felt to be able to measure the optimal condition, i.e., optimal physiological age, of seed potatoes in relation to the intended time of planting (van der Zaag and van Loon 1987, and coworkers cited therein). Such measures would be valuable to evaluate storage practices and seed treatments for cultivars widely differing in length of dormancy and rate of progression of physiological age. While suitable (bio)chemical parameters were not found, the sprouting capacity seems the most appropriate method at the moment; the definition is: "the sprout weight of uniform tubers, expressed in g (fresh weight) sprouts per tuber, after a standard sprouting procedure whereby tubers that were stored at X °C were then desprouted, and sprouted again by storing for 4 weeks at 18 to 20 °C and approximately 80% relative humidity in darkness" (Hartmans and van Loon 1987). This definition leaves open the exact temperature and growing conditions for the test. The user can select these, taking into account the technical facilities available. The essence is that the test conditions need to be reproducible in order to be able to compare results from different series of tests.

Seed can be either too young or too old at planting, resulting in negative effects on crop performance and yield. However, there is much confusion in the literature on the effects of physiological age for the intermediate ranges. Within this range emergence, tuber initiation and senescence occur somewhat faster with increase in physiological age. The general idea (discussed in detail by van der Zaag and van Loon 1987) is that effects on yield can be larger: (1) as conditions near emergence are less favourable for rapid crop growth, (2) when harvests are made in early stages of crop growth (e.g., seed potato crops) and (3) for early-maturing cultivars in comparison to late-maturing ones.

4 Interrelations Between Numbers of Component Plant Parts

Tuber yield is the product of the number of tubers per unit area and mean tuber weight. However, mean tuber weight is agronomically and economically less important than yields in specific tuber size classes or grades. Specific grades are required for specific market outlets, e.g., at harvest of a seed potato crop the grower generally wants a smaller size grade than for the processing industry. An important practical problem is to apply agronomic measures that result in maximization of the required grade. Much scientific work has been undertaken to develop and evaluate such measures and to explore their physiological background. Important issues are the determination of the number of emerging sprouts per seed tuber, the number of tuber-bearing stems per unit area, the number of stolons per stem, the number of tubers per stolon and tuber size. Figure 2 shows an example of associations between sprout, stem, stolon and tuber numbers per plant (four plants m^{-2}; data from Haverkort et al. 1990). The differences in initial sprout numbers on seed tubers resulted from eight different temperature × light treatments during the last 0 to 13 weeks during storage of the seed. However, whatever the cause of that variation, the graph clearly illustrates that manipulation of sprout numbers can affect stem

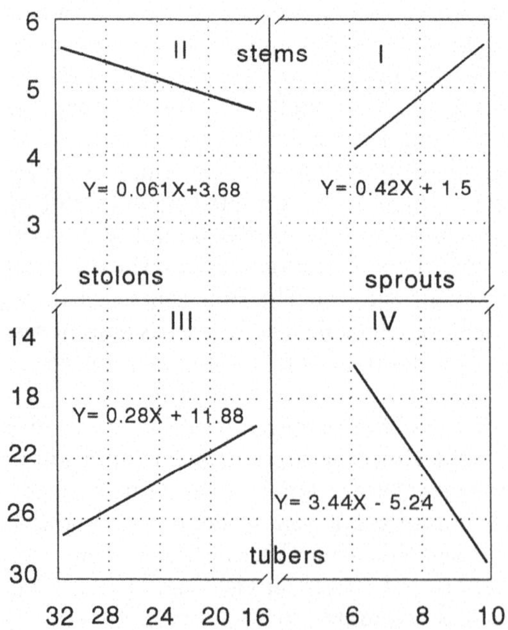

Fig. 2I–IV. Four-quadrant diagram showing an example of the relations between **I** number of sprouts on the seed tuber and number of emerging stems per plant, **II** the number of stems and the number of stolons per plant, **III** the number of stolons and the number of harvested tubers per plant, **IV** the number of harvested tubers per plant and the initial number of sprouts per seed tuber. Sprout numbers per seed tuber varied as a result of storage treatments. (After Haverkort et al. 1990)

and tuber numbers. Linear regressions were adequate to summarize the experimental data per season. The r-values of the associations in quadrants I–IV were 0.71, 0.79, 0.82 and 0.74, respectively. Yet the lines in Fig. 2 should primarily be read as trends and as illustrations of the ranges in which the numbers are often found. Pooled data on five seasons gave different parameters of the regressions and the four r-values dropped to 0.61, 0.01, 0.68 and 0.10, respectively. If Fig. 2 were the general case, apart from cultivar dependency of the parameters, then tuber number could be adequately controlled via manipulation of the initial number of sprouts, but apparently the association between the number of stems and the number of stolons is very variable, destroying the association between initial sprout number per seed tuber and final tuber number per plant. With choices with respect to seed treatment during the storage period, seed size and planting pattern the grower can influence the size distribution of the crop, although the control is far from perfect. In this context the finding of Schepers and Hoogland (1968) is relevant: these authors found the following relation between the number of stems (Y, m^{-2}) and the number of planted sprouts showing clear root initials (X, m^{-2}): $Y = 1.78 + 0.83X$. Although there are no examples where this equation is applied, there might be scope to further test its validity.

5 The Process of Tuber Formation

A sequence of events is necessary to form tubers on a potato plant (Vreugdenhil and Struik 1989; Struik et al. 1990; Ewing and Struik 1992). These include stolon induction and stolon initiation, the elongation and branching of the stolons, cessation of the longitudinal growth of stolons, tuber induction and tuber initiation. Although these steps occur in chronological order they do not occur simultaneously at each node, and some of the steps, e.g., tuber induction, can be reversed. Stolon initiation starts at the nodes closest to the mother tuber and progresses acropetally (Plaisted 1957). This results in about 2 weeks overlap in the periods of formation of stolons and tubers (Fig. 3). Warm temperatures and long photoperiods generally stimulate stolon formation (Ewing and Struik 1992).

Short days and cool temperatures stimulate tuberization and higher temperatures and long photoperiods retard tuberization. There is genetic variation in the critical photoperiod, i.e., the longest photoperiod that will permit tuberization of the cultivar. *Solanum tuberosum* ssp. *andigena* (cultivated in the Andean regions nowadays) generally fails to tuberize at photoperiods exceeding 12 to 13 h even under cool temperatures. The potato cultivars used in the temperate zones show longer critical photoperiods and may tuberize eventually even under continuous illumination. Cultivars with a long critical photoperiod generally still respond to temperature. Snyder and Ewing (1989) discussed the case where cultivar Norchip was induced to tuberize both under 16-h and 10-h days at day/night temperatures of 20/15 °C, but when tempera-

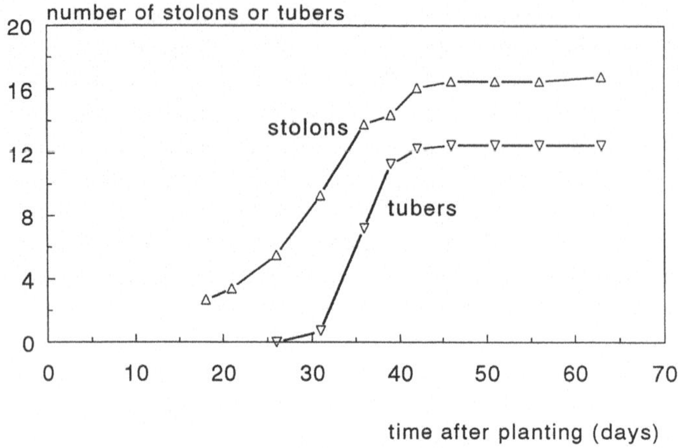

Fig. 3. Changes with time in the number of stolons and the number of tubers per plant. Data taken from the Struik et al. (1988); non-destructive measurements could be made frequently using a special pot assembly

tures were raised to 30/25 °C the longer photoperiods gave much less induction. Snyder and Ewing used single-node stem cuttings to analyze induction: if the plant from which the cutting was taken was induced to tuberize, a tuber-like structure develops rapidly from the bud in the axil of the cutting. For cultivars developed and grown in temperate climates, the influence of the photoperiod on time to tuberization may be small (Firman et al. 1991), but the response and its genetic variation is important when such cultivars are grown in other climates (Kooman and Haverkort 1995).

Environmental changes that increase induction to tuberize cause decreased gibberellic activity of shoots. In the 1980s progress was made in identifying compounds that are involved in the triggering of tuber induction (e.g., Koda et al. 1991; Ewing 1995). Under inductive conditions signals are produced or made transportable in the leaves and transported to the stolon tips, triggering the cessation of longitudinal stolon growth and the start of radial growth. Differential sensitivity in tuberization for photoperiod and temperature partly determines the length of the growth cycle. Early cultivars with a relatively short growing season tuberize faster and start tuber growth at a faster rate than late-maturing cultivars.

For a long time it has been a matter of debate whether the first *Solanum tuberosum* species that were introduced in Europe belonged to the subspecies *andigena* or *tuberosum*. Early drawings that were preserved suggest the former. Through a process of selection of tuberizing material the cultivars probably acquired more and more the characteristics of the subspecies *tuberosum*, which is less sensitive to photoperiod (Hawkes 1992). Recently, progress has been made in identifying loci on the genome that are involved in the photoperiodic response (van den Berg et al. 1996). Knowledge of the genetics will facilitate breeding and selection and should aid in understanding the physiological regulation of tuberization.

6 Rate and Duration of Development Processes

An objective of crop ecophysiology is to quantify the responses of crops to environmental factors. A collection of such quantitative relations covering the key aspects of growth and development is called a model. With such quantifications one can explain the variations in crop performance under a range of contrasting conditions. It is helpful to divide the cycle of crop development into particular stages. Subdivision in stages can be based on changes in the pattern of dry matter distribution, for instance the appearance of new sinks, and on changes in reaction to environmental factors. Suitable stages of development of a potato crop (Kooman and Haverkort 1995) are the periods (1) from planting to emergence, (2) from emergence to the start of tuber growth, (3) from the start of tuber growth till cessation of leaf growth and (4) from cessation of leaf growth till senescence.

Temperature and photoperiod are the main driving factors of plant development. The concept of thermal time is often an appropriate way to quantify effects of temperature on the time that is required to complete a process or development stage. Calling this period of time d (from duration, dimension: day) it is often found that 1/d, which is a rate, is related to mean daily temperature, T (°C) by:

$$1/d = a + bT, \tag{1}$$

where a and b are parameters with dimensions day^{-1} and $day^{-1}\ °C^{-1}$, respectively. The intercept with the X-axis is called the base temperature for development (T_b). Equation (1) can be reworked to yield:

$$1/d = b(T - T_b). \tag{2}$$

Usually there is an optimal temperature (T_0) above which the rate of progress in development declines, while development is completely halted at the maximum or ceiling temperature (T_c).

Responses complying with Eqs. (1) and (2) have been found to govern rates of development and expansion, for instance sprout growth and emergence, and rate of primordia production. Values for T_b generally range between 0 and 4 °C, with 2 °C as a fair average. The variation in T_b among authors may result from the fact that T_b is obtained by extrapolation below the experimentally covered range of temperatures. Differential temperature sensitivity of processes may also play a role. T_0 differs with the type of process: often it is in the range of 20 to 25 °C, but it is lower for the rate of tuber induction and higher for the rate of initiation of the (first 20) leaf primordia on sprouts Kirk et al. (1985; cf. Squire 1995). Kirk and Marshall (1992) observed that a new leaf appeared every 0.032 °C day (T_b = 0 °C and T_0 25 °C); the data in Vos (1995a) corroborate this rate.

In the previous section it was indicated that both temperature and photoperiod affect the number of days required to reach tuberization. One objective of modelling is summarizing the current understanding of a process. Many models on potato have functions describing the effects of temperature and photoperiod on tuberization (see Haverkort and MacKerron 1995 and Kabat et al. 1995 for various examples). Figure 4 illustrates such a quantification, taken from Kooman and Haverkort (1995) and pertaining to cv. Spunta. The relative temperature effect for a given temperature represents the multiplication factor of the actual temperature to calculate 1/d [Eq. (2)]. To be able to work with Fig. 4 one needs to know the maximum rate (1/d) for the temperature range of 19 to 23 °C. In this particular case that value is 0.0672 day^{-1}, meaning that the period of time from emergence to start of tuber growth is the reciprocal of 0.0672 = 14.88 days for temperatures in the optimum range and a photoperiod at emergence of 11 h.

The initiation of flower primordia can be affected by photoperiod and temperature in a similar way as tuberization. Although initiation of flower primordia generally occurs before tuberization, there is experimental evidence in support of the hypothesis that some of the hormonal signals involved in flowering and tuberization are similar (Almekinders and Struik 1996). On the other hand, the linkage between both processes is not very intimate since flower primordia can develop before planting (Firman et al. 1991) and flowering can occur without tuberization and vice versa.

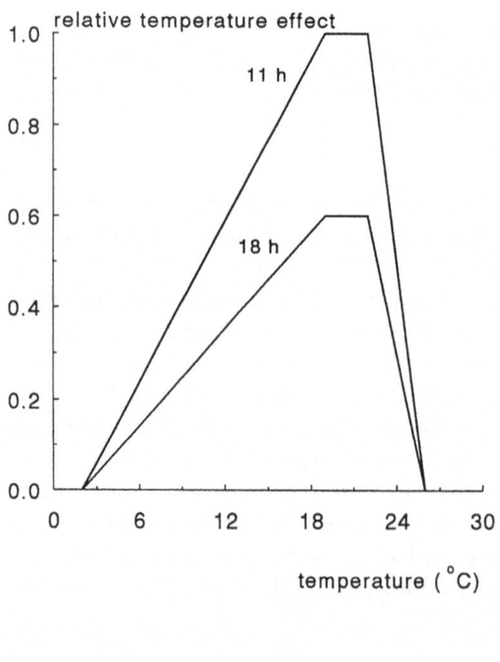

Fig. 4. The relative effect of mean daily temperature on the rate of progress towards start of tuber growth for photoperiods at emergence of 11 and 18 h. See text for explanation. (After Kooman and Haverkort 1995)

Shorter photoperiod and increase in temperature (up to 20 °C) were found to shorten the duration of the period between start of tuber growth and cessation of leaf growth; the fourth stage of development is shorter for warmer temperatures and higher levels of incident radiation. Of crucial importance are the effects of temperature and photoperiod on the duration of the period between emergence and the beginning of tuber growth because the durations of the subsequent stages (and hence, the duration of the total growth cycle and yield) vary more or less in proportion (Kooman et al. 1996).

7 Radiation Interception, Growth Rate and Total Dry Matter Production

A common device to analyze crop production is Eq. (3), relating the crop growth rate (δY; $g\,m^{-2}$) to the quantity of incident radiation received at the top of the canopy during the period under consideration (I; $MJ\,m^{-2}$), the fraction of light intercepted, F, and the radiation conversion efficiency (E; $g\,MJ^{-1}$):

$$\delta Y = I\,F\,E. \tag{3}$$

The factor for conversion from energy intercepted to dry matter produced, E, is called different names in the literature, among which radiation- or light-use efficiency (with acronyms RUE and LUE). The numerical value of E depends on the units in which I is expressed: most authors use PAR (photosynthetically active radiation), but also global radiation (GR) is used (GR is about 2 PAR). When radiation is expressed as PAR, E for potato varies between 2 and 3 $g\,MJ^{-1}$ with 2.5 $g\,MJ^{-1}$ as a fair average value in the absence of stress. Such values are common for C_3 species.

E generally declines with growth temperature; above 20 °C E declines with about 0.02 $g\,MJ^{-1}\,°C^{-1}$ (Kooman and Haverkort 1995). Under high radiation loads E is lower than under moderate radiation (Haverkort and Harris 1987). To account for this effect, Kooman and Haverkort (1995) suggested discarding I in excess of 12 MJ (PAR) day^{-1} in resolving Eq (3) [I > 12 MJ (PAR) day^{-1} rarely occurs in temperate climates].

Since the harvest index of mature crops (i.e., the fraction of the total dry weight in tubers), grown under suitable conditions, generally ranges from about 0.7 to 0.8, and since the dry matter concentration of mature tubers is fairly conservative for a given cultivar in a given climate, the approach given with Eq. (3) can be extended to obtain estimates of fresh yields in particular environments. Tuber dry matter concentrations typically range from 18 to 26%.

Leaf area index, L, is an important determinant of F. The rule of thumb applicable to many crop species is that F increases little for L values higher than 3. Spitters (1987) found the following relation for PAR: $F = 1 - e^{(-1.01L)}$. For L = 3, F is 0.95. Haverkort et al. (1991) also reported full interception of PAR at L values of 3 and higher. However, Burstall and Harris (1983) found as equation of best fit for GR: $F = (L - 0.28)/[1.46 + 0.486(L - 0.28)]$ and Firman and Allen (1989) found for GR: $F = 0.12 + 0.43\,L - 0.067\,L^2$ (valid up to L values of 4). The last two equations return F values of 0.72 and 0.79, respectively for $L = 3$. The extinction coefficient of PAR is higher than that for GR and this can explain lower fractions of interception at $L = 3$ for the last two equations in comparison with the data from Spitters (1987) and Haverkort et al. (1991). The difference in extinction between PAR (400–700 nm) and the longer wavelengths (700–3000 nm) explains the decline in red:far-red ratio as the light travels down the canopy. Potatoes are planted in rows with distances of 75 cm (Europe) to 90 cm or more (e.g., USA) at within row distances of ca. 30 cm between plants. Due to this pattern, some degree of clustering of stems is likely, especially early in the season. For this reason it seems plausible that in potato the value for L for complete light interception would be somewhat higher than in crop species with shorter interplant distances.

To obtain high yields, the objective is to increase the period of complete light interception and to prevent the decline of E below the optimum value. The duration of the period of complete light interception is a function of development and growth of leaves on stems and branches, which, in turn, depends on the change during the growth cycle in allocation of carbon and nitrogen in the plant.

8 Temperature and the Rate of Growth Processes

The growth rate (dry weight) of a potato crop is sensitive to the ambient temperature. Figure 5 illustrates relative temperature effects taken from two models – again assumed to be summaries of the current understanding (or the perception of that by the modellers!). The lines 1 and 2 illustrate the relative effect of mean daily temperature on the growth rate of the vines (vegetative above-ground parts) and of the tubers, respectively in the SUBSTOR potato (Ritchie et al. 1995). Lines 3 and 4 are from LINTUL-Potato by Kooman and Haverkort (1995) and denote the effect on E (which is not exactly equivalent to vine growth represented in line 1) while line 4 pertains to the tuber growth rate.

The overall growth response of the crop is the combined result of responses of component processes which differ to some extent in temperature response. Important processes in this respect are the growth rate of sprouts, the rate of foliar expansion, the rates of photosynthesis and respiration, the rate of tuber growth and the rate of leaf senescence. In principle, the temperature response of the crop will depend on the growth stage (Squire 1995); this is not generally considered in models (several contributions in Haverkort and MacKerron 1995 and in Kabat et al. 1995).

According to Feddes (1987), for instance, 116 °C days are required for emergence under optimal water availability ($T_b = 2.9\,°C$). The thermal time required for emergence is modified by water availability (Feddes 1987) and by other factors such as physiological age and planting depth (Klemke and Moll 1990);

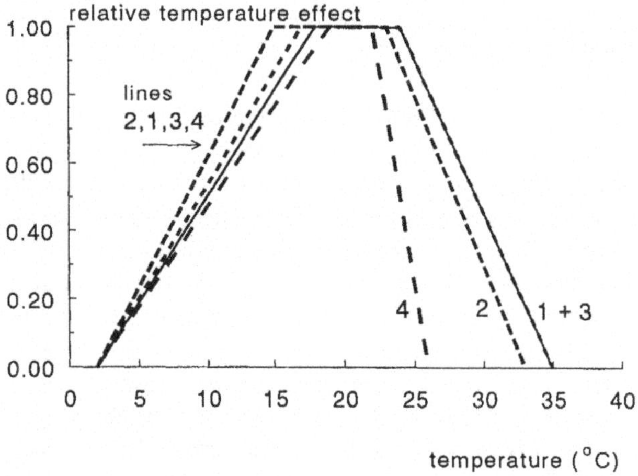

Fig. 5. Relative effects of mean daily temperature on growth rates. Lines *1* and *2* illustrate the relative effect of mean daily temperature on the growth rate of the vines (vegetative above-ground parts) and of the tubers, respectively in SUBSTOR-Potato (Ritchie et al. 1995). Lines *3* and *4* denote the effect on the radiation conversion factor (E) and tuber on the growth rate, respectively, in the LINTUL-POTATO model. (Kooman and Haverkort 1995)

in the latter example, physiological age itself was described as a function of thermal time.

Kirk and Marshall (1992) found that the thermal duration of expansion was 170 °C day of leaf numbers 4 to 10 on the main stem (range of observations: 9 to 29 °C with T_b = 0 °C and T_0 25 °C). Equal thermal durations of expansion do not mean equal leaf sizes: (potential) leaf sizes depend on leaf number; they increase to about leaf number 10 to 13 and decline for higher main stem leaf numbers (Kirk and Marshall 1992; Vos and Biemond 1992). Equal thermal duration of expansion implies that larger leaves become larger because they expand faster than smaller leaves. In principle it is possible to estimate foliar expansion from information on leaf sizes and the thermal requirements of leaf appearance and duration of leaf expansion.

Observations such as those made by Ku et al. (1977) corroborated the established opinion that the rate of leaf photosynthesis was optimal in the range of 15/16 to 25 °C. However, Wolf et al. (1990b) showed that the photosynthetic rate was not reduced up to 38 °C, provided plants had been adapted to high temperatures. Interestingly, they included in their study cvs. Desirée and Up-to-Date which were bred in the temperate zone (northwest Europe).

Particularly important is the sensitivity of tuber growth to temperature: a rapid decline in the relative rate is assumed to occur at temperatures exceeding 21 to 23 °C (Fig. 5). Under such temperatures the potato plant may regress in its development and return to some extent to a more vegetative type of growth (as will be discussed in more detail in Sect. 12).

9 Dry Matter Distribution

The temporal pattern of allocation of carbon and nitrogen to sink organs is of decisive significance for the length of the growth cycle, the yield level and harvest index. Figure 6 shows a schematic representation of the change with thermal time in the partitioning coefficients of dry matter. When, in the example shown, thermal time is 200 °C, the fraction of the (daily) increment in dry matter which is allocated to leaves is 0.22, while the fractions allocated to stems and tubers are 0.20 and 0.58, respectively. Genetically controlled differences in maturity class and photothermal influences determine the onset of allocation to tubers and the rate at which tubers gain absolute dominance in dry matter allocation, i.e., these factors affect the intercept on the thermal time axis and the slope of line t in Fig. 6. Wolf et al. (1990a) showed these photothermal effects with $^{14}CO_2$ feeding of leaves. Poor supply of nitrogen advances and ample supply delays tuberization and thus the position of line t (cf. Oparka et al. 1987; Biemond and Vos 1992).

Figure 6 is not complete because allocation to roots is not considered. Vos and Groenwold (1986) found maxima in root dry weight of 40 to 70 $g m^{-2}$, but assuming about 30% of the dry weight was lost during recovery, these maxima are probably some 30% higher. Lesczynski and Tanner (1976) also observed maximum root dry weight over 100 $g m^{-2}$. Vos and Groenwold found that root expansion in the deeper soil horizons continues till advanced stages of growth.

A process not accounted for in Fig. 6 is the relocation of material from the tops to the tubers. Relocation accounts for 10 to perhaps 20% of tuber yield.

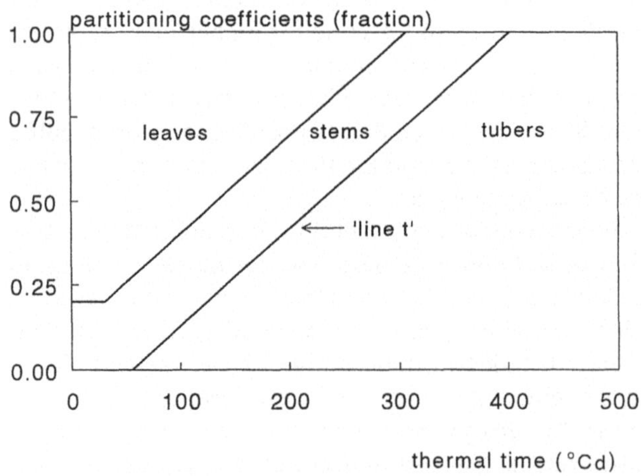

Fig. 6. Cumulative partitioning coefficients for (daily) increments in dry matter of leaves, stems and tubers plotted against thermal time (T_b = 7 °C). See text for explanation of *line t*. (After van Heemst 1986)

10
Nitrogen Nutrition and Potato Production

The period of maximal light interception depends strongly on the amount of nitrogen that is available during the season (Millard and Marshall 1986; Vos 1995b). The degree of apical branching at which foliar growth stops depends on nitrogen supply as is illustrated by Fig. 7a,b for two treatments: one without fertilizer nitrogen and one with $30\,g\,m^{-2}$ nitrogen. In the zero-N treatment foliar development of most of the main stems was halted in the first level of apical branching (Fig. 1); less than 5% of the main stems completed the first level of apical branching and showed some development of the second level of

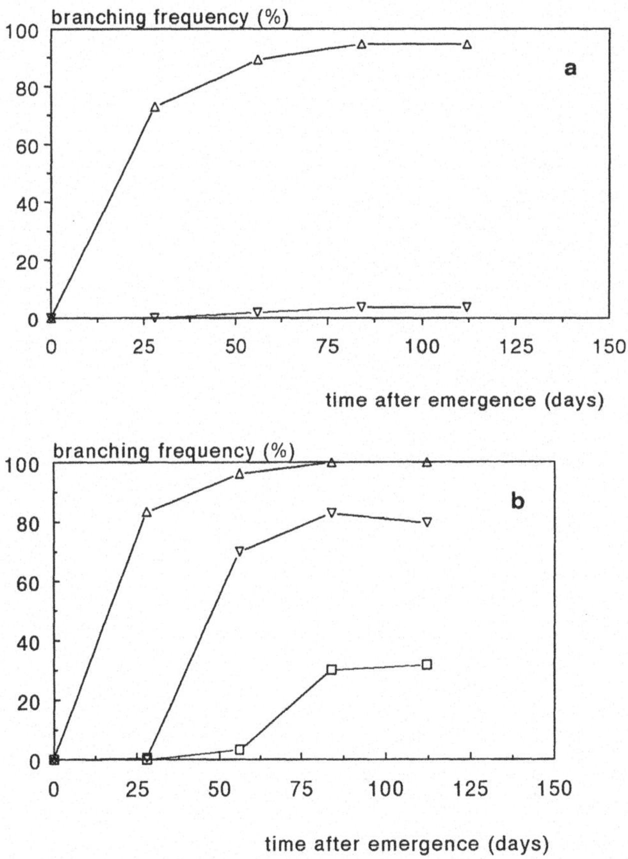

Fig. 7a,b. Effect of nitrogen supply on frequencies of apical branching. **a** Treatment with zero fertilizer nitrogen supply. **b** Treatment with $30\,g\,m^{-2}$ fertilizer nitrogen. *Arrows up* First level apical branches (see Fig. 1); *arrows down* second level of apical branches; *squares* third level of branches. A branch was counted when some leaves were present, not when it was complete; the sum of counts was expressed as percent of the potential number (J. Vos, unpubl.). Emergence = May 12, 1993; latitude of location: 52°N branching. For most stems of the high-N treatment, development was halted in the second level of apical branching; about 30% of the stems even showed some development of the third level of branching.

Young leaves show initially a high photosynthetic capacity, P_{max}. Therefore, prolonged appearance of new leaves with more nitrogen extends the period with a high growth rate. Since P_{max} declines with leaf age (Vos and Oyarzún 1987; Firman and Allen 1988; Olesinski et al. 1989), the growth rate of the crop is bound to decline gradually from the moment on that the plants cease to produce new leaves. These points are illustrated in Fig. 8a which shows the change with time in total dry weight and in tuber dry weight for the same treatments as in Fig. 7. Figure 8 shows fairly representative values for dry matter production and nitrogen uptake for well-managed crops in temperate zones.

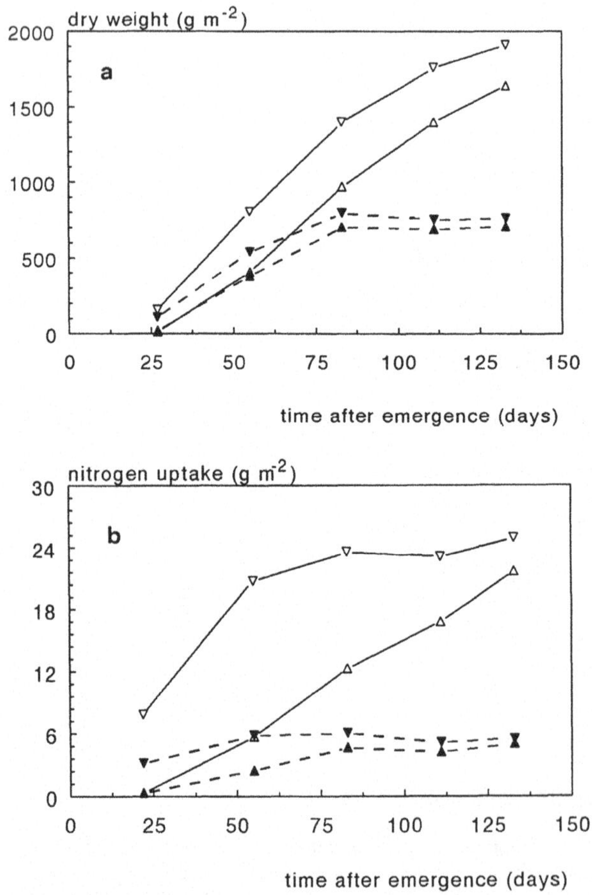

Fig. 8a,b. Effect of nitrogen supply on dry matter production (**a**) and amount of nitrogen in the crop (**b**); *broken lines* zero fertilizer nitrogen supply; *full drawn line* 30 g m^{-2} fertilizer nitrogen; *arrows down* total plant (foliage plus tubers); *arrows up* tubers. (J. Vos, unpubl.)

Relocation from tops to tubers is particularly important for nitrogen, as is illustrated in Fig. 8b. The total amount of nitrogen in the crop changed comparatively little after 60 days from emergence, but the amount in the tubers kept on increasing due to relocation. From the time when little additional nitrogen is taken up from the soil, the relative sink strengths for nitrogen of the foliage and the tubers determine the quantity of additional carbon that can be produced: rapid relocation to the tubers will lead to faster senescence, whereas higher seasonal dry matter production can be achieved the longer nitrogen is kept in photosynthesis systems.

The most important effects of nitrogen supply regard the degree of apical branching, the rate of leaf expansion, leaf size and the final degree of apical branching. Effects on specific leaf area are negligible (Vos and Biemond 1992; Vos 1995b,c). In spaced plants leaf longevity may be increased to some extent by high rates of nitrogen supply (Vos and Biemond 1992), but in a crop situation leaf longevity appears to decline with increase in nitrogen supply (Millard and MacKerron 1986; Firman and Allen 1988), enhanced shading being the probable cause (cf. Marshall and Vos 1991). Seasonal E is little affected by nitrogen supply, except under severe nitrogen deficiency (Millard and Marshall 1986; J. Vos, unpubl.).

11 Drought and Potato Production

Potato shows generally a linear relation between water use and dry matter production. The slope of the line being dependent on the (average) water vapour saturation deficit of the atmosphere, V (mbar). Plots of dry matter production versus W/V yield a linear relation with a characteristic slope of 27.4 g (dry matter) $m^{-2} mm^{-1}$ mbar, where W is transpiration (mm) (Feddes 1987).

Potato is claimed to be a relatively drought-sensitive species due to its relatively shallow, sparse and weak root system. However, direct comparative evidence in support of this view seems to be lacking. Root length density of potato in upper horizons is in the order of 2 to 3 cm cm^{-3}, which is lower than in Gramineae, for instance (Vos and Groenwold 1986). Yet it is questionable whether such root length densities are limiting water uptake. If rooting depth is somewhat shallow, this could perhaps be due to a greater sensitivity of expansion growth to mechanical impedance than in other species. Little attention has been paid to the water relations of potato, and perhaps here lies a clue to the potato's apparent drought sensitivity. Gandar and Tanner (1976) concluded that expansion of leaves and tubers ceased at water potentials of about −0.4 to −0.5 MPa. Vos and Groenwold (1988) corroborated that turgor loss could occur at such or slightly lower water potentials and reported osmotic potentials at saturation (full turgor) of −0.54 to −0.70 MPa. Although the water relations of roots have not been studied, it can be assumed that roots have osmotic potentials in the same order as leaves. Such high osmotic potentials and turgor loss at relatively high water potentials in roots could have two unfavourable consequences. The first is a higher wilting point for potato than the conventional value for soil water potential of pF = 4.2 (−1.6 MPa). This would mean that the potato would leave more water unused in the soil than species with a lower wilting point. The second consequence could be that the maximum growth pressure that potato roots can exert to penetrate the soil is relatively low (because that pressure is determined by the maximum turgor, which, in turn, depends on the osmotic potential). Hence, root growth could be inhibited by relatively small external forces.

12 Second Growth

Suppression or stimulation of outgrowth of buds is important in the reactions of potato to the environment. This also holds for phenomena called second growth. A normal sequence of events is that plants emerge, develop foliage and stolons and build up the stimuli to stop stolon growth and start tuber growth. At some point in time during the last stage, growth of the foliage is halted and the crop matures eventually as a result of ageing of the foliage. This normal sequence of processes can be interrupted. In the field, such an interruption often occurs during a period of drought stress. Upon rewatering phenomena of second growth can be observed, including changes in shape due to knobs, chain tuberization, i.e., chains of interconnected tubers with short stolons, and sprouting of the tubers, possibly leading to the appearance of secondary plants. Bodlaender et al. (1964) and Lugt et al. (1964) showed that the high temperature, not drought itself, is the causal environmental factor, high temperatures in this context being in excess of ca. 25 °C for some period of time. It is difficult to specify the critical conditions more exactly, one reason being the genetic variation in sensitivity for second growth. The essence of second growth is the loss of dormancy of buds in the eyes of growing tubers. Depending on the strength of the stimuli, growth of the buds (primarily in the apical region) results in a more or less pronounced change in shape of the tuber to the more severe expressions just mentioned. Second growth, whatever its degree, can result in the resorption of starch from the original tuber part. Thus at harvest, a tuber can have a primary part of low dry matter content and a secondary part with a higher dry matter content. It stands to reason that severe incidences of second growth are detrimental to the cooking and processing quality of the crop.

13 Tuber Size Distribution

The size distribution of a harvested crop is one of the factors determining its economic value. Disregarding the smallest tubers, tuber size shows a *more or less* normal distribution, both on a weight basis (Sands and Regel 1983; Marshall 1986) and as sizes, obtained by sieving using square-meshed sieves

(Travis 1987). The distribution can be characterized by two parameters μ (mu, the mean) and σ (sigma, the measure of the spread). Evaluation of the size distributions in many crops led to the conclusion that μ is a function of the total yield (Y, $g\,m^{-2}$) and the number of tubers (N, m^{-2}):

$$\mu = a + b(Y/N), \tag{4}$$

where a and b are parameters. Regel and Sands found values for a and b of 17 g and 1.6, respectively, and Marshall and Thompson (1986) reported values for a of −4.65 (g) and for b 1.417 (tubers through 15-mm sieve were disregarded for estimates of Y and N). In both studies 98% of the variance in μ could be explained with Eq. (4). Marshall (1986) defined a relative variability parameter, Rv, as the ratio of $\sigma:\mu$, that is analogous to the coefficient of variation. MacKerron et al. (1988) showed a relation between Rv and the number of tubers per m^{-2}: for 20 tubers m^{-2} Rv was about 0.6 while Rv declined to about 0.4 for 70 tubers m^{-2}. Comparable values for Rv were found in the other studies mentioned.

There are several possible applications of Eq. (4). MacKerron et al. used the analysis to answer the question whether drought had other effects on tuber-size distribution other than those mediated by Y and N. The same approach can be applied to evaluate effects of other factors or cultural practices. The final tuber number is set well in advance of the harvest. When N and an estimate of Y are known at a particular point in time during tuber growth, additional growth can be estimated using Eq. (3) and the change in tuber-size distribution can be estimated from Eq. (4). Although N is not under perfect control by the grower, it is in principle possible to advise growers with such devices on tactic decisions regarding plant and stem number and on operational decisions on the optimal time of harvest (or rather time of killing) of the crop.

Acknowledgement. The author is grateful to Dr. B. Marshall (SCRI, Dundee), Dr. A.J. Haverkort (AB-DLO, Wageningen) and Prof. Dr P.C. Struik (Agronomy-WAU, Wageningen) for helpful comments on the manuscript.

References

Almekinders CJM, Struik PC (1994) Photothermal response of sympodium development and flowering in potato (*Solanum tuberosum* L.) under controlled conditions. Neth J Agric Sci 42:311–329

Almekinders CJM, Struik PC (1996) Shoot development and flowering in potato (*Solanum tuberosum* L.). Potato Res 39:581–607

Biemond H, Vos J (1992) Effects of nitrogen on the development and growth of the potato plant. 2. The partitioning of dry matter, nitrogen and nitrate. Ann Bot 70:37–45

Bodlaender KBA, Lugt C, Marinus J (1964) The induction of second growth in potato tubers. Eur Potato J 7:57–71

Burstall L, Harris PM (1983) The estimation of percentage light interception from leaf area index and percentage ground cover in potatoes. J Agric Sci Camb 100:241–244

Danert S (1957) Ueber den Sprossaufbau und die Blattentwicklung bei der Kartoffel. Züchter 27:22–33 van der Zaag DE, van Loon CD (1987) Effect of physiological age on growth vigour of seed potatoes of two cultivars. 5. Review of literature and integration of some experimental results. Potato Res 30:451–472

van Heemst HDJ (1986) The distribution of dry matter during growth of a potato crop. Potato Res 29:55–66

van Ittersum MK (1992a) Relation between growth conditions and dormancy of seed potatoes. 1. Effects of nitrogen. Potato Res 35:355–364

van Ittersum MK (1992b) Relation between growth conditions and dormancy of seed potatoes. 3. Effects of light. Potato Res 35:377–387

van Ittersum MK (1993) Advancing growth vigour of seed potatoes by storage temperature regimes. Neth J Agric Sci 41:23–36

van Ittersum MK, Scholte K (1992) Relation between growth conditions and dormancy of seed potatoes. 2. Effects of temperature. Potato Res 35:365–375

Vos J (1995a) Foliar development of the potato plant and moulations by environmental factors. In: Kabat P, Marshall B, van den Broek BJ, Vos J, van keulen H (eds) Modelling and parameterization of the soil-plant-atmosphere system. A comparison of potato growth models. Wageningen Pers, Wageningen, pp 21–38

Vos J (1995b) Nitrogen and the growth of potato crops. In: Haverkort AJ, MacKerron DKL (eds) Potato ecology and modelling of crops under conditions limiting growth. Current issues in production ecology, vol 3. Kluwer, Dordrecht, pp 115–128

Vos J (1995c) The effects of nitrogen supply and stem density on leaf attributes and stem branching in potato (*Solanum tuberosum* L.). Potato Res 38:271–279

Vos J, Biemond H (1992) Effects of nitrogen on the development and growth of the potato plant. 1. Leaf appearance, expansion growth, life span of leaves and stem branching. Ann Bot 70:27–35

Vos J, Groenwold J (1986) Root growth of potato crops on a marine-clay soil. Plant Soil 94:17–33

Vos J, Groenwold J (1988) Water relations of potato leaves. I. Diurnal changes, gradients in the canopy and effects of leaf-insertion number, cultivar and drought. Ann Bot 62:363–371.

Vos J, Groenwold J (1989) Genetic differences in water-use efficiency, stomatal conductance and carbon isotope fractionation in potato. Potato Res 32:113–121

Vos J, Oyarzún PJ (1987) Photosynthesis and stomatal conductance of potato leaves – effects of leaf age, irradiance, and leaf water potential. Photosynth Res 11:253–264

Vreugdenhil D, Struik PC (1989) An integrated view of the hormonal regulation of tuber formation in potato (*Solanum tuberosum* L.). Physiol Plant 75:525–531

Wolf S, Marani A, Rudich J (1990a) Effects of temperature and photoperiod on assimilate partitioning in potato plants. Ann Bot 66:513–520

Wolf S, Olesinski AA, Rudich J, Marani A (1990b) Effect of high temperature on photosynthesis in potatoes. Ann Bot 65:179–185

Ku SB, Edwards GE, Tanner CB (1977) Effects of light, carbon dioxide, and temperature on photosynthesis, oxygen inhibition of photosynthesis, and transpiration in *Solanum tuberosum.* Plant Physiol 59:868–872

Lesczynski DB, Tanner CB (1976) Seasonal variation in root distribution of irrigated field-grown Russet Burbank potato. Am Potato J 53:69–78

Lovell PH, Booth A (1969) Stolon initiation and development in *Solanum tuberosum* L. New Phytol 68:1175–1185

Lugt C, Bodlaender KBA, Goodijk G (1964) Observations on the induction of second growth in potato tubers. Eur Potato J 7:219–227

MacKerron DKL, Marshall B, Jefferies RA (1988) The distributions of tuber sizes in droughted and irrigated crops of potato. II. Relation between size and weight of tubers and the variability of tuber-size distribution. Potato Res 31:279–288

Marshall B (1986) Tuber size. Aspects Appl Biol 12:393–396

Marshall B, Thompson R (1986) Tuber-size distribution. Potato Res 29:261–262

Marshall B, Vos J (1991) The relation between the nitrogen concentration and photosynthetic capacity of potat (*Solanum tuberosum* L.) leaves. Ann Bot 68:33–39

Millard P, MacKerron DKL (1986) The effects of nitrogen application on growth and nitrogen distribution within the potato canopy. Ann Appl Biol 109:427–437

Millard P, Marshall B (1986) Growth, nitrogen uptake and partitioning within the potato (*Solanum tuberosum* L.) crop, in relation to nitrogen application. J Agric Sci Camb 107:421–429

Olesinski AA, Wolf S, Rudich J, Marani A (1989) Effect of leaf age and shading on photosynthesis in potatoes (*Solanum tuberosum*). Ann Bot 64:643–650

Oparka KJ, Davies HV, Prior DAM (1987) The influence of applied nitrogen on export and partitioning of current assimilate by field-grown potato plants. Ann Bot 59:311–323

Pallais N (1991) True potato seed: changing potato propagation from vegetative to sexual. Hort Science 26:239–241

Plaisted PH (1957) Growth of the potato tuber. Plant Physiol 32:445–452

Ritchie JT, Griffin TS, Johnson BS (1995) SUBSTOR: functional model of potato growth, development and yield. In: Kabat P, Marshall B, van den Broek BJ, Vos J, van Keulen H (eds) Modelling and parameterization of the soil-plant-atmosphere system. A comparison of potato growth models. Wageningen Pers, Wageningen, pp 401–435

Sands PJ, Regel PA (1983) A model for the development and bulking of potatoes (*Solanum tuberosum* L.). Field Crops Res 6:25–40

Schepers A, Hoogland RF (1968) Relation between sprout development in seed potatoes and the number of main stems in the crop. Mededeling 337 van het IBS, Wageningen (Comm 337 of the Institute for Biological and Chemical Research on Crops and Herbage, Wageningen), pp 47–53

Snyder RG, Ewing EE (1989) Interactive effects of temperature, photoperiod, and cultivar on tuberization of potato cuttings. HortScience 24:336–338

Spitters CJT (1987) An analysis of variation in yield among potato cultivars in terms of light absorption, light utilization and dry matter partitioning. Acta Hortic 214:71–84

Squire GR (1995) Linkages between plant and weather in models of crop production. In: Kabat P, Marshall B, van den Broek BJ, Vos J, van Keulen H (eds) Modelling and parameterization of the soil-plant-atmosphere system. A comparison of potato growth models. Wageningen Pers, Wageningen, pp 57–76

Struik PC, van Heusden E, Burger-Meijer K (1988) Effects of short periods of long days on the development, yield and size distribution of potato tubers. Neth J Agric Sci 36:11–22

Struik PC, Haverkort AJ, Vreugdenhil D, Bus CB, Dankert R (1990) Manipulation of tuber-size distribution of a potato crop. Potato Res 33:417–432

Travis KZ (1987) Use of a simple model to study factors affecting the size distribution of tubers in potato crops. J Agric Sci Camb 109:563–571

van den Berg JH, Ewing EE, Plaisted RL, McMurry S, Bonierbale MW (1996) QTL analysis of potato tuberization. Theor Appl Genet 93:307–316 van der Zaag DE, van Loon CD (1987) Effect of physiological age on growth vigour of seed potatoes of two cultivars. 5. Review of literature and integration of some experimental results. Potato Res 30:451–472

van Heemst HDJ (1986) The distribution of dry matter during growth of a potato crop. Potato Res 29:55–66

van Ittersum MK (1992a) Relation between growth conditions and dormancy of seed potatoes. 1. Effects of nitrogen. Potato Res 35:355–364

van Ittersum MK (1992b) Relation between growth conditions and dormancy of seed potatoes. 3. Effects of light. Potato Res 35:377–387

van Ittersum MK (1993) Advancing growth vigour of seed potatoes by storage temperature regimes. Neth J Agric Sci 41:23–36

van Ittersum MK, Scholte K (1992) Relation between growth conditions and dormancy of seed potatoes. 2. Effects of temperature. Potato Res 35:365–375

Vos J (1995a) Foliar development of the potato plant and moulations by environmental factors. In: Kabat P, Marshall B, van den Broek BJ, Vos J, van keulen H (eds) Modelling and parameterization of the soil-plant-atmosphere system. A comparison of potato growth models. Wageningen Pers, Wageningen, pp 21–38

Vos J (1995b) Nitrogen and the growth of potato crops. In: Haverkort AJ, MacKerron DKL (eds) Potato ecology and modelling of crops under conditions limiting growth. Current issues in production ecology, vol 3. Kluwer, Dordrecht, pp 115–128

Vos J (1995c) The effects of nitrogen supply and stem density on leaf attributes and stem branching in potato (*Solanum tuberosum* L.). Potato Res 38:271–279

Vos J, Biemond H (1992) Effects of nitrogen on the development and growth of the potato plant. 1. Leaf appearance, expansion growth, life span of leaves and stem branching. Ann Bot 70:27–35

Vos J, Groenwold J (1986) Root growth of potato crops on a marine-clay soil. Plant Soil 94:17–33

Vos J, Groenwold J (1988) Water relations of potato leaves. I. Diurnal changes, gradients in the canopy and effects of leaf-insertion number, cultivar and drought. Ann Bot 62:363–371.

Vos J, Groenwold J (1989) Genetic differences in water-use efficiency, stomatal conductance and carbon isotope fractionation in potato. Potato Res 32:113–121

Vos J, Oyarzún PJ (1987) Photosynthesis and stomatal conductance of potato leaves – effects of leaf age, irradiance, and leaf water potential. Photosynth Res 11:253–264

Vreugdenhil D, Struik PC (1989) An integrated view of the hormonal regulation of tuber formation in potato (*Solanum tuberosum* L.). Physiol Plant 75:525–531

Wolf S, Marani A, Rudich J (1990a) Effects of temperature and photoperiod on assimilate partitioning in potato plants. Ann Bot 66:513–520

Wolf S, Olesinski AA, Rudich J, Marani A (1990b) Effect of high temperature on photosynthesis in potatoes. Ann Bot 65:179–185

Cowpea

A.E. Hall[1]

1 Introduction

Cowpea (*Vigna unguiculata* L. Walp.) is an annual legume that is adapted to warm conditions and sensitive to chilling. Consequently, it is cultivated either in the tropics or in the subtropics during the warm season. Cowpea is grown to produce dry grain, fresh peas, edible pods, edible leaves, hay or forage, or as a green manure crop. This chapter will consider the production of dry grain by cowpea. Cowpea has been defined as including four cultigroups (Baudoin and Maréchal 1985), and two of these cultigroups are grown to produce dry grain: unguiculata, which is the major group; and biflora or catjang, which is differentiated mainly by its small erect pods and is grown in Southeast Asia. Cowpea is highly self-pollinated in most production environments, although significant outcrossing can occur in some environments, possibly associated with pollen transfer by bumblebees.

Cowpea is grown both as a sole-crop and as an intercrop. I will examine cowpea as a sole-crop to facilitate the delineation of the relationships between plant physiology and grain yield. Sole-crop production environments for this crop range from being either near-optimal, such as under irrigation during the summer in subtropical zones of California, or often suffer from severe drought, heat, and other stresses, such as under rainfed conditions in the semiarid tropical Sahelian zone of Senegal. Most examples discussed here are chosen from studies in these two contrasting environments, but they have relevance to many cowpea production zones.

Progress has been slow in breeding cowpea cultivars with either enhanced yield potential or enhanced adaptation to dry environments. The most practical approach is to yield-test advanced lines over several locations and years, and select lines that have consistently greater yield than check cultivars. This approach has been successful under irrigated conditions in California (Helms et al. 1991a,b), and rainfed conditions in the semiarid zone of Senegal (Cisse et al. 1995). However, yield-testing is more difficult for cowpea than for many other grain crops. Large plots are needed and to date most cowpeas are not combine-harvested as a standing crop but are harvested either by hand or

[1] Botany and Plant Sciences Department, University of California, Riverside, California 92521-0124, USA

Crop Yield, Physiology and Processes
D. L. Smith, C. Hamel (Eds.)
© Springer-Verlag Berlin Heidelberg 1999

using cumbersome procedures involving cutting the plants, drying them in windrows, and then separating the grains using special multiple-drum bean threshers (Hall and Frate 1996). In addition, yield-testing takes considerable time and effort, and interpreting yield data is particularly difficult when the tests are conducted in stressful environments. Consequently, it is useful to consider physiological, morphological, and phenological criteria that would contribute to yield potential or adaptation to drought, and could be selected in early generations, thereby complementing the empirical approach to breeding improved cultivars. This chapter will focus on ways that information concerning plant physiology can be used to increase grain yield by either plant breeding or changes in agronomic practices. Further information on the physiology of cowpea is presented in comprehensive reviews by Wien and Summerfield (1984), and Summerfield et al. (1985). I will begin by presenting models for cowpea performance under optimal conditions and drought to identify and discuss some of the factors that influence grain yield.

2 Grain Yield in Optimal Environments

In optimal environments, grain yield of cowpea per unit ground area (Y) can be modeled using the following equation:

$$Y = \sum_{i=n,}^{i=o} PFD_i \times GC_i \times Q_i \times PG_i \tag{1}$$

where PFD_i is the flux density of photosynthetically active photons per day, GC_i is the ratio of plant cover to ground area on that day, Q_i is the efficiency for the conversion of intercepted photons to plant dry biomass, and PG_i is a ratio for the partitioning of plant dry biomass to grain. This equation applies over the period of days (n) when carbon is being partitioned to grain. The days from first flowering to the time when 95% of the pods are mature is an effective value for n. This should be considered an approximation because some of the carbohydrate produced during the vegetative stage may be translocated to the developing grain but the amount appears to be small (Wien and Summerfield 1984).

The length of the reproductive period (n) has a strong influence on grain yield. Over 7 t/ha of grain has been obtained on large guarded plots in subtropical conditions in California when the cowpea cultivars were permitted to grow until 145 days from sowing and produced two flushes of pods (e.g., Sanden 1993). In this experiment, the period to first flowering was about 50 days, and the first pod flush had a duration of about 50 days and produced 5 t/ha (1 t/ha every 10 days). The second pod flush began about 15 days after the first flush, lasted about 30 days, and produced 2 t/ha (2/3 t/ha every 10 days). When the same California cultivars are grown in tropical zones, such as Senegal, they have the potential to produce about 2 t/ha from one flush of pods

(Hall and Patel 1985). Due to the high night temperature in Senegal, they begin flowering early, about 30 days from sowing, and exhibit 95% pod maturity about 60 days from sowing. This is a production rate of 2/3 t/ha every 10 days. Locally adapted cultivars that flower about 35 days after sowing and have a reproductive duration of about 40 days have a yield potential in Senegal of 3 t/ha, which is a production rate of 3/4 t/ha every 10 days. The main point is that, under optimal conditions, grain yield variations from 2 to 7 t/ha can occur and be largely explained by variations in reproductive duration from 30 to 80 days, but with some variation in grain production rate from 2/3 to 1 t/ha every 10 days. The tendency for higher night temperature to accelerate reproductive development by decreasing the period from sowing to first flowering (Grantz and Hall 1982), and by decreasing the development period of individual pods (e.g., 21 days from pod set to maturity under cool night temperatures to 14 days under tropical night temperatures; Nielsen and Hall 1985b), strongly reduces grain yield by reducing overall reproductive duration. Wien and Ackah (1978) observed that genotypic variation exists for length of the pod development period and that it is positively associated with individual seed weight.

Current cowpea cultivars grown in California (CB5, CB46, and CB88) and some breeding lines that flower early but are not determinate have the potential to produce two discrete flushes of flowers and pods under optimal growing conditions with a period of 10 to 20 days between each flush when they produce few flowers (Kwapata and Hall 1990a; Gwathmey et al. 1992b). This inactive period in the middle of the season when few flowers or pods are produced represents an inefficiency whose removal by improved management or plant breeding could enhance grain yields. Unfortunately, at this time no genes or management methods have been identified that can eliminate or reduce the inactive period. The inactive period appears to be internally programmed in that it still occurs when young or older pods produced during the first flush are continually removed (Kwapata and Hall 1990a; Gwathmey et al. 1992b). A few cowpea genotypes have been discovered that have the ability to produce alternating reproductive and vegetative nodes on the main stem (Ehlers and Hall 1996). This trait may be useful for breeding early-flowering cultivars that have a long but less synchronous first flush of pods. Also, genotypes have been identified with delayed leaf senescence that have substantial ability to survive after producing the first flush of pods, which enables the crop to produce a greater second flush of pods than the more typical genotypes where leaves senesce and most plants may die after maturing the first flush (Gwathmey et al. 1992a). The delayed leaf senescence genes could enhance grain yield potential in many environments and are consistently expressed in both subtropical California and tropical Senegal (Hall et al. 1997).

The photon flux density per day (PFD_i) can be about 50 mol photon/m^2 on a sunny, summer day in California with a day length of about 10 h. In the tropics PFD_i can be substantially less than this due to shorter days and clouds. Solar radiation data are often provided in terms of irradiance which can be converted to PFD using a locally determined conversion factor (e.g., 2.1 μmol photon/J for summer conditions in California from Kwapata et al. 1990). As a first approximation, the accumulation of carbohydrate by cowpea will be proportional to intercepted PFD (Littleton et al. 1979), as has been observed with several crop species (Monteith 1977). It should be noted that the photosynthetic response of the canopy probably is linear over a greater range of PFD than that of individual leaves which achieves 95% saturation at about 50% of full sunlight (Kirchhoff et al. 1989c).

When attempting to achieve maximum grain yield per hectare, plant spacings should be chosen that enable the plants to achieve 100% ground cover ($GC_i = 1$) by first flowering and maintain complete ground cover during most of the reproductive period. A canopy of this type would also be effective in suppressing weeds that emerge late in the season. Sole-crops of cowpea in Senegal often are grown at wide spacings that do not permit the achievement of 100% ground cover at first flowering. As a first approximation, grain yield of cowpea is proportional to the amount of PFD intercepted and therefore GC_i during the reproductive period.

Under optimal conditions, the efficiency by which canopies of cowpea leaves convert intercepted photons into shoot dry matter (Q_i) is about 400 to 500 mg/mol photon (Kwapata et al. 1990), which is equivalent to crop growth rates of 2.0 to 2.5 t/ha over a 10-day period with PFD_i of 50 mol photon m^{-2} day^{-1} and complete ground cover. In tropical conditions, Littleton et al. (1979) observed maximum daily crop growth rates of 16 to 25 g/m^2, which is equivalent to 1.6 to 2.5 t/ha over a 10-day period. The high grain yields achieved in California of 1 t/ha over 10 days would require PG_i values, including partitioning to roots, of 0.4 to 0.5. Apparent harvest index (HI = grain yield/total shoot biomass at harvest) values, which did not include partitioning to roots or dead leaves, were between 0.41 and 0.47 for high-yielding grain-type cultivars (Kwapata and Hall 1990b). Surprisingly, a genotype which had many pods above the leaves was 54% less efficient at converting intercepted photons into shoot dry matter than a genotype which had its pods within the canopy (Kwapata et al. 1990). The authors hypothesized that this was due to the absorption of PFD by pods which had a much lower Q_i than leaves. Studies by Littleton et al. (1981) have shown that green pods of cowpea experience a net loss of CO_2 even in full sunlight. Removal of young pods would be expected to reduce pod interference with PFD interception by leaves and was shown to substantially increase Q_i (Kwapata et al. 1990). Canopies where the pods are displayed above the leaves may have advantages in terms of providing fewer oviposition sites for insect pests, such as pod borer, and ease of harvesting, but it is likely that they have substantially reduced ability to produce biomass compared with canopies in which pods are covered with leaves. California cultivars that have produced very high yields of about 1 t/ha for every 10 days of the reproductive period retain most of their pods within the canopy.

The canopy of cowpea may not have the highest possible photosynthetic efficiency for converting PFD into carbohydrate due to nonuniform distribution of PFD within the canopy. However, Wien (1982) compared genotypes having conventional broad leaves with genotypes having smaller narrow leaves and observed that they had similar light distribution within their canopies and similar shoot dry matter production. In principle, Q_i could be increased by breeding to increase photosynthesis per unit leaf area. It is not known whether this is possible because there has been little progress with other crop species; but it has recently been observed that genotypic differences in grain yield potential can be positively associated with carbon isotope discrimination in cowpea (Hall et al. 1997). This positive association has been observed in some other crops with C_3 photosynthetic metabolism (reviewed by Hall et al. 1994a), and it has been suggested that it may result from more open stomata and higher photosynthesis per unit leaf area in the more productive genotypes (Condon et al. 1987). The reasons for the positive associations between grain yield potential and carbon isotope discrimination in cowpea are not known. However, if the objective is simply to change gas exchange properties, then selection based upon carbon isotope composition may be more effective than selection based upon measurements of photosynthetic rates, because carbon isotope composition provides a time-integrated measure of gas exchange properties and is influenced less by environmental conditions at the time of measurement (Hall et al. 1992). There appears to be no advantage in selecting for high chlorophyll content per unit leaf area in that a cowpea mutant with yellow leaves and 36% lower chlorophyll content (Kirchhoff et al. 1989b) had similar quantum efficiencies (Habash et al. 1994), rates of leaf photosynthesis (Kirchhoff et al. 1989c), and grain yield (Kirchhoff et al. 1989a) as its parent.

Attaining high grain yield requires optimal partitioning of carbohydrate to grain. This partitioning may be estimated by HI (Kwapata and Hall 1990b). In selection experiments, grain yield has been positively correlated with HI for cowpea under either well-irrigated or dry conditions (Hall et al. 1997). Ranking of cowpea genotypes for HI was consistent across a broad range of plant densities, indicating that selection for HI may be effective with widely spaced plants in the F_2 generation (Kwapata and Hall 1990b). This possibility is supported by the observation that realized heritabilities for HI can be moderate (Menéndez and Hall 1996). Obviously, there are limits to the extent that grain yield can be increased by selecting for greater HI, and plants with excessively high HI tend to be very early in flowering and stunted. Early-generation selection for HI is most likely to be useful for populations from crosses involving at least one exotic parent, which typically have relatively low yields and low HI. It has been proposed that grain yield of some bush-type cowpeas may be increased by selecting for high HI and then growing them at high densities approaching 400 000 plants/ha (Kwapata and Hall 1990b). A tradeoff is present between high yield potential and ability to withstand stresses in that the cultivars with high HI tend to be sensitive to midseason stresses. This sensitivity to midseason stresses can be partially overcome by incorporating genes for delayed leaf senescence, as will be described in the next section.

An overall strategy for increasing grain yield potential of cowpea is to breed cultivars and use management methods that maximize the length of the effective reproductive period, within the constraints set by climatic, biotic and agronomic conditions. This strategy could involve developing early-flowering types with delayed leaf senescence genes that produce two flushes of flowers or later-flowering types that produce one large flush of flowers over a long period. Plant spacings should be used that result in complete canopy coverage of the ground by first flowering. The efficiency of conversion of intercepted PFD to plant biomass could be enhanced by developing cultivars that maintain pods within the canopy of leaves, and by breeding for enhanced photosynthesis/ground area, if this is possible. The HI should be optimized and early generation (e.g., F_2) selection for HI may speed up breeding for appropriate levels of this trait, especially when exotic parents are used in the crosses. Cowpea often is grown as a minor crop in cropping systems; consequently, an important strategy is to develop cultivars that suit specific niches in these systems. Cowpea cultivars have been developed that fit into specific niches in tropical cereal-based cropping systems. Under tropical conditions, these cowpea cultivars are extra-early and some begin flowering 30 days after sowing and are ready for harvest after only 60 days and can produce 2 t/ha of grain (Singh 1987). The extremely short cycle length of these cultivars enables them to fit into niches within cropping systems that other crop species cannot exploit in an effective manner.

3 Grain Yield Under Drought

Studies have been conducted in which cowpea was subjected to drought during the vegetative stage (prior to floral buds reaching the stage where they are obvious without using a hand lens) and the reproductive stage (Turk et al. 1980; Ziska and Hall 1983a; Ziska et al. 1985). These studies demonstrated that grain yield of cowpea is strongly dependent upon the water supply during the reproductive stage, with relatively little influence of vegetative-stage drought on grain yield. It was shown that grain yield of cowpea can be maximized while reducing irrigations and water use by imposing a planned water deficit during the vegetative stage, providing adequate soil water is available beginning with the first appearance of floral buds (Ziska and Hall 1983a,b; Ziska et al. 1985). A model has been proposed (Hall et al. 1994a) based upon these and other observations with annual grain crops.

$$Y = \sum_{i=n,}^{i=o} E_i \times (T_i/E_i) \times W_i \times PG_i \tag{2}$$

where grain yield/area (Y) depends upon the length of the period in days from the first appearance of macroscopic floral buds (i = o) to physiological maturity of the latest-maturing pods (n), total water use per day (E_i), the proportion of this water that is transpired per day (T_i/E_i), transpiration efficiency per day (W_i), and the partitioning of carbohydrate to grain per day (PG_i). A model of this type would be very useful in plant breeding and developing improved management practices if the various components (n, E_i, T_i, W_i, and PG_i) were relatively independent or positively associated, such that increases in any of them would increase grain yield. Unfortunately, this appears not to be the case, as will be discussed in the following paragraphs.

The amount of water available in the root zone influences E_i and possibly n and depends on rainfall, soil physical characteristics, and rooting depth. Cultivars with greater rooting depth, that access more soil moisture, would have greater T_i and E_i than cultivars with shallower roots which have insufficient water in their root zone. Cowpea accessions have been evaluated for depth of rooting by growing them in tubes (Singh 1993). Extent of rooting of cowpea accessions has been evaluated under natural soil conditions by placing an herbicide deep in the soil and scoring plant symptom responses as an indicator of the time taken for roots to grow and reach the herbicide (Robertson et al. 1985). Genotypic differences among cowpea accessions in extent of rooting (Hall and Patel 1985), which were positively associated with soil-water extraction (Robertson et al. 1985), were detected using the herbicide technique. The influence of rooting on soil-water extraction and plant water relations is complex, however, and the uniformity of root distribution may have greater impact on the maintenance of plant water status than the overall density of the roots (Petrie and Hall 1992a,b,c; Petrie et al. 1992). It should also be recognized that deeper rooting has costs to plant adaptation associated with a greater investment of carbohydrates in the root system that could result in decreases in PG_i, depending on whether investment in rooting affects partitioning of carbohydrate during the reproductive stage. In favorable soil conditions some cowpea cultivars can develop root systems that extract substantial soil moisture to 1.5 to 2 m so there may be many circumstances where breeding cultivars with even deeper roots would not enhance adaptation.

Depending on the environmental circumstances, adaptation to water-limited conditions may be enhanced by selecting for either greater or less T_i. Transpiration rate depends on stomatal conductance, percent ground cover and the evaporative demand of the atmosphere. Cultivars with deeper rooting, that are accessing more soil water, would tend to have more open stomata and greater ground cover and, therefore, greater transpiration and photosynthesis than shallow-rooted cultivars. In contrast, cultivars with more sensitive stomatal closing responses to soil drying would have lower T_i, greater ability to survive severe drought, and greater W_i than cultivars with more open stomata. Sensitive stomatal responses to soil drying would appear to be adaptive for indeterminate cultivars of cowpea growing on unreliable summer rainfall in a semiarid zone. However, many cowpea cultivars already have very sensitive stomatal responses to soil drying (Bates and Hall 1981; Shackel and Hall 1983; Petrie and Hall 1992a) that result in higher W_i (Hall and Schulze 1980; Kirchhoff et al. 1989c; Ismail and Hall 1992, 1993a; Ismail et al. 1994). Consequently, there appears to be little value in breeding cultivars with stomata that are even more sensitive to soil drying.

The proportion of crop water use that is transpired (T_i/E_i), rather than being evaporated from the soil surface, is higher when the percent ground cover is higher, such as with more prostrate plants or with higher plant densities. However, when plants are selected that have early first flowering they usually have erect main stems. Agronomic trials under rainfed, semiarid conditions in Senegal with erect or semierect cultivars demonstrated that the highest grain yields were obtained with relatively dense plant spacings of about 25 × 50 cm per plant (C. Dancette, unpubl. data 1985).

There has been considerable interest in recent years in attempting to increase W_i through breeding (Hall et al. 1994a). Seasonal transpiration efficiency (W) is the ratio of seasonal plant biomass production to transpiration. Unfortunately, measurement of transpiration and root biomass usually is done with potted plants, which requires substantial labor, and the soil environment is not natural and can influence plant responses (Ismail et al. 1994). A major breakthrough occurred in the early 1980s when Farquhar et al. (1982) demonstrated that, for C_3 plants, W should be related to the extent that the plants discriminate against the heavy stable isotope ^{13}C compared with the more abundant isotope ^{12}C (Δ) in the photosynthetic uptake of CO_2. A negative correlation between W and Δ was predicted due to their mutual dependence upon the [CO_2] inside leaves. Empirical studies with many C_3 species have shown that, as expected based on theory, there is a strong negative correlation between W and Δ for plants growing in the same aerial environment (Hall et al. 1994a). The value of this correlation to plant breeding is that Δ can be measured on plants in field nurseries, whereas W cannot, and Δ can be measured without destroying the plant. In principle, large numbers of genotypes could be evaluated since it is simply necessary to take leaf samples (preferably from the same nodal position), dry them, grind them, and measure Δ using an isotope ratioing mass spectrometer (assuming a specific constant carbon isotope composition of the air). Unfortunately, the measurements are still relatively expensive (about US $15/sample); however, methods are available for reducing costs (Hall et al. 1994a).

For cowpea, genotypic differences in Δ have been described under field conditions (Hall et al. 1990, 1992, 1994b). The genotypic differences in Δ are negatively correlated with W as expected based on theory (Ismail and Hall 1992, 1993a). The genotypic differences in Δ were associated with differences in both stomatal conductance and mesophyll components of photosynthesis (Ismail and Hall 1993b). Studies with reciprocal crosses indicated that inheritance of both Δ and W are determined by nuclear genes (Ismail and Hall 1993a). Realized heritabilities were low for Δ (0.25 and 0.31) compared with values for days to flowering (0.89 and 0.96; Menéndez and Hall 1995). The data indicated that selection for earliness of flowering could be effective with single F_2 plants, whereas selection for Δ may be most efficient with family selection in advanced generations. Genetic correlations were observed between Δ and days to flowering which were negative and low to intermediate in magnitude (−0.14 to −0.66; Menéndez and Hall 1995). This indicates that selection for earliness in the F_2 generation could result in some undesirable, indirect selection for high Δ (and therefore low W). In addition, in other experiments, positive genetic correlations were observed between Δ and HI (0.36 and 0.38; Menéndez and Hall 1996). This indicates that selection for low Δ could result in some undesirable, indirect selection for low HI. However, by simultaneously imposing selection pressure for these three traits it should be possible to select plants with appropriate levels of earliness, HI and Δ. What is not known at this time is whether breeders should select for low Δ (and high W) or high Δ (and low W), even for dry environments, as will be discussed in the next paragraph.

Theory and the model [Eq. (2)] suggest that for water-limited environments, one should select for low Δ to obtain high W. However, studies in two water-limited environments in northern Senegal indicated a tendency for well-adapted local cultivars to have high Δ and poorly adapted exotics to have low Δ (Hall et al. 1994b). Also, well-adapted cultivars and advanced lines selected for high grain yield under well-watered conditions in California have high Δ (Hall et al. 1993). Selection experiments with cowpea have confirmed that general adaptability under wet or dry conditions is associated with high Δ (Hall et al. 1993, 1997). A further complication for breeding is that substantial genotype x environment interaction was observed for Δ when comparing the same set of cowpea genotypes under tropical and subtropical conditions (Hall et al. 1994b). This interaction would not necessarily constrain cowpea breeding programs that are developing improved cultivars for specific target production environments. In fact, genotypic rankings for Δ were quite consistent under drier or wetter conditions within either the tropical or the subtropical zone. However, cowpea performance with respect to Δ and W does not appear to be transferrable between radically different production zones. The overall conclusion is that a clear case has not yet been made to justify selection for Δ in cowpea breeding programs. Selecting for high Δ might enhance grain yield in many environments but would it be more effective than selecting for high HI and high grain yield? Measurement of Δ is useful for more basic studies and germplasm evaluation.

The utility of selecting for greater HI was discussed in the previous section. The comments also have relevance for breeding for adaptation to drought, although the optimum HI for dry conditions may be lower than the HI that is optimal for well-watered environments, since plants with higher HI tend to be more sensitive to midseason stresses.

Developing cultivars or management methods that result in cowpea having a more plastic reproductive period (a variable value for n) may be an effective means for enhancing adaptation to variable semiarid environments. Early erect cowpea cultivars have proved to be useful in some dry environments and years due to their ability to escape drought (Hall and Grantz 1981; Hall and Patel 1985). Unfortunately, early erect cowpea cultivars can be more sensitive to mid-season drought than medium-cycle spreading cultivars (Thiaw et al. 1993). Grain yield of cowpea is more sensitive to soil-water deficits during flowering and pod filling than during the vegetative stage (Turk et al. 1980; Ziska and Hall 1983a; Ziska et al. 1985). Indeterminate cowpeas have been discovered that are early but have delayed leaf senescence (DLS) after producing the first flush of pods and the ability to produce a second substantial flush of pods (Gwathmey et al. 1992a). The DLS trait results from a higher proportion of plants surviving after producing the first flush of pods and probably results from the maintenance of root viability (Gwathmey et al. 1992b). The DLS trait was shown to enhance adaptation to midseason drought by permitting recovery of yield through the production of a substantial second flush of pods after the drought (Gwathmey and Hall 1992). The combination of early flowering and DLS may increase adaptation of cowpea to environments that frequently experience either premature terminal droughts or midseason droughts. An early cultivar would be able to produce a substantial single flush of pods when the rainy season is short, whereas the DLS character would enable an early cultivar to be productive when the rainy season is longer but has an extreme drought at midseason. Cultivars with DLS also often have enhanced production and quality of forage, compared with the more typical senescent cultivars. Dual purpose cultivars with the ability to produce substantial quantities of both grain and hay are useful in many parts of Africa where cowpea are extensively cultivated.

Field nurseries, where the seasonal supply of water through irrigation or rainfall is adequately long, have been shown to be effective in screening for DLS in subtropical (California) and tropical (Senegal) zones (Hall et al. 1997). It is necessary to have an adequate supply of water because all genotypes will senesce if they are subjected to an extended drought during pod maturation. It should be noted that in making plant selections, DLS must be combined with earliness and substantial production of pods during the first flush of pods. Genotypes that produce no or few pods during the first flush, due to genetic or nongenetic causes, will exhibit, a "false" DLS that has little agronomic value. An example of this are the male-sterile plants which can occur in cowpea breeding nurseries and stay green as long as water is available in the root zone, providing temperatures are favorable and there are no significant pests or diseases. Selection studies indicate that the DLS trait has adequate heritability for breeding and is not substantially affected by genotype x environment interaction in that DLS was consistently expressed in subtropical California and tropical Senegal over 2 years (Hall et al. 1997). Advanced cowpea breeding lines have been developed that have both early flowering and DLS. In tropical conditions in Senegal, with high night temperatures, these lines can produce a first flush of pods in about 60 days after planting, and the second flush of pods is mature after about 90 days. Consequently, early cultivars with DLS should be adapted to the broad range of conditions that can occur in the Sahelian zone of Africa, where the annual rainfall can vary from 200 mm with a short growing season of less than 60 days to wetter years with as much as 400 mm of rain and a growing season as long as 90 days, but with the possibility of midseason droughts (Dancette and Hall 1979; Khalfaoui 1991). In subtropical conditions in California, with moderate night temperatures, the same cowpea lines produce a first flush of pods in about 100 days, and obtaining the second flush of pods requires a season of about 140 days (Gwathmey et al. 1992a).

An alternative approach to increasing the plasticity of the reproductive period in a population of cowpea plants is to plant a varietal intercrop consisting of alternating rows of cowpea cultivars that have different earliness of flowering and lengths of the reproductive period. Varietal intercrops of cowpea were shown to have higher and more stable yields of grain and hay than the most productive sole crops, under the conditions of infertile soil and drought that are prevalent in the Sahelian zone of Africa (Thiaw et al. 1993). In this study, cultivars were used that differed in both phenology and morphology. One cultivar in the varietal intercrop flowered early and was erect, whereas the other cultivar flowered later and was spreading. In years and locations where midseason stresses occurred, the later spreading cultivar produced most of the grain and hay, and compensated by growing into the space left by the senescing early erect cultivar. In contrast, in years and locations where the growing season was short due to terminal drought, the early erect cultivar produced most of the grain in the varietal intercrop.

4 Photoperiod and Temperature Effects on Phenology and Responses to Stress

The previous discussion illustrated how the adaptation of cowpea to either optimal or water-limited environments depends on the time of first flowering and duration of the reproductive period. The time of first flowering depends on the nodal position at which the first floral buds are initiated (which varies among cultivars depending on the duration of their juvenile periods and sensitivity to photoperiod), the rate at which nodes are produced (which is positively correlated with temperature), and the extent to which floral bud development of the cultivar is suppressed by high temperatures (Ehlers and Hall 1996).

Cowpea is a short-day (SD) species and some cultivars have an obligate or highly sensitive response to photoperiod and do not initiate floral buds when subjected to day lengths that are longer than a critical value (Summerfield et al. 1985; Ehlers and Hall 1996). Other cultivars have a quantitative response to photoperiod such that under long days flowering is delayed, while others are day-neutral in that the initiation of floral buds is not influenced by day length (Summerfield et al. 1985; Ehlers and Hall 1996). Useful quantitative models for the effects of day length and temperature on time to first flowering in cowpea have been developed by Summerfield et al. (1985). Unfortunately, these models are not valid at high night temperatures for cultivars that exhibit suppression of floral bud development (Patel and Hall 1990). A more complete classification system has been developed for cowpea by Ehlers and Hall (1996) that includes the effects of photoperiod, high temperature, and juvenility on reproductive development. The classification systems of Patel and Hall (1990) and Ehlers and Hall (1996) have agronomic significance. According to these systems, cultivars developed for different regions of the world and seasons fall into discrete classification groups, suggesting that their phenological and stress responses to photoperiod and temperature are optimized for the different climates.

Heat-induced reductions in grain yield in cowpea can occur that are probably often caused by reductions in pod set (Warrag and Hall 1983; Nielsen and Hall 1985b) and, possibly, damage during early floral bud development (Ahmed and Hall 1993). In both cases high night temperatures appear to be more damaging than high day temperatures. Cowpea production zones in the tropics typically have high night temperatures (Nielsen and Hall 1985a; Hall and Patel 1987), and short days during all or parts of the growing season. Heat-induced suppression of floral bud development only occured under long days (Mutters et al. 1989b) but it is influenced by the quality of light during the day (Ahmed et al. 1993b). In growth chambers, low red/far-red quantum flux ratios of 1.3 to 1.6 were required to obtain the same heat-induced suppression of floral bud development under long days as was observed under field conditions, whereas a ratio of 1.9 resulted in normal floral bud development. These experiments have implications for methods in that growth chambers with fluorescent lights and only a small wattage of tungsten lights would have high red/far-red ratios and may not produce plant responses to high temperature that occur in field conditions. These experiments also indicated that dense canopies may have enhanced heat-induced suppression of floral bud development in that red/far-red ratios within canopies are usually very low (Smith 1995). Damage to floral buds due to heat causes them to appear brown and necrotic, which can be confused with damage to floral buds by flower thrips or lygus bugs. It has been speculated that phytochrome genes, which are different from the ones responsible for the classical SD photoperiod effects on floral bud initiation in cowpea, are responsible for the heat-induced suppression of floral bud development under long days (Ahmed et al. 1993b). Crosses with cowpea genotypes that exhibit day-neutral responses to day length with respect to floral bud initiation, but differences in heat-induced suppression of floral buds, demonstrated that this sensitivity to heat can be explained by the presence of a single gene with complete dominance (Hall 1993). Heat tolerance during floral bud development can be effectively selected beginning in the F_2 generation using either field nurseries or glasshouses that have high night temperatures and long days.

High night temperatures can cause male sterility in cowpea (Warrag and Hall 1984b). The stage of floral development most sensitive to high night temperature occurs 9 to 7 days before anthesis and is after meiosis and involves premature degeneration of the tapetal tissue and lack of endothecial development (Ahmed et al. 1992). Transfer of proline from anther cell walls to pollen is inhibited in sensitive genotypes (Mutters et al. 1989a). Interestingly, pollen is damaged more by late-night heat than early-night heat suggesting that the sensitive processes are under circadian control (Mutters and Hall 1992). Sensitivity to high night temperature during pod set appears to be determined by a single gene but in this case the gene is recessive and there can be substantial environmental effects, such that the realized heritability is only about 0.26 (Marfo and Hall 1992). Consequently, tolerance to high night temperature during pod set is conferred by a dominant gene, and this gene appears to be effective under either long or short days (Ehlers and Hall 1996). Selection for heat tolerance during pod set, by selecting for large numbers of pods per peduncle, requires selection over several generations and is more effective with families than single plants, and can be more effective in glasshouses where night temperature can be controlled than in field nurseries.

Some other stages of reproductive development are sensitive to heat (Hall and Patel 1990; Hall 1992). Pods of different cowpeas produce 9 to 20 ovules with many cultivars having 15, but they rarely produce this many seeds per pod. Under optimal conditions two thirds of the ovules may produce seed, on a per-plant basis, whereas with high temperature pods can have fewer seeds (Warrag and Hall 1983). For most cultivars and stresses it is the ovules at the blossom end of the pod that do not produce seed. High day temperatures can cause cowpea to produce seed with asymetrical twisted cotyledons and more wrinkled seed coats, but without influencing germinability (Warrag and Hall 1984a). Under tropical conditions, certain cowpea cultivars (TVx 3236) and accessions (TVu 4552) can exhibit deep yellow or brown patches on the seed coat which decreases their acceptability to consumers. This seed-coat discoloration is more pronounced where night temperatures are high (Nielsen and Hall 1985b) and is caused by a single dominant gene (Patel and Hall 1988).

Breeding can be used to overcome problems caused by heat stress during reproductive development (Hall 1992, 1993). The following procedure has been shown to be effective in breeding cowpea cultivars with enhanced heat tolerance under a broad range of photoperiods (Hall 1993). Parents with the most effective documented heat tolerance during reproductive development are advanced breeding lines developed by P.N. Patel and J.D. Ehlers (H8-8-27 and H8-9) which are available from the University of California, Riverside. They are early-flowering, day-neutral and semierect with large blackeye seed type and resistances to certain biotypes of *Fusarium* wilt and root knot nematode. Heat-tolerant lines should be crossed with a locally adapted cultivar and, possibly, immediately backcrossed with the same cultivar (Ehlers and Foster 1993). Progeny from the first segregating generation should be screened for ability to produce flowers and set pods under long-day environments with very high night temperatures. In subsequent generations, families should be screened in breeding nurseries in the target production environment and single plants selected for agronomic traits that can be fixed in early generations, such as seed size and quality, plant habit, and resistance to certain diseases. In a more advanced generation, families should be screened again in the long-day environment with very high night temperatures. The few families that do not have heat tolerance during early floral development and produce few if any flowers should be eliminated. The remaining families should be carefully evaluated to select those having heat tolerance during pod set and ability to produce three to four pods per peduncle on the main stem nodes. Cultivars developed by this program would have heat tolerance under a broad range of photoperiods but they would be extremely early and day neutral, but they would not have the photoperiod responses that regulate time of flowering and enhance adaptation in some tropical rainfed environments.

In subtropical zones, early sowing enables cowpea to escape heat at flowering, but this may result in cool soils and chilling damage during emergence (Hall and Frate 1996). Average soil temperatures greater than 19 °C are needed for about 3 days after sowing or plant emergence will be delayed and reduced. Differences among cowpea genotypes have been observed in rate of germination under chilling conditions (El-kholy et al. 1997). These genotypic effects were detected using a constant temperature of 15 °C and they mainly involved differences in rate of germination and not maximal percentage germination; but under field conditions, seed that germinate slowly have a high probability of not emerging from the soil, and genotypic differences in maximal percentage emergence were observed in chilling field conditions (A.M. Ismail and A.E. Hall, unpubl 1995). Genotypes with slower rates of germination under chilling conditions had greater rates of leakage of electrolytes and carbohydrates from the seed, indicating that chilling was damaging membrane integrity (El-kholy et al. 1997). No genetic associations were observed between chilling tolerance during germination and heat tolerance during reproductive development. Consequently, the opportunity exists for breeding cowpea cultivars that can germinate and emerge from chilling soils and yet have heat tolerance during reproductive development, which would be valuable in the subtropics.

In tropical zones, emergence of cowpea may be decreased by high soil temperature. Depth of sowing can be critical for cowpea under hot soil conditions. With soil temperature of 35 °C emergence was only 46% for seed sown at a depth of 5 cm, whereas it was 100% for seed sown at a depth of 2.5 cm, and emergence was high at both depths with soil temperatures of 30 or 25 °C (Warrag and Hall 1984a). Onwueme and Adegoroye (1975) also observed less emergence under heat stress for seeds sown deeper in the soil, and that seeds of cowpea were most sensitive to heat during the first day after sowing. Genotypic differences in maximal percentage germination have been detected in cowpea using a constant temperature of 38 °C, which were not associated with genotypic differences in heat tolerance during reproductive development (El-kholy et al. 1997). Consequently, for tropical environments it may be useful, in some cases, to incorporate genes for heat tolerance during germination and additional genes for heat tolerance during reproductive development.

5 Conclusions

There is a critical need to develop and use physiological information in breeding and agronomic programs involved in increasing the performance and productivity of cowpea production systems because conventional empirical approaches are making slow progress with this crop. The main opportunities for improvement arise from the choice of sowing dates and plant spacings, and development of cultivars that result in: (1) adequate plant establishment, (2) attainment of full canopy ground cover by first flowering, and (3) an adequately long and effective reproductive period during the part of the season when temperatures, solar radiation, and soil moisture are optimal for photosynthesis and grain production. Consideration should be given to breeding cultivars of cowpea, and other crop species, for the climatic conditions of the next century, especially the projected increases in atmospheric (CO_2) and temperature. Elevated (CO_2) increases productivity of cowpea, and most other C_3 species, but it does not overcome the detrimental effects of high night temperature on reproductive development (Ahmed et al. 1993a). Hall and Allen (1993) have hypothesized that cultivars of C_3 species with heat tolerance during reproductive development, high HI, high photosynthetic capacity per unit leaf area, small leaves and low leaf area per unit ground area, under present levels of (CO_2), will be most responsive to elevated atmospheric (CO_2) under both hot and intermediate temperatures.

References

Ahmed FE, Hall AE (1993) Heat injury during early floral bud development in cowpea. Crop Sci 33:764–767

Ahmed FE, Hall AE, DeMason DA (1992) Heat injury during floral development in cowpea (*Vigna unguiculata*, Fabaceae). Am J Bot 79:784–791

Ahmed FE, Hall AE, Madore MA (1993a) Interactive effects of high temperature and elevated carbon dioxide concentration on cowpea [*Vigna unguiculata* (L.) Walp.]. Plant Cell Environ 16:835–842

Ahmed FE, Mutters RG, Hall AE (1993b) Interactive effects of high temperature and light quality on floral bud development in cowpea. Aust J Plant Physiol 20:661–667

Bates LM, Hall AE (1981) Stomatal closure with soil water depletion not associated with changes in bulk leaf water status. Oecologia 50:62–65

Baudoin JP, Maréchal R (1985) Genetic diversity in Vigna. In: Singh SR, Rachie KO (eds) Cowpea research, production and utilization. John Wiley, New York, pp 3–9

Cisse N, Ndiaye M, Thiaw S, Hall AE (1995) Registration of Mouride cowpea. Crop Sci 35:1215–1216

Condon AG, Richards RA, Farquhar GD (1987) Carbon isotope discrimination is positively correlated with grain yield and dry matter production in field-grown wheat. Crop Sci 27:996–1001

Dancette C, Hall AE (1979) Agroclimatology applied to water management in the Sudanian and Sahelian zones of Africa. In: Hall AE, Cannell GH, Lawton HW (eds) Agriculture in semi-arid environments. Ecological Studies, vol 34. Springer, Berlin Heidelberg New York, pp 98–118

Ehlers JD, Foster KW (1993) Introgression of agronomic characters from exotic cowpea germplasm into blackeye bean. Field Crops Res 35:43–50

Ehlers JD, Hall AE (1996) Genotypic classification of cowpea based on responses to heat and photoperiod. Crop Sci 36:673–679
El-kholy AS, Hall AE, Mohsen AA (1997) Heat and chilling tolerance during germination, and heat tolerance during flowering are not associated in cowpea. Crop Sci 37:456–463
Farquhar GD, O'Leary MH, Berry JA (1982) On the relationship between carbon isotope discrimination and the intercellular carbon dioxide concentration in leaves. Aust J Plant Physiol 9:121–137
Grantz DA, Hall AE (1982) Earliness of an indeterminate crop, *Vigna unguiculata* (L.) Walp., as affected by drought, temperature, and plant density. Aust J Agric Res 33:531–540
Gwathmey CO, Hall AE (1992) Adaptation to midseason drought of cowpea genotypes with contrasting senescence traits. Crop Sci 32:773–778
Gwathmey CO, Hall AE, Madore MA (1992a) Adaptive attributes of cowpea genotypes with delayed monocarpic leaf senescence. Crop Sci 32:765–772
Gwathmey CO, Hall AE, Madore MA (1992b) Pod removal effects on cowpea genotypes contrasting in monocarpic senescence traits. Crop Sci 32:1003–1009
Habash DZ, Genty B, Baker NR (1994) The consequences of chlorophyll deficiency for photosynthetic light use efficiency in a single nuclear gene mutation of cowpea. Photosynth Res 42:17–25
Hall AE (1992) Breeding for heat tolerance. Plant Breed Rev 10:129–168
Hall AE (1993) Physiology and breeding for heat tolerance in cowpea, and comparisons with other crops. In: Kuo CG (ed) Adaptation of food crops to temperature and water stress. Asian Vegetable Research and Development Center, Taipai, Taiwan, pp 271–284
Hall AE, Allen LH (1993) Designing cultivars for the climatic conditions of the next century. In: Buxton DR, Shibles R, Forsberg RA, Blad BL, Asay KH, Paulsen GM, Wilson RF (eds) International crop science I, Crop Science Society of America, Madison, Wisconsin, pp 291–297
Hall AE, Frate CA (eds) (1996) Blackeye bean production in California. University of California, Oakland, Division of Agriculture and Natural Resources, Publ 21518
Hall AE, Grantz DA (1981) Drought resistance of cowpea improved by selecting for early appearance of mature pods. Crop Sci 21:461–464
Hall AE, Patel PN (1985) Breeding for resistance to drought and heat. In: Singh SR, Rachie KO (eds) Cowpea research, production and utilization. John Wiley, New York, pp 137–151
Hall AE, Patel PN (1987) Cowpea improvement for semiarid regions of sub-Saharan Africa. In: Menyonga JM, Bezuneh T, Youdeowei A (eds) Food grain production in semi-arid Africa. OAU/STRC-SAFGRAD Ouagadougou, Burkina Faso, pp 279–290
Hall AE, Patel PN (1990) Breeding heat-tolerant cowpeas. In: Miller JC, Miller JP, Fery RL (eds) Cowpea research – US perspective. Texas Agricultural Experiment Station, College Station, Texas, pp 41–46
Hall AE, Schulze ED (1980) Stomatal response to environment and a possible interrelation between stomatal effects on transpiration and CO_2 assimilation. Plant Cell Environ 3:467–474
Hall AE, Mutters RG, Hubick KT, Farquhar GD (1990) Genotypic differences in carbon isotope discrimination by cowpea under wet and dry field conditions. Crop Sci 30:300–305
Hall AE, Mutters RG, Farquhar GD (1992) Genotypic and drought-induced differences in carbon isotope discrimination and gas exchange of cowpea. Crop Sci 32:1–6
Hall AE, Ismail AM, Menéndez CM (1993) Implications for plant breeding of genotypic and drought-induced differences in water-use efficiency, carbon isotope discrimination, and gas exchange. In: Ehleringer JR, Hall AE, Farquhar GD (eds) Stable isotopes and plant carbon-water relations. Academic Press, San Diego, pp 349–369
Hall AE, Richards RA, Condon AG, Wright GC, Farquhar GD (1994a) Carbon isotope discrimination and plant breeding. Plant Breed Rev 12:81–113
Hall AE, Thiaw S, Krieg DR (1994b) Consistency of genotypic ranking for carbon isotope discrimination by cowpea grown in tropical and subtropical zones. Field Crops Res 36:125–131
Hall AE, Thiaw S, Ismail AM, Ehlers JD (1997) Water-use efficiency and drought adaptation of cowpea. In: Singh BB, Mohan Raj DR, Dashiell KE, Jackai LEN (eds) Advances in cowpea

research. Copublication of International Institute of Tropical Agriculture, Japan, and International Research Center for Agricultural Sciences, Ibadan, Nigeria, pp 87–98

Helms D, Panella L, Buddenhagen IW, Tucker CL, Gepts PL (1991a) Registration of California Blackeye 46 cowpea. Crop Sci 31:1703

Helms D, Panella L, Buddenhagen IW, Tucker CL, Foster KW, Gepts PL (1991b) Registration of California Blackeye 88 cowpea. Crop Sci 31:1703–1704

Ismail AM, Hall AE (1992) Correlation between water-use efficiency and carbon isotope discrimination in diverse cowpea genotypes and isogenic lines. Crop Sci 32:7–12

Ismail AM, Hall AE (1993a) Inheritance of carbon isotope discrimination and water-use efficiency in cowpea. Crop Sci 33:498–503

Ismail AM, Hall AE (1993b) Carbon isotope discrimination and gas exchange of cowpea accessions and hybrids. Crop Sci 33:788–793

Ismail AM, Hall AE, Bray EA (1994) Drought and pot size effects on transpiration efficiency and carbon isotope discrimination of cowpea accessions and hybrids. Aust J Plant Physiol 21:23–35

Khalfaoui JLB (1991) Determination of potential lengths of the crop growing period in semi-arid regions of Senegal. Agric For Meteorol 55:251–263

Kirchhoff WR, Hall AE, Isom WH (1989a) Phenotypic expression of a chlorophyll mutant in cowpea [*Vigna unguiculata* (L.) Walp.]: environmental influences and effects on productivity. Field Crops Res 21:19–28

Kirchhoff WR, Hall AE, Roose ML (1989b) Inheritance of a mutation influencing chlorophyll content and composition in cowpea. Crop Sci 29:105–108

Kirchhoff WR, Hall AE, Thomson WW (1989c) Gas exchange, carbon isotope discrimination, and chloroplast ultrastructure of a chlorophyll-deficient mutant of cowpea. Crop Sci 29:109–115

Kwapata MB, Hall AE (1990a) Response of contrasting vegetable-cowpea cultivars to plant density and harvesting of young green pods. I. Pod production. Field Crops Res 24:1–10

Kwapata MB, Hall AE (1990b) Determinants of cowpea (*Vigna unguiculata*) seed yield at extremely high plant density. Field Crops Res 24:23–32

Kwapata MB, Hall AE, Madore MA (1990) Response of contrasting vegetable-cowpea cultivars to plant density and harvesting of young green pods. II. Dry matter production and photosynthesis. Field Crops Res 24:11–21

Littleton EJ, Dennett MD, Monteith JL, Elston J (1979) The growth and development of cowpeas (*Vigna unguiculata*) under tropical field conditions. 2. Accumulation and partition of dry weight. J Agric Sci Camb 93:309–320

Littleton EJ, Dennett MD, Elston J, Monteith JL (1981) The growth and development of cowpeas (*Vigna unguiculata*) under tropical field conditions. 3. Photosynthesis of leaves and pods. J Agric Sci Camb 97:539–550

Marfo KO, Hall AE (1992) Inheritance of heat tolerance during pod set in cowpea. Crop Sci 32:912–918

Menéndez CM, Hall AE (1995) Heritability of carbon isotope discrimination and correlations with earliness in cowpea. Crop Sci 35:673–678

Menéndez CM, Hall AE (1996) Heritability of carbon isotope discrimination and correlations with harvest index in cowpea. Crop Sci 36:233–238

Monteith JL (1977) Climate and the efficiency of crop production in Britain. Philos Trans R Soc Lond Ser B 281:227–294

Mutters RG, Hall AE (1992) Reproductive responses of cowpea to high temperatures during different night periods. Crop Sci 32:202–206

Mutters RG, Ferreira LGR, Hall AE (1989a) Proline content of anthers and pollen of heat-tolerant and heat-sensitive cowpea subjected to different temperatures. Crop Sci 29:1497–1500

Mutters RG, Hall AE, Patel PN (1989b) Photoperiod and light quality effects on cowpea floral development at high temperatures. Crop Sci 29:1501–1505

Nielsen CL, Hall AE (1985a) Responses of cowpea [*Vigna unguiculata* (L.) Walp.] in the field to high night air temperature during flowering. I. Thermal regimes of production regions and field experimental system. Field Crops Res 10:167–179

Nielsen CL, Hall AE (1985b) Responses of cowpea [*Vigna unguiculata* (L.) Walp.] in the field to high night air temperature during flowering. II. Plant responses. Field Crops Res 10:181–196

Onwueme IC, Adegoroye SA (1975) Emergence of seedlings from different depths following high temperature stress. J Agric Sci Camb 84:525–528

Patel PN, Hall AE (1988) Inheritance of heat-induced brown discoloration in seed coats of cowpea. Crop Sci 28:929–932

Patel PN, Hall AE (1990) Genotypic variation and classification of cowpea for reproductive responses to high temperature under long photoperiods. Crop Sci 30:614–621

Petrie CL, Hall AE (1992a) Water relations in cowpea and pearl millet under soil water deficits: I. Contrasting leaf water relations. Aust J Plant Physiol 19:577–589

Petrie CL, Hall AE (1992b) Water relations in cowpea and pearl millet under soil water deficits: II. Water use and root distribution. Aust J Plant Physiol 19:591–600

Petrie CL, Hall AE (1992c) Water relations in cowpea and pearl millet under soil water deficits: III. Extent of predawn equilibrium in leaf water potential. Aust J Plant Physiol 19:601–609

Petrie CL, Kabala ZJ, Hall AE, Simunek J (1992) Water transport in an unsaturated medium to roots with different geometries. Soil Sci Soc Am J 56:1686–1694

Robertson BM, Hall AE, Foster KW (1985) A field technique for screening for genotypic differences in root growth. Crop Sci 25:1084–1090

Sanden B (1993) Blackeye varietal and irrigation cutoff trial. University of California Dry Bean Research - 1993 Progress Report, California Dry Bean Advisory Board, Dinuba, California, pp 120–121

Shackel KA, Hall AE (1983) Comparison of water relations and osmotic adjustment in sorghum and cowpea under field conditions. Aust J Plant Physiol 10:423–435

Singh BB (1987) Breeding cowpea varieties for drought escape. In: Menyonga JM, Bezuneh T, Youdeowei A (eds) Food grain production in semi-arid Africa. OAU/STRC-SAFGRAD Ouagadougou, Burkina Faso, pp 299–306

Singh BB (1993) Grain legume improvement program. Part I. Cowpea breeding. Archival Report (1988–1992). International Institute of Tropical Agriculture, Ibadan, Nigeria, p 70

Smith H (1995) Physiological and ecological function within the phytochrome family. Ann Rev Plant Physiol Plant Mol Biol 46:289–315

Summerfield RJ, Pate JS, Roberts EH, Wien HC (1985) The physiology of cowpeas. In: Singh SR, Rachie KO (eds) Cowpea research, production and utilization. John Wiley, New York, pp 66–101

Thiaw S, Hall AE, Parker DR (1993) Varietal intercropping and the yields and stability of cowpea production in semiarid Senegal. Field Crops Res 33:217–233

Turk KJ, Hall AE, Asbell CW (1980) Drought adaptation of cowpea. I. Influence of drought on seed yield. Agron J 72:413–420

Warrag MOA, Hall AE (1983) Reproductive responses of cowpea to heat stress: genotypic differences in tolerance to heat at flowering. Crop Sci 23:1088–1092

Warrag MOA, Hall AE (1984a) Reproductive responses of cowpea [*Vigna unguiculata* (L.) Walp.] to heat stress. I. Responses to soil and day air temperatures. Field Crops Res 8:3–16

Warrag MOA, Hall AE (1984b) Reproductive responses of cowpea [*Vigna unguiculata* (L.) Walp.] to heat stress. II. Responses to night air temperature. Field Crops Res 8:17–33

Wien HC (1982) Dry matter production, leaf area development and light interception of cowpea lines with broad and narrow leaflet shape. Crop Sci 22:733–737

Wien HC, Ackah EE (1978) Pod development period in cowpeas: varietal differences as related to seed characters and environmental effects. Crop Sci 18:791–794

Wien HC, Summerfield RJ (1984) Cowpea [*Vigna unguiculata* (L.) Walp.]. In: Goldsworthy PR, Fisher NM (eds) The physiology of tropical field crops. John Wiley, New York, pp 353–383

Ziska LH, Hall AE (1983a) Seed yields and water use of cowpeas *Vigna unguiculata* (L.) Walp., subjected to planned-water-deficit irrigation. Irrig Sci 3:237–245

Ziska LH, Hall AE (1983b) Soil and plant measurements for determining when to irrigate cowpeas *Vigna unguiculata* (L.) Walp., grown under planned-water deficits. Irrig Sci 3:247–257
Ziska LH, Hall AE, Hoover RM (1985) Irrigation management methods to reduce water use of cowpea [*Vigna unguiculata* (L.) Walp.] and lima beans (*Phaseolus lunatus* L.) while maintaining seed yield at maximum levels. Irrig Sci 6:223–239

Soybean [*Glycine max* (L.) Merr.] Physiology and Symbiotic Dinitrogen Fixation

F. ZHANG and D.L. SMITH[1]

1 Origin and Introduction

The soybean, a native of eastern Asia, is one of the oldest crops of that area and is considered to be a vital grain crop. The soybean belongs to the family Leguminosae, subfamily Papilionideae, and the genus *Glycine*. *Glycine* is composed of two subgenera, *glycine* and *soja*. The cultivated soybean, *Glycine max* (L.) Merr., and its wild counterpart, *G. soja*, are now classified as species of the subgenus *soja*. *G. max*, *G. soja*, and most species of the subgenus *glycine* are diploid (2n = 40) (Hymowitz and Singh 1987). Soybean was first domesticated in China from the 11th to the 7th century B.C as understood by pictographic evidence of the ancient Chines word "shu". The wild annual soybean *G. soja* is the ancestor of *G. max*. It has been crossed with *G. tomentella*, a wild perennial relative of the subgenus *glycine*. *G. tomentella* and *G. tabacina*, both perennials, overlap geographically with *G. soja* and are thought to be the closest relatives of *G. max* from the subgenus *glycine* (Hymowitz and Singh 1987). The genus of *Glycine* is characterized by trifoliolate leaves, flowers inserted singly at each node of the raceme, a five-toothed calyx with the upper pair of teeth not well united, a glabrous corolla with long clawed petals, a keel which is shorter than the wings, and estrophilate seed (Hymowitz and Newell 1981). There are two types of stem growth habit and floral initiation in soybean. One type is the indeterminate stem, in which the terminal bud continues vegetative activity during most of the reproductive flowering period. The second type is the determinate stem, in which the vegetative activity of the terminal bud ceases when it becomes an inflorescence.

Soybean is one of the most important oilseed crops. Since 1970, soybean production has been at least double that of any of the other oilseed crops, such as peanut (*Arachis hypogaea* L.), sunflower (*Helianthus* L.), and rapeseed (*Brassica napus* L.) (Johnson 1987). The share of soybean in world oilseed production has increased from 32% in 1965 to over 50% in the 1980s. Soybean oil accounts for 20 to 25% of the total world fat and oil production, and 30 to 35% of the total edible vegetable oil production (Smith and Huyser 1987). Another important product of soybean is protein. Soybean meal plays a key role as a protein ingredient in feeds in the USA. Poultry feeds are the largest outlet for

[1] McGill University, Natural Resource Department, Macdonald Campus, 21, 111 Lakeshore road, Ste. Anne de Bellevue, QC, Canada H9X3V9

Crop Yield, Physiology and Processes
D. L. Smith, C. Hamel (Eds.)
© Springer-Verlag Berlin Heidelberg 1999

meal, followed by swine feeds; the two account for 78% of total usage. Soybean meal has been estimated to make up more than 90% of the oilseed meals consumed in poultry feeds. Smaller quantities of soybean meal are fed to beef and dairy cattle. Production of edible protein products is small as compared to soybean meal for feed uses (Mounts et al. 1987). Soybean cultivars grown in the United States generally contain 20–30% oil and 39–45% protein (Hymowitz and Singh 1987).

Soybean requires at least 2400 Corn Heat Units (Brown 1981) in one growing season. Cultivated soybean is adapted to a wide range of latitudes, from 52°N to 50°S, and to a range of seasonal durations (90 to 180 days) (Shibles 1980). The four major producers, USA, Brazil, China, and Argentina, together account for 90 to 95% of the world's production. Canada is at the northern limit of the North American soybean growth range, but soybean production there has risen sharply in recent times. Temperature is usually the limiting climatic factor in short growing season areas, such as Canada.

2 Plant Development

2.1 Germination and Vegetative Stages

Seeds of most soybean cultivars imbibe water rapidly following planting. Under favourable conditions, the radicle emerges from the seed in 1 or 2 days. Downward growth of the primary root is rapid and by the 4th or 5th day the first secondary roots emerge 4 to 5 cm behind the primary roots apex (Lersten and Carlson 1987). The cotyledons emerge in 3 to 4 days, and are pulled out of the soil by rapid growth in the doubled-over crook in the upper region of the hypocotyl. Further growth and straightening of the hypocotyl elevates the pale cotyledons, which soon become green and photosynthetic in addition to supplying stored reserves to the seedling. The microbodies, glyoxysomes, appear in cotyledon cells about 48 h after planting, while others, assumed to be peroxisomes, appear as the cotyledons emerge and chloroplasts become functional. This suggests that the glyoxysomes help to convert stored lipids to hexose sugar, whereas the peroxisomes are associated with the chloroplasts (Liu et al. 1971). After emergence of the cotyledons, the two primary leaves expand from the apical bud and mature within a few days. Further growth of the seeding and plant before flowering involves the formation of trifoliolate leaves.

2.2 Reproductive Stage

Following the period of vegatative growth, which varies depending upon cultivar and environmental conditions such as daylength and temperature, the plant enters the reproductive stage, during which axillary buds develop into flower clusters of 2 to 35 flowers each. The node of the first flower is related to the development stage of the plant. Since nodes of the cotyledons, the primary leaves, and the first two or three trifoliolate leaves are usually vegetative, the first flowers appear at nodes five or six and sometimes higher. Flowers form progressively toward the tip of the main stem and also toward the tip of the branches. Under favourable conditions, the period of bloom may extend from 3 to more than 5 weeks (Carlson and Lersten 1987). Several investigators have reported that a soybean plant produces many more flowers than can develop into pods. From 20 to 80% of the flowers are reported to abscise for various cultivars (Wiebold et al. 1981). Most cultivars with many flowers per node have a higher percentage of flower abscission than those with few flowers per node. Abscission can occur at the time of bud initiation, during the development of floral organs, at the time of fertilization, during the early embryo state, or at any stage of cotyledon development. Flower abscission occurs most often from 1 to 7 days after flowering, while pod abscission must be later than 7 days after flowering (Carlson and Lersten 1987). In general, the earliest and latest flowers produced tend to abscise most often.

Many crops are affected by the light-dark cycle, or photoperiod regime, in which they are grown. An important photoperiod response is the difference in time required for flower initiation under various day lengths. Soybean is a short-day crop, which begins flowering much earlier under short than under long days. When inflorescences are initiated, there are marked changes from the normal vegetative development of the axillary buds. The two opposite prophyll primordia are differentiated as usual, but instead of the typical two-rowed arrangement of the trifoliolate leaf primordia, the pattern of initiation changes to a spiral two fifths phyllotaxy. These primordia do not develop into typical trifoliolate leaves with paired stipules, but instead become simple bracts that lack stipules. Growth of both prophylls and bracts is somewhat slower than their vegetative counterparts, but their axillary buds develop much more rapidly, each becoming the primordium of a flower. When a bud in the axil of a trifoliolate leaf develops into an inflorescence, the stalk of that inflorescence remains stem-like, with typical stem anatomy, including epidermis, cortex, endodermis, vascular tissue, and considerable secondary growth from a vascular cambium. In the development of an inflorescence, the bract of each flower is homologous to a trifoliolate leaf, and the two bracteoles are homologous to the prophylls that normally develop at the base of every branch. After forming the primordia of the bracteoles, the apical meristem of the flower gives rise directly to the floral organs.

Patterns of dry matter accumulation in soybean plant leaves are similar in both determinate and indeterminate types of soybean (Beaver et al. 1985). Dry matter increases from the vegetative to the reproductive stage until a maximum value is attained, and decreases thereafter, during the rapid development of the seeds. The middle leaves show the most characteristic changes in the patterns of dry matter accumulation in blades of trifoliolate leaves. These changes are initiated by the transport of material to the developing seeds and coincide with the time of increased leaf fall, as the rate of abscission of leaves from the plants exceeds the production of new leaves (Drossopoulos et al. 1994). In the middle leaves, the rate of dry matter accumulation per trifoliolate leaf during the vegetative stage is large and leads, at the end of this stage, to a level of accumulation 3.45 times higher than the initial value. During the reproductive stage, dry matter continues to accumulate although at a lesser rate.

Fehr et al. (1971) reported stage development descriptions for soybean. Vegetative stages are determined by counting the number of nodes on the main stem, beginning with the unifoliolate node. Reproductive stages are based on flower, pod, and seed development, and on plant maturation (Table 1).

There is a major difference in plant development stages between indeterminate and determinate soybean plants. Indeterminate cultivars begin flowering when about half the nodes on the main stem have developed, and flowering proceeds upward on the plant as additional nodes are produced. Therefore, flower, pod, and seed development are more advanced on the bottom portion of the main stem than on the top. For determinate cultivars, flowering frequently does not occur until all of the nodes on the main stem have developed. In this situation, flower, pod, and seed development are similar throughout the plant.

Table 1. Summary of soybean plant development stages. (Fehr et al. 1971)

Stage number	Description
	Vegetative stages
VE	Cotyledons above the soil surface
VC	Unifoliolate leaves unrolled sufficiently so the leaf edges are not touching
V1	Completely unrolled leaf at the unifoliolate node
V2	Completely unrolled leaf at the first node above the unifoliolate node
V3	Three nodes on main stem beginning with the unifoliolate node are fully developed
V(N)	N nodes on the main stem beginning with the unifoliolate node are fully developed
	Reproductive stages
R1	One flower at any node
R2	Flower at node immediately below the uppermost node with a completely unrolled leaf
R3	Pod 0.5 cm long at one of the four uppermost nodes with a completely unrolled leaf
R4	Pod 2 cm long at one of the four uppermost nodes with a completely unrolled leaf
R5	Beans beginning to develop at one of the four uppermost nodes with a completely unrolled leaf
R6	Pod containing full size green beans at one of the four uppermost nodes with a completely unrolled leaf
R7	Pods yellowing; 50% of leaves yellow. Physiological maturity
R8	95% of pods brown. Harvest maturity

3 Plant Physiology

3.1 Carbon Assimilation and Metabolism

Soybean is a C_3 plant, those that fix CO_2 initially into three carbon sugars via the carboxylation of ribulose-1,5-bisphosphate (RuBP). The primary products of photosynthesis are partitioned between two major pathways, one leading to chloroplastic starch synthesis, the other to cytoplasmic sucrose synthesis (Shibles et al. 1987). Starch and sucrose are the most heavily labelled products from either pulse-chase or steady-state $^{14}CO_2$ uptake, with starch often accounting for most of the radioactivity (Fisher 1970a,b). Monosaccharides such as glucose, fructose, and hexose phosphates are present, but usually in concentrations of <1% (Streeter and Jeffers 1979). There is a daily rhythm in leaf starch content, with accumulation occurring during daytime and hydrolysis and translocation resulting in a progressive reduction at night. Under a normal photoperiod, the diurnal increase in leaf starch content occurs at a rate of approximately 3 to 3.5 $mg\,dm^{-2}h^{-1}$ regardless of nodal position (Shibles et al. 1987). Total carbon assimilation in the plant system is dependent in part on the maximum photosynthetic rate per unit leaf area and in part on interception of photosynthetically active radiation (PAR) by the total leaf area of the plant. While the maximum photosynthetic rate depends on age and nitrogen level of the leaves, and on water status, temperature, and CO_2 concentration in the environment, interception of PAR is affected both by flux density of radiation above the canopy and distribution within the canopy. The optimal environmental conditions for leaf CO_2 assimilation in soybean include saturating photosynthetic photon flux density, moderate temperature (20 to 30 °C), and ambient CO_2 and O_2 concentration (340 $\mu l\,CO_2\,l^{-1}$ and 21% O_2). Of course, faster rates would be expressed in high CO_2 and/or low O_2 environments, but a leaf is not normally subjected to such conditions (Shibles et al. 1987).

Soybean leaves also adapt to their growth regime primarily by adjusting leaf dimensions; for example, high irradiance during growth results in thicker leaves (Gourdon and Planchon 1982) with more photosynthetic apparatus per unit leaf area. Each leaf on a soybean plant follows a similar ontogenetic pattern with respect to CO_2 assimilation. During leaf expansion, the apparent photosynthetic (AP) rate per unit leaf area increase. Because both AP rate and leaf area growth are linear functions of time. During rapid leaf expansion the total CO_2-fixing ability of the leaf increases exponentially. Greatest photosynthetic capacity is attained at about the same time as the leaf reaches full expansion, and is then maintained for several days before beginning to decline gradually (Shibles et al. 1987). The total field dry matter accumulation is dependent not only upon AP per unit leaf area, but also on canopy AP per unit of land area (CAP). The light compensation point for soybean canopies is about 200 $\mu mol\,m^{-2}s^{-1}$, although it increases for maturing canopies (Larson et al.

1981). The CAP seems insensitive to temperatures between 25 and 32 °C (Baldocchi et al. 1981). Temperatures outside of this range can suppress CAP. Depending on plant developmental stage, plant and row spacing, and photosynthetic photon flux density, the critical leaf area index required for maximum CAP ranges between 3 and 6 (Baldocchi et al. 1981). Seasonal maximum CAP occurs during the late vegetative or early reproductive period. The time most likely depends on when the canopy closes; delays will occur for sparse populations. The seasonal maximum CAP will be maintained until the commencement of seed filling. The CAP rates of soybean during this period are commonly between 22 and 38 CO_2 $\mu mol\,m^{-1}\,s^{-1}$ (Shibles et al. 1987). When rapid seed filling commences, daily maximum CAP begins a steady decline, reaching zero at physiological maturity.

Dark (mitochondrial) respiration has been regarded as an obligatory carbon expenditure associated with metabolism essential for production of the complex compounds needed for growth and maintenance of tissue. In darkness, photosynthetically active leaves respire at about 3 $\mu mol\ CO_2\,m^{-2}\,s^{-1}$, or about 10 to 20% of the photosynthetic rate (Secor et al. 1983). In field-grown plants, the respiration rate, on a dry weight basis, was most rapid in leaf tissue, slowest in the stem, and intermediate in the root. Soybean plants show a diurnal pattern in dark CO_2 evolution, not totally associated with changes in temperature, that peaks between noon and late afternoon, or early in the dark period (Warrington et al. 1977). The peak in respiration may be a consequence of increased carbohydrate substrate or a circadian rhythm. Respiration rate and carbohydrate content are well correlated in actively growing tissues, such as young leaves and meristems, but are poorly related in more mature tissues, such as older leaves and full-sized pods before rapid seed growth begins (Coggeshall and Hodges 1980). Pods with rapidly growing seeds contribute more to respiration than do leaves, contributing 35 to 50% of whole-plant respiration (Kishitani and Shibles 1986). Under natural conditions, intact soybean pods respire more CO_2 during the day because the increase in pod temperature due to solar radiation hastens respiration more than the increase in solar radiation increases pod photosynthesis.

Soybean, as a subtropical crop, requires temperatures in the range of 25 to 30 °C for growth and physiological activities. The work of Thomas et al. (1981) suggests that there is an inverse relationship between temperature and soluble-carbohydrate concentration in leaves. For individual leaves, photosynthesis increases with increasing temperature to between 35 to 40 °C and then begins to decline, while respiration usually continues to increase with temperature above the optimum for photosynthesis (Raper and Kramer 1987). However, total plant dry matter accumulation begins to decline as air temperature increases above 28 to 30 °C (Wann and Raper 1979). Changes in temperature affect plant growth also through altered partitioning of photosynthate between organs. For example, low temperature decreases translocation of photosynthate into soybean seed both by decreased unloading into the seed coat and by decreasing embryo growth (Thorne 1982).

Photosynthesis is dependent on incoming light both for energy and for light-dependent activation of regulatory enzymes of assimilation. Declines in photon flux density levels within a canopy result in immediate decreases in net carbon uptake. The induction phase of photosynthesis has been found to be dependent on three separate processes that operate on different time scales (Pearcy 1990). A fast phase that activates and deactivates rapidly as photon flux density increases and decreases are associated with limitations in RuBP regeneration. Limitation of enzymes in this part of the carbon reduction cycle by the light activation state is most evident after relatively short low-light periods. The slow phase of induction is dependent on ribulosebiphosphate carboxylase/oxygenase (Rubisco) activation, which requires longer illumination at high photon flux density and exhibits a slower deactivation upon return to low photon flux density than the RuBP-regenerating enzymes. Changes in stomatal conductance also contribute to the slow phase of photosynthetic induction.

As a result of fossil fuel combustion and deforestation, the ambient air concentration of CO_2 is increasing. The increase in CO_2 results in an increase in C_3 plant photosynthesis. Soybean plant photosynthesis has been reported to increase as the CO_2 level increases from 330 to $1000\,\mu l\,l^{-1}$ (Wittwer 1979). When plants are exposed to higher CO_2 concentration conditions, some physiological events change; for example, the amount of both chlorophyll a and b increases in soybean plants on exposure to 400 and $450\,\mu l\,l^{-1}$ CO_2, compared to $350\,\mu l\,l^{-1}$ (Gupta and Li 1994).

3.2 Water Requirement and Water Stress

Raper and Kramer (1987) have indicated that the seasonal water used for soybean can range from 450 to 825 mm of water where the growing season ranges from 100 days at low altitudes to up to 190 days in the higher altitudes. These water-use values for soybean indicate a fairly broad range that is highly dependent on the rainfall distribution, the total evaporative demand, and soil water-holding characteristics for each of the locations. A typical seasonal water-use pattern consists of both soil evaporation and plant transpiration. Water-use rate is generally low during the germination and seedling stages due to partial canopy cover and with a large portion of the water lost due to soil evaporation. However, as the plant moves into the rapid growth stage from V3 to V6, there is a rapid increase in water use. Maximum water use occurs during R1 to R6 when full canopy is attained. Once the plant has started to mature and the pods are filled, there is a rapid decrease in water use associated with leaf senescence at the end of the season, and a concurrent reduction in evaporation demand (Doren and Reicosky 1987).

Water stress can be caused either by an excess of water that injures root systems by deficient aeration or by deficit in available soil moisture that results in dehydration of the shoots. Water stress, whether caused by flooding or drought, affects every aspect of plant growth, including anatomy, morphology, physiology, and biochemistry. The process of germination requires imbibition of water and is accompanied by a higher rate of respiration. Excessive soil moisture limits the oxygen available for the respiratory processes of germination. A combination of cold soil and excess water may be especially damaging to germination and emergence (Willing and Leopold 1983). Germination and emergence are decreased more often by dry soil than by flooding. The water content of soybean seeds must reach about 500 $g\,kg^{-1}$ of dry weight to insure germination (Raper and Kramer 1987), in contrast to corn seeds which require only about 300 $g\,kg^{-1}$. Soybean germinates at a soil moisture tension of −0.6 MPa, while corn germinates at −1.25 MPa.

During plant vegetative growth, total photosynthesis of water-stressed plants decreases due to decreases in rate of carbon fixation per unit leaf area resulting from premature stomatal closure and non-stomatal inhibition of the photosynthetic machinery, and due to decreases in photosynthetic surface area caused by decreased leaf enlargement and hastened leaf senescence. The effects of water stress on photosynthetic rates of soybean leaves are readily detectable at leaf water potentials of about −1.0 to −1.2 MPa (Cure et al. 1983). Photosynthetic rate declines rapidly with further decreases in leaf-water potential to about −1.8 MPa, and then continues to decline gradually with decreasing potential. The effects of water stress on photosynthetic rate are reversible down to leaf-water potentials of about −1.6 MPa or less (Cure et al. 1983). Respiration also decreases with leaf-water potential, however, this decrease is less than that of photosynthesis (Raper and Kramer 1987). Translocation continues at water stresses lower than those inhibiting photosynthesis. The velocity of translocation is not greatly decreased until water potential drops as low as −2.0 to −3.0 MPa.

The reproductive stage of plant growth is particularly sensitive to water stress, and soybean is no exception. A major portion of the variations in soybean yield can be attributed statistically to seasonal variations in rainfall during flowering and fruiting (Thompson 1970). The yield component associated with yield decreases in response to water stress varies with growth stage; the number of seeds and pods is affected by stress during flowering and early pod-fill stages, and seed size is affected by stress during seed fill (Constable and Hearn 1978). Abscission of flowers, pods, and seeds of water-stressed plants, as well as the later decreases in seed size, may be at least partially a response to water stress through the effects of leaf-water deficits on photosynthetic rates with resulting decreases in the concentration of assimilates in vegetative pools. If the stress is relieved prior to completion of the seed-filling period, photosynthesis can recover, growth can resume, and the remaining seeds can continue to develop to normal size (Patterson et al. 1979). If, however, photosynthesis is decreased by short periods of water stress in the range of −1.0 and −1.6 MPa after pod and seed set, the size of the reproductive sink demand is not adjusted during the stress period (Cure et al. 1983). Comparison of water relations between soybean and other legumes, such as white bean, were summarized by Smith et al. (1988a): (1) white bean conserves more water than soybean by virtue of (a) greater leaf diffusive resistance at the same plant water potential, (b) lesser depletion of soil water, and (c) a greater degree of parahelionastic response to water-deficit stress; (2) water deficit has a greater effect on N_2 fixation in white bean than in soybean, initially through a greater restriction of photosynthesis and, under extreme conditions, through drought-induced nodule senescence.

3.3 Mineral Nutrition and Metabolism

Soybean offers a number of advantages over other crops, e.g., its ability to fix N_2, its low phosphorus requirements, its tolerance to low pH and high levels of Al, and its tolerance to high soil moisture content (Tanaka 1985). Patterns of macronutrient accumulation in blades of trifoliolate leaves for most of the most of the macronutrients are in accordance with patterns of dry matter accumulation, while patterns of concentration are different (Drossopoulos et al. 1994). Total N, P, and K concentrations in leaves increase and those of Ca and Mg decrease from the lower to the middle and upper leaves during the vegetative stage. Nitrogen and P accumulation increase parallel to dry matter accumulation, with rates higher than those for the dry matter accumulation (Olson 1978). During the reproductive stage, both total N and P concentration and content in blades are decreased. These changes are attributed to the mobilization and transport of N and P to the reproductive sink observed during this period (Weaver and Morris 1982). Potassium concentration in blades of middle leaves is higher before flowering than afterwards. At the same time, K accumulation in leaves also increases. During the reproductive stage, the K concentration is stable and relatively low, although the accumulation of this nutrient in leaves decreases, especially during flowering and pod development, but increases again after the development seeds, at the end of the life cycle (Drossopoulos et al. 1994). In upper leaves, K concentration declines and in lower leaves K continues to increase during the reproductive stage. Magnesium concentration in the middle leaves is decreased with the approach of flowering, while after the appearance of pods, it increases again to about the same levels as at the beginning of the vegetative period. Magnesium accumulation increases during the early vegetative stage, but remains constant until the appearance of pods and then increases thereafter (Drossopoulos et al. 1994).

With the exception of Fe, the accumulation of all micronutrient elements in blades of the middle and upper leaves increases from the vegetative to the reproductive stage until a maximum value is attained and subsequently decreases during the rapid development of the seeds (Drossopoulos et al. 1994). Iron, Cu, and Zn concentrations in leaf blades before flowering are greater than later in the life cycle. Iron levels present the most impressive difference among blades of lower, middle, and upper leaves during the vegetative stage. Patterns of Fe concentration are the same in blades at every position on the plant and show their highest values at the onset of the early vegetative stage, and decline rapidly as flowering begins. The Fe concentration in blades during the reproductive stage increases slightly after the development of the pods.

The interdependence and interaction of nitrate (NO_3^-) metabolism and symbiotic dinitrogen (N_2) fixation provides the basis for consideration of overall N metabolism of soybean plants. The symbiotic association of *Bradyrhizobium japonicum* with roots of soybean provides one N input system to the plant. Alternatively, NO_3^- is readily taken up from the soil solution by soybean roots and provides a second N input system. The net rate of NO_3^- uptake by intact plants depends on (1) availability of NO_3^- to the uptake sites and (2) the relative rates of influx and efflux processes across the root cell plasmalemma (Harper 1987). The former is dependent on NO_3^- level in the soil solution, water availability and rate of flow to roots, mineralization rates, and root penetration into previously unexploited soil area. Since NO_3^- is freely mobile, the moisture status of a soil is likely the predominant factor controlling NO_3^- availability to the root (Olson and Kurtz 1982). Nitrate uptake by soybean occurs under both light and dark conditions (Nicholas and Harper 1985). Uptake of NO_3^- appears to be saturated at a relatively low (0.5 mM) solution NO_3^- level during the seedling and early vegetative growth stages (Gibson and Harper 1985). Peak uptake rates per plant occur at the time of early to mid-pod fill under conditions of adequate water supply and continuous NO_3^- availability (Harper 1987). Increasing rates of NO_3^- uptake with plant age are most likely due in large part to increasing root mass rather than to an increase in specific uptake rate. Nitrate which is taken up may be either reduced in the root, stored in the roots, or translocated to the shoots, and reduced there. Nitrate translocation has been primarily studied through monitoring of xylem sap composition, whereas nitrate does not appear to move in the phloem of soybean (Ohyama and Kawai 1983). McClure and Israel (1979) reported that NO_3^- comprised 58% of the total nitrogen in the xylem sap of non-inoculated soybean supplied with 20 mM NO_3^-. This verifies that a portion of the NO_3^- taken up is translocated to the shoots in an unreduced state, but also indicates that active NO_3^- reduction most like occurs in the roots before translocation. Nitrate which remains in the root system and which is not directly reduced is most likely stored in the vacuole (Oaks 1979). The concentration of NO_3^- in soybean roots exceeds that found in either the leaf or the stem fractions, indicating that roots may serve as temporary storage sites for NO_3^- (Harper 1987). The initial step of NO_3^- reduction is catalyzed by nitrate reductase, which is considered a cytoplasmic enzyme in higher plants. The subsequent reduction of NO_2 to ammonium (NH_4^+) is mediated by nitrite reductase. The major product of ammonia assimilation is usually considered to be amino nitrogen. Asparagine plays a key role in the metabolic pathway of amino nitrogen and is the major amino acid involved in xylem transport in soybean (Harper 1987). The greatest amount of ^{15}N label from $^{15}N\text{-}NO_3^-$ was also shown to be incorporated into asparagine in soybean roots (Ohyama and Kawai 1983).

Compared to white bean, soybean N accumulation and distribution were summarized as follows: (1) soybean seed always contained more N than white bean seed, and a large portion of the total soybean plant N (excluding the senesced leaves) was in seeds; (2) the addition of fertilizer N increased the concentration of N in most tissues most of the time, and always increased the concentration in seeds; (3) irrigation generally increased the concentration of N in plant tissues on a heavier soil, but decreased the concentration on a light soil; (4) white bean is much more variable in its response to N fertilizer and irrigation (Smith et al. 1988b).

The amount of dry matter accumulated per unit of assimilated N by soybean is about half that for cereal crops (Shinano et al. 1994). The production cost of a particular plant component is calculated as the amount of the material formed from a unit weight of glucose, and on this basis the ascribed values are 0.40 for synthesis of protein, 0.33 for lipid, and 0.83 for carbohydrate (Shanano et al. 1993a). As soybean accumulates a large amount of protein and lipid in its reproductive organs, respiratory losses associated with such synthesis are also assumed to be higher. Based on studies of growth efficiency in the harvested organs and in the whole plant during the vegetative growth stage, the low productivity of soybean is mainly due to the higher respiratory loss of carbohydrates in leaves and stems (Shinano et al. 1993b).

Phosphorous deficiency decreases whole-plant photosynthate production by decreasing both photosynthetic rate per unit area and expansion of photosynthetic surface area. The decreased photosynthetic rates under phosphorus deficiency have been associated with decreases in ribulose 1,5-bisphosphate carboxylase/oxygenase activity, ribulose 1,5-bisphosphate regeneration, CO_2 carboxylation, stomatal conductance, and an increase in mesophyll resistance (Qiu and Israel 1994). Photosynthetic rates can also be regulated by leaf cytoplasmic P concentration through the phosphate translocator, by the rate of P release from sucrose phosphate, and by the conversion of phosphorylated intermediates into starch (Preiss 1982). Alterations in plant phosphorus status influence soluble carbohydrate utilizations of hexose within a plant. The initial response to the onset of phosphorus deficiency is an increase in the concentration of hexose in all organs of the soybean plant (Qiu and Israel 1994). As phosphorus deficiency becomes more severe, hexose in stems and roots cannot be metabolized efficiently through the glycolysis pathway and is eventually converted into starch. As phosphorus deficiency progresses, a decrease in relative growth rate and decreases in day and night leaf elongation rates either coincide with or precede significant increases in leaf, stem, and root starch concentration. Thus, metabolic functions associated with growth are also disrupted, causing decreased utilization of the available nonstructural carbohydrate (Qiu and Israel 1994).

The role of micronutrients (such as Mo, Fe) in symbiotic dinitrogen fixation is well known. Soybean plants need additional Mo and Fe, which act as an electron acceptor and donor, respectively, to make nitrogenase and leghaemoglobin. Increased nodule activity has been reported in plants after increasing Mo and Fe concentrations in the plant rhizosphere. Dalton et al. (1988) review the role of nickel as plant micronutrient. In large amounts the element has a detrimental effect. However, Ni is required in small amounts for normal growth and development of soybean. Both urease and hydrogenase are Ni-requiring enzymes present in nitrogen-fixing symbionts and their presence suggests that optimum growth of nodulated legumes and actinorhizal plants may depend on an adequate supply of Ni to meet the requirements of these Ni-requiring enzymes in host plants and endophytes (Sprent and Sprent 1990).

4
Symbiotic Dinitrogen Fixation

4.1
Dinitrogen Fixation

Nitrogen is a major component of this planet occurring either free as N_2 or bound in various inorganic or organic forms. However, 98% of this global N occurs in primary rock and is unavailable to the biosphere. By far the largest part of the remainder is found in the atmosphere as N_2. It has been estimated that the reservoir of atmospheric N_2 is 3.9×10^{15} tonnes. Terrestrial living systems contain a total of only 1×10^{10} tonnes, or about 25×10^{-6}% of that in the atmosphere (Gallon and Chaplin 1987). The ability to reduce atmospheric dinitrogen is limited to prokaryotes. Legumes and a few other plant species have the ability to fix atmospheric N_2 through symbiotic relationships; for legumes the N_2-fixation is carried out in nodules located on the plant root by prokaryotes, generally *Rhizobium* or *Bradyrhizobium* (Sprent and Sprent 1990). Biological N_2 fixation is estimated to convert 175×10^6 t of N from dinitrogen to ammonia every year, whereas global industrial processes fix about 60×10^6 t. Soybean plants, by forming a symbiosis with the bacteria *B. japonicum*, can fix up to $200\,kg\,N\,ha^{-1}\,yr^{-1}$ of atmospheric N_2 (Smith and Hume 1987).

The family *Rhizobiaceae* consists of a heterogeneous group of Gram-negative, aerobic, non-spore-forming rods that can invade and form nodules on the root and, in some instances, on the stem of leguminous plants (McCardell et al. 1993). The slow-growing nodulation bacteria which have specific association with soybean are referred to *as Bradyrhizobium*. Currently, *Bradyrhizobium* has only one designated species. *B. japonicum* (Hollis et al. 1981). Some soybean roots can also nodulate with a fast grower named *Rhizobium fredii* (Sprent and Sprent 1990).

4.2
Nodule Formation

Nodulation begins when rhizobia attach themselves to epidermal cells. Epidermal cells with immature or as yet unformed root hairs are the usual sites for bacterial penetration (Bhuvaneswari et al. 1980). Prior to attachment, communication between the two symbiotic partners, soybean plant and *B. japonicum* bacteria, is required and a certain minimum period of contact is needed. Infected hairs are always shorter than mature intact hairs, due to marked curling upon infection. At the point of infection, the root hair wall forms a depression that invaginates deeply, forming an infection thread lined by a continuation of the root hair cell wall and membrane. Infection threads may branch within a root hair (Turgeon and Bauer 1982). The infection thread, with its included dividing bacteria, grows 60 to 70 μm to the base to the root hair cell. The cortex adjacent to infected root hairs becomes meristematic and produces a wedge-shaped area of dividing cells even before any infection threads enter (Turgeon and Bauer 1982). These mitoses increase cell number in the cortical layer, which then becomes the main area of infected cells (Newcomb et al. 1979). The combination of multiple threads and branching of threads in the cortex results in penetration of many, but not all, of these cells. The peripheral uninfected area becomes the nodule cortex, which includes a scleroid layer and several vascular bundles. At some time during or following mitotic activity, rhizobia are released into cortical cells through thin areas on the tips of the infection threads. Bradyrhizobia are called bacteroids after their release into the host cell. Their cell walls have been considerably modified or, in the case of peanut entirely removed (Werner and Mörschel 1978). Mitosis in infected cortical cells ceases about 14 days after infection. Subsequent increases in the volume of infection tissue are due entirely to cell enlargement. As the nodule matures, oxygen-binding leghaemoglobin develops gradually in the host tissue and the nodule becomes pink, remaining so until it begins to senesce. As leghaemoglobin forms, bacteria cease dividing, and dinitrogen fixation commences (Lersten and Carlson 1987).

The time-course of each these stages has been described by Turgeon and Bauer (1982). Bacterial attachment to root hairs occurs within minutes of inoculation and is followed, within 12 h, by marked curling of root hairs. Infection threads, first visible within 24 h of infection, reach the base of the root hair by 48 h after inoculation. Anticlinal divisions of the adjacent cortical cells has already occurred, giving rise to nodule primordia. Infection thread penetration of this extensively dividing meristem is not observed until between 48 to 96 h after inoculation. Bassett et al. (1977) noted that the bacteria are released from the infection thread to form bacteroids within 7 to 10 days after inoculation. A spherical mass of cytoplasmically rich cells, which have been invaded by infection threads, divide and differentiate into the central zone of N_2-fixing cells within 12 to 18 days after inoculation (Newcomb et al. 1979).

4.3 Recognition Between Symbiotic Partners

The molecular mechanisms for recognition between (*Brady*)*rhizobium* and soybean can be considered as a form of interorganismal cell-to-cell communication. A precise exchange of molecular signals between the host plant and bradyrhizobia over space and time is essential to the development of effective root nodules. The first apparent exchange of signals involves the secretion of phenolic compounds (flavonoids, flavones and isflavones) by soybean plants (Peters and Verma 1990). These signal compounds are often excreted by the portion of the root with emerging root hairs, a region that is highly susceptible to infection by bradyrhizobia (Verma 1992). These compounds activate the expression of *nod* genes in bradyrhizobia, stimulating production of the bacterial nod factor (Kondorosi 1992). This nod factor has been identified as a lipo-oligosaccharide (Carlson et al. 1993), able to induce many of the early events in nodule development, including deformation and curling of plant root hairs, the initiation of cortical cell divisions, and induction of root nodule meristems (Dénarié and Roche 1992). The isoflavones daidzein and genistein are the major components of soybean root excretions responsible for inducing the *nod* genes of *B. japonicum* (Kosslak et al. 1987). These substances are active at very low concentrations (10^{-7} to 10^{-8} M) and stimulate bacterial *nod* gene expression within minutes.

Plant lectins play a major role in the initiation of infection. Lectins are carbohydrate-binding proteins produced by legumes and are recognized by bacterial receptor molecules. Legumes of different cross-inoculation groups make lectins with different sugar-binding specificities. Lectins are found in seeds, roots, leaves, and stems. Those found on the root are often concentrated in the area where nodule initiation occurs. Lectins are thought to have two major roles in N_2-fixation symbioses. First, lectins are important in accumulating rhizobia on root hairs through adsorption of bacterial cells to the plant (non-biovar-specific). The second function plays a host-specific role in infection. In previous work, the mechanism of recognition in the *B. japonicum*-soybean symbiosis was investigated by using a mutant of *B. japonicum* strain HS111, which exhibits a delayed-nodulation phenotype (Halverson and Stacey 1984, 1985). The nodulation phenotype of mutant strain HS111 is the result of its inability to promptly initiate infection leading to subsequent nodulation. The defect in initiation of nodulation in HS111 can be phenotypically reversed by stimulating the plant rhizosphere through preinoculation with soybean root exudate, soybean seed lectin, or root-secreted components prior to inoculation.

4.4 Biochemistry and Physiology

Nitrogenase, which comprises 30% of total protein in infected cells, has been purified from all known types of N_2-fixing organisms, with the exception of archaebacteria (Sprent and Sprent 1990). Nitrogenases are known to be made up of two fairly distinct parts, dinitrogenase reductase (Fe-protein), which is an electron carrier, and dinitrogenase (Mo-Fe protein), which is the enzyme responsible for the reduction of N_2.

The process of N_2 fixation is very costly in terms of plant energy requirement. It has been estimated that 5 to 10 g of carbon are required for 1 g N fixed from N_2. In addition, carbon substrates are required for the subsequent assimilation of ammonium into organic compounds and for nodule growth and maintenance (Day and Copeland 1991). A number of studies have shown that current photosynthate, translocated into nodules as sucrose, is preferentially used to support N_2 fixation (Kouchi and Nakaji 1985). The active uptake of disaccharides into bacteroids has been demonstrated for fast-growing *Rhizobium* species, whereas slow-growing *Bradyrhizobium* accumulate disaccharides only by passive diffusion. Feeding experiments carried out in several laboratories with ^{13}C or ^{14}C-labelled precursors have shown that label is converted rapidly from sucrose into the dicarboxylic acids malate and succinate in nodules and that these are the compounds supplied to the bacteroid in quantity.

The nitrogenase enzyme complex inside the bacteroid is extremely sensitive to inactivation by O_2. On the other hand, O_2 is required to support the highly active respiratory processes that take place aerobically in the plant and bacteroid compartments. A nodule gaseous diffusion barrier, composed of water-filled intercellular pores located within the nodule cortex, regulates oxygen flux from the rhizosphere into the nodule and prevents nitrogenase inhibition by O_2. Leghaemoglobin, which occurs in the plant cytoplasm of infected cells, plays a key role in maintaining a low concentration of free O_2. This suggests that leghaemoglobin acts as an O_2 buffer in the nodule, and facilitates the transport of O_2 at a strictly controlled concentration to rapidly respiring bacteroids (Day and Copeland 1991). The free O_2 concentration inside nodules is around 10 nM, a concentration at which oxidative phosphorylation is no longer possible. Mitochondria of infected cells appear morphologically modified and are clustered around the cell periphery where O_2 concentrations of 20 to 26 mM are postulated to exist, allowing some oxidative phosphorylation (Sheehy et al. 1985). Weisz and Sinclair (1988) showed that soybean nodule activity and nodule gas permeability are both closely correlated to, and decreased by, low soil temperature, probably due to temperature-mediated changes in the turgor of cortical cells. Under these conditions, the decreased flux of oxygen into the nodules limits nodule respiration, results in inhibition of host carbohydrate conversion to ATP, and limits availability of reductant required for nitrogenase activity (Weisz and Sinclair 1988).

Ammonium, the product of nitrogenase activity, is excreted and moves freely across membranes. Thus, it is assumed to flow out of the bacteroid by diffusion down the concentration gradient caused by its continual removal from the host cytoplasm (Millor and Werner 1990). In almost all green plants the glutamine synthetase (GS)/glutamate synthase (GOGAT) pathway is the preferred route of ammonia assimilation (Groat and Vance 1981). Glutamine may be further used for asparagine or purine synthesis depending on whether the plant is an amide-transporting legume (most of this group are temperate legumes) or a ureide-transporting legume (most of this group are tropical legumes). The amide exporters transport asparagine and glutamine, while ureide exporters transport either allantoin and allantoin acid or citrulline. Soybean is a ureide-exporting legume, in which glutamine, glycine, and aspartate are used for purine synthesis in plastids of infected cells. Purine nucleotides are transformed to xanthine and uric acid, which are transformed by peroxisomes in neighbouring uninfected cells to allantoin. This is further transformed in the smooth endoplasmic reticulum of uninfected cells to allantoic acid (Millor and Werner 1990).

4.5 General Genetic Characteristics

As N_2 fixation is a symbiosis, the genetic aspects of both organisms (bacteria and higher plant) are involved. Plant nodulin genes can, according to the timing of their expression during nodule development, be divided into early and late nodulin genes (Govers et al. 1987). Early nodulins are involved in root-hair deformation, infection, and nodule morphogenesis. Later nodulins are detectable after the nodule has developed and bacterial release has taken place. The activity of late nodulin genes is associated with nodule function.

The only early nodulin that has been characterized in sufficient detail to warrant discussion about its function is the soybean nodulin Ngm-75 which is encoded by GmENOD2 (Nap and Bisseling 1990). It is a hydroxyproline-rich protein that is most likely a cell-wall component (Franssen et al. 1987) and is formed in nodule parenchyma cells. The Ngm-75 component is expressed in soybean roots within 6 to 7 days of inoculation by *B. japonicum* (Dickstein et al. 1988). The function of N-75 is most likely associated with nodule development (Franssen et al. 1987). Other soybean nodulins characterized so far, GmENOD13 and GmENOD55, are also related to nodule morphogenesis. Leghaemoglobin (Lb) transcript detection was used as a marker for the expression of late nodulin genes. Leghaemoglobin is the most abundant late nodulin, comprising up to 25% of the total soluble nodule protein, and is an O_2 carrier in the infected cells of the nodule that facilitates oxygen diffusion toward the bacteroid (Appleby 1984). Leghaemoglobin is detectable just before nitrogenase activity can be measured (Verma et al. 1979). Uricase is the second most abundant protein in legume root nodules and turns out to be the product of Ngm-35 in the cytoplasm of soybean. This protein is a key enzyme in the ureide biosynthetic pathway used to assimilate ammonium (Bergmann et al. 1985), and is specifically located in the peroxisomes of the uninfected cells. In nodules, glutamine synthetase is located in the cytoplasm of infected cells (Verma et al. 1986), and is the first enzyme involved in assimilation of ammonium. With respect to nodule carbon metabolism it has been shown that the subunit of the sucrose synthase that acts in soybean nodules is nodulin Ngm-100 (Franssen et al. 1992). Several other nodulins (termed peribacteroid membrane nodulins) have been identified and are localized in the peribacteroid membrane which surrounds bacteroids in infected cells. Since the peribacteroid membrane is the physical and metabolic interface between *(Brady)rhi-*

zobium and its eukaryotic partner, nodulins may function in the exchange of metabolites between the symbionts. Nodulin Ngm-24 is exclusively located in the peribacteroid membrane (Fischer and Hennecke 1987). Another peribacteroid membrane nodulin, Ngm-26, probably is a transmembrane protein exposed both to the bacteroid space and the plant cytoplasm, and may have a role in small molecule transport across the peribacteroid membrane (Fortin et al. 1987).

The bacterial genome has been characterized as occurring in series of functional groups. The bacterial genes involved can be categorized as (1) the *nod* genes, involved in the induction of nodule tissue, and subdivided into the common *nod* genes (*nod*A, B, C, I, and J) and host specific *nod* genes (*nod*E, F, D, and H), (2) the *fix* genes, which are involved in the process of N_2 fixation and (3) the *nif* genes, about 20 in number, which are really a subset of *fix* genes and code for nitrogenase and related proteins. In the *Rhizobium* species the *nod*, *fix*, and *nif* genes are all contained on a sym plasmid, while in *Bradyrhizobium* species these genes are on the chromosome (Appelbaum 1990). The common *nod* region is a cluster of *nod* genes that are structurally and functionally conserved in many *Rhizobium* and *Bradyrhizobium* strains (Appelbaum 1990). DNA sequencing and gene production analysis have led to the identification of four genes in two transcriptional units, *nodD* and *nodABC* (Nieuwkoop et al. 1987). Mutations in *nodABC* prevent marked root hair curling (Hac^-) and formation of infection threads (Inf). Mutations in *nodD* also result in a Hac^- phenotype, as expected due to the role *nodD* has in regulation of *nodABC* transcription. The *nodABC* genes also are involved in morphogenesis and encode functions that stimulate cell division without direct contact between the bacteria and the stimulated plant cells. Host-specific *nod* genes influence root hair curling, infection thread formation, and nodule initiation in various ways. The host-specific *nod* gene products are involved in the construction of an altered bacterial cell surface that is recognized by the host plant at the stages of root hair curling and infection thread initiation. The *fix* and *nif* genes contribute to the late stages of nodule development and nodule function. In *K. pneumoniae* about 20 contiguous *nif* genes are required for nitrogenase activity and growth on N free medium (Appelbaum 1990). The functions of the *fix* genes are less clear. It is now established, at least in *R. meliloti*, that the *fixL* gene product is a protein that spans the membrane of the bacteroid and senses the oxygen concentration inside the nodule (Earl et al. 1987).

4.6 Factors Affecting Soybean Dinitrogen Fixation

The capacity of soybean to fix N_2 is affected by general ecological factors such as temperature, macronutrients, micronutrients and water regime.

As a subtropical legume, soybean requires temperatures in the range of 25 to 30 °C for optimal symbiotic activity. When soil temperature drops below this range, legume nodulation and dinitrogen fixation are negatively affected. Low soil temperatures decrease both soybean nodulation and dinitrogen fixation. When the soil temperature drops below 25 °C, but remains above 17 °C, the time between inoculation and onset of dinitrogen fixation is delayed by 2 to 3 days for each °C decrease in temperature, whereas soil temperatures between 17 and 15 °C were more strongly inhibitory, and each °C delayed the onset of dinitrogen fixation about 1 week (Zhang and Smith 1994; Zhang et al. 1995a). When the plants are grown under 10 °C, nodulation fails (Matthews and Hayes 1982). All stages of nodule formation and function are affected by low temperature. Experiments involving the transfer of soybean plants, at various stages of their life cycle, between optimal and low temperature have generally indicated that infection and early nodule development processes are the most sensitive based on microscopic observation (Lynch and Smith 1993a; Zhang and Smith 1994). Zhang and Smith (1995) found that low temperature decreases early infection and nodule initiation by disruption of the interorganism signal exchange. Low temperature decreases both the biosynthesis of genistein and the excretion of genistein from plant root cells to soil rhizosphere (Zhang and Smith 1996a). The effects of low temperature on the function of nitrogen-fixing nodules may be due to the changes in nodule oxygen permeability (Walsh and Layzell 1986). The solubility of ureide is low and decreases sharply as temperature declines, therefore low temperature may also limit the rate of export of symbiotically fixed nitrogen from the nodule (Sprent 1982). In turn, higher nitrogen concentrations inside the nodule inhibits dinitrogen fixation. The effect of low temperature on dinitrogen fixation and nitrogen assimilation may also be mediated via effects on photosynthesis or translocation (Walsh and Layzell 1986). Different genotypes of soybean or strain of *B. japonicum* respond differently to low soil temperatures; for example, the reports from Lynch and Smith (1993b, 1994) show that the strains obtained from the cold soils of Hokkaido, northern Japan, result in increased plant biomass and symbiotically fixed nitrogen level at the plant-flowering stage under short-season conditions. Recently, some studies have suggested that the following factors could improve nodulation, dinitrogen fixation, and grain and protein yield: preincubation of the plant molecular signal with *B. japonicum* (Zhang and Smith 1996b, 1997); coinoculation of *B. japonicum* with other soil organisms, such as vesicular arbuscular mycorrhizal fungi (Zhang et al. 1995b); adding growth-promoting rhizobacteria (Zhang et al. 1996, 1997).

Soybean needs the same macro- and micronutrients as other green plants. Soybean plants need additional Mo and Fe to make nitrogenase and leghaemoglobin. Combined nitrogen (NO_3^-, NH_4^+, and urea) has been demonstrated to influence symbiotic dinitrogen fixation from the initial bidirectional signal exchange between the symbionts through subsequent nodule development and nodule senescence. Although the inhibition of nodulation by nitrate has been known for over 100 years, little is known about mechanisms involved. An early study indicated that greater delays in nodule formation were caused by nitrate than by either ammonium or urea (Darbyshire 1966). At nitrate concentrations greater than perhaps 2 mM, both nodule growth and nodule activity are depressed (Eardly et al. 1984). Nitrate affects a broad range of infection events, including a decrease in root-hair deformation, a decrease in the binding of rhizobia to root hairs, a decrease in the number of infection threads, and an increase in the number of aborted infection events (Streeter 1988). With long-term exposure of plants to 4 to 8 mM nitrate, effects of nitrate on nodule growth are clearly greater than effects of nitrate on specific nitrogenase activity (Streeter 1988). Studies with field-grown plants also indicate much greater effects on nodule growth than on specific nodule activity (Miller et al. 1982). In a recent study, the inhibition of symbiotic dinitrogen fixation by the presence of mineral nitrogen in the soil was generally attributed to a shortage of carbohydrates and or reducing power necessary for the competing metabolic processes of NO_3^- assimilation and symbiotic dinitrogen fixation (Waterer and Vessey 1993). Depending on the site of NO_3^- reduction and point of assimilation, it has been estimated that it is four to eight times more efficient for a plant to assimilate NO_3^- than to fix atmospheric dinitrogen (Postgate 1982).

Water supply has a major effect on nodulation and dinitrogen fixation. The relationship between soil moisture and nodulation has been recognized; nodulation increases with soil-water content until water logging occurs. Irrigation is reported to increase rates of N_2 fixation (Smith and Hume 1985), photosynthesis, and soil N mineralization. Water stress significantly decreases the number of infection threads formed and inhibits nodulation. This effect on infection is reversible when the plants are rewatered; immature root hairs can resume normal growth and become infected, Following successful infection, decreased water supply can retard nodule development (Sprent and Sprent 1990). Lower water potential in the nodules may decrease dinitrogen-fixing activity (Pankhurst and Sprent 1976), with a corresponding decrease in nodule respiration, decrease in transport of fixed nitrogen out of nodules, and impairment of photosynthate supply from a stressed shoot system (Huang et al. 1975). Dinitrogen fixation activity of nodules is decreased as water potential decreases and ceases irreversibly in detached nodules when fresh weight drops below about 80% of the fully turgid weight (Raper and Kramer 1987).

Low pH generally decreases nodulation. This may be due to decreased growth and multiplication of rhizobia in the soil (Rice et al. 1977), increased numbers of ineffective rhizobia or sensitivity of the infection process (Date 1988). Acid soil may also be low in levels of available calcium, magnesium, phosphorus, and molybdenum and also contains levels of aluminium and manganese toxic to the host plant (Sprent and Sprent 1990). With high saline concentration in the water, nodule activity is negatively affected by osmotic withdrawal from nodules. Bacterial colonization and root-hair curling of plants maintained at 100 mM NaCl are both decreased when compared to those plants grown at 50 mM NaCl conditions and the proportion of root hairs containing infection threads is decreased by about 30% (Zahran and Sprent 1986). Three weeks after saline treatment, nodule numbers are decreased by about 50%, and nodule weight by more than 60%.

When dinitrogen-fixing organisms are afflicted with pests or pathogens the fixation process is usually affected. Conversely, the dinitrogen-fixing symbiosis may affect the plant disease either by its effect on plant nutrition status or directly (Sprent and Sprent 1990). Plant viruses affect the structure, physiology and biology of both root nodules and host plants. Tu and Ford (1984) point out that a virus-affected plant may, via root exudate, selectively affect rhizobial strains. Nodules may have levels of virus that affect nodule activity directly. Root nodules may be eaten by some insects or insect larvae.

References

Appelbaum E (1990) The *Rhizobium/Bradyrhizobium*-legume symbiosis. In: Gresshof PM (ed) Molecular biology of symbiotic nitrogen fixation. CRC Press, Boca Raton, pp 131–158

Appleby CA (1984) Leghemoglobin and *Rhizobium* respiration. Annu Rev Plant Physiol 35:443–478

Baldocchi JO, Higgs LR, Paulson WH, Jacobs JA, Schrader WD (1981) Legume and mineral N effects on crop yields in several crop sequences in the upper Mississippi Valley. Agron J 73:885–890

Bassett B, Gloudeman RN, Novacky A (1977) Ultrastructure of soybean nodule I. release of rhizobia from the infection thread. Can J Microbiol 23:537–582

Beaver JS, Cooper RL, Martin RJ (1985) Dry matter accumulation and seed yield of determinate and indeterminate soybeans. Agron J 77:675–679

Bergmann H, Preddlie E, Verma DPS (1985) Nodulin-35: a subunit of specific uricase induced and localized in the uninfected cells of soybean nodules. EMBO J 2:2333–2339

Bhuvaneswari TV, Goodman RN, Bauer WD (1980) Early events in the infection of soybean [*Glycine max* (L.) Merr.] by *Rhizobium japonicum* 1. location of infectible root cells. Plant Physiol 66:1027–1031

Brown DM (1981) Heat units for corn in Southern Ontario. Ontario Ministry of Agriculture and Food Factsheet #78-063 Ontario

Carlson JB, Lersten NR (1987) Reproductive morphology. In: Wilcox JR (ed) Soybean: improvement, production, and uses, 2nd edn. #16 Agron Ser ASA-CSSA-SSSA, Madison, Wisconsin, pp 95–134

Carlson RW, Sanjuan J, Bhat R, Glushka J, Spaink HP, Wijfjes AHM, van Brussel AAN, Stokkermans TJW, Peters NK, Stacey G (1993) The structures and biological activities of the lipo-oligosaccharide nodulation signals produced by type I and II strains of *Bradyrhizobium japonicum*. J Biol Chem 268:18372–18381

Coggeshall BM, Hodges HF (1980) The effect of carbohydrate concentration on the respiration rate of soybean. Crop Sci 20:86–90

Constable GA, Hearn (1978) Agronomic and physiological responses of soybean and sorghum crops to water deficits. I. Growth, development, and yield. Aust J Plant Physiol 5:159–167

Cure JD, Patterson RP, Raper CD, Jackson WA (1983) Water stress recovery of soybeans as affected by photoperiod during seed development. Crop Sci 23:110–114

Dalton DA, Russell SA, Evans HJ (1988) Nickel as a micronutrient for plants. BioFactors 1:11–16

Darbyshire JF (1966) Studies on the physiology of nodule formation. IX. The influence of combined nitrogen, glucose, light intensity and day length on root hair infection in clover. Ann Bot 30:623–638

Date RA (1988) Nodulation difficulties related to low pH. In: Gibson AH, Newton WE (eds) Current perspectives in nitrogen fixation. Australian Academy of Sciences, Canberra, pp 216–262

Day DA, Copeland L (1991) Carbon metabolism and compartmentation in nitrogen-fixing legume nodules. Plant Physiol Biochem 29:185–201

Dénarié RM, Roche P (1992) *Rhizobium* nodulation signals. In: Verma DPS (ed) Molecular signals in plant-microbs communication. CRC Press, Boca Raton, Florida, pp 295–324

Dickstein R, Bisseling T, Reinhold UN, Ausubel F (1988) Expression of nodule-specific genes in alfalfa root nodule blocked at an early stage of development. Genes Dev 2:677–687

Doren DMV, Reicosky DC (1987) Tillage and irrigation. In: Wilcox JR (ed) Soybean: improvement, production, and uses. 2nd edn. #16 Agron Ser. ASA-CSSA-SSSA, Madison, Wisconsin, pp 391–428

Drossopoulos JB, Bouranis DL, Bairaktari BD (1994) Patterns of mineral nutrient fluctuations in soybean leaves in relation to their position. J Plant Nutr 17:1017–1135

Eardly BD, Hannaway DB, Bottomley PJ (1984) Nitrogen nutrition and yield of seedling alfalfa as affected by ammonium nitrate fertilization. Agron J 77:57–62

Earl CD, Ronson CW, Ausubel FM (1987) Genetic and structural analysis of the *Rhizobium meliloti fixA, fixB, fixC,* and *fixX* genes. J Bacteriol 169:1127–1136

Fehr WR, Caviness CE, Burmood DT, Pennington JS (1971) Stages of development descriptions for soybeans, *Glycine max* (L.) Merrill. Crop Sci 11:929–930

Fisher DB (1970a) Kinetics of C-14 translocation in soybean. I. Kinetics in the stem. Plant Physiol 45:107–113

Fisher DB (1970b) Kinetics of C-14 translocation in soybean. I. Kinetics in the leaf. Plant Physiol 45:114–118

Fischer HM, Hennecke H (1987) Direct response of *Bradyrhizobium japonicum nif*A-mediated *nif* gene regulation to cellular oxygen status. Mol Gen Genet 209:621–626

Fortin MG, Morrison NA, Verma DPS (1987) Nodulin-26, a peribacteroid membrane nodulin is expressed independently of the development of the peribacteroid component. Nucleic Acids Res 15:813–824

Franssen HJ, Nap JP, Gloudemans T, Stiekema W, van Dam H, Govers F, Louwerse J, Van Kammen A, Bisseling T (1987) Characterization of cDNA for nodulin-75 of soybean: a gene product involved in early stages of root nodule development. Proc Natl Acad Sci USA 84:4495–4499

Franssen HJ, Nap JP, Bisseling T (1992) Nodulins in root nodule development. In: Stacey G, Burris RH, Evans HJ (eds) Biological nitrogen fixation. Chapman and Hall, New York, pp 598–624

Gallon JR, Chaplin AE (1987) An introduction to nitrogen fixation. Cassel Education Limited, London

Gibson AH, Harper JE (1985) Nitrate effect on nodulation of soybean by *Bradyrhizobium japonicum.* Crop Sci 25:497–501

Gourdon F, Planchon C (1982) Responses of photosynthesis to irradiance and temperature in soybean, *Glycine max* (L.) Merr. Photosynth Res 3:31–43

Govers F, Nap JP, van Kammen A, Bisseling T (1987) Nodulin in the developing root nodule. Plant Physiol Biochem 25:309–322

Groat RG, Vance CP (1981) Root nodule enzymes of ammonia assimilation in alfalfa (*Medicago sativa* L.). Plant Physiol 67:1198–1203

Gupta G, Li Y (1994) Soybean response to carbon dioxide and molybdenum. Commun Soil Sci Plant Anal 25:2571–2581

Halverson LJ, Stacey G (1984) Host recognition in the *Rhizobium* soybean symbiosis: detection of a protein factor in soybean root exudate which is involved in the nodulation process. Plant Physiol 74:84–89

Halverson LJ, Stacey G (1985) Host recognition in the *Rhizobium* soybean symbiosis: evidence for the involvement of lectin in nodulation. Plant Physiol 77:621–625

Harper JE (1987) Nitrogen metabolism. In: Wilcox JR (ed) Soybean: improvement, production, and uses. 2nd edn. #16 Agron Ser. ASA-CSSA-SSSA, Madison, Wisconsin, pp 497–533

Hollis AB, Kloos WE, Elkan GH (1981) DNA:DNA hybridization studies of *B. japonicum* and related Rhizobiaceae. J Gen Microbiol 123:215–222

Huang CY, Boyer JS, Vanderhoef LN (1975) Limitation of acetylene reduction (nirogen fixation) by photosynthesis in soybean having low water potentials. Plant Physiol 56:228–232

Hymowitz T, Newell CA (1981) Taxonomy, speciation, domestication, dissemination, germplasm resources and variation in the genus *Glycine*. In: Summerfield RJ, Bunting AH (eds) Advances in legume science Royal Botanical Garden, Kew, England, pp 251–277

Hymowitz T, Singh RJ (1987) Taxonomy and speciation. In: Wilcox JR (ed) Soybean: improvement, production, and uses. 2nd edn. #16 Agron Ser. ASA-CSSA-SSSA, Madison, Wisconsin, pp 23–48

Johnson RR (1987) Crop management. In: Wilcox JR (ed) Soybean: improvement, production, and uses. ASA-CSSA-SSSA, Madison, Wisconsin, pp 355–390

Kishitani S, Shibles R (1986) Respiration rate of soybean cultivars. Crop Sci 26:580–583

Kondorosi A (1992) Regulation of nodulation genes in rhizobia. In: Verma DPS (ed) Molecular signals in plant-microbe communication. CRC Press, Boca Raton, Florida, pp 325–340

Kosslak RM, Bookland R, Barkei J, Paaren H, Appelbaum ER (1987) Induction of *Bradyrhizobium japonicum* common *nod* genes by isoflavones isolated from *Glycine max*. Proc Natl Acad Sci USA 84:7428–7432

Kouchi H, Nakaji K (1985) Utilization and metabolism of photoassimilated ^{13}C in soybean roots and nodules. Soil Sci Plant Nutr 31:323–334

Larson EM, Hesketh JD, Woolley JT, Peters DB (1981) Seasonal variations in apparent photosynthesis among plant stands of different soybean cultivars. Photosynth Res 2:3–20

Lersten NR, Carlson JB (1987) Vegetative morphology. In: Wilcox JR (ed) Soybean: improvement, production, and uses. 2nd edn. #16 Agron Ser. ASA-CSSA-SSSA, Madison, Wisconsin, pp 49–94

Liu KC, Pappelis AJ, Kaplan HM (1971) Microbodies of soybean cotyledon mesophyll. Trans Ill State Acad Sci 64:136–141

Lynch DH, Smith DL (1993a) Soybean [*Glycine max* (L.) Merr.] nodulation and N_2 fixation as affected by period of exposure to a low root zone temperature. Physiol Plant 88:212–223

Lynch DH, Smith DL (1993b) Early seedling and seasonal N_2-fixing symbiotic activity of two soybean [*Glycine max* (L.) Merr.] cultivars inoculated with *Bradyrhizobium* strains of diverse origin. Plant Soil 157:289–303

Lynch DH, Smith DL (1994) The effect of low root zone temperature stress on two soybean (*Glycine max*) genotypes when combined with *Bradyrhizobium* strains of varying geographic origin. Physiol Plant 90:105–113

Matthews DJ, Hayes P (1982) Effect of root zone temperature on early growth, nodulation and nitrogen fixation in soybeans. J Agric Sci Camb 98:371–376

McCardell A, Sadowsky MJ, Cregan PB (1993) Genetics and improvement of biological nitrogen fixation. In: Metting FB Jr (ed) Soil microbial ecology Marcel Dekker, New York, pp 151–176

McClure PR, Israel DW (1979) Transport of nitrogen in the xylem of soybean plants. Plant Physiol 64:411–416

Miller WJ (1979) Dairy cattle feeding and nutrition. Academic Press, New York

Miller JC, Scott JS, Zary KW, O'Hair SK (1982) The influence of available nitrate levels on nitrogen fixation in three cultivars of cowpea. Agron J 74:14–18

Millor RB, Werner D (1990) Legume nodule biochemistry and function. In: Gresshoff PM (ed) Molecular biology of symbiotic nitrogen Fixation. CRC Press, Boca Raton, Florida, pp 111–130

Mounts TL, Wolf WJ, Martinez WH (1987) Processing and utilization. In: Wilcox JR (ed) Soybean: improvement, production, and uses. 2nd edn. #16 Agron Ser. ASA-CSSA-SSSA, Madison, Wisconsin pp 819–866

Nap JP, Bisseling T (1990) Nodulin function and nodulin regulation in root nodule development. In: Gresshoff PM (ed) Molecular biology of symbiotic nitrogen fixation. CRC Press, Boca Raton, Florida, pp 181–230

Newcomb W, Sippell D, Peterson RL (1979) The early morphogenesis of *Glycine max* and *Pisum sativum* root nodule. Can J Bot 57:2603–2616

Nicholas JC, Harper JE (1985) Nitrate uptake and reduction by soybean seedlings during dark periods. Plant Physiol 77:365–369

Nieuwkoop AJ, Banfalvi Z, Deshmande N, Schell MG, Sirotkin KM, Stacey G (1987) Locus encoding host range is linked to the common nodulation genes of *Bradyrhizobium jpaonicum*. J Bacteriol 169:2631–2638

Oaks A (1979) Nitrate reductase in roots and its regulation. In: Hewitt EJ, Cutting CV (eds) Nitrogen assimilation in plants. Academic Press, New York, pp 217–226

Ohyama T, Kawai S (1983) Nitrogen assimilation and transport in soybean leaves: Investigation by petiole girdling treatment. Soil Sci Plant Nutr 29:227–231

Olson RA (1978) The indispensable role of nitrogen in agricultural production. In: Pratt PF (ed) Natural conference. Management of nitrogen in irrigated agriculture. US National Foundation. USEPA and University of California, Riverside, California, pp 1–31

Olson RA, Kurtz LT (1982) Crop nitrogen requirements, utilization, and fertilization. In: Nitrogen in agricultural soils. Agronomy 22:567–604

Pankhurst CE, Sprent JI (1976) Effects of temperature and oxygen tension on the nitrogenase and respiratory activities of turgid and water-stressed soybean and French bean root nodules. J Exp Bot 27:1–9

Patterson RP, Raper CD, Gross HD (1979) Growth and specific nodule activity of soybean during application and recovery of a leaf moisture stress. Plant Physiol 64:551–556

Pearcy RW (1990) Sunflecks and photosynthesis in plant canopies. Annu Rev Plant Physiol Plant Mol Biol 41:421–453

Peters NK, Verma DPS (1990) Phenolic compounds as regulators of gene expression in plant-microbe interactions. Mol Plant-Microbe Inter 3:4–8

Postgate JR (1982) The fundamentals of nitrogen fixation. Cambridge University Press, New York

Preiss J (1982) Regulation of the biosynthesis and degradation of starch. Annu Rev Plant Physiol 333:431–454

Qiu J, Israel W (1994) Carbohydrate accumulation and utilization in soybean plants in response to altered phosphorus nutrition. Physiol Plant 90:722–728

Raper CD, Kramer PJ (1987) Stress physiology. In: Wilcox JR (ed) Soybean: improvement, production, and uses. 2nd edn. #16 Agron Ser. ASA-CSSA-SSSA, Madison, Wisconsin, pp 589–642

Rice WA, Penney DC, Nyborg M (1977) Effects of soil acidity on rhizobia numbers, nodulation and fixation by alfalfa and red clover. Can J Soil Sci 57:197–203

Sassenrath-Cole GF, Pearcy RW (1994) Regulation of photosynthetic induction state by the magnitude and duration of low light exposure. Plant Physiol 105:1115–1123

Secor J, Shibles R, Stewart CR (1983) Metabolic changes in senescing soybean leaves of similar plant ontogeny. Crop Sci 23:106–110

Sheehy JE, Minchin FR, Witty ST (1985) Control of nitrogen fixation in a legume nodule: an analysis of the role of oxygen diffusion in relation to nodule structure. Ann Bot 55:549–562

Shibles R (1980) Adaptation of soybeans to different seasonal duration. In: Summerfield RJ, Bunting AH (eds) Advances in legume science. Royal Botanic Garden, Kew, England, pp 287–296

Shibles R, Secor J, Ford DM (1987) Carbon assimilation and metabolism. In: Wilcox JR (ed) Soybean: improvement, production, and uses. 2nd edn. #16 Agron Ser. ASA-CSSA-SSSA, Madison, Wisconsin, pp 497–588

Shinano T, Osaki M, Tadano T (1993a) Comparison of production efficiency among field crops related to nitrogen nutrition and application. Plant Soil 155/156:207–210

Shinano T, Komatsu M, Tadano T (1993b) Comparison of production efficiency of the harvesting organs among field crops. I. Growth efficiency of the harvesting organs. Soil Sci Plant Nutr 39:269–280

Shinano T, Osaki M, Tadano T (1994) ^{14}C-allocation of ^{14}C-compounds introduced to a leaf to carbon and nitrogen components in rice and soybean during ripening. Soil Sci Plant Nutr 40:199–209

Smith DL, Hume DJ (1985) Effects of irrigation and fertilizer N on N_2 (C_2H_2) fixation and yield of white bean and soybean. Can J Plant Sci 65:307–316

Smith DL, Hume DJ (1987) Comparison of assay methods for nitrogen fixation utilizing white bean and soybean. Can J Plant Sci 67:11–19

Smith DL, Dijak M, Hume DJ (1988a) Effect of irrigation and fertilizer N on N accumulation and partitioning in white bean and soybean. Can J Plant Sci 68:31–39

Smith DL, Dijak M, Hume DJ (1988b) Effect of water deficit on N_2 (C_2H_2) fixation by white bean and soybean. Can J Plant Sci 67:11–19
Smith KJ, Huyser W (1987) World distribution and significance of soybean. In: Wilcox JR (ed) Soybean: improvement, production, and uses. 2nd edn. #16 Agron Ser. ASA-CSSA-SSSA, Madison, Wisconsin, pp 1–22
Sprent JI, Sprent P (1990) Nitrogen-fixing organisms. Chapman and Hall. New York
Sprent JI (1982) Nitrogen fixation by grain legumes in U.K. Philos Trans R Soc Lond B 296:387–395
Streeter JG (1988) Inhibition of legume nodule formation and N_2 fixation by nitrate. CRC Crit Rev Plant Sci 7:1–23
Streeter JG, Jeffers DL (1979) Distribution of total nonstructural carbohydrates in soybean plants having increased reproduction load. Crop Sci 19:729–734
Tanaka A (1985) The physiology of soybean yield improvement. In: Shanmugasundaram S, Sulzberger EW (eds) Soybean in tropical and subtropical cropping systems. Fortune Printing CO, Taiwan, pp 324–331
Thomas JF, Raper CD, Weeks WW (1981) Day and night temperature effects on nitrogen and soluble carbohydrate allocation during early reproductive growth of soybean. Agron J 73:577–582
Thompson JH (1970) Weather and technology in the production of soybeans in the central United States. Agron J 62:232–236
Thorne JH (1982) Characterization of the active sucrose transport system of immature soybean embryos. Plant Physiol 70:953–958
Tu JC, Ford RE (1984) Plant virus interaction in nitrogen-fixing nodules. Z Pflanzenkr Pflanzenschutz 91:200–212
Turgeon BG, Bauer WD (1982) Early events in the infection of soybean by *Rhizobium jpaonicum*. Time course and cytology of the initial infection process. Can J Bot 60:152–161
Verma DPS (1992) Signals in root nodule organogenesis and endocytosis of *Rhizobium*. Plant Cell 4:373–382
Verma DPS, Ball S, Guérin C, Wanamker L (1979) Leghemoglobin biosynthesis in soybean root nodule. Characterization of the synthesis of the major leghemoglobin. Biochemistry 18:476–483
Verma DPS, Fortin MG, Stanley J, Mauro VP, Purohit S, Morrison N (1986) Nodulins and nodulin genes of *Glycine max*, a perspective. Plant Mol Biol 7:51–61
Walsh KB, Layzell DB (1986) Carbon and nitrogen assimilation and partitioning in soybeans exposed to low root temperatures. Plant Physiol 80:249–255
Wann M, Raper CD (1979) A dynamic model for plant growth: adaptation for vegetative growth of soybean. Crop Sci 19:461–467
Warrington IJ, Peet M, Patterson DT, Bunce J, Haslemore RM, Hellmers H (1977) Growth and physiological responses of soybean under various thermoperiods. Aust J Plant Physiol 4:371–380
Waterer JG, Vessey JK (1993) Effect of low static nitrate concentrations on mineral nitrogen uptake, nodulation, and nitrogen fixation in field pea. J Plant Nutr 16:1775–1789
Weaver RW, Morris DR (1982) Mobilization of labelled N in soybean leaves during early pod filling. J Plant Nutr 5:1099–1108
Weisz PR, Sinclair TR (1988) Soybean nodule gas permeability, nitrogen fixation and diurnal cycles in soil temperature. Plant Soil 109:227–234
Werner D, Mörschel E (1978) Differentiation of nodules of *Glycine max*. Planta 141:169–177
Wiebold WJ, Ashley DA, Boerma HR (1981) Reproductive abscission levels and patterns for eleven determinate soybean cultivars. Agron J 73:43–46
Willing RP, Leopold AC (1983) Cellular expansion at low temperature as a cause of membrane lesions. Plant Physiol 71:118–121
Wittwer SH (1979) Future technological advances in agriculture and their impact on the regulatory environment. BioScience 29:603–610

Zahran HH, Sprent JI (1986) Effects of sodium chloride and polyethylene glycol on root hair infection and nodulation and NH_4NO_3 fertilized *Vicia faba* L. plants by *Rhizobium leguminosarum*. Planta 167:303–309

Zhang F, Smith DL (1994) Effects of low root zone temperatures on the early stages of symbiosis establishment between soybean [*Glycine max* (L.) Merr.] and *Bradyrhizobium japonicum*. J Exp Bot 279:1467–1473

Zhang F, Smith DL (1995) Preincubation of *Bradyrhizobium japonicum* with genistein accelerates nodule development of soybean [*Glycine max* (L.) Merr.] at suboptimal root zone temperatures. Plant Physiol 108:961–968

Zhang F, Smith DL (1996a) Genistein accumulation in soybean [*Glycine max* (L.) Merr.] root systems under suboptimal root zone temperatures. J Exp Bot 47:785–792

Zhang F, Smith DL (1996b) Inoculation of soybean [*Glycine max* (L.) Merr.] with genistein-preincubated *Bradyrhizobium japonicum* or genistein directly applied into soil and soybean protein and dry matter yield under short season conditions. Plant Soil 179:233–241

Zhang F, Smith DL (1997) Preincubation of *Bradyrhizobium japonicum* with genistein accelerates nodulation, and increases seasonal nitrogen fixation activity of early planted soybean [*Glycine max* (L.) Merr.] under short season conditions. Plant Soil 192:141–151

Zhang F, Lynch DH, Smith DL (1995a) Impact of low root zone temperatures in soybean [*Glycine max* (L.) Merr.] on nodulation and nitrogen fixation. Environ Exp Bot 35:279–285

Zhang F, Hamel C, Kianmhr H, Smith DL (1995b) Root-zone temperature and soybean [*Glycine max* (L.) Merr.] vesicular-arbuscular mycorrhizae: development and interactions with the nitrogen fixation symbiosis. Environ Exp Bot 35:287–298

Zhang F, Dashti N, Hynes RK, Smith (1996) Plant growth promoting rhizobacteria and soybean [*Glycine max* (L.) Merr.] nodulation and nitrogen fixation at suboptimal root zone temperatures. Ann Bot 77:453–459

Zhang F, Dashti N, Hynes RK, Smith (1997) Plant growth promoting rhizobacteria and soybean [*Glycine max* (L.) Merr.] growth and physiology at suboptimal root zone temperatures. Ann Bot 79:243–249

Physiological Control of Growth and Yield in White Clover

J.R. CARADUS and M.J.M. HAY[1]

1 Growth Form of White Clover

1.1 Clonal Growth

White clover is a stoloniferous plant of indeterminate size that perennates by forming clonal patches of rooted stolons. As a seedling it consists of a weakly taprooted crown with radiating stolons (Erith 1924). Taproot death occurs within 1 to 2 years. Growth of plants then continues through the production of new nodes at stolon apices and death of nodes at the base (Beinhart et al. 1963). Each node has a leaf, two root primordia and an axillary bud which may remain dormant or develop into either a branch stolon or an inflorescence (Chapman 1983; Sackville Hamilton and Harper 1989). The development of branch stolons is critical for the perennation and spread of the plant. Eventual death of the primary stolon from which a branch is formed will result in the formation of an independent plant of the same genotype as the parent.

The rate at which a stolon apex moves is determined by leaf appearance rate and internode length (Sackville Hamilton 1990). Both leaf appearance rate and internode length are influenced by a number of environmental stimuli such as shading (Simon et al. 1989), defoliation (Mitchell 1956b; King et al. 1978; Grant et al. 1991; Hay and Newton 1996), and temperature (Beinhart 1963; Sackville Hamilton and Harper 1989), and by genetic effects (Caradus and Chapman 1996). Internode length is also influenced by the number, length and proximity of roots, and to a lesser extent, by the number of leaves supporting the elongating internode (Sackville Hamilton 1990). Although the potential for movement of apices through swards is high (Burdon 1983; Turkington and Burdon 1983; Chapman 1983), susceptibility of faster-growing stolons to grazing and treading damage limited elongation rates to 18 cm yr^{-1} or less (Chapman 1983).

Apical dominance has a strong regulating effect on rate of branching from nodes (Chapman 1983; Thomas 1987a). Hormones such as gibberellin depress total stolon numbers but increase the number of nodes per stolon and internode length (Norris 1988). Recently formed nodes support most of the branches

[1] AgResearch Grasslands, Private Bag 11008, Palmerston North, New Zealand

Crop Yield, Physiology and Processes
D. L. Smith, C. Hamel (Eds.)
© Springer-Verlag Berlin Heidelberg 1999

on a plant, such that nodes 1 to 12 and 1 to 25 support 50 and 90%, respectively, of total branches (Hay et al. 1991).

1.2 Exceptions to Clonal Growth

White clover seedlings are rarely observed in mature temperate pastures (Turkington et al. 1979; Chapman 1987). However, in grasslands close to the high-temperature (subtropics) or low-moisture limits of the distribution of the species it persists as an annual or short-lived perennial (Gibson and Hollowell 1966; Sackville Hamilton 1980; Jones 1982; Macfarlane and Sheath 1984; Archer and Robinson 1989). While white clover seeds are small, seedlings develop leaves quickly and have a relatively high photosynthetic rate (Hur and Nelson 1985).

1.3 Plagiotropic Habit

The plagiotropic habit of white clover is a feature of its clonal growth in pastures in that it confers ability for lateral spread and limits the loss of apices and axillary buds by defoliation events (Forde et al. 1988). In white clover plagiotropism reflects a balance between photo- and geotropic responses in that the strength of the geotropic response decreases as light intensity increases (Thomas 1987a). Hence, under high light intensities, stolons are adpressed to the ground and at low intensities stolons may grow upwards. As application of gibberellic acid to plants growing in full light will induce stolons to grow upwards; this suggests that gibberellic acid is implicated in the control of growth responses to darkness (Fletcher and Martin 1962; Mange 1963; Thomas 1987a).

1.4 Flowering

Once formed, an apical meristem is potentially immortal (Sackville Hamilton and Harper 1989) and continues to produce a succession of modular units, referred to here as phytomers. A phytomer is the repeating vegetative unit of growth, differentiated from an apical meristem, composed of a leaf, a node, an axillary bud which may be reproductive or vegetative, two root primordia and an internode (Robin et al. 1994a). When exposed to certain environmental conditions, the axillary buds of forming phytomers, rather than remaining vegetative, are committed to reproductive development or flowering (Thomas 1987b).

Mechanisms controlling flowering are complex, with some genotypes exhibiting a requirement for cold, and others for a defined photoperiod, both short- and long-day types occurring, as well as short-/long-day types (Beatty and Gardner 1961; Britten 1961; Thomas 1961; Norris 1985b). Hormones such as gibberellin do not induce vegetative plants to flower. However, in flowering plants gibberellin increases the number of reproductive buds per stolon and peduncle length (Norris 1988). Larger-leaved types tend to produce more reproductive buds per stolon at temperatures above 18 °C than smaller-leaved types (Norris 1985a).

The optimum temperature for nectar secretion varies with genotype from 10 °C for material of Icelandic origin to 18 °C for a cultivar of Danish origin (Jakobsen and Kristjansson 1994). Additionally, the amount of nectar secreted at optimum temperatures can vary by as much as fourfold (Norris 1985a). The optimum day temperature for nectar secretion was higher when plants were exposed to low night temperatures, presumably a result of decreased night respiration. The major sugars in nectar are sucrose, fructose and glucose, the proportion of which can vary with temperature. Reabsorption of nectar in senescing florets has been observed.

2 Physiological Control of Growth

All physiological responses affect growth by influencing, (1) production of new phytomers (modules) through node appearance rate, apical dominance and branching, (2) size of organs, (3) distribution of dry matter within phytomers, and (4) reproductive growth. The smallest viable unit of growth is a node with a leaf and root attached, and has been referred to as a module (Sackville Hamilton and Harper 1989) phytomer (Robin et al. 1994a), or metamer (Chapman 1989).

The importance of young nodes in the growth and physiology of white clover in grazed swards has been demonstrated. Nodes 1 to 10 account for more than 60% of all nodes in a population (Hay et al. 1991); starch storage is concentrated in stolons between nodes 4 and 10 (Hay et al. 1989b); effective branch initiation ceases at node position 10 (Chapman 1983; Davies 1989; Hay et al. 1991); and most leaves are present on nodes 1 to 4 (Brock et al. 1988). Older nodes (>10) bear well-developed, secondarily thickened roots, referred to as strong adventitious roots (Chapman 1983) or pseudo-taproots (Sackville Hamilton and Harper 1989), and provide a reservoir of viable vegetative axillary buds in the eventuality of younger nodes being removed or damaged (Newton et al. 1992).

Size of organs is influenced by both environment and genetics. In the absence of defoliation, processes that determine differentiation and organ size appear mainly dependent on the light environment and in particular the balance or proportions of red to far-red radiation (Brougham et al. 1978). Defoliation has an overriding effect such that with more frequent defoliation, organ sizes decrease (Hay and Newton 1996) within the genetically defined limits of phenotypic plasticity (Brock 1988).

Young leaves are carbon sinks until they have expanded to between 25 and 45% of their final surface area and reached between 45 and 55% of their final weight (Chapman et al. 1990a). The transition from carbon sink to carbon source coincides with carbon fixation exceeding the leaf's requirement for respiration during the dark. The smooth transition between export and import of carbon is mediated by the stolon from which reserve carbohydrate may be remobilized and utilized (Danckwerts and Gordon 1989).

There is reciprocal exchange of carbohydrate between parent stolon and branches to the extent that up to 41% of the carbon exported from leaves on the parent stolon moves to branches, while branches simultaneously export between 25% (from old source branches) to 54% (for young source branches) of the carbon they assimilate to the parent plant (Chapman et al. 1992a). If plants were either shaded or defoliated just before a young leaf was about to change from being a carbohydrate sink to a source there was no observed effect on the point at which this transition occurred (Chapman et al. 1990b). Thus the leaves exported a similar proportion of the carbohydrate they fixed irrespective of stresses imposed upon the rest of the plant. However, the primary destination of the carbohydrate changed from stolon to stolon apex as imposed stresses increased.

Stolon branches are largely independent of carbohydrate supply from the main stolon after they have developed about three mature leaves of their own (Chapman et al. 1989). After this stage the carbohydrate reserve in the main stolon is maintained by inputs not only from mature leaves on that stolon but also from leaves on branch stolons (Chapman et al. 1989). The pattern of carbohydrate distribution on the main stolon is very similar to that from equivalent positioned leaves on a basally positioned branch stolon (Kemball and Marshall 1995).

The most critical period in the annual seasonal growth cycle of white clover in moist temperate climates occurs in midspring. At this time death and decay rates of older stolon tissue exceeds new stolon formation at the apex, resulting in the rapid loss of stolon biomass, the fragmentation of plants, and a reduction in the overall complexity and size of plants (Hay et al. 1983, 1987; Brock et al. 1988). Although white clover herbage production and growing point density is usually unaffected, many weaker plants die (Hay et al. 1989a). Energy reserves of stored starch are rapidly mobilized to meet the demands of increased leaf, stolon and root growth (Hay et al. 1989b). An associated loss of axillary bud activity also causes reduced branching at this time of the year (Newton et al. 1990, 1992). Total non-structural carbohydrate (TNC) reserves are highest in stolons in autumn, constituting 47 and 37% of stolon dry matter in unshaded and shaded plants, respectively (Lüscher and Nösberger 1988). Irrespective of treatment, TNC decreases in spring to 10% of stolon dry matter. In autumn, TNC are mostly starch but in winter they are made up of both starch and sucrose in similar proportions.

Annually, an average of 54% of nodes have non-viable or dormant buds, 25% have axillary buds that were viable but non-active and 0.1% bear reproductive buds. The remaining 21% of nodes initiated branch growth but only 8.5% of nodes bore live branches (Newton et al. 1992).

In undefoliated plants the death of old stolon material may be associated with the withdrawal of carbon previously supplied to stolon tissue by leaves at the apex. Stolon death was greater when stolons were covered rather than fully illuminated, perhaps as a result of the elimination of direct carbon assimilation by stolon tissue accentuating the carbohydrate deficit (Chapman and Robson 1992).

2.1 Environmental Effects

2.1.1 *Temperature*

Optimum temperatures for growth of the whole shoot are between 22 and 25 °C, when supplied with 16h of light per day (Mitchell and Lucanus 1962). Temperature affects plant leaf area by influencing both the rate at which leaves are produced from already active meristems, and through the activation of meristems to form new branches (Beinhart 1963). Different processes of shoot growth have different temperature optima. For example, Beinhart (1963) demonstrated an inverse relationship between branching frequency and temperature, but a positive curvilinear response for leaf appearance rate. These differences may cancel each other out and mean that white clover grows well in temperatures that range from 18 to 30 °C (Mitchell 1956a; Beinhart 1962, 1963) and some growth can occur over a much wider range, 5 to 40 °C. Increasing temperature from 10 to 20 °C causes an increase in leaf appearance rate, petiole length and internode length but not leaf size and proportion of dry matter allocated to leaf (Arnott and Ryle 1982; Eagles and Othman 1988). At low temperature (8 °C day/4 °C night) growth of plants dependent for nitrogen on nitrogen fixation was lower than that of plants supplied with abundant nitrate (Woledge and Calleja 1983). This was due largely to a large root/shoot ratio and small leaf area of nodulated plants compared with nitrate fed plants.

2.1.2 *Light*

There are four aspects of the light regime that influence plant growth: photoperiod, intensity, spectral quality and total irradiation. Unfortunately, the independent action of each of these variables remains largely undefined for white clover. Many studies confound the effects of photoperiod and total irradiation (Mitchell and Lucanus 1962; Boller and Nösberger 1983; Nösberger et al. 1983). Both light quality and quantity have appreciable effects on white clover growth and yield. In a sward, white clover leaves are formed in shade. Each new lamina is raised to the top of the sward and into full sunlight by petioles of successive leaves growing longer as sward height increases (Dennis and

Woledge 1982). As long as the lamina is no more than stage 0.5 (Carlson 1966) when it reaches the top of the sward, it has photosynthetic capacities as high as those of leaves unshaded throughout their entire development (Dennis and Woledge 1983). When leaf area index becomes greater than 3, leaf emergence rapidly decreases and branching terminates, but elongation at the stolon apex may accelerate (Simon et al. 1989). Doubling day length from 8 to 16h accelerates leaf production and expansion, and increases petiole length and the final area of leaves (Arnott and Ryle 1982).

Canopy-filtered radiation causes a reduction in both photosynthetically active radiation and the red to far-red light ratio, which results in reduced node appearance rate, branching and rooting of nodes, but greatly increases the length of petioles (Solangaarachchi and Harper 1987; Thompson and Harper 1988; Kemball et al. 1992), seedling hypocotyl and cotyledon lengths (Robin et al. 1994a).

The quality of incident light may be sensed by several plant organs. For example, branch appearance rate is affected when the apical meristem alone receives a low ratio of red to far-red light (Robin et al. 1994a). Additionally, reducing light quantity reaching an established basal branch caused axillary branching to cease although stolon elongation was unaffected (Beinhart 1963; Kemball et al. 1992).

Shaded stolons benefit from having connections with unshaded stolons from which they can receive photosynthate. The extra carbon does not reduce etiolation or increase branching, but is used to extend the main axis and thereby allow the stolon apex to move into a more favourable light environment (Robson and Polniaszek 1984). On the other hand, shading a basal branch had little impact on the proportion of assimilate exported to such a branch (Kemball et al. 1992). The rate of photosynthesis of leaves decreased faster and death occurred sooner when leaves were shaded rather than unshaded (Woledge 1986).

Shading has an adverse effect on maintenance of root and nodule tissue (Butler et al. 1959; Chu and Robertson 1974; Kemball et al. 1992). There is a linear relationship between photoperiod duration and total nodule respiration associated with N-fixation during the photoperiod and the following night (dark) period (Kessler et al. 1988; Ryle et al. 1988). Thus, N-fixation potential is closely linked to current photosynthesis during a photoperiod, and can be limited during the night by the quantity of assimilate generated in the previous photoperiod (Ryle 1988). Shading also reduced nodulated root respiration (Ryle et al. 1985b).

Photoperiod extension has a positive effect on leaf size and petiole length but no significant effect on leaf appearance rate, internode length and distribution of dry matter to leaf (Eagles and Othman 1988). Day length can affect floral development, but also interacts with cultivar origin. Cultivars from lower latitudes had a greater spread of response to day length (Norris 1987). Long days increased peduncle length and nectar production.

2.1.3 Nutrients

White clover distribution is limited to relatively fertile soils. In association with grasses, white clover is a poor competitor for phosphorus (P) (Mouat and Walker 1959; Jackman and Mouat 1972a,b), sulphur (S) (Walker and Adams 1958), and nitrogen (N) (Walker et al. 1956). This is due in large part to root morphology, with white clover having thicker roots, shorter root hairs, and a lower frequency of root hairs than grasses (Evans 1977), and apparently in spite of white clover roots being mycorrhizal (Crush 1976).

Phosphorus absorbed through nodal roots moves acropetally through the stolon with the principal sink being the closest components to the supplied root (Chapman and Hay 1993; Kemball and Marshall 1994). Branch stolons at the node of the supplied root are the dominant sink which increases with age and size of the branch (Kemball and Marshall 1994). However, the extent of lateral movement across stolons varies with genotype, and in some genotypes the distribution of nutrients is restricted by the vascular connections (architecture) within stolons (Hay and Sackville Hamilton 1996; Lötscher and Hay 1996a,b). Roots of white clover genotypes may export less than a third of the P absorbed to the shoot within 24 h when P supply is limited, whereas at a higher rate of P supply more than half of the recently assimilated P may be exported (Chapman and Hay 1993). Essentially, translocation patterns of P are consistent with the same source-sink principles that govern carbohydrate translocation in a clonally growing species, although in some genotypes this is constrained by the vascular architecture of the genotype (Lötscher and Hay 1996b).

Increasing P supply increases the concentration of all P fractions in nodules and leaves of white clover (Hart 1989). P concentrations were higher in nodules than leaves. Inorganic P showed the greatest overall increase in concentration compared with other P fractions, although at low levels of supply lipid and residue P appeared to be the main contributors to the rise in total P concentration in the nodules. This coincided with an increase in nitrogen fixation rate.

The greater mobility of P from the inorganic to organic pool may explain the higher yields of white clover at high P levels compared with species, such as *Stylosanthes*, which are more tolerant of low P soils (Chisholm 1982). During periods of high P availability species such as *Stylosanthes* (Hart and Jessop 1982) and pigeon pea (*Cajanus cajan*) (Abu-Gyamfi et al. 1990) that are adapted to low-P soil can accumulate large amounts of inorganic P in their tissues. Subsequently, in times of P deficiency, this stored P is utilized and tissue P concentrations can fall to very low levels. In white clover this fluctuation of inorganic P is less extreme. Phosphorus deficiency increased root/shoot ratio of *Agrostis* but had little effect on that of white clover (Mouat 1983a). The minimum P concentration required for maximum growth was 5 μm P for white clover compared with 3 μm P for *Agrostis*, which may exacerbate the competitive disadvantage of white clover in mixed species swards (Mouat 1983b).

Stolons can absorb P directly but at a much lower rate than roots and do not show enhanced P uptake following P deprivation, as shown by roots (Hay et al. 1986).

Rate of photosynthesis is reduced only at very deficient levels of P supply whereas leaf area and whole plant growth is affected over a wide range of P supply levels. Thus, factors other than photosynthesis per se determine the extent of growth in response to P supply (Hart and Greer 1988).

Mycorrhizal infection of roots of white clover enhances uptake of nutrients such as P and improves plant growth, particularly at deficient levels of P supply (Crush 1976; Powell 1977a,b). However, there seems to be no consistent relationship between proportion of root system infected by mycorrhiza and P inflow or dry matter production, suggesting that VA mycorrhiza do not always function mutualistically (Crush 1978; McGonigle and Fitter 1988).

White clover in the absence of interspecific competition will respond to added N. Uptake of mineral N is a more energy-efficient process than dinitrogen fixation (Ryle et al. 1979) and can result in increased growth rate (Arnott and Ryle 1982). However, with added N in the presence of interspecific competition, clover content in the sward may decrease due to a reduced leaf appearance rate and branching incidence (Barthram et al. 1992).

Potassium supply ameliorates water stress in terms of effects on root length and internode weight (Robin et al. 1992).

2.1.4 Nitrogen Fixation

Temperature has a proportionately smaller effect on N-fixation per unit nodule weight than on unit absorption rate of nitrate (Macduff et al. 1989). Thus the percentage of total N requisition by N-fixation when nitrate is also supplied decreased from 77% at 5 °C to 11% at 25 °C. Marriott et al. (1988) observed that nodule mass per plant and nitrogen fixed per unit nodule dry weight was greater at 8/5 °C than at 20/15 °C day/night temperatures.

Maximum nitrogen fixation rate occurs at 60–80% of field capacity, depending on plant type. In the absence of water stress N-fixation is positively associated with amount of leaf material (Marriott 1988). Large-leaved types require lower soil moisture content for maximum nitrogen fixation (Noor et al. 1989). N-fixation and nodulation are reduced as applied nitrate (Davidson and Robson 1986a) and intensity of defoliation increase (Lee 1993). Low light levels and short photoperiod reduced nitrogenase activity per unit root weight (Kessler et al. 1988). There is a linear relationship between photoperiod duration and total nodule respiration associated with N-fixation during the photoperiod and the following night (dark) period (Ryle et al. 1988). Thus, N-fixation potential is closely linked to current photosynthesis during a photoperiod, and limited by assimilate generated in the previous photoperiod during the night (Ryle 1988). Removal of all leaves and petioles causes an 80% reduction in nitrogen-fixation and nitrogenase-linked respiration after 3h, reaching a 100% reduction by 24h. Recovery to low levels occurred after 3 days, reaching complete recovery after 15 days. Nitrogen fixation may never recover if plants are continuously defoliated (Gordon et al. 1990).

Growth of plants receiving N from nitrate is generally greater than that of plants receiving N from fixation at both low (9 °C) and high (15 °C) temperatures (Glendining and Mytton 1989). Increased reliance on nitrate will result in plants with lower root/shoot ratios (Davidson and Robson 1986b). Source of N, either as nitrate-N or symbiotic-N, has no effect on shoot P concentration (Breeze and Hopper 1987). However, roots of plants dependent on symbiotic-N have higher P uptake than those supplied with nitrate due largely to the need of a cation-charge-balancing anion (Breeze and Hopper 1987) and high organic-P concentrations in nodules (Hart 1989). Total nodulated root respiration is unaffected by supply of nitrate, but analysis of its components showed an 81% decrease in nitrogenase-linked respiration and a 340% increase in growth and maintenance respiration (Minchin et al. 1986).

The specific respiration rates of roots lacking nodules and utilizing nitrate-N were 20–30% greater than the equivalent rates of roots from nodulated plants (Ryle et al. 1983). The contribution of mineral-N to total plant N was 50% more in defoliated than in undefoliated plants (Marriott and Haystead 1992).

2.1.5 Moisture

Density of white clover plants is reduced by low soil moisture levels, particularly when in combination with mean weekly maximum temperatures that exceed 20 °C (Archer and Robinson 1989). In species comparisons, the relatively poor performance of white clover under water-deficit stress has been associated both with a less extensive root system (Burch and Johns 1978; Guobin et al. 1992) and physiological mechanisms such as ineffective stomatal control and an inability to roll leaves to regulate water balance (Burch and Johns 1978). Increasing atmospheric water vapour pressure deficit (VPD) decreased both stomatal conductance and rate of leaf photosynthesis, irrespective of light saturation (Woledge et al. 1989). However, during a typical day the rate of light-saturated photosynthesis would remain stable because any depression caused by increased VPD could be offset by increasing temperatures. Immediate impacts of water-deficit stress are a rapid decrease in dawn water potential and net assimilation rate (Noor et al. 1989).

Waterlogging causes a marked decrease in nitrogenous activity (Pugh et al. 1995). There is increased aerenchyma production in roots and around nodules of waterlogged plants. Infected cells of submerged nodules were larger and had larger vacuoles than those from nodules of normally watered plants. When anaerobiosis is severe due to waterlogging there may be a depletion of available nitrate through denitrification and an increase in the levels of manganese causing toxicity (Adams and Akhtar 1994). A component of anaerobic energy production in white clover, alcohol dehydrogenase (ADH) activity, increases markedly in response to flooding (Chan and Burton 1992). This higher ADH activity during flooding may be associated with anaerobic conditions.

2.1.6 Physical Impedance

Growth of stolon apices is often physically impeded by the presence of other plants, uneven soil surface and soil (if the stolon apex is buried). White clover shows characteristic responses to physical impedance of, in particular, a decrease in stolon elongation rate and an increase in the radial growth of stolons (Newton 1986). Such responses have been termed thigmomorphogenetic (Jaffe 1973). In addition, obstruction of the primary stolon apex induced a striking increase in the frequency of branching and an increase in the dry weight of branches (Newton 1986). It has been concluded that ethylene is the active agent initiating thigmomorphogenetic responses (Biro et al. 1980).

2.1.7 Defoliation

As a forage legume, white clover is well adapted to withstand the vicissitudes of defoliation, whether it be grazing or cutting. For the majority of the year, except in summer when the soil surface is dry, stolons are firmly rooted close to the soil surface, or, as often occurs in wet winters, they are buried (Hay et al. 1987). Thus, stolons are located below the grazing height of animals, allowing regrowth to occur from both main and branch apical meristems. The proportion of white clover leaves lower in the sward profile increases as cutting frequency increases (Winkler and Nösberger 1985). Leaf appearance rate appears to be little affected by grazing intensity measured as continuous grazing to different sward heights, although with infrequent grazing in taller swards white clover leaves take longer to complete their expansion (Parsons et al. 1991). Severe defoliation resulting in less than two leaves remaining per stolon, which may occur in continuously grazed sheep pastures, causes a significant reduction in axillary branch production and an increase in proportion of dead stolon (Jones and Davies 1988). There is a concomitant reduction in leaf size, petiole length, internode length, stolon diameter (Hay and Newton 1996) and increase in proportion of stolon to total shoot dry matter (Ryle et al. 1989). Defoliation causes a reduction in nitrogen fixation rate, photosynthesis and respiration (Ryle et al. 1985a,b, 1986).

The immediate response of white clover to defoliation is an increase in specific leaf area (i.e., leaf area per dry weight of leaf) of new leaves, so that the assimilation rate (net photosynthesis × area) of these leaves remains unchanged (Chapman et al. 1990b). Any leaves remaining after defoliation export more of their fixed carbon to support stolon apical growth (Chapman and Robson 1988) and branch stolon growth (Chapman et al. 1992b). Carbon fluxes, in all combinations, between parent stolon and branches increase with defoliation of the sink at the expense of carbon supply to stolon tissue or roots of the source module (Chapman et al. 1992a). Only after continued severe defoliation was reduced apical support from old leaves apparent and apical dominance increased so as to delay branch formation (Hay and Newton 1996). These responses should largely buffer the effects of different defoliation frequencies and severities that may occur, for example, with continuous (set-stocked) or rotational grazing, so that overall growth is maintained. However, in reality, this is not the case because defoliation does not occur in isolation from other effects such as interspecific competition and climate. Under rotational grazing, white clover populations are less stable and are influenced by a wider, more complex range of factors where plant competition and physiological processes of regrowth are important as compared to set stocking where populations are more stable, and are influenced more directly by climate (principally soil temperature) than by physiological processes of growth recovery following defoliation (Brock and Hay 1993).

Rotational grazing by sheep results in a higher percentage of nodes with roots compared with plants grazed continuously by set-stocking (Newton and Hay 1994). Root presence is highly correlated with the establishment and survival of branches, but not with branch initiation or the viability of axillary buds. This association between nodal roots and branches is strongest when resources are limited. With frequent defoliation, a modified balance between death of old roots and regeneration of new roots is established (Butler et al. 1959; Evans 1973; Chu and Robertson 1974). As older roots have greater potential to be larger, frequent defoliation can reduce the root weight of field-grown white clover, although it may have little effect on perennial ryegrass root weight (Whitehead 1983). After prolonged regrowth, defoliation causes extensive root and nodule tissue death (Butler and Bathurst 1956; Butler et al. 1959), and reduction in nitrogen fixation (Moustafa et al. 1969).

2.1.8
Biotic Influences

When white clover is grown at high plant density the development of individual plants is reduced as node appearance rate, branching frequency and stolon elongation rate decrease. Solangaarachchi and Harper (1989) suggest that intraspecific effects by plants of the same or differing genotypes suppress the development of a plant via strongly localized control of axillary bud development mediated by alteration of the red to far-red ratio of radiation within the canopy (Solangaarachchi and Harper 1987).

Interspecific effects of an associated grass species on white clover development vary with species of grass. White clover grows well with *Lolium perenne* L. as the neighbouring grass species (Kershaw 1959; Harris and Brougham

1968; Harris 1973; Turkington and Harper 1979a). The compatibility of a grass species with white clover increases when it colonizes deeper soils (Kershaw 1959), has an asynchronous growth rhythm (Turkington and Harper 1979b; Harris and Hoglund 1980), has an open rather than dense or mat-like habit (Jackman and Mouat 1972b; Schmid and Harper 1985) and a high red to far-red ratio of light transmitted or reflected by its foliage (Thompson and Harper 1988).

Symbioses formed with *Rhizobium trifolii* (review Crush 1987) and vesicular-arbuscular mycorrhizes (Dunlop and Hart 1987) improve growth respectively when there is a low soil supply of nitrogen or other nutrients (e.g., phosphorus, sulphur). Negative influences on white clover growth result from acute pathogenic attack by fungal, bacterial, viral, mycoplasma-like organisms (review Latch and Skipp 1987) and/or nematodes (Skipp and Gaynor 1987), as well as from damage through pest attack (Gaynor and Skipp 1987).

2.1.9 Ozone and Ultraviolet Light

Ozone exposure, at a rate of 60 ppb, causes foliar injury and premature senescence, reductions in root and shoot growth, and decreases in foliar chlorophyll concentration (Ommen et al. 1994) and nitrogen fixation (Montes et al. 1983). Ozone-susceptible genotypes of white clover have higher leaf conductance and higher rates of transpiration and net carbon assimilation than leaves of ozone-resistant genotypes when exposed to ozone (100 ppb), although there was no difference in water-use efficiency (Fuhrer et al. 1993).

2.1.10 Carbon Dioxide Concentration

Carbon dioxide (CO_2) concentrations in the atmosphere have doubled in the last 100 years and projections are that concentrations will continue to increase. In response to this, several studies have measured the impact of elevated CO_2 concentrations on white clover growth and physiology.

Elevated CO_2 concentration (twice ambient) had no effect on either number or size of plant organs (Scheidegger and Nösberger 1984) or leaf appearance rate of white clover, and did not change the relationship between growth and senescence per meristem (Clark et al. 1995). Absolute, but not specific, nitrogen-fixation activity increases due to elevated CO_2 concentration due to increased nodule mass attributed to an increase in nodule number (Murphy 1986). Therefore, excess carbon supply is used solely to promote further nodulation.

In ryegrass/white clover swards exposure to elevated CO_2 concentrations should result in an increase in white clover content (Newton et al. 1994, 1995, 1996). Comparison of canopy level with individual leaf functioning showed that elevated CO_2 concentration (600 ppm) caused greater photosynthetic stimulation at the canopy level as a consequence of accelerated development of foliage area, and this resulted in increased water consumption (by 16%) despite strong stomatal closure (Nijs et al. 1989a). Thus, while internal CO_2 concentration in the leaf (Ci) increased with rising atmospheric CO_2 concentration (Ca), Ci/Ca decreased. Higher canopy CO_2 loss at night and faster light extinction in the longer high CO_2 canopy prevented further dry matter production increase. Yet plants grown at low CO_2 partial pressures have a higher affinity for CO_2 than those grown at high CO_2 partial pressures (Lehnherr et al. 1985).

Elevated CO_2 concentration can significantly increase water-use efficiency, favouring the survival of white clover when exposed to severe, rapidly developing drought stress (Nijs et al. 1989b). While at ambient CO_2 growth stops in response to moisture stress, at elevated CO_2 growth can continue with fixed carbon primarily allocated below ground (Newton et al. 1996).

2.2 Correlative Effects

Correlative effects refer to those factors within the plant that affect the processes determining growth and yield. The current status of a plant reflects previous environments and determines the ability the plant has to respond to environmental change, e.g., the number and location of viable axillary buds within a plant at a point in time is a function of past environments and these factors can influence the branching response to subsequent environmental change (Newton and Hay 1995).

2.2.1 Age of Phytomer

The position of a phytomer relative to the apical bud (age), because of ontogenetic effects, influences the potential of any given phytomer to respond to environmental change. An adequate supply of resources to the apical bud determines the upper limits of subsequent growth potential of the organs of phytomers developing within the apical bud (Harvey 1979; Thomas 1987a; Hay and Newton 1996). Subsequently, during the growth stages of an organ's developmental pathway the microenvironment at a phytomer can modify growth of various organs within the limit of the previously set potential. It is pertinent to note that organs differ in their ontogeny (Thomas 1987a). For instance, the second phytomer proximal to the apical bud has lamina area of mature size, although petiole, root, internode and axillary bud development may continue to be responsive to the local environment. However, a phytomer more than three from the apical bud only has the option of varying root and axillary bud development along with secondary thickening of the internode as developmental possibilities.

Perhaps apical dominance is the characteristic most commonly associated with correlative effects. In white clover apical dominance is seen as an inhibiting of outgrowth of axillary buds near the apical bud so that branching occurs at older nodes (Thomas 1987a). The number of nodes proximal to the apical bud at which outgrowth of axillary buds is inhibited increases as the extent of apical dominance increases. The extent of apical dominance varies with genotype (Erith 1924), shading (Harvey 1979; Caradus and Chapman 1991), nutrient supply (Harvey 1979; Thomas 1987a; Hay 1994) and season (Thomas 1987a).

Increasing phytomer position relative to the apical bud (age) decreases the viability of both root primordia (Newton and Hay 1994) and axillary buds (Newton et al. 1992) such that viability is respectively <1 and <4% at phytomer positions >20. Successful branch development from phytomers with position >10 is relatively rare in the field (Chapman 1983; Davies and Evans 1990b; Hay et al. 1991).

2.2.2
Presence of Root, Leaf, Flower or Branch at a Phytomer

The survival, development and productivity of a branch have consistently been found to be enhanced by the presence of a root at the phytomer of origin of the branch (Knight 1953; Trautner and Gibson 1966; Chapman 1983; Newton and Hay 1994). However, axillary bud viability and initiation of outgrowth of an axillary bud are only weakly correlated with root presence at the phytomer (Newton and Hay 1994). There is a close physiological relationship between the root and branch of a phytomer involving the supply of nutrients (Hoshino 1974; Chapman and Hay 1993; Kemball and Marshall 1994; Hay and Sackville Hamilton 1996; Lötscher and Hay 1996b) and water from the root and translocation of photosynthates from the branch to the root (Chapman et al. 1992a,b). As up to 84% of carbon translocated to a root may come from the branch at the same phytomer (Chapman et al. 1992c), the survival and growth of a nodal root benefits from the presence of a branch at the phytomer (Newton and Hay 1994).

Leaf/petiole presence inhibits development of the axillary bud at the phytomer (Carlson 1966; Newton and Hay 1992) although under shaded conditions leaf presence stimulated bud activity (Davies and Evans 1990a). Low R:FR ratios of incident light at a leaf can delay branch development (Robin et al. 1994b) and reduce leaf appearance rates of the branch originating from the same phytomer (M.J.M. Hay, P.C.D. Newton, A. Cresswell, C. Robin, unpubl. data).

An increased proportion of vegetative axillary buds of phytomers adjacent to a phytomer with a flower initiate branch development (N.R. Sackville-Hamilton, 1982, pers. comm.); this may reflect earlier initiation of axillary bud primordia with the apical bud when conditions are conducive to flowering (Thomas 1962). However, as young branch survival is variable, differences in the frequency of branches at these phytomers have as yet not been noted. The presence of a flower can increase the length of the proximal internode and decrease the length of the distal internode (de Booysen and Laude 1964).

2.2.3
Genotype

There is considerable genetic variation in white clover. Cultivars are classified by leaf size and there is a positive association between leaf size and yield, although this can be masked by defoliation regime (Caradus and Mackay 1991). As a result, it is usual for small-leaved types with high stolon growing-point densities to predominate where defoliation is severe and for large-leaved types with low stolon growing-point densities to predominate where defoliation is lax (Brock 1988). Large-leaved cultivars may have a greater reliance on stored carbohydrate for regrowth than smaller-leaved types (Kang and Brink 1995).

Genetic correlations between plant characters may influence growth of both the whole plant and plant components. For example, stolon elongation rate has been shown to have an inverse association with leaf size (Kang and Brink 1995). Alternatively, there is a positive association between stolon diameter and leaf size. A direct relationship between root and shoot type is evident, with large-leaved types being more taprooted than small-leaved types (Caradus 1977; Pederson 1989).

Genotype may affect P nutrition since storage of P in stolons and roots was of greater importance in small-leaved than in large-leaved plants (Chapman and Hay 1993). Intraspecific variation for P response (Caradus et al. 1992), P uptake and N accumulation has been described (Mackay et al. 1990).

Genotypes with a high degree of stolon branching have a higher root mass than those with less stolon branching, leading to a potential improvement in competitive ability and population stability in mixed swards (Chapman et al. 1989).

Photosynthetic properties may differ between genotypes. Plants from high-altitude sites can have higher photosynthetic rates, especially when measured at low temperature, than plants from lower altitudes (Mächler and Nöberger 1977).

References

Abu-Gyamfi J, Fujita K, Ogata S (1990) Phosphorus fractions in relation to growth in pigeon pea [*Cajanus cajan* (L) Millsp.] at various levels of P supply. Soil Sci Plant Nutr 36:531–543

Adams WA, Akhtar N (1994) The possible consequences for herbage growth of waterlogging compacted pasture soils. Plant Soil 162:1–17

Archer KA, Robinson GG (1989) The role of stolons and seedlings in the persistence and production of white clover (*Trifolium repens* L. cv. Huia) in temperate pastures on the Northern Tablelands, New South Wales. Aust J Agric Res 40:605–616

Arnott RA, Ryle GJA (1982) Leaf surface expansion on the main areas of white and red clovers. Grass Forage Sci 37:227–233

Barthram GT, Grant SA, Elston DA (1992) The effects of sward height and nitrogen fertiliser application on changes in sward composition, white clover growth and the stock carrying capacity of an upland perennial ryegrass/white clover sward grazd by sheep for four years. Grass Forage Sci 47:326–341
Beatty DW, Gardner FP (1961) Effect of photoperiod and temperature on the flowering of white clover, *Trifolium repens* L. Crop Sci 1:323–326
Beinhart G (1962) Effects of temperture and light intensity on carbon dioxide uptake, respiration and growth of white clover. Plant Physiol 37:709–715
Beinhart G (1963) Effects of environment on meristematic development, leaf area, and growth of white clover. Crop Sci 3:209–213
Beinhart G, Gibson PB, Halpin JE, Hollowell EA (1963) Selection and evaluation of white clover clones. III. Clonal differences in relation to leaf area production and persistence. Crop Sci 3:89–92
Biro RL, Hart ER, Erner Y, Jaffe MJ (1980) Thigmomorphogenesis: changes in cell division and elongation in the internodes of mechanically perturbed or ethrel-treated bean plants. Ann Bot 45:555–664.
Boller BC, Nösberger J (1983) Effects of temperature and photoperiod on stolon characteristics, dry matter partitioning, and nonstructural carbohydrate concentration of two white clover ecotypes. Crop Sci 23:1057–1062
Breeze VG, Hopper MJ (1987) The uptake of phosphate by plants from flowering nutrient solution. J Exp Bot 189:618–630
Britten EJ (1961) The influence of genotype and temperature on flowering in *Trifolium repens*. Agron J 53:11–14
Brock JL (1988) Evaluation of New Zealand-bred white clover cultivars under rotational grazing and set stocking with sheep. Proc NZ Grassl Assoc 49:203–206
Brock JL, Hay RJM (1993) An ecological approach to forage management. Proc XVII Int Grassl Congr Massey University, Palmerston North, New Zealand, pp 837–842
Brock JL, Hay MJM, Thomas VJ, Sedcole JR (1988) Morphology of white clover (*Trifolium repens* L.) plants in pastures under intensive sheep grazing. J Agric Sci 111:273–283
Brougham RW, Ball PR, Williams WM (1978) The ecology and management of white clover-based pastures. In: Wilson JR (ed) Plant relations in pastures. CSIRO, East Melbourne, pp 309–324
Burch GJ, Johns GG (1978) Root absorption of water and physiological responses to water deficits by *Festuca arundinacea* Schreb and *Trifolium repens* L. Aust J Plant Physiol 5:859–871
Burdon JJ (1983) Biological flora of the British Isles. J Ecol 71:307–330
Butler GW, Bathhurst NO (1956) The underground transference of nitrogen from clover to associated grass. Proc 7th Int Grassl Congr, Massey University, Palmerston North, New Zealand, pp 168–178
Butler GW, Greenwood RM, Soper K (1959) Effects of shading and defoliation on the turnover of root and nodule tissue of plants of *Trifolium repens, Trifolium pratense* and *Lotus uliginosus*. NZ J Agric Res 2:415–426
Caradus JR (1977) Structural variation of white clover root systems. NZ J Agric Res 20:213–219
Caradus JR, Chapman DF (1991) Variability of stolon characteristics and response to shading in two cultivars of white clover (*Trifolium repens* L.). NZ J Agric Res 34:239–247
Caradus JR, Chapman DF (1996) Selection for heritability of stolon characteristics in two cultivars of white clover. Crop Sci 36:900–904
Caradus JR, Mackay AC (1991) Performance of white clover cultivars and breeding lines in a mixed species sward. 2. Plant characters contributing to differences in clover proportion in swards. NZ J Agric Res 34:155–160
Caradus JR, Mackay AD, Wewala S, Dunlop J, Hart AL, van den Bosch J, Lambert MG, Hay MJM (1992) Inheritance of phosphorus response in white clover. Plant Soil 146:199–208
Carlson GE (1966) Growth of clover leaves – developmental morphology and parameters at ten stages. Crop Sci 6:293–294

Chan JWY, Burton RS (1992) Variation in alcohol dehydrogenase activity and flood tolerance in white clover, *Trifolium repens*. Evolution 46:721–734

Chapman DF (1983) Growth and demography of *Trifolium repens* stolons in grazed hill pastures. J Appl Ecol 20:597–608

Chapman DF (1987) Natural re-seeding and *Trifolium repens* demography in grazed hill pastures II. Seedling appearance and survival. J Appl Ecol 24:1037–1043

Chapman DF (1989) The assimilation and distribution of carbon in white clover (*Trifolium repens* L.). PhD Thesis, University of Reading, Reading, UK

Chapman DF, Hay MJM (1993) Translocation of phosphorus from nodal roots in two contrasting genotypes of white clover (*Trifolium repens* L.). Physiol Plant 89:323–330

Chapman DF, Robson MJ (1988) Effects of defoliation on carbon assimilation and partitioning in white clover stolons Proc 12th General Meeting European Grassl Fed, Dublin, Ireland, 4–7 July 1988, pp 277–281

Chapman DF, Robson MJ (1992) The physiological role of old stolon material in white clover (*Trifolium repens* L.). New Phytol 122:53–62

Chapman DF, Robson MJ, Snaydon RW (1989) Interactions between parent stolons and branches in assimilate partitioning in white clover (*Trifolium repens* L.). Proc XV1 Int Grassl Congr 4–11 Oct 1989, Nice, France, pp 1047–1048

Chapman DF, Robson MJ, Snaydon RW (1990a) The carbon economy of developing leaves of white clover (*Trifolium repens* L.). Ann Bot 66:623–628

Chapman DR, Robson MJ, Snaydon RW (1990b) Short-term effects of manipulating the source: sink ratio of white clover (*Trifolium repens* L.) plants on export of carbon from, and morphology of, developing leaves, Physiol Plant 80:262–266

Chapman DF, Robson MJ, Snaydon RW (1992a) Physiological integration in the clonal perennial herb *Trifolium repens* L. Oecologia 89:338–347

Chapman DF, Robson MJ, Snaydon RW (1992b) Interactions between defoliation and the nitrogen nutrition of white clover (*Trifolium repens* L.): effects on carbon utilisation in clonal plants. Plant Soil 139:159–165

Chapman DF, Robson MJ, Snaydon RW (1992c) The carbon economy of clonal plants of *Trifolum repens* L. J Exp Bot 43:427–434

Chisholm RH (1982) Phosphorus utilisation in contrasting pasture species. J Aust Inst Agric Sci 48:90

Chu ACP, Robertson AG (1974) The effects of shading and defoliation on nodulation and nitrogen fixation by white clover. Plant Soil 41:509–519

Clark H, Newton PCD, Bell CC, Glasgow EM (1995) The influence of elevated $C0_2$ and simulated seasonal changes in temperature on tissue turnover in pasture turves dominated by perennial ryegrass (*Lolium perenne*) and white clover (*Trifolium repens* L.). J Appl Ecol 32:128–136

Crush JR (1976) Endomycorrhizas and legume growth in some soils of the Mackenzie Basin, Canterbury, New Zealand. NZ J Agric Res 19:473–476

Crush JR (1978) Changes in effectiveness of soil endomycorrhizal fungal populations during pasture development. NZ J Agric Res 21:683–685

Crush JR (1987) Nitrogen fixation. In: Baker MJ, Williams WM (eds) White clover. CAB International, Wallingford, UK, pp 185–201

Danckwerts JE, Gordon AJ (1989) Long-term partitioning, storage and remobilisation of ^{14}C assimilated by *Trifolium repens* (cv. Blanca). Ann Bot 64:533–544

Davidson IA, Robson MJ (1986a) Effect of contrasting patterns of nitrate application on the nitrate uptake, N_2-fixation, nodulation and growth of white clover. Ann Bot 57:331–338

Davidson IA, Robson MJ (1986b) Effect of temperature and nitrogen supply on the growth of perennial ryegrass and white clover. 2. A comparison of monocultures and mixed swards. Ann Bot 57:709–719

Davies A (1989) The structure of the grass/clover sward and its implications in sward management. Proc XVI Int Grassl Congr, Nice, France, pp 1065–1066

Davies A, Evans ME (1990a) Axillary bud development in white clover in relation to defoliation and shading treatments. Ann Bot 66:349–357
Davies A, Evans ME (1990b) Effects of spring defoliation and fertiliser nitrogen on the growth of white clover in ryegrass/clover swards. Grass Forage Sci 45:345–356
de Booysen PV, Laude HM (1964) Influence of flower initiation and development on internode growth in the ladino clover stolon. Crop Sci 4:520–524
Dennis WD, Woledge J (1982) Photosynthesis by white clover leaves in mixed clover-ryegrass swards. Ann Bot 49:627–635
Dennis WD, Woledge J (1983) The effect of shade during leaf expansion on photosynthesis by white clover leaves. Ann Bot 51:111–118
Dunlop J, Hart AL (1987) Mineral nutrition. In: Baker MJ, Williams WM (eds) White clover. CAB International, Wallingford, UK, pp 153–183
Eagles CF, Othman OB (1988) Variation in growth of overwintered stolons of contrasting white clover populations in response to temperature, photoperiod and spring environment. Ann Appl Biol 112:563–574
Erith AG (1924) White clover (*Trifolium repens* L.). A monograph. Duckworth, London
Evans PS (1973) The effect of the repeated defoliation to three different levels on root growth of five pasture species. NZ J Agric Res 16:13–34
Evans PS (1977) Comparative root morphology of some pasture grasses and clovers. NZ J Agric Res 20:331–335
Fletcher WW, Martin DJ (1962) The effects of gibberellic acid on the germination, growth and development of *Trifolium repens* L. J Agric Sci 58:235–241
Forde MB, Hay MJM, Brock JL (1988) Development and growth characteristics of temperate perennial legumes. In: Marten GC, Matches AG, Barnes RF, Brougham RW, Clements RJ, Sheath GW (eds) Persistence of Forage Legumes. Proc Trilateral Workshop, Honolulu, Hawaii, 18–22 July 1988, pp 91–110
Fuhrer J, Parker R, Shariat-Madari H (1993) Growth and gas exchange characteristics of two clones of white clover *Trifolium repens* L differing in ozone sensitivity. Angew Bot 67:163–167
Gaynor DL, Skipp RA (1987) Diseases. In: Baker MJ, Williams WM (eds) White Clover. CAB International, Wallingford, UK, pp 421–460
Gibson PB, Hollowell EA (1966) White clover. Agriculture Handbook, Agric Res Service, USDA 314, 33 pp
Glendining MJ, Mytton LR (1989) The response of white clover (*Trifolium repens* L.) seedlings to spring root temperatures: the relative roles of the plant and the *Rhizobium* bacteria. Plant Soil 113:147–154
Gordon AJ, Kessler W, Minchin FR (1990) Defoliation-induced stress in nodules of white clover. I. Changes in physiological parameters and protein synthesis J Exp Bot 41:1245–1253
Grant SA, Torrell L, Sim EM, Small J (1991) The effect of stolon burial and defoliation early in the growing season on white clover performance. Grass For age Sci 46:173–182
Guobin L, Kemp DR, Lin GB (1992) Water stress affects the productivity, growth components, competitiveness and water relations of phalaris and white clover growing in a mixed pasture. Aust J Agric Res 43:654–672
Harris W (1973) Ryegrass genotype-environment interactions in response to density, cutting height, and competition with white clover. NZ J Agric Res 16:207–222
Harris W, Brougham RW (1968) Some factors affecting change in botanical composition in a ryegrass-white clover pasture under continuous grazing. NZ J Agric Res 11:15–38
Harris W, Hoglund JH (1980) Influence of seasonal growth periodicity and N-fixation on competitive combining abilities of grasses and legumes. Proc 13th Int Grassl Congr, pp 239–243
Hart AL (1989) Nodule phosphorus and nodule activity in white clover. NZ J Agric Res 32:145–149
Hart AL, Greer DH (1988) Photosynthesis and carbon export in white clover plants grown at various levels of phosphorus supply. Physiol Plant 73:46–51
Hart AL, Jessop D (1982) Concentrations of total, inorganic, and lipid phosphorus in leaves of white clover and Stylosanthes. NZ J Agric Res 25:69–76

Harvey HJ (1979) The regulation of vegetative reproduction. PhD Thesis, University of Wales, UK
Hay MJM (1994) Autecology of white clover (*Trifolium repens* L.) with special reference to the effect of stolon burial on branch formation. PhD Thesis, Massey University, Palmerston North, New Zealand
Hay MJM, Newton PCD (1996) Effect of severity of defoliation on the viability of reproductive and vegetative axillary buds of *Trifolium repens* L. Ann Bot 78:117–123
Hay MJM, Sackville Hamilton NR (1996) Influence of xylem vascular architecture on the translocation of phosphorus from nodal roots in a genotype of *Trifolium repens* during undisturbed growth. New Phytol 132:575–582
Hay MJM, Brock JL, Fletcher RH (1983) Effect of sheep grazing management on distribution of white clover stolons among three horizontal strata in ryegrass/white clover swards. NZ J Exp Agric 11:215–218
Hay MJM, Saker LR, Drew MC (1986) Phosphate uptake and transport by roots and stolons of intact white clover (*Trifolium repens* L.). J Exp Bot 37:577–588
Hay MJM, Chapman DF, Hay RJM, Pennell CGL, Woods PW, Fletcher RH (1987) Seasonal variation in the vertical distribution of white clover stolons in grazed swards. NZ J Agric Res 30:1–8
Hay MJM, Brock JL, Thomas VJ (1989a) Density of *Trifolium repens* plants in mixed swards under intensive grazing by sheep. J Agric Sci 113:81–86
Hay MJM, Chu ACP, Knighton MV, Wewala S (1989b) Variation with season and node position in carbohydrate content of white clover stolons. Proc 16th Int Grassl Congr, Nice, France, pp 1059–1060
Hay MJM, Newton PCD, Thomas VJ (1991) Nodal structure and branching of *Trifolium repens* in pastures under intensive grazing by sheep. J Agric Sci 116:221–228
Hoshino M (1974) Translocation and accumulation of assimilates and phosphorus in ladino clover. Bull Natl Grassl Res Inst 5:35–84
Hur SN, Nelson CJ (1985) Cotyledon and leaf development associated with seedling vigour of six forage legumes. Proc XV Int Grassl Congr, August 24–31, 1985, Kyoto, Japan, pp 374–376
Jackman RH, Mouat MCH (1972a) Competition between grass and clover for phosphate. I. Effect of browntop (*Agrostis tenuis* Sibth.) on white clover (*Trifolium repens* L.) growth and nitrogen fixation. NZ J Agric Res 15:653–666
Jackman RH, Mouat MCH (1972b) Competition between grass and clover for phosphate. II. Effect of root activity, efficiency of response to phosphate, and soil moisture. NZ J Agric Res 15:667–675
Jaffe M (1973) Thigmomorphogenesis: The response of plant growth and development to mechanical stimulation with special reference to *Bryonia dioica*. Planta 114:143–157
Jakobsen HB, Kristjansson K (1994) Influence of temperature and floret age on nectar secretion in *Trifolium repens* L. Ann Bot 74:327–334
Jones DR, Davies A (1988) The effects of simulated continuous grazing on development and senescence of white clover Grass Forage Sci 43:421–425
Jones RM (1982) White clover (*Trifolium repens*) in subtropical southeast Queensland. I. Some effects of site, season and management practices on the population dynamics of white clover. Trop Grassl 16:118–127
Kang JH, Brink GE (1995) White clover morphology and physiology in response to defoliation interval. Crop Sci 35:264–269
Kemball WD, Marshall C (1994) The significance of nodal rooting in *Trifolium repens* L.: 32 P distribution and local growth responses. New Phytol 127:83–91
Kemball WD, Marshall C (1995) Clonal integration between parent and branch stolons in white clover: a developmental study. New Phytol 129:513–521
Kemball WD, Plamer MJ, Marshall C (1992) The effects of local shading and darkening on branch growth, development and survival in *Trifolium repens* and *Galium aparine*. Oikos 62:366–375
Kershaw KA (1959) An investigation of the structure of a grassland community. II. The pattern of *Dactylis glomerata*, *Lolium perenne* and *Trifolium repens*. III. Discussion and conclusions. J Ecol 47:31–53

Kessler W, Boller BC, Nösberger J (1988) Einfluss des Lichtangebotes auf das Wachstum und die biologische Stickstoff-Fixierung von Weissklee (*Trifolium repens* L.). J Agron Crop Sci 160:250–259

King J, Lamb WIC, McGregor MT (1978) Effect of partial and complete defoliation on regrowth of white clover plants. J Br Grassl Soc 33:49–55

Knight WE (1953) Interrelationships of some morphological and physiological characteristics of ladino clover. Agron J 45:197–199

Latch G, Skipp R (1987) Diseases. In: Baker MJ, Williams WM (eds) White clover. CAB International, Wallingford, UK, pp 421–460

Lee HW (1993) Studies on the nitrogen fixation of legumes. I. Effects of nitrogen fertilisation level on nitrogen fixation and growth of Ladino clover. J Korean Soc Grassl Sci 13:16–22

Lehnherr B, Machler F, Nösberger J (1985) Effect of $C0_2$ concentrating mechanism in white clover as predicted from differential $^{14}C0_2/^{12}C0_2$ uptake. J Exp Bot 36:1835–1841

Lötscher M, Hay MJM (1996a) Distribution of phosphorus and calcium from nodal roots of *Trifolium repens* L.: the relative importance of transport via xylem or phloem. New Phytol 133:445–452

Lötscher M, Hay MJM (1996b) Distribution of mineral nutrient from nodal roots of *Trifolium repens*: genotypic variation in intra-plant allocation of ^{32}P and ^{45}Ca. Physiol Plant 97:269–276

Lüscher A, Nösberger J (1988) Überwinterung von Weissklee - Veränderungen der Kohlenhydratreserven und der Wachstumsparameter. Mitt Ges Pflanzenb 1:91–92

Macduff JH, Gordon AJ, Ryle GJA, Powell CE (1989) White clover N_2-fixation in response to root temperature and nitrate. I. Growth and uptake of nitrate from flowering nutrient solutions. J Exp Bot 40:517–526

Macfarlane MJ, Sheath GW (1984) Clover – what types for dry hill country? Proc NZ Grassl Assoc 45:140–150

Mächler F, Nösberger J (1977) Effect of light intensity and temperature on apparent photosynthesis of altitudinal ecotypes of *Trifolium repens* L. Oecologia 31:73–78

Mackay AD, Caradus JR, Hart AL, Wewala GS, Dunlop J, Lambert MG, van den Bosch J, Mouat MCH (1990) Phosphorus uptake characteristics of a world collection of white clover (*Trifolium repens* L.) cultivars. In: van Beusichem ML (ed) Plant nutrition – physiology and applications. Proc 11th Int Plant Nutr Coll, Wageningen, Netherlands, 30 July – 4 August 1989, pp 655–658

Mange M (1963) Action de l'acide gibbérellique sur *Trifolium repens* L.: observations sur quelques modifications – morphologies, physiologiques et anatomiques. Ann Sci Univ de bescançon, 2° Séné, Botanique 19:43–53

Marriott CA (1988) Seasonal variation in white clover content and nitrogen fixation (acetylene reduction) activity in a cut upland sward. Grass Forage Sci 43:253–262

Marriott CA, Haystead A (1992) The effect of lenient defoliation on the nitrogen economy of white clover: the contribution of mineral nitrogen to plant nitrogen accumulation during regrowth. Ann Bot 69:429–435

Marriott CA, Thomas RJ, Smith MA, Logan KAB, Baird MA, Ironside AD (1988) The effect of temperature and nitrogen interactions on growth and nitrogen assimilation of white clover. Plant Soil 111:43–51

McGonigle TP, Fitter AH (1988) Growth and phosphorus inflows of *Trifolium repens* L. with a range of indigenous vesicular-arbuscular mycorrhizal infection levels under field conditions. New Phytol 108:59–65

Minchin FR, Minguez MI, Sheehy JE, Witty JF, Skot L (1986) Relationship between nitrate and oxygen supply in symbiotic nitrogen fixation by white clover. J Exp Bot 37:1103–1113

Mitchell KJ (1956a) Growth of pasture species under controlled environment. I. Growth at various levels of constant temperature. NZ J Sci Tech Sect A38:203–216

Mitchell KJ (1956b) Growth of pasture species. III. White Clover (*Trifolium repens*), red clover (*T. pratense*), and lotus major (*L. uliginosus*). NZ J Sci Tech Sect A37:395–413

Mitchell KJ, Lucanus R (1962) Growth of pasture species under controlled environment III. Growth at rarious levels of constant temperature with 8 and 16 hours of uniform light per day. NZ J Agric Res 5:135–144
Montes RA, Blum U, Heagle AS, Volk RJ (1983) The effects of ozone and nitrogen fertiliser on tall fescue, ladino clover, and a fescue-clover mixture. II. Nitrogen content and nitrogen fixation. Can J Bot 61:2159–2168
Mouat MCH (1983a) Competitive adaptation by plants to nutrient shortage through modification of root growth and surface change. NZ J Agric Res 26:327–332
Mouat MCH (1983b) Phosphate uptake from extended soil solutions by pasture plants. NZ J Agric Res 26:483–487
Mouat MCH, Walker TW (1959) Competition for nutrients between grasses and white clover. I. Effects of grass species and nitrogen supply. Plant Soil 11:30–40
Moustafa E, Ball R, Field TRO (1969) The use of acetylene reduction to study the effect of nitrogen fertiliser and defoliation on nitrogen fixation by field grown white clover. NZ J Agric Res 12:691–696
Murphy PM (1986) Effect of light and atmospheric carbon dioxide concentration on nitrogen fixation by herbage legumes. Plant Soil 95:399–409
Newton PCD (1986) The establishment, growth and fate of white clover plants with special reference to the physiology of stolon growth. PhD Thesis, University of North Wales, Bangor, UK
Newton PCD, Hay MJM (1992) Technique for evaluating the potential for growth of shoot and root buds of white clover (*Trifolium repens*). J Agric Sci 119:179–183
Newton PCD, Hay MJM (1994) Patterns of nodal rooting in *Trifolium repens* (L.) and correlations with stages in the development of axillary buds. Grass Forage Sci 49:270–276
Newton PCD, Hay MJM (1995) Non-viability of axillary buds as a possible constraint on effective foraging of *Trifolium repens* L. Abstr Bot 19:83–88
Newton PCD, Hay MJM, Thomas VJ, Glasgow EM, Dick HB (1990) Patterns of axillary bud activity in white clover. Proc NZ Grassl Assoc 52:247–249
Newton PCD, Hay MJM, Thomas VJ, Dick HB (1992) Viability of axillary buds of white clover (*Trifolium repens* L.) in grazed pasture. J Agric Sci 119:345–354
Newton PCD, Clark H, Bell CC, Glasgow EM, Campbell BD (1994) Effects of elevated $C0_2$ and simulated seasonal changes in temperature on the species composition and growth rates of pasture turves. Ann Bot 73:53–59
Newton PCD, Clarke H, Bell CC, Glasgow EM, Tate KR, Ross DJ, Yeates GW (1995) Plant growth and soil processes in temperate grassland communities at elevated $C0_2$. J Biogeogr 22:235–240
Newton PCD, Clarke H, Bell CC, Glasgow EM (1996) Interaction of soil moisture and elevated CO_2 on the above-ground growth rate, root length density and gas exchange of turves from temperate pasture. J Exp Bot 47:771–779
Nijs I, Impens I, Behaeghe T (1989a) Effects of different $C0_2$ environments on the photosynthesis – yield relationship and the carbon and water balance of a white clover (*Trifolium repens* L. cv. Blanca) sward. J Exp Bot 40:353–359
Nijs I, Impens I, Behaeghe T (1989b) Effects of long-term elevated atmospheric CO_2 concentration on *Lolium perenne* and *Trifolium repens* canopies in the course of a terminal drought stress period. Can J Bot 67:2720–2725
Noor LS, Robin C, Schontz D, Heulin T, Guckert A (1989) Effect d'un deficit hydrique sur le trèfle blanc (*Trifolium repens* L.). I. Importance du cultivar. Agronomie 9:251–257
Norris IB (1985a) Flowering of contrasting white clover varieties in relation to temperature in controlled environments. Ann Bot 56:317–322
Norris IB (1985b) Flowering of contrasting white clover varieties in relation to daylength in controlled environments and in the field. Ann Appl Biol 107:301–308
Norris IB (1987) Effects of daylength on floral characteristics of six white clover (*Trifolium repens* L.) varieties. Ann Bot 57:255–261
Norris IB (1988) Gibberellin, daylength and temperature effects on floral and vegetative development in white clover. Ann Appl Biol 112:359–366

Nösberger J, Machler F, Boller BC (1983) Photosynthesis and dry matter distribution in altitudinal ecotypes of white clover. Proc XIV Int Grassl Congr, 15–24 June, Lexington, USA, pp 410–413

Ommen OE, Bender J, Weigel HJ (1994) Effects of ozone on white clover: protection of ozone damage by ethylenediurea. Angew Bot 68:146–150

Parsons AJ, Harvey A, Woledge J (1991) Plant-animal interactions in a continuously grazed mixture. I. Differences in the physiology of leaf expansion and the fate of leaves of grass and clover. J Appl Ecol 28:619–634

Pederson GA (1989) Taproot and adventitious root growth of white clover as influenced by nitrogen nutrition. Crop Sci 29:764–768

Powell CL (1977a) Mycorrhizas in hill country soils. I. Spore-bearing mycorrhizal fungi in 37 soils. NZ J Agric Res 20:53–58

Powell CL (1977b) Mycorrhizas in hill country soils II. Effect of several mycorrhizal fungi on clover growth in sterilised soils. NZ J Agric Res 20:59–62

Pugh R, Witty JF, Mytton LR, Minchin FR (1995) The effect of waterlogging on nitrogen fixation and nodule morphology in soil-grown white clover (*Trifolium repens* L.). J Exp Bot 46:284–290

Robin C, Muljanto D, Guckert A, Scaife A (1992) Effect of potassium on the growth and development of roots and nodules of white clover subjected to a PEG-induced water stress. Proc 2nd Congr European Soc Agron, Warwick University, 23–28 August 1992, pp 300–301

Robin C, Hay MJM, Newton PCD, Greer DH (1994a) Effect of light quality (red: far-red ratio) at the apical bud of the main stolon on morphogenesis of *Trifolium repens* L. Ann Bot 74:119–123

Robin C, Hay MJM, Newton PCD (1994b) Effect of light quality (red: far-red ratio) and defoliation treatments applied at a single phytomer on axillary bud outgrowth in *Trifolium repens*. Oecologia 100:236–242

Robson MJ, Polniaszek T (1984) Effect of contrasting light regimes on the growth and morphology of individual plants of white clover, and of different stolons of the same plant. In: Forage Legumes. Proc Br Grassl Soc, pp 189–190

Ryle GJA (1988) The influence of host plant energy supply on nitrogen fixation. In: O'Gara F, Manion S, Drevon JJ (eds) Physiological limitations and the genetic improvement of symbiotic nitrogen fixation. Kluwer, Dordrecht, pp 3–10

Ryle GA, Arnott RA, Powell CE, Gordon AJ (1983) Comparisons of the respiratory effluxes of nodules and roots in six temperate legumes. Ann Bot 52:469–477

Ryle GJA, Powell CE, Jackson JP (1979) The respiratory costs of nitrogen fixation in soyabean, cowpea, and white clover. I. Nitrogen fixation and the respiration of the nodulated root. J Exp Bot 114:133–144

Ryle GJA, Powell CE, Gordon AJ (1985a) Defoliation in white clover: regrowth, photosynthesis and N_2 fixation. Ann Bot 56:9–18

Ryle GJA, Powell CE, Gordon AJ (1985b) Short-term changes in $C0_2$ evaluation associated with nitrogenase activity in white clover in response to defoliation and photosynthesis. J Exp Bot 36:634–643

Ryle GJA, Powell CE, Gordon AJ (1986) Defoliation in white clover: nodule metabolism, nodule growth and maintenance, and nitrogenase functioning during growth and regrowth. Ann Bot 567:263–271

Ryle GJA, Powell CE, Gordon AJ (1988) Responses of N_2 fixation linked respiration to host-plant energy status in white clover acclimated to a controlled environment. J Exp Bot 39:879–887

Ryle GJA, Powell CE, Timbrell MK, Jackson JP (1989) Carbon and nitrogen yield, and N_2 fixation in white clover plants receiving simulated continuous defoliation in controlled environments. Ann Bot 63:675–686

Sackville Hamilton NR (1980) Variation and adaptation in wild populations of white clover (*Trifolium repens* L.) in East Anglia. PhD Thesis, University of Cambridge, Cambridge, UK

Sackville Hamilton NR (1990) The dynamics of *Trifolium repens* in a permanent pasture. II. Internode length and the movement of shoot axes. Proc R Soc Lond B239:359–381

Sackville Hamilton NR, Harper JL (1989) The dynamics of *Trifolium repens* in a permanent pasture. I. The population dynamics of leaves and nodes per shoot axis. Proc R Soc Lond B237:133–173

Scheidegger UC, Nösberger J (1984) Influence of carbon dioxide concentration on growth, carbohydrate content, translocation and photosynthesis of white clover. Ann Bot 54:735–742

Schmid B, Harper JL (1985) Clonal growth in grassland perennial. I. Density and pattern-dependent competition between plants with growth forms. J Ecol 73:793–808

Simon JC, Gastal F, Lemaire G (1989) Compétition pour la lumière et morphologie du trèfle blanc (*Trifolium repens* L.): émission des feuilles et des remifications. Agronomie 9:383–389

Skipp RA, Gaynor DL (1987) Pests-nematodes. In: Baker MJ, Williams WM (eds) White clover. CAB International, Wallingford, UK, pp 493–512

Solangaarachchi SM, Harper JL (1987) The effect of canopy filtered light on the growth of white clover *Trifolium repens*. Oecologia 72:372–376

Solangaarachchi SM, Harper JL (1989) The growth and asymmetry of neighbouring plants of white clover (*Trifolium repens* L). Oecologia 78:208–213

Thomas RG (1961) Flower initiation in *Trifolium repens* L.: a short-long-day plant. Nature 190:1130–1131

Thomas RG (1962) The initiation an growth of axillary bud primordia in relation to flowering in *Trifolium repens* L. Ann Bot 26:329–344

Thomas RG (1987a) The structure of the mature plant. In: Baker MJ, Williams WM (eds) White clover. CAB International, Wallingford, UK, pp 1–29

Thomas RG (1987b) Reproductive development. In: Baker MJ, Williams WM (eds) White clover. CAB International, Wallingford, UK, pp 63–123

Thompson L, Harper JL (1988) The effects of grasses on the quality of transmitted radiation and its influence on the growth of white clover. *Trifolium repens*. Oecologia 75:343–347

Trautner JL, Gibson PB (1966) Fate of white clover axillary buds at five intensities of sunlight. Agron J 58:557–558

Turkington R, Burdon JJ (1983) The biology of Canadian weeds. 57. *Trifolium repens* L. Can J Plant Sci 63:243–266

Turkington R, Harper JL (1979a) The growth, distribution and neighbour relationships of *Trifolium repens* in a permanent pasture. I. Ordination, pattern and contact. J Ecol 67:201–218

Turkington R, Harper JL (1979b) The growth, distribution and neighbour relationships of *Trifolium repens* in a permanent pasture. II. Inter- and intra-specific contact. J Ecol 67:219–230

Turkington R, Cahn MA, Vardy A, Harper JL (1979) The growth, distribution and neighbour relationships of *Trifolium repens* in a permanent pasture. J Ecol 67:231–243

Walker RW, Adams AFR, Orchiston HD (1956) Fate of labelled nitrate and ammonium nitrogen when applied to grass and clover grown separately and together. Soil Sci 8:339–351

Walker TW, Adams AFR (1958) Competition for sulphur in a grass-clover association. Plant Soil 9:353–366

Whitehead DC (1983) The influence of frequent defoliation and of drought on nitrogen and sulphur in the roots of perennial ryegrass and white clover. Ann Bot 52:931–934

Winkler L, Nösberger J (1985) Einfluss der Schnitthäufigkeit und N-Düngung auf die Bestandsstruktur und die vertikale Verteilung von Weissklee (*Trifolium repens* L.) in einer Dauerwiese. Z Acker Pflanzenb 155:43–50

Woledge J (1986) The effect of age and shade on the photosynthesis of white clover leaves. Ann Bot 57:257–262

Woledge J, Calleja SA (1983) The growth and photosynthesis of seedling plants of white clover at low temperature. Ann Bot 52:239–245

Woledge J, Bunce JA, Tewson V (1989) The effect of air humidity on photosynthesis of ryegrass and white clover at three temperatures. Ann Bot 63:271–279

Physiological Control of Alfalfa Growth and Yield

J.J. Volenec[1]

1 Introduction

Unlike grain crops, the economic yield of alfalfa (*Medicago sativa* L.) is herbaceous, composed predominantly of leaves and stems. Understanding physiological control of alfalfa yield, therefore, requires careful evaluation of factors influencing leaf and stem initiation and growth. Alfalfa yield is determined by the total amount of dry matter produced and the partitioning of dry matter. Partitioning of dry matter to crowns and roots also influences the survival of this perennial species. Natural selection may favor alfalfa survival over herbage production in most ecosystems. Finally, recent emphasis on production of high-quality alfalfa forage results in frequent defoliation at immature stages of development. As a result, increasing yield of high-quality forage without compromising alfalfa survival presents a challenging opportunity.

Holland and Bingham (1994) compared herbage yield of alfalfa cultivars released over an 87-year period, some dating back to the turn of the century. They reported an average herbage yield increase of 0.18%/year. This indicates that genetic improvement of alfalfa herbage yield has been a mere 15% during the 87 years spanned by this study. This relatively meager yield enhancement underscores the complexity and difficulty in improving herbage production and plant survival simultaneously in this perennial species. A similar study comparing changes in alfalfa persistence resulting from selection has not been published, but would be of value in understanding how plant persistence has been impacted by selection for improved agronomic performance.

The goal of this chapter will be to examine limitations to growth and yield of alfalfa. The initial discussion will focus on alfalfa yield component analysis in an effort to understand factors that determine herbage yield at the whole-plant level. The remainder of the chapter will examine physiological and morphological characteristics that influence individual yield components. From these discussions selection criteria will be identified that may enhance future efforts to improve alfalfa yield and persistence.

[1]Department of Agronomy, Purdue University, 1150 Lilly Hall of Life Sciences, West Lafayette, Indiana 47907-1150, USA

Crop Yield, Physiology and Processes
D. L. Smith, C. Hamel (Eds.)
© Springer-Verlag Berlin Heidelberg 1999

2 Alfalfa Yield Component Analysis

Initial attempts to understand whole-plant morphological factors influencing herbage yield used correlation analysis to evaluate a wide array of traits. These studies showed that crown width, stem length, and number of shoots per plant were associated with herbage yield (Rumbaugh 1963). Path coefficient analysis was an improvement over correlation analysis for identifying direct effects of whole-plant morphological traits on yield. Of the traits they examined, Liang and Riedl (1964) determined that plant height and number of stems per plant had the greatest direct effect on yield. Although these characters are associated with alfalfa herbage yield, they unfortunately cannot be used mathematically in a detailed analysis of alfalfa yield determination.

In a manner similar to yield component analysis of cereals, alfalfa herbage yield can be calculated as the mathematical product of three plant characteristics:

$$\text{Herbage yield/area} = \text{plants/area} \times \text{shoots/plant} \times \text{mass/shoot}.$$

Bolger and Meyer (1983) examined the influence of seeding rate on alfalfa herbage yield and yield components. Herbage yield was reduced at plant populations of 19 or fewer plants/m^2. As plant population increased from 16 to 31 plants/m^2 the decline in number of shoots per plant offset the increase in plants/area, resulting in no change in number of shoots per area. Mass/shoot was largely unaffected as plant populations increased from 16 to 89 plants/m^2, but increased significantly at very low plant densities.

Yield components differ among cultivars, and cultivar differences can interact with plant population in determining final herbage yield (Volenec et al. 1987). At low plant populations Vernal alfalfa produced more shoots/plant than Hi-Phy or BIC, an unselected germplasm. This advantage was lost, however, at high plant populations where interplant competition limited shoot production per plant. Mass per shoot was always lowest in Vernal regardless of plant population. Path coefficient analysis revealed that the direct effects of shoots per plant and mass per shoot on yield per plant were equal at low plant populations, whereas mass/shoot was 30% more important than shoots/plant in determining herbage yield per plant at high plant densities. Kephart et al. (1992) also used path analysis of yield components of alfalfa established at different populations to conclude that mass/shoot was the principal barrier to increased herbage yield at high plant populations.

As alfalfa stands age, the contribution of plants/area and shoots/plant to herbage yield changes dynamically (Fig. 1). Although young stands have high initial plant populations, there is a rapid decline during the first 2 to 3 years after seeding, presumably caused by high interplant competition (Nelson et al. 1986). Thereafter, a slower decrease in plant population occurs, during which time there is a compensatory increase in shoots/plant. The net effect is little

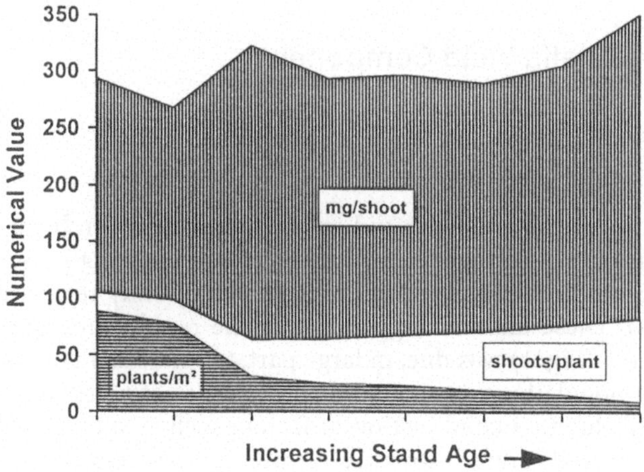

Fig. 1. Changes in alfalfa yield components with increasing stand age. (After plant population data of Bolger and Meyer 1983)

change in shoots/m^2 (plants/m^2 $\times$ shoots/plant) and yield until plant populations decline to a point at which crown width and shoot production/plant can no longer compensate for plant losses. This suggests that increasing crown width and shoot production/crown are useful approaches for enhancing stand longevity. However, in situations where alfalfa is grown in 3 to 5 year rotations, the need for spreading crowns and exceptional shoot production/crown may not be necessary. Rotation will most likely occur before genetic differences in crown development can be exploited. Despite large changes in plants/m^2 and shoots/plant, mass/shoot remains relatively constant with stand age (Fig. 1). This indicates that genetic improvement in mass/shoot could result in greater herbage yield at most plant populations and in stands of varying ages.

Initial genetic studies show that alfalfa yield components are heritable traits (Winsett 1991). Analysis of parents and progeny over 2 years revealed narrow-sense heritabilities for shoots/plant ranging from 0.10 $\pm$ 0.14 for selections from within the cultivar Hi-Phy to 0.89 $\pm$ 0.38 for selections from Beltsville-6, a fall nondormant germplasm. Narrow-sense heritabilities for mass/shoot were 0.54 $\pm$ 0.19 and 0.43 $\pm$ 0.16 for these respective germplasms over 2 years. The moderate heritability values suggest that altering these yield components through selection is possible, although progeny testing may be necessary to make reasonable gains from selection. Simultaneously increasing mass/shoot without altering shoots/plant may also be possible because phenotypic correlations between these traits are low (r values ranged from -0.26 to 0.10).

3 Factors Influencing Alfalfa Yield Components

3.1 Alfalfa Persistence

Although both seedling establishment and plant persistence influence plants/area, this discussion will focus primarily on factors influencing persistence of established plants. Forage legume persistence was recently reviewed (Beuselinck et al. 1994). These authors suggested that the rapid thinning of stands during years 1 to 3 (Fig. 1) was due, in large part, to interplant competition. Thereafter, plant populations decline gradually because of increases in "cumulative stress level" due to a combination of factors such as insects and diseases, poor soil fertility, temperature stress, water stress, and winter stresses in temperate regions.

Understanding the seasonality of alfalfa plant death may help us identify the principal stress(es) causing alfalfa plant death. In California, no consistent seasonal pattern of plant death was observed when disease and insect pests were controlled, whereas a greater proportion of plants died in summer when diseases and insects were allowed to develop (Summers and Gilchrist 1991). Stand age was the most important factor influencing alfalfa persistence in this study. By comparison, data from Missouri (Nelson et al. 1986) indicated that alfalfa plant death was 50% greater over winter than that observed during summer. These authors suggested that some plants which died during summer might have been weakened by winter injury, but they were unable to quantify the magnitude of this effect. These findings agree with Sheaffer et al. (1992), who indicated that the traits most often associated with alfalfa persistence in temperate regions are disease resistance and winter hardiness.

The biotic components of "cumulative stress" can be minimized with prophylactic application of pesticides and their effects on plant persistence separated from abiotic factors. Application of insecticides, fungicides, or both pesticides in combination had no effect on alfalfa plant populations during the first 2 production years (Fig. 2) (Summers and Gilchrist 1991). After year 2, plant populations declined 36% faster when fungal disease pests were not suppressed than if fungicides had been applied. Control of insect pests had less influence on persistence than control of fungal pathogens. From year 2 through year 5 alfalfa plant populations declined 91% on plots where fungal diseases and insects were not suppressed, but also declined 71% on plots receiving preventative applications of fungicides and insecticides. This suggests that stress other than these biotic pests accounted for 78% of alfalfa plant losses observed in this California environment. This conclusion is supported by results from other studies, where persistence of established stands was not shown to be closely associated with genetic differences in disease resistance (Jones et al. 1984; Smith and Bouton 1993). Genetic pathogen resistance may be more important for survival of alfalfa seedlings during initial stand establishment.

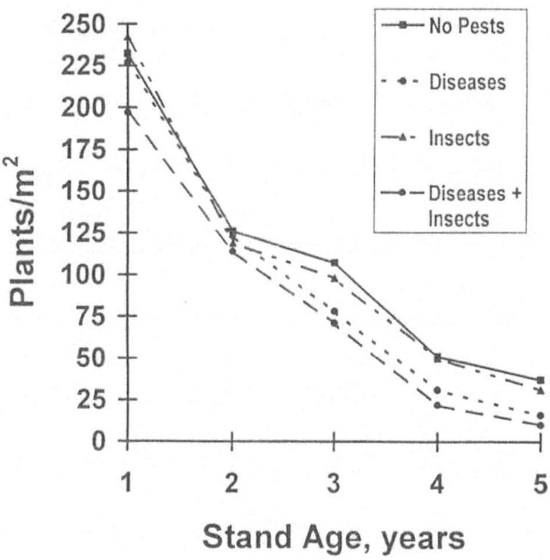

Fig. 2. Influence of stand age and biotic stress on alfalfa plant populations in California. Plots were kept free of both insects and fungal diseases (*No Pests* treatment) with preventative applications of pesticides. Plots in the *Diseases* treatment received preventative insecticide applications, while the *Insects* treatment received preventative applications of fungicides. *Diseases and Insects* treatment were not sprayed with pesticides. (After Summers and Gilchrist 1991)

Wiersma et al. (1995) reported improved establishment of alfalfa seedlings possessing resistance to Phytophthora root rot and Aphanomyces root rot in soils where these organisms are prevalent.

The physiological basis for abiotic stress tolerance of alfalfa is traditionally associated with accumulation of root total nonstructural carbohydrates (TNC). Graber et al. (1927) were among the first to show a correlation between alfalfa shoot regrowth or plant persistence, and root TNC levels. Others have reported correlations between alfalfa forage yield and root TNC concentrations (Feltner and Massengale 1965). However, as pointed out by May (1960), the cause-effect nature of this popular concept has rarely been critically evaluated. For example, Smith and Marten (1970) used ^{14}C-labeling of roots prior to defoliation to study mobilization of root TNC to regrowing shoots. They concluded that carbohydrates initially stored in roots play an active role in shoot regrowth after harvest. They failed to recognize that ^{14}C-labeling would have labeled all organic constituents in roots (sugars, starch, *and* amino acids, proteins, lipids, etc.). Therefore, they could not accurately assess the role of root TNC in shoot regrowth based on their data.

Recent studies suggest that the association between root TNC levels and alfalfa persistence is more complex than originally thought. Suzuki (1991) reported that persistent alfalfa plants actually had lower root TNC concentrations than that found in alfalfa plants that died. Jones et al. (1984) studied persistence of alfalfa grown in yield performance trials and reported no association between root TNC concentrations and alfalfa survival. Smith et al. (1992) reported higher TNC concentrations in roots of grazing-tolerant alfalfa germplasms. They attributed part of this increase to the observation that grazing-tolerant plants had greater basal leaf areas on crowns than grazing-intolerant plants. Other studies showed that thin stems with decumbent growth were morphological adaptations associated with the greater crown leaf area but, unfortunately, these traits are associated with low forage yield (Brummer and Bouton 1991, 1992).

Understanding the mechanistic role of root TNC in alfalfa persistence has been difficult. For example, winter survival of an alfalfa genotype with low root starch was similar to that of plants with normal root starch levels suggesting that high root starch levels were not essential for winter survival (Boyce and Volenec 1992). However, plants in this study were not isogenic, and most likely differed in other traits in addition to root starch concentration. Reductions in root TNC can be induced by frequent harvest in autumn, but are not consistently associated with alfalfa winter killing (Edminsten et al. 1988; Brink and Marten 1989). However, results from such studies are complicated by differences in stand age and variation in the prevailing environment in the year(s) the study was conducted. Even without these complications, the physiological responses of alfalfa in experiments where frequent harvesting is used as a treatment are difficult to interpret. This is because, like root TNC, complete defoliation is likely to alter numerous interacting physiological and biochemical processes. Nevertheless, it appears root TNC concentrations alone cannot adequately describe differences in alfalfa winter hardiness and persistence. This agrees with findings of Bowley and McKersie (1990), where selection for winter hardiness depended on several things including plant age and the nature of the prevailing environmental stress (ice encasement, freezing, anaerobiosis, flooding) to which plants were exposed. They suggested that different resistance mechanisms are operative in alfalfa, depending on the stress to which plants are exposed, and that selection for tolerance to one stress is not likely to confer tolerance to other stresses associated with winter.

Significant genetic variation in winter hardiness exists among cultivars. An enigma in alfalfa winter survival is the association of winter hardiness with fall dormancy reaction. Fall dormancy is a relative measure of shoot length and shoot decumbancy in autumn. Fall-dormant cultivars with short, prostrate shoots are often winter-hardy, whereas very fall-nondormant cultivars that grow tall and upright in autumn often winter kill. Up to one-half of the genetic variation in winter hardiness and winter injury is associated with fall dormancy (Sheaffer et al. 1992). This makes greater fall dormancy attractive as a selection criterion for improving alfalfa winter hardiness (Stout et al. 1992). However, fall dormancy is also associated with slow shoot regrowth after harvest and reduced herbage yield in spring and summer (Stout and Hall 1989). Therefore, creating high-yielding alfalfa cultivars that persist over winter may be difficult if fall-dormancy reaction is used as a selection criterion for winter hardiness. There is a clear need for an improved understanding of the physiological and biochemical mechanisms conferring high winter hardiness to alfalfa. Genes and gene products associated with enhanced freezing tolerance have been identified and cloned, that may prove insightful to our understanding as they emerge from the laboratory for field evaluation (McKersie et al. 1993; Monroy et al. 1993a,b).

The physiological basis for fall-dormancy reaction is poorly understood. Heichel and Henjum (1990) used reciprocal grafting between plants differing in fall-dormancy reaction to show that fall dormancy is conditioned mainly by factors located in shoots. Herbage production of fall nondormant plants was reduced in dormancy-inducing conditions whenever fall-dormant tissues were used in the grafts either as scion or rootstock. Morphological changes toward greater fall-dormancy reaction did not improve crown-freezing tolerance of fall nondormant plants. This indicates that fall-dormancy reaction alone is not causal in its effects on alfalfa freezing tolerance, and that selection for winter-hardy fall-nondormant plants may be possible. However, selection for winter hardiness is difficult because it is conditioned by numerous plant traits, many of which interact with the environment. For example, Philippot et al. (1991) concluded that daily irradiance alone did not mediate fall-dormancy reaction in alfalfa, and that the effects of temperature and its interaction with irradiance on morphology and carbon partitioning should be investigated.

Selection for contrasting fall-dormancy reaction from within a cultivar can alter winter hardiness (Table 1). Three cycles of selection for contrasting fall-dormancy reaction using the fall-nondormant cultivar CUF 101 as a starting population were conducted at Davis, CA. Winter survival, and taproot and bud physiology of CUF 101 and the selections from it were evaluated for two winters at West Lafayette, IN. Two check cultivars were included; the very fall-

Table 1. Fall-dormancy reaction, winter survival, and root and crown bud composition of alfalfa selected for contrasting fall-dormancy reaction. Norseman and Wadi Qurayat are very fall-dormant and -nondormant cultivars, respectively, and are included for comparison. Selections from CUF 101, a fall-nondormant cultivar, were for greater or less fall dormancy reaction. Data are expressed relative to values obtained for Norseman, and are averaged over 2 years. Roots and buds were sampled in December for compositional analysis (S.M. Cunningham, J.J. Volenec, and L.R. Teuber, unpubl.)

Trait	Norseman (very fall-dormant)	CUF 101 Selected for greater fall dormancy	CUF 101 (nondormant)	CUF 101 Selected for less fall dormancy	Wadi Qurayat (very nondormant)
Fall dormancy	100	189	261	263	275
Winter survival	100	95	1	8	0
Root TNC, mg/g	100	119	112	112	117
Root starch, mg/g	100	153	163	160	161
Root sugar, mg/g	100	72	43	47	58
Root protein, mg/g	100	92	76	73	70
Bud TNC, mg/g	100	95	69	58	43
Bud starch, mg/g	100	146	114	63	40
Bud sugar, mg/g	100	86	61	58	43
Bud protein, mg/g	100	110	102	108	90

dormant Norseman; and the extremely fall-nondormant cultivar Wadi Qurayat. Selection for greater fall dormancy was effective in reducing height of CUF 101, and markedly improved winter survival to near that observed for Norseman (S.M. Cunningham, J.J. Volenec, and L.R. Teuber, unpubl.). The greater winter hardiness of the more fall-dormant CUF 101 was associated with higher sugar concentrations in roots and crown buds, and high buffer-soluble protein concentrations in roots. Results indicate that fall-dormant alfalfa cultivars accumulate higher levels of raffinose and stachyose in crowns when compared to CUF 101, differences that were more closely associated with winter hardiness than the accumulation of sucrose alone (Castonguay et al. 1995). Direct selection for high root and bud sugar (stachyose and raffinose) concentrations and/or high root protein levels in autumn, with subsequent winter hardiness evaluation, would be useful in ascertaining the direct effects of these constituents on winter hardiness and their association with fall dormancy reaction.

3.2 Mass per Shoot

Genotypic differences in mass per shoot have been ascribed to several morphological and physiological characteristics of alfalfa. Rapid leaf-area expansion and shoot elongation after defoliation are positively associated with shoot mass accumulation (Volenec 1985). Rapidly regrowing shoots originate from large buds located on crowns (versus buds present on bases of previously harvested stems) that resume growth quickly after defoliation (Leach 1968). Rapid shoot morphogenesis hastens the conversion of vegetative shoots to reproductive development, which, in turn, stimulates more extensive bud development on crowns (Leach 1969). The resultant larger buds develop into rapidly regrowing shoots after defoliation that again exhibit greater rates of morphogenesis, and in doing so, continue the pattern of rapid shoot regrowth and acquisition of mass per shoot. Hodgkinson (1973) speculated that development of large shoots inhibited morphogenesis of smaller buds on alfalfa crowns, implying that growth regulators may have a role on controlling alfalfa bud development. A critical assessment of growth-regulator status of alfalfa crown buds as influenced by shoot morphogenesis and defoliation has not been published. In addition, photomorphogenic control of bud development has been reported for several species (Volenec and Nelson 1995), but its role in controlling crown bud development of alfalfa remains to be demonstrated.

The importance of crown bud development prior to defoliation in determining shoot growth rates and mass per shoot is consistent with several observations. Nelson and Smith (1968) reported that initial shoot growth in spring originates from crown buds, while shoots comprising herbage of the second and subsequent harvests originate largely from axillary buds of previously cut stems. These later shoots are slower to develop, smaller in size, and result in lower herbage yields than those observed in spring (Singh and Winch 1974).

Defoliation prior to the transition of shoots from vegetative to reproductive development reduces morphogenesis of buds on crowns, and slows subsequent shoot regrowth rates (Nelson and Smith 1968). These effects, however, have traditionally been confounded with reductions in root TNC concentrations, making it difficult to discern the exact cause of the slower shoot regrowth rates caused by frequent harvesting. The morphometric bases for threefold differences in mass per shoot between near-isogenic alfalfa populations differing in ploidy were investigated using growth analysis (Volenec 1988). The relative growth rates (RGR, mg/mg/day) of shoots were similar, whereas the tetraploid population had 12-fold greater bud mass at defoliation when compared to the diploid population. This difference in initial bud mass resulted in faster shoot regrowth and higher herbage yield in the tetraploid plants.

Musgrave and Langer (1977) reported that shoots originating from lower positions on crowns regrew faster and attained greater mass than did shoots originating from axillary buds located on recently cut stems. Two-thirds or more of harvestable yield was derived from shoots originating from low positions on crowns. This agrees with findings of Gosse et al. (1988), who reported that shoots originating from low crown positions (their type 2 stems) grew taller, intercepted more light, and accounted for over 75% of shoot dry matter production when compared to shoots originating from axillary positions high on crowns (their type 1 stems). They speculated that persistence was poor for plants that produced few type 2 stems because inability to compete effectively for light would reduce root reserves and result in slow regrowth after cutting. Later work showed that nitrogen was distributed preferentially to shoots intercepting the most light (Lemaire et al. 1991). These shoots, and the plants from which they developed, survived better than did shoots developing in low light. The role of nitrogen reserves in alfalfa regrowth is discussed later in this chapter.

Genotypes selected for contrasting shoot elongation rates exhibit differences in both origin of buds and shoot development rates after defoliation (Table 2). The genotype selected for slow shoot elongation rate (SSER) had one-half the number of shoots developing from crown buds when compared to the genotype selected for rapid shoot elongation rate (RSER) (Hendershot 1986). Shoots developing from crown buds on the RSER genotype grew rapidly after defoliation when compared to crown buds developing on the SSER genotype, and resulted in production of substantially greater mass per shoot. Although shoot mass of buds originating from stubble positions was similar between genotypes, overall shoot mass of the RSER genotype was greater due to the rapid growth of its greater number of crown buds. Shoots developing on the SSER genotype had similar growth rates, irrespective of bud origin. This indicates that, while bud origin can be associated with genetic or management-induced differences in shoot mass accumulation, bud origin alone is not the sole determinant of shoot growth rate and mass per shoot.

Our understanding of the physiological and biochemical cause(s) for differences in shoot regrowth rates, and ultimately mass per shoot is incomplete.

Table 2. Influence of crown bud origin on shoot growth and development of alfalfa genotypes selected for slow (SSER) or rapid (RSER) shoot elongation rate. Shoots developing from crown buds were defined as those attached directly to the base of the crown, whereas shoots from stubble buds were those developing from axillary buds of recently cut stems. (Hendershot 1986)

Characteristic	SSER genotype		RSER genotype	
	Crown buds	Stubble buds	Crown buds	Stubble buds
Shoot growth rate, mg/day	6.4	4.0	19.0	5.3
Mass per shoot, mg[a]	131	78	376	102
Shoots per plant, no[b]	2.3	20	4.6	14
Shoot mass, mg[a]	459	1609	1882	1492

[a] Obtained after 19 days of regrowth.
[b] Averaged over values obtained on days 5, 12, and 19 of regrowth.

A popular concept is that defoliation-induced reductions in shoot growth rates are due to low root TNC levels (Constable et al. 1977). This concept has found favor because of the obvious influence of defoliation on photosynthate production. May (1960) questioned the cause-effect relationship between depletion of root carbohydrates and shoot regrowth of defoliated forages including alfalfa. He argued that "carbohydrate-driven" shoot regrowth is most likely an oversimplification of the physiological and biochemical processes necessary for herbage regrowth. Leach (1969) examined cultivars that differed in shoot growth rate and mass per shoot and found no evidence supporting the commonly held notion that accumulation of root carbohydrate reserves is closely associated with shoot regrowth rate. Other work with genetic selections differing up to fourfold in root-starch concentration were used to evaluate the influence of root-starch concentration on shoot regrowth rate after defoliation (Habben and Volenec 1990; Boyce and Volenec 1992). Contrary to expectations, early shoot regrowth rate of the low-starch genotype equaled or exceeded that of the high-starch genotype. Using reciprocal grafting to interchange roots and shoots between the RSER and SSER genotypes described above, we showed no consistent effect of rootstock on shoot regrowth rate and mass per shoot (Fankhauser and Volenec 1989). Plants created using RSER scions generally had rapid regrowth rates and high mass per shoot, irrespective of rootstock. This suggests that shoot and not root factors controlled shoot regrowth rates of these genotypes. Surprisingly, significant scion effects were noted for root TNC accumulation. Plants possessing RSER scions exhibited more rapid reaccumulation of root TNC when compared to plants constructed using SSER as scions. This indicates that, while rootstock had little effect on shoot growth and development, shoot development (as determined by scion source) can alter carbon partitioning and root carbohydrate accumulation.

Other work has brought into question the availability of root TNC as a carbon source for shoot development. By extending the dark period slightly we have shown marked reductions in both rates of expansion and dark respiration in expanding alfalfa leaves (Hendershot and Volenec 1989). Reductions occurred despite the fact that roots contained approximately 200 mg/g TNC. These results agree with earlier findings of Hodgkinson (1969), who used ^{14}C-labeling of roots prior to defoliation to determine the role of root organic reserves in shoot regrowth. Expanding leaves on regrowing shoots received very little labeled C from roots during the early phases of regrowth, suggesting that their growth was largely independent of root TNC. Results using ^{13}C-labeling to quantify carbon movement from roots to shoots showed that only 5% of the carbon lost by roots is recovered in regrowing shoots (Avice et al. 1996a). Most of the carbon disappearing from roots after defoliation was attributed to root and shoot respiratory losses. These results, which are contrary to the conventional wisdom regarding the role of root TNC in alfalfa shoot regrowth, suggest that critical reassessment of the relationship between root TNC and shoot regrowth rate is needed.

In contrast to root TNC, results are emerging that suggest that root nitrogen (N) can influence shoot regrowth rate and mass per shoot (Volenec et al. 1996). The dependency of shoot regrowth on endogenous root N pools may result from the influence of defoliation on nodule activity. Vance et al. (1979) reported that dinitrogen fixation of alfalfa declines 88% within 24 h after defoliation and remains very low for 13 days. Early regrowth, composed largely of N-rich leaves (Volenec 1985), precedes recovery of dinitrogen fixation (Vance et al. 1979; Cralle and Heichel 1981). Addition of N as nitrate immediately after defoliation has no effect on herbage regrowth rate (Vance and Heichel 1981). These authors suggested that N remobilization from nodules and from storage pools in roots must be sufficient to meet the N needs of regrowing shoots during early regrowth until nodule function is restored.

Recent results using ^{15}N labeling have confirmed that substantial N is transferred from roots to regrowing shoots after defoliation (Kim et al. 1991; Ourry et al. 1994; Barber et al. 1996). Extensive transfer of root N to regrowing shoots occurs during the initial 10 days after defoliation. While no relationship between root-starch concentration and shoot regrowth rate was observed, highly significant correlations between root-N levels at defoliation and subsequent shoot dry matter production were obtained. These researchers concluded that the main factor limiting alfalfa shoot regrowth after cutting was availability of N reserves in roots and crowns.

Nitrogen can be a factor limiting shoot growth rates of alfalfa. Hodgkinson (1973) reported that apices of slow-growing shoots had about one-half the N concentration of fast-growing shoots. He speculated that redistribution of nutrients from roots to shoots during early regrowth may be significant in meeting the nutrient requirements of regrowing shoots. Our preliminary results using ineffective nodulating alfalfa demonstrate the importance of root N on alfalfa shoot regrowth (Fig. 3). In these experiments the N and protein levels of roots of ineffective nodulating alfalfa are readily increased to near-normal levels using fertilizer N. After defoliation and transfer of crowns and

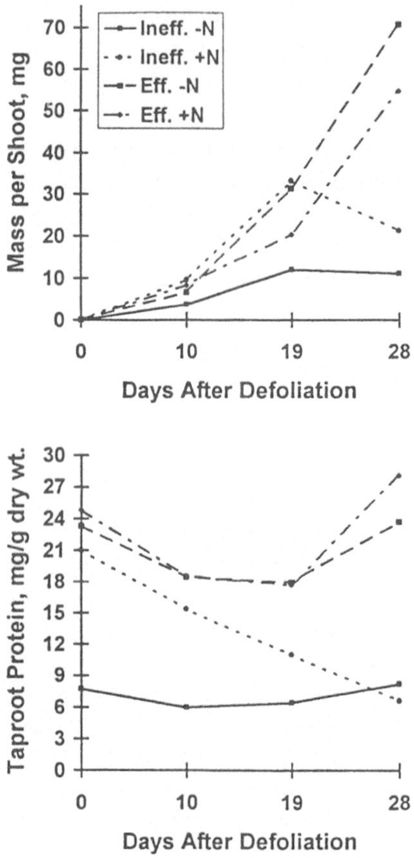

Fig. 3. Influence of root protein level on shoot dry weight accumulation in alfalfa. Well-established plants of ineffective nodulating Saranac (*Ineff.*) or effectively nodulating Saranac (*Eff.*) were provided fertilizer nitrogen (*+N*) or were not provided nitrogen (*−N*) during a previous regrowth cycle. After defoliation, crowns and taproots were carefully washed free of soil and transplanted into coarse sand to remove the +N plants from their N-rich environment. Preapplication of N did not influence root-protein levels or shoot regrowth of effective alfalfa, but these plants exhibited a cyclic pattern of root-protein depletion after defoliation followed by reaccumulation after day 19. Preapplication of N to ineffective nodulating alfalfa increased root-protein levels. Shoot regrowth of the ineffective plants +N was similar to that of effective plants until after day 19, when their root-protein levels declined to values similar to those observed for the ineffective plants −N. (B.S. Ruff, S.M. Cunningham, and J.J. Volenec, unpubl.)

roots to N-free sand culture conditions, shoot regrowth rates of ineffective nodulating plants with high root protein equals that of effectively nodulated alfalfa (with or without prior N fertilization), and exceeds that of ineffective plants possessing low root-protein levels. The cessation of shoot mass accumulation of ineffective plants that were fertilized with N prior to transplanting coincides with depletion of root protein to levels found in normal, ineffectively nodulating alfalfa. Field studies also have confirmed extensive transfer of root N to regrowing shoots (Lemaire et al. 1992; Hendershot and Volenec 1993a), and its importance in winter hardening and initial herbage growth in spring (Bula and Smith 1954; Bula et al. 1956; Volenec et al. 1991; Hendershot and Volenec 1993b).

The form(s) of root N available as an N source for regrowing shoots has recently been investigated. Alfalfa roots accumulate substantial quantities of N as protein and amino acids. Relative abundance of these N pools declines when shoot growth is initiated in spring and when herbage regrowth occurs after defoliation (Volenec et al. 1991; Hendershot and Volenec 1993a,b). Barber et al. (1996) labeled root N pools with ^{15}N and determined that preferential mobilization of N occurred from the amino acid pool and a pool enriched in three putative vegetative storage proteins. Because they comprise up to one-third of the buffer-soluble protein pool and appear to be preferentially mobilized from roots to regrowing shoots, we have characterized these vegetative storage proteins (Cunningham and Volenec 1996). These proteins accumulate to high concentrations in alfalfa roots, but are not present in leaves, stems, nodules, or seeds. They are localized in ray parenchyma cells of wood tissues, and throughout bark tissues (Avice et al. 1996b). Within alfalfa root cells, these storage proteins accumulate in vacuoles, and are associated with the periphery of starch grains. They are abundant in roots of all perennial *Medicago* species examined to date, but are found in trace amounts or not at all in roots of annual *Medicago* species. Roots of red clover (*Trifolium pratense* L.), birdsfoot trefoil (*Lotus corniculatus* L.), and sweetclover (*Melilotus officinalis* L.) also do not contain these proteins even though these are biennial or perennial species are used for forage. This is despite the fact that root-protein concentration in these forage legume species declines sharply in spring when shoot growth is initiated and, with the exception of trefoil, after defoliation in summer (Li et al. 1996). Our work is continuing to focus on the molecular mechanisms controlling synthesis and degradation of alfalfa root vegetative storage proteins.

3.3 Shoots per Plant

Environment interacts with plant genotype in determining shoots per plant. Genetic differences in shoots per plant, that might be expressed at low plant populations characteristic of old stands (Fig. 1), are generally not expressed in young stands where competition among plants is high (Volenec et al. 1987). Low competition allows crowns to spread, generally increasing shoots per plant, and in doing so, maintains herbage yield per area (Miller et al. 1969; Chatterton et al. 1977). For example, mechanical thinning of alfalfa stands reduced plant populations by one-third, but resulted in only a 20% reduction in yield the 1st year after thinning and a mere 9% reduction in the 2nd year (Stout and Thompson 1994). The ability to compensate for plant loss by producing more shoots per plant is an important adaptation because intensive harvest schedules (that eliminate flower production) and allelopathy prevent alfalfa from reseeding itself (Beuselinck et al. 1994). Crown spreading also plays a critical role in adaptation of alfalfa to continuous stocking by maintaining high leaf area below the grazing heights accessible to livestock (Smith et al. 1992).

Few studies have evaluated how shoot production per plant is controlled. Management influences shoots per plant, and in doing so, brings about increases in herbage yield. For example, shoots per plant are enhanced with proper soil fertility, especially phosphorus and potassium, and are reduced as harvest frequency increases from 6- to 2-week intervals (Nelson et al. 1986). Little is known concerning regulation of bud dormancy and bud development of alfalfa despite the obvious agricultural importance of this information. Inference from work where analysis of crowns was conducted is inadequate because segments of previously harvested stems represent the majority of the crown mass and would most likely mask important differences in bud physiology and biochemistry. Further complicating our understanding of bud and early shoot development is the need to identify and analyze separately the physiological features of developing buds and those of quiescent or dormant buds. Hodgkinson (1973) suggested that rapidly developing buds which begin growth soon after defoliation may exert apical dominance over other buds, thereby reducing their growth rates. Growth substances like auxins may control the number of shoots becoming "dominant" by altering the partitioning of carbohydrates, amino acids, and minerals among buds. The continued supply of these nutrients facilitated rapid growth of dominant shoots. By restricting the number of shoots developing per crown from eight to two, Leach (1971) was able to demonstrate intershoot competition for growth substrates from roots. However, he did not identify the nature of the factor(s) limiting shoot development.

4 Conclusions and Future Prospects

Alfalfa herbage yield is determined by three components: plants per area; shoots per plant; and mass per shoot. Available data suggest that increases in mass per shoot will translate to improved yield, whereas compensation between plants per area and shoots per plant often occurs, resulting in little change in shoots per area. Shoot growth rate determines mass acquisition per shoot, and has potential as a selection criterion for improved yield. Historically, many of the adaptive responses of alfalfa have been associated with the accumulation of root TNC. While the conventional wisdom that root TNC depletion causes slow regrowth and poor persistence may be a useful model because it is easily understood, data supporting this concept at the mechanistic level are however, lacking. New information on root N reserves will help clarify the role of roots in alfalfa stress tolerance, and may lead to innovative strategies for alfalfa improvement.

Genetic selection continues to be the primary mechanism by which we increase yield and stress tolerance of alfalfa cultivars. However, as in many crop species, future improvements by genetic manipulation, especially using modern genetic techniques, will depend on new insights into basic physiological, biochemical, and morphogenic plant processes. Knowledge is critically deficient in several areas of alfalfa research including: plant vs. environmental factors altering crown bud dormancy and bud development; physiological mechanisms regulating fall-dormancy reaction; mechanisms conditioning genetic differences in winter hardiness; plant aging and alfalfa persistence;

clarification of the role of root TNC reserves in shoot regrowth and plant persistence; and the physiological limits to shoot regrowth after defoliation. New insights into these long-standing problems related to alfalfa growth and persistence are critically needed if we are to alter the very slow rate of genetic improvement in alfalfa yield attained during the last century.

Acknowledgments. The insightful reviews of S.M. Cunningham, K.D. Johnson, and H.R. Koller were appreciated. This is contribution No. 14734 from the Purdue University Agricultural Research Programs office.

References

Avice JC, Ourry A, Lemaire G, Boucaud J (1996a) Nitrogen and carbon flows estimated by ^{15}N and ^{13}C pulse-chase labeling during regrowth of alfalfa. Plant Physiol 112:281–290
Avice JC, Ourry A, Volenec JJ, Lemaire G, Boucaud J (1996b) Defoliation-induced changes in abundance and immuno-localization of vegetative storage proteins in taproots of *Medicago sativa*. Plant Physiol Biochem 34:561–570
Barber LD, Joern BC, Volenec JJ, Cunnigham SM (1996) Supplemental nitrogen effects on alfalfa regrowth and nitrogen mobilization from roots. Crop Sci 36:1217–1223
Beuselinck PR, Bouton JH, Lamp WO, Matches AG, McCaslin MH, Nelson CJ, Rhodes LH, Sheaffer CC, Volenec JJ (1994) Improving legume persistence in forage crop systems. J Prod Agric 7:311–322
Bolger TP, Meyer DW (1983) Influence of plant density on alfalfa yield and quality. In: Proc 1983 Forage and Grassl Conf, 23–26 Jan 1993, Eau Claire WI., Am Forage and Grassl Council, Lexington, Kentucky
Bowley SR, McKersie BD (1990) Relationship among freezing, low temperature flooding, and ice encasement tolerance in alfalfa. Can J Plant Sci 70:227–235
Boyce PJ, Volenec JJ (1992) Taproot carbohydrate concentrations and stress tolerance of contrasting alfalfa genotypes. Crop Sci 32:757–761
Brink GE, Marten GC (1989) Harvest management of alfalfa – nutrient yield vs. forage quality, and relationship to persistence. J Prod Agric 2:32–36
Brummer EC, Bouton JH (1991) Plant traits associated with grazing-tolerant alfalfa. Agron J 83:996–1000
Brummer EC, Bouton JH (1992) Physiological traits associated with grazing-tolerant alfalfa. Agron J 84:138–143
Bula RJ, Smith D (1954) Cold resistance and chemical composition in overwintering alfalfa, red clover, and sweetclover. Agron J 46:397–401
Bula RJ, Smith D, Hodgson HJ (1956) Cold resistance in alfalfa at two diverse latitudes. Agron J 48:153–156
Castonguay Y, Nadeau P, Lechasseur P, Chouinard L (1995) Differential accumulation of carbohydrates in alfalfa cultivars of contrasting winterhardiness. Crop Sci 35:509–516
Chatterton NJ, Akao S, Carlson GE, Hungerford WE (1977) Physiological components of yield and tolerance to frequent harvests in alfalfa. Crop Sci 17:918–923
Constable GA, Sheridan KP, Gleeson AC (1977) Effects of sequential defoliation on lucerne (*Medicago sativa* L.). Aust J Agric Res 28:769–776
Cralle HT, Heichel GH (1981) Nitrogen fixation and vegetative regrowth of alfalfa and birdsfoot trefoil after successive harvests or floral debudding. Plant Physiol 67:898–905
Cunningham SM, Volenec JJ (1996) Purification and characterization of vegetative storage proteins from alfalfa (*Medicago sativa* L.) taproots. J Plant Physiol 147:625–632
Edminsten KL, Wolf DD, Lentner M (1988) Fall harvest management of alfalfa. I. Date of fall harvest and length of growth period prior to fall harvest. Agron J 80:688–693

Fankhauser JJ Jr, Volenec JJ (1989) Root vs. shoot effects on herbage regrowth and carbohydrate metabolism of alfalfa. Crop Sci 29:735–740

Feltner KC, Massengale MA (1965) Influence of temperature and harvest management on growth, level of carbohydrates in roots, and survival of alfalfa (*Medicago sativa* L.). Crop Sci 5:585–588

Gosse G, Lemaire G, Chartier M, Balfourier F (1988) Structure of a lucerne population (*Medicago sativa* L.) and dynamics of stem competition for light during regrowth. J Appl Ecol 25:609–617

Graber LF, Nelson NT, Luekel WT, Albert WB (1927) Organic food reserves in relation to growth of alfalfa and other perennial herbaceous plants. Research Bull 80, Univ of Madison, Wisconsin

Habben JE, Volenec JJ (1990) Starch grain distribution in taproots of defoliated *Medicago sativa* L. Plant Physiol 94:1056–1061

Heichel GH, Henjum KI (1990) Fall dormancy response of alfalfa investigated with reciprocal cleft grafts. Crop Sci 30:1123–1127

Hendershot KL (1986) Shoot origin, shoot growth, and carbohydrate metabolism of two contrasting alfalfa genotypes. Msc Thesis, Purdue University, West Lafayette, Indiana

Hendershot KL, Volenec JJ (1989) Shoot growth, dark respiration, and nonstructural carbohydrates of contrasting alfalfa genotypes. Crop Sci 29:1271–1275

Hendershot KL, Volenec JJ (1993a) Nitrogen pools in taproots of *Medicago sativa* L. after defoliation. J Plant Physiol 141:129–135

Hendershot KL, Volenec JJ (1993b) Taproot nitrogen accumulation and use in overwintering alfalfa. J Plant Physiol 141:68–74

Hodgkinson KC (1969) The utilization of root organic compounds during the regeneration of lucerne. Aust J Biol Sci 22:1113–1123

Hodgkinson KC (1973) Establishment and growth of shoots following low and high cutting of lucerne in relation to the pattern of nutrient uptake. Aust J Agric Res 24:497–510

Holland JB, Bingham ET (1994) Genetic improvement for yield and fertility of alfalfa cultivars representing different eras of breeding. Crop Sci 34:953–957

Jones ER, Carroll RB, Swain RH, Bell KW (1984) Evaluation of factors affecting yield reduction in mature alfalfa (*Medicago sativa* L.). Can J Plant Sci 64:901–907

Kephart KD, Twidwell EK, Bortnem R, Boe A (1992) Alfalfa yield component responses to seeding rate several years after establishment. Agron J 84:827–831

Kim TH, Ourry A, Boucaud J, Lemaire G (1991) Changes in source-sink relationship for nitrogen during regrowth of lucerne (*Medicago sativa* L.) following removal of shoots. Aust J Plant Physiol 18:593–602

Leach GJ (1968) The growth of the lucerne plant after cutting: the effects of cutting at different stages of maturity and at different intensities. Aust J Agric Res 19:517–530

Leach GJ (1969) Shoot numbers, shoot size, and yield of regrowth in three lucerne cultivars. Aust J Agric Res 20:425–434

Leach GJ (1971) The relation between shoot growth and temperature. Aust J Agric Res 22:49–59

Lemaire G, Onillon B, Gosse G, Chartier M, Allirand JM (1991) Nitrogen distribution within a lucerne canopy during regrowth: relation with light distribution. Ann Bot 68:483–488

Lemaire G, Khaity M, Onillon B, Allirand JM, Chartier M, Gosse G (1992) Dynamics of accumulation and partitioning of N in leaves, stems and roots of lucerne (*Medicago sativa* L.) in a dense canopy. Ann Bot 70:429–435

Li R, Volenec JJ, Joern BC, Cunningham SM (1996) Seasonal changes in nonstructural carbohydrates, proteins, and macronutrients in roots of alfalfa, red clover, sweetclover, and birdsfoot trefoil. Crop Sci 36:617–623

Liang GHL, Riedl WA (1964) Agronomic traits influencing forage and seed yield in alfalfa. Crop Sci 4:394–396

May LH (1960) The utilization of carbohydrate reserves in pasture plants after defoliation. Herb Abstr 30:239–245

McKersie BD, Chen Y, de Beus M, Bowley SR, Bowler C, Inze D, D'alluin K, Botterman J (1993) Superoxide dismutase enhances tolerance of freezing stress in transgenic alfalfa (*Medicago sativa* L.). Plant Physiol 103:1155–1163

Miller DA, Shrivastava JP, Jackobs JA (1969) Alfalfa yield components in solid seedings. Crop Sci 9:440–443

Monroy AF, Sarhan F, Dhindsa RS (1993a) Cold-induced changes in freezing tolerance, protein phosphorylation, and gene expression. Evidence for a role for calcium. Plant Physiol 102:1227–1235

Monroy AF, Castonguay Y, Laberge F, Sarhan F, Vezina LP, Dhindsa RS (1993b) A new cold-induced alfalfa gene associated with enhanced hardening at sub-zero temperature. Plant Physiol 102:873–879

Musgrave DJ, Langer RHM (1977) Crown development of two diverse genotypes of lucerne. NZ J Agric Res 20:453–458

Nelson CJ, Smith D (1968) Growth of birdsfoot trefoil and alfalfa. II. Morphological development and dry matter distribution. Crop Sci 8:21–25

Nelson CJ, Buchholz DD, Rausch DL, Coutts JH (1986) Managing alfalfa and alfalfa-grass mixtures for persistence. Proc 16th Nat Alf Symp, Fort Wayne, Indiana, 5–6 March 1986. Certified Alfalfa Seed Council, Woodland, California

Ourry A, Kim TH, Boucaud J (1994) Nitrogen reserve mobilization during regrowth of *Medicago sativa* L. Relationships between availability and regrowth yield. Plant Physiol 105:831–837

Philippot S, Allirand JM, Chartier M, Gosse G (1991) The role of different daily irradiations on shoot growth and root/shoot ratio in lucerne (*Medicago sativa* L.). Ann Bot 68:329–335

Rumbaugh MD (1963) Effects of population density on some components of yield of alfalfa. Crop Sci 3:423–424

Sheaffer CC, Barnes DK, Warnes DD, Lueschen WE, Ford HJ, Swanson DR (1992) Seeding year cutting affects winter survival and its association with fall growth score in alfalfa. Crop Sci 32:225–231

Singh Y, Winch JE (1974) Morphological development of two alfalfa cultivars under various harvesting schedules. Can J Plant Sci 54:79–87

Smith LH, Marten GC (1970) Foliar regrowth of alfalfa utilizing ^{14}C-labeled carbohydrates stored in roots. Crop Sci 10:146–150

Smith SR Jr, Bouton JH (1993) Selection within alfalfa cultivars for persistence under continuous stocking. Crop Sci 33:1321–1328

Smith SR Jr, Bouton JH, Hoveland CS (1989) Alfalfa persistence and regrowth potential under continuous grazing. Agron J 81:960–965

Smith SR Jr, Bouton JH, Hoveland CS (1992) Persistence of alfalfa under continous grazing in pure stands and in mixtures with tall fescue. Crop Sci 32:1259–1264

Stout DG, Hall JW (1989) Fall growth and winter survival of alfalfa in interior British Columbia. Can J Plant Sci 69:491–499

Stout DG, Thompson DT (1994) Yield and compensatory growth of lucerne following simulated winter-kill. Grass Forage Sci 49:401–404

Stout DG, Acharya SN, Huang HC, Hanna MR (1992) Alfalfa plant death during the summer versus the winter in interior British Columbia. Can J Plant Sci 72:931–936

Summers CG, Gilchrist DG (1991) Temporal changes in forage alfalfa associated with insect and disease stress. J Econ Entomol 84:1353–1363

Suzuki M (1991) Effects of stand age on agronomic, morphological and chemical characteristics of alfalfa. Can J Plant Sci 71:445–452

Vance CP, Heichel GH (1981) Nitrate assimilation during vegetative regrowth of alfalfa. Plant Physiol 68:1052–1056

Vance CP, Heichel GH, Barnes DK, Bryan JW, Johnson LE (1979) Nitrogen fixation, nodule development, and vegetative regrowth of alfalfa (*Medicago sativa* L.) following harvest. Plant Physiol 64:1–8

Volenec JJ (1985) Leaf area expansion and shoot elongation of diverse alfalfa germplasms. Crop Sci 25:822–827

Volenec JJ (1988) Herbage growth and carbohydrate metabolism of diploid and tetraploid alfalfa. Crop Sci 28:128–132

Volenec JJ, Nelson CJ (1995) Forage crop management: application of emerging technologies. In: Barnes RF, Miller DA, Nelson CJ (eds) Forages vol II: The science of grassland agriculture. Iowa State University Press, Ames Iowa, pp 3–20

Volenec JJ, Cherney JH, Johnson KD (1987) Yield components, plant morphology, and forage quality of alfalfa as influenced by plant population. Crop Sci 27:321–326

Volenec JJ, Boyce PJ, Hendershot KL (1991) Carbohydrate metabolism in taproots of *Medicago sativa* L. during winter adaptation and spring regrowth. Plant Physiol 96:786–793

Volenec JJ, Ourry A, Joern BC (1996) A role for nitrogen reserves in forage regrowth and stress tolerance. Physiol Plant 97:185–193

Wiersma DW, Grau CR, Undersander DJ (1995) Alfalfa cultivar performance with differing levels of resistance to Phytophthora and Aphanomycetes root rots. J Prod Agric 8:259–264

Winsett BL (1991) The inheritance of shoot weight, shoot number and yield per plant in two diverse alfalfa germplasms. Msc Thesis, Purdue University, West Lafayette, Indiana

Physiological Control of Forage Grass Yield and Growth

A.R. Overman and D.M. Wilson[1]

1 Introduction

Forage grasses are used in a variety of ways in both managed and natural systems. These uses include animal feed (hay, greenchop, haylage, and pasture), crop rotations, erosion control, soil improvement, and waste management. Forage grass management has been discussed by numerous authors (Mays 1974; Buckner and Bush 1979; Heath et al. 1985; Ball et al. 1991; Fageria et al. 1991; Dovrat 1993). Other related topics include plant nutrition (Marschner 1986; Mengel and Kirkby 1987; Glass 1989; Jones 1994) and soil erosion (Schwab et al. 1993).

A primary focus of this chapter is on simplified models of forage production. The subject of crop modeling has been dealt with extensively (Hunt 1982; Ratkowsky 1983; France and Thornley 1984; Charles-Edwards et al. 1986; Tilman 1988; Cerrato and Blackmer 1990; Hanks and Ritchie 1991; Keen and Spain 1992). The subject of population models has a long history (Verhulst 1838; Lotka 1925; Pearl 1925; Volterra 1926; Ginzburg and Golenberg 1985; Kingsland 1985; Caswell 1989; Thompson 1992; Murray 1993; Huxley 1993). These models have included the logistic equation, first proposed by Verhulst (1838) to correct the simple Malthus model of geometric growth, and later popularized by Pearl (1925). The controversy surrounding various models of population dynamics has been reviewed by Kingsland (1985). People interested in modeling may find these references very informative. The general subject of models is treated by numerous authors (Davis and Hersh 1981; Eigen and Winkler 1981; Kramer 1981; Will 1986; Gleick 1987; Rucker 1987; Pagels 1988; Bunch 1989; Freund and Littell 1991; Lindley 1993; Aris 1994). Logistic regression has been discussed by Hosmer and Lemeshow (1989) and Kleinbaum (1994). The weakness of polynomial models has been treated in detail (Rucker 1987; Lanczos 1988; Freund and Littell 1991). In this analysis we view a model as a simplification of reality (Gleick 1987) and apply Occam's razor to simplify to the essentials required for description (Hunt 1982; Will 1986; Davies 1992). In fact, in the field of physics a functional approach has replaced the strictly

[1] Agricultural & Biological Engineering Department, Rogers Hall, Museum Road, University of Florida, P.O. Box 110570, Gainesville, Florida, 32611-0570, USA

Crop Yield, Physiology and Processes
D. L. Smith, C. Hamel (Eds.)
© Springer-Verlag Berlin Heidelberg 1999

mechanistic one (Einstein and Infeld 1938; Walker 1963; Feynman 1986; Weinburg 1993).

2 Dynamic Models

In the present context, a dynamic model is one which contains time explicitly (France and Thornley 1984). The first step will be to relate dry matter accumulation (harvested portion) to time through an empirical model using the probability function. This simple model is extended to include accumulation of plant nutrient. A phenomenological model is then developed which combines an environmental function with a growth function to describe dry matter accumulation over the season. This model is related to partitioning of dry matter between stems and leaves for a warm-season grass.

2.1 Probability

Examination of field data for the warm-season perennial Coastal bermudagrass [*Cynodon dactylon* (L.) Pers.] led Overman (1984) to use the probability model

$$Y = A/2\{1 + \mathrm{erf}[(t - \mu)/\sqrt{2}\sigma]\}, \tag{1}$$

where Y is cumulative dry matter yield, t is time in the season, A is maximum dry matter, μ is time to the mean of the dry matter distribution, and σ is the standard deviation of the distribution. The error function in Eq. (1) is given by

$$\mathrm{erf}\, z = 2/\sqrt{\pi} \int_0^z \exp(-u^2)du. \tag{2}$$

By defining the fractional (normalized) yield as F = Y/A, Eq. (1) can be written

$$F = 1/2\{1 + \mathrm{erf}[(t - \mu)/\sqrt{2}\sigma]\}. \tag{3}$$

Data which follow Eq. (3) produce a straight line on normal probability paper. Overman (1984) showed this to be true for bermudagrass grown at Watkinsville, Georgia (Mays et al. 1980). This applied for fixed harvest intervals of 4 and 6 weeks, as well as for irrigated and nonirrigated treatments. Since this analysis simply used nonlinear regression to determine the parameters A, μ, and σ, the approach was classified as an empirical model. The high quality of the probability plot suggested a possible rational basis for this approach, which led to the phenomenological model discussed below.

2.2
Extended Probability

This simple idea was then expanded by Overman et al. (1988a) to include plant N accumulation (harvested portion) as well as for dry matter. This leads to the extension

$$N_u = A_n/2\{1 + \mathrm{erf}[(t - \mu_n)/\sqrt{2}\sigma_n]\}, \quad (4)$$

where N_u is cumulative plant N, A_n is maximum plant N, μ_n is time to the mean of the plant N distribution, and σ_n is the standard deviation of the distribution. Fractional (normalized) plant N, $F_n = N_u/A_n$, is given by

$$F_n = 1/2\{1 + \mathrm{erf}[(t - \mu_n)/\sqrt{2}\sigma_n]\}. \quad (5)$$

Since it was shown that normalized plant N vs. normalized dry matter produced a straight line for bermudagrass data, it was concluded that $\mu_n = \mu$ and $\sigma_n = \sigma$. Further analysis showed that the extended model was applicable to various treatments including applied N, harvest interval, and available water (irrigated and nonirrigated).

Further evaluation of the empirical model (Overman et al. 1988b) showed that it applied to data for bermudagrass from several sites with different soils, which provided even greater confidence in this simple approach. Model parameter A was shown to depend upon applied N.

2.3
Phenomenological

The success of the empirical model in describing accumulation of dry matter and plant N by bermudagrass led to speculation about a possible rational basis (Overman et al. 1989). This model is based on three postulates: (1) an environmental function, (2) an intrinsic growth function, and (3) net growth rate which is the product of (1) and (2). Assume that the environmental function, E, follows a Gaussian distribution over the season and is given by

$$E = \mathrm{constant}\ \exp\left\{-[(t - \mu)/\sqrt{2}\sigma]^2\right\}, \quad (6)$$

where μ and σ are as defined above. We postulate a quadratic intrinsic growth function given by

$$dY'/dt = a + b(t - t_i), \quad (7)$$

where a and b are linear parameters for the growth function and t_i is the time of the previous cutting. The net growth rate, dY/dt, at time t is then assumed to be given by

$$dY/dt = \mathrm{constant}\ [a + b(t - t_i)]\ \exp\left\{-[(t - \mu)/\sqrt{2}\sigma]^2\right\}. \quad (8)$$

Two reference times are apparent in Eq. (8), viz. t_i and μ, which seems intuitively reasonable. Equation (8) is piecewise continuous due to harvests at discrete times, t_i. We now define the dimensionless variable

$$x = (t - \mu)/\sqrt{2}\sigma \tag{9}$$

from which it follows that

$$(t - t_i)/\sqrt{2}\sigma = (t - \mu)/\sqrt{2}\sigma - (t_i - \mu)/\sqrt{2}\sigma = x - x_i = \Delta x_i \tag{10}$$

Equation (8) can now be written in the form

$$dY/dt = A[2/\sqrt{\pi} + 4/\sqrt{\pi}C(x - x_i)] \exp(-x^2), \tag{11}$$

where A and $C = \sigma b/\sqrt{2}\ a$ are constants. The quantity x_i is fixed for a particular growth interval, but changes in discrete steps from one interval to the next.

Equation (11) can be integrated (Abramowitz and Stegun 1965) to give the yield increment, ΔY_i,

$$\Delta Y_i = AQ_i, \tag{12}$$

where

$$\begin{aligned} Q_i = {} & (1 - 2Cx_i)[\mathrm{erf}(x_i + \Delta x_i) - \mathrm{erf}(x_i)] \\ & -2/\sqrt{\pi}\, C\{\exp[-(x_i + \Delta x_i)^2] - \exp[-(x_i)^2]\}. \end{aligned} \tag{13}$$

Now the cumulative yield, Y_n, for n harvests is given by

$$Y_n = \Sigma \Delta Y_i = A\Sigma Q_i = AQ_n. \tag{14}$$

Normalized yield fraction for n harvests, F_n, is then given by

$$F_n = Y_n/Y_T = Q_n/Q_T, \tag{15}$$

where Y_T is the total dry matter for the season, and Q_n and Q_T are summed for n harvests and over the season, respectively. Overman et al. (1989) have shown that for a fixed interval, $\Delta x_i = \Delta x$, Eq. (15) produces a straight line on normal probability paper. It was further shown that

$$Q_T = 2 + 2C\Delta x. \tag{16}$$

Since field data (Prine and Burton 1956) for seasonal dry matter yield vs. harvest interval, Δt, have been shown to follow a straight line (for harvest intervals up to 6 weeks) then

$$Y_T = \alpha + \beta\Delta t, \tag{17}$$

where α and β are linear parameters, and the model is consistent with field data for bermudagrass. Finally, it was shown that both model parameters A and AC followed logistic response to applied N (Overman et al. 1990a).

What is the physical basis for this model, i.e., to what do the model parameters A and AC correspond in the plant system? Overman and Wilkinson

(1989) used data from Tifton, Georgia (Burton et al. 1963), and College Station, Texas (Holt and Conrad 1986), on partitioning of dry matter between leaves and stems for bermudagrass to show that A and AC correspond to leaves and stems, respectively, for harvest intervals up to 6 weeks. For longer intervals, leaf loss occurs due to shading of lower leaves.

The phenomenological model contains four parameters, viz. μ, σ, A, and AC. The first two are evaluated from the probability distribution of dry matter over the growing season, while the last two come from the dependence of seasonal total dry matter on harvest interval. Overman et al. (1989) described growth patterns between harvests as the season passes. The model could be used to describe yield data for a variable harvest interval, but the procedure would be somewhat more involved than discussed here.

3 Seasonal Models

In the previous section model parameters A and AC were found to follow logistic response to applied N. Is the relationship somewhat general, or is it unique to the data sets chosen for analysis? In many cases managers and others may only be interested in the dependence of seasonal total yield and plant nutrient uptake upon applied nutrients, and perhaps other factors. More data of this type have been reported in the literature. And what about annual crops which are harvested only once during the season? In this section we expand upon this topic in considerable detail.

3.1 Logistic

The logistic equation can be written as

$$Y = A/[1 + \exp(b - cN)], \tag{18}$$

where Y is the total dry matter yield for the season, N is applied nitrogen, A is estimated maximum yield, b is the intercept parameter (N = 0) for dry matter, and c is the nutrient response coefficient. Model parameters A, b, and c are determined from data by nonlinear regression. Model characteristics and parameter estimation have been discussed by Overman et al. (1990b). The function exhibits sigmoid behavior and is bounded by 0 and A, with an inflection point at N = b/c and Y = A/2. Overman et al. (1990c) showed by analysis of variance that dependence of yield upon harvest interval and water availability (year-to-year variations) could be accounted for in the linear parameter A, with b and c constant. In fact, parameter A showed linear dependence upon harvest interval for bermudagrass at Tifton, Georgia. This greatly simplifies data analysis and makes it legitimate to average data over years in many cases. Field studies with bahiagrass and bermudagrass (Overman et al. 1992) showed that the effect of overseeding with clover was to increase available soil N and decrease the intercept parameter b with A and c unchanged.

3.2 Extended Logistic

Plant nutrient accumulation is of interest to many different groups, including producers, managers, engineers, and regulators. Hence, a procedure is needed to relate nutrient uptake to factors such as applied nutrient, water availability, and harvest interval. An extension of the logistic model has been proposed by Overman et al. (1994a)

$$N_u = A_n/[1 + \exp(b_n - cN)], \tag{19}$$

where N_u is plant nutrient uptake, A_n is estimated maximum plant nutrient uptake, and b_n is the intercept parameter for plant nutrient uptake. Response coefficient c is assumed to be the same for dry matter and plant nutrient accumulation. A consequence of Eqs. (18) and (19) with common c is the hyperbolic relationship between dry matter yield and plant nutrient uptake

$$Y = Y_m N_u/(K' + N_u), \tag{20}$$

where maximum potential yield, Y_m, is given by

$$Y_m = A/[1 - \exp(b - b_n)] \tag{21}$$

and the system response coefficient, K', by

$$K' = A_n/[\exp(b_n - b) - 1]. \tag{22}$$

It also follows that plant nutrient concentration, $N_c = N_u/Y$, is linearly related to plant nutrient uptake by

$$N_C = (K'/Y_m) + (1/Y_m)N_u. \tag{23}$$

Dependence of plant nutrient concentration on applied N is given by the ratio of Eqs. (19) to (18), so that

$$N_C = N_{cm}[1 + \exp(b - cN)]/[1 + \exp(b_n - cN)], \tag{24}$$

where $N_{cm} = A_n/A$ = maximum plant nutrient concentration.

Data from Louisiana (Robinson et al. 1988) for dallisgrass (*Paspalum dilatatum* Poir.) response to applied N were used by Overman et al. (1994a) to illustrate the validity of this model, with a correlation coefficient of 0.9990. Equations (20) and (23) were followed very closely. Maximum potential yield of 27.5 Mg ha^{-1}, calculated from Eq. (21), appears to be in the range reported in the literature. The model showed plant N concentration bounded by $11.8 < N_c < 27.5$ g kg^{-1}. Further analysis by Overman et al. (1995a) has shown that the extended model can be used to couple dry matter and plant nutrient uptake for applied P and applied K, as well as applied N. It has also been applied to cool-season grass (Overman et al. 1995b).

The logistic model has been criticized as not having a mechanistic basis. On the other hand, Overman (1995a) has shown that it does have a rational basis. A brief summary of this reasoning is now given. The logistic differential equation can be written as

$$dY/dN = kY(A - Y) \quad \text{with } Y = Y_0 \quad \text{at } N = 0, \tag{25}$$

where k is the response coefficient. The solution to Eq. (25) is given by Eq. (18), where $c = kA$ and $b = \ln(A/Y_0 - 1)$. By defining the dimensionless variables $\varphi = Y/A$ and $\xi = cN - b$, Eq. (18) can be written as

$$\varphi = 1/[1 + \exp(-\xi)] \tag{26}$$

and Eq. (25) as

$$d\varphi/d\xi = \varphi(1 - \varphi) = \exp(-\xi)/[1 + \exp(-\xi)]^2 . \tag{27}$$

Now Eq. (27) can be shown to exhibit bilateral symmetry of $d\varphi/d\xi$ on ξ, similar to the Gaussian bell-shaped curve. In physics, symmetry in models has proven very significant and related to conservation of various quantities (Eigen and Winkler 1981; Kramer 1981; Pagels 1985; Rothman 1985; Ferris 1988; Krauss 1993). Then we might ask as to the significance of symmetry in Eq. (27)? What in the system is conserved? The answer lies in Eq. (25). If we view Y as the filled capacity of the system and (A − Y) as the unfilled capacity, then the total capacity is given by the sum $Y + (A - Y) = A$, which remains constant (is conserved) with variations in applied N. This assumes that other factors (such as other nutrients, harvest interval, water availability, etc.) remain constant. Since dY/dN represents incremental increase in dry matter with applied N, it seems reasonable that increase would depend upon amount present and upon unused resources available.

Equation (20) suggests that factors other than uptake of the particular nutrient in question ultimately limit potential production. Overman et al. (1994b) showed this true for maize (*Zea mays* L.) grown in North Carolina. Maximum potential grain production was estimated to be 25.9 Mg ha^{-1}, in excellent agreement with the estimate of 26.5 Mg ha^{-1} (500 bu $acre^{-1}$) made by Tollenaar (1986) for conditions in the midwestern United States. It was also shown that the hyperbolic relationship was followed for both total plant and grain components. Linear dependence of model parameter A on seasonal water has been demonstrated for maize (Reck and Overman 1996).

3.3 Multiple Logistic

The next extension of the logistic model is to include response to multiple nutrients such as N, P, and K. If response to each nutrient is logistic and if we can treat each nutrient as independently applied, then logically the combined function would have the product form

$$Y = A/[1 + \exp(b_n - c_nN)][1 + \exp(b_p - c_pP)] \\ [1 + \exp(b_k - c_kK)], \quad (28)$$

where subscripts n, p, and k refer to applied N, P, and K, respectively. This model was applied by Overman et al. (1991) to factorial data for Coastal bermudagrass grown at Watkinsville, Georgia (Carreker et al. 1977), with a resulting correlation coefficient of 0.9884. As is often observed, dry matter accumulation was less responsive to applied P and K than to applied N due to high levels of soil P and K.

Several characteristics of Eq. (28) may be noted. With fixed levels of P and K, it reduces to the simple logistic model with a modified maximum parameter A, as observed from field studies. The model allows evaluation of the effect of combinations of N, P, and K on dry matter production. Applied nutrient required to achieve half of maximum yield is given by $N_{1/2} = b_n/c_n$, $P_{1/2} = b_p/c_p$, and $K_{1/2} = b_k/c_k$. A negative b value means that the soil already contains more than enough nutrient to achieve half of maximum yield.

3.4 Extended Multiple Logistic

The multiple logistic model is now extended in a manner similar to the simple logistic equation. It is assumed that the multiple logistic model applies to plant nutrient uptake as well as to dry matter yield, so that

$$N_u = A_n/[1 + \exp(b_n' - c_nN)][1 + \exp(b_p' - c_pP)] \\ [1 + \exp(b_k' - c_kK)] \quad (29)$$

where A_n is maximum plant nutrient uptake (N, P, or K) and b′ are intercept parameters for each nutrient (N, P, or K). As with the simple extended model, response coefficient c is assumed common for plant nutrient uptake and dry matter yield. Overman and Wilkinson (1995) used the simplifying assumption that, to first approximation at least, plant nutrient concentration for a given element is independent of the applied level of the other two. For example, for plant N only A_n and b_n are different for plant N uptake and dry matter, while $b_p' = b_p$ and $b_k' = b_k$. For plant N concentration, N_c, this leads to

$$N_c = N_{cm}[1 + \exp(b_n - c_nN)]/[1 + \exp(b_n' - c_nN)], \quad (30)$$

where $N_{cm} = A_n/A$ = maximum plant N concentration. By similar reasoning, it follows that

$$P_c = P_{cm}[1 + \exp(b_p - c_pP)]/[1 + \exp(b_p' - c_pP)] \quad (31)$$

$$K_c = K_{cm}[1 + \exp(b_k - c_kK)]/[1 + \exp(b_k' - c_kK)] \quad (32)$$

These procedures were shown to apply rather well with data from a full factorial experiment ($N \times P \times K = 4 \times 4 \times 4$) with Coastal bermudagrass at Watkinsville, Georgia.

A further simplification is possible if the b and c parameters are known from response of dry matter to applied N, P, and K. Response of plant N uptake to applied N is used for illustration. Define standardized yield, Y'_n, as

$$Y'_n = Y[1 + \exp(b_p - c_pP)][1 + \exp(b_k - c_kK)] \quad (33)$$

$$= A/[1 + \exp(b_n - c_nN)] \quad (34)$$

and standardized plant N uptake, N'_u, as

$$N'_u = N_u[1 + \exp(b'_p - c_pP)][1 + \exp(b'_k - c_kK)] \quad (35)$$

$$= A_n/[1 + \exp(b'_n - c_nN)] \quad (36)$$

With b and c known for applied P and K, values of N_u' can be calculated at each N level for all applied P and K. Average N_u' values are then used to estimate A_n and b'_n. The hyperbolic relationship [Eq. (20)] now becomes

$$Y'_n = Y'_m N'_u/(K' + N'_u). \quad (37)$$

Similar steps can be taken for applied P and applied K. While this analysis appears very complicated, Overman and Wilkinson (1995) have discussed the procedures step-by-step to illustrate practical application. Unpublished results by our lab show that this model also applies to data for other sites and crops. We should remember that, as a general rule, as a model becomes more comprehensive it becomes more mathematically complex (Lindley 1993; Lorenz 1993).

3.5 Harvest Interval and Water Availability

There are cases where various decision makers need information on forage response to harvest intervals beyond the 6 weeks mentioned above. Since it is known that dry matter yield peaks and declines for longer and longer intervals, Eq. (17) is modified to the form

$$Y = (\alpha + \beta\Delta t)\exp(-\gamma\Delta t), \quad (38)$$

where α, β, and γ are to be determined from data analysis. In a similar manner, plant nutrient uptake, N_u, is assumed to follow

$$N_u = (\alpha_n + \beta_n\Delta t)\exp(-\gamma\Delta t). \quad (39)$$

A common γ has been assumed to apply to both dry matter yield and plant nutrient uptake. Plant nutrient concentration, $N_c = N_u/Y$, is given by

$$N_c = (\alpha_n + \beta_n\Delta t)/(\alpha + \beta\Delta t). \quad (40)$$

It should be noted that N_c is bounded by $\alpha_n/\alpha \geq N_c \geq \beta_n/\beta$.

Equations (38) and (39) can be linearized by

$$Y' = Y\exp(\gamma\Delta t) = \alpha + \beta\Delta t \tag{41}$$

and

$$N'_u = N_u\exp(\gamma\Delta t) = \alpha_n + \beta_n\Delta t. \tag{42}$$

According to Eqs. (41) and (42) plots of Y' and N'_u vs. Δt should produce straight lines. The trick is to optimize the choice of γ. In the section on examples, the procedure for parameter evaluation will be discussed for both warm-season and cool-season grasses.

4 Examples

The models and procedures discussed above are now applied to fresh examples which have not been published previously in journal articles, using data from the literature. This approach is intended to complement articles already published on these models and to provide further documentation.

4.1 Warm-Season Perennial

The data for this analysis are taken from Overman and Blue (1991) for Pensacola bahiagrass (*Paspalum notatum* Flügge) in Florida. Field studies were conducted by Blue (1987) on an excessively drained Entisol Astatula sand and a poorly drained Spodosol Myakka fine sand. Experiments were conducted over the period 1979–1983. Average values are used in this analysis. Harvest interval averaged 6.6 weeks during the season. Applied N levels were 0, 100, 200, 300, and 400 kg ha^{-1}. Measurements included both dry matter and plant N uptake.

Cumulative dry matter with time (reference to Jan. 1) is shown in Fig. 1 for N = 400 kg ha^{-1} for the two soils. The curves are drawn from Eq. (1) with A = 11.6 Mg ha^{-1}, μ = 28.9 weeks, and σ = 7.3 weeks for the Entisol and A = 19.3 Mg ha^{-1}, μ = 25.6 weeks, and σ = 8.6 weeks for the Spodosol. The probability plots for these same data are shown in Fig. 2 to lend further support to the model. The effect of applied N on response is shown in Fig. 3 for the Spodosol. Data are described well by the probability function. Dependence of μ and σ on N are summarized in Fig. 4 for the two soils. Greater applied N and water availability enhance production early in the season, and both appear to increase availability of soil N to plant roots as shown in Fig. 5, where the line is given by

$$\sigma = 19.6 - 0.428\mu. \tag{43}$$

The scatter diagrams of cumulative plant N uptake vs. cumulative dry matter for the Entisol (Fig. 6) and for the Spodosol (Fig. 7) suggest that μ and σ values are the same for plant N and dry matter for each soil and each applied N, provided that harvest interval is constant over the season. This simplifies data analysis somewhat.

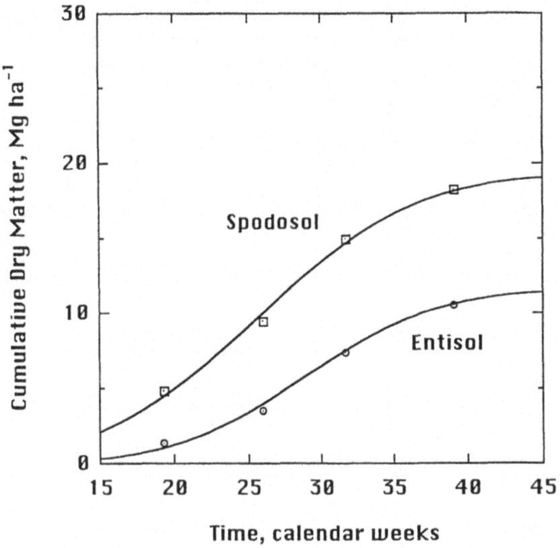

Fig. 1. Accumulation of dry matter (harvested portion) with time for bahiagrass with N = 400 kg ha^{-1} on two soils in Florida

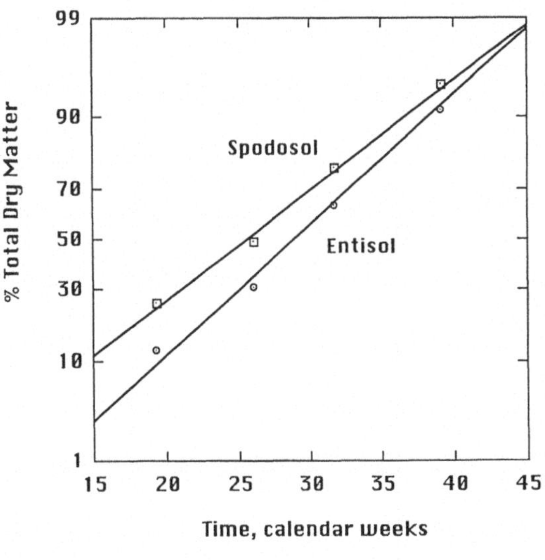

Fig. 2. Probability distribution of dry matter for bahiagrass with N = 400 kg ha^{-1} on two soils in Florida

Response of seasonal total dry matter, plant N uptake, and plant N concentration to applied N is shown in Fig. 8 for both soils, where the curves are drawn from Eqs. (18), (19), and (24) with parameters listed in Table 1. The extended logistic model appears to describe the data quite well. Both dry matter yields and plant N uptake are greater for the wetter soil (Spodosol) than the drier soil (Entisol). The slightly higher plant N concentration for the drier soil reflects a higher N in the soil solution. To verify that intercept and response parameters are the same for the two soils, data are plotted in linearized form

$$\ln(A/Y - 1) = b - cN = 1.45 - 0.0120N \qquad (44)$$

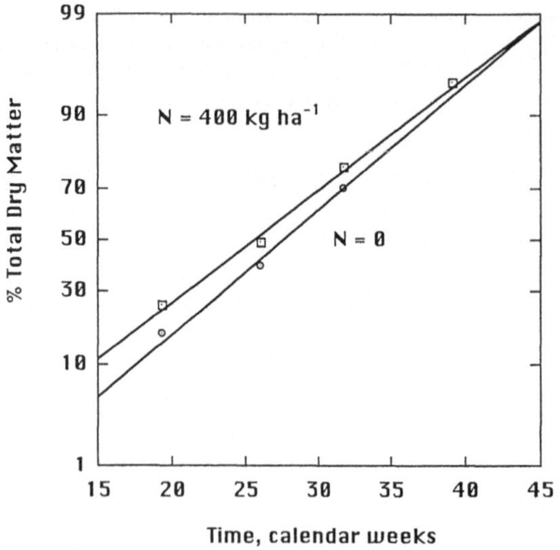

Fig. 3. Probability distribution of dry matter for bahiagrass at N = 0 and 400 kg ha^{-1} in Florida

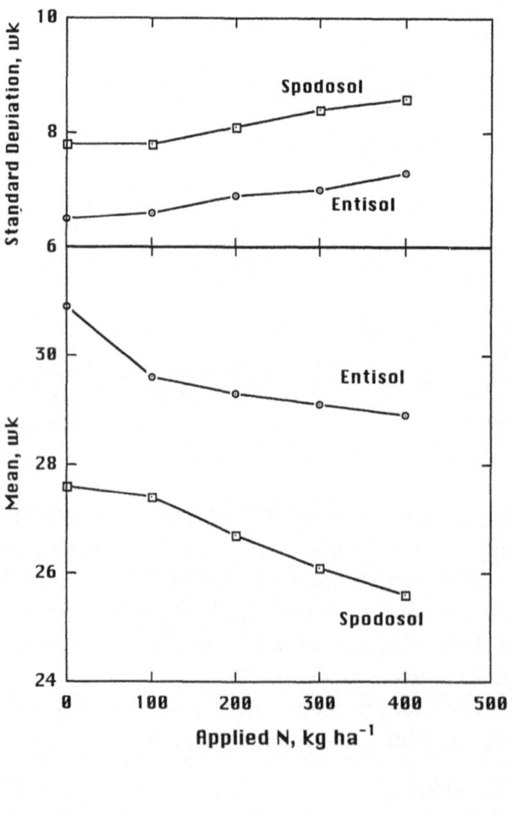

Fig. 4. Dependence of time to mean (μ) and standard deviation (σ) of dry matter distribution on applied N for bahiagrass on two soils in Florida

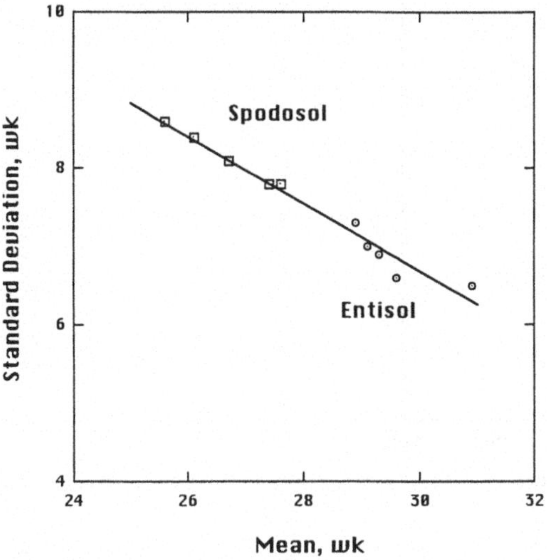

Fig. 5. Correlation of standard deviation (σ) with time to mean (μ) of dry matter distribution for bahiagrass on two soils in Florida

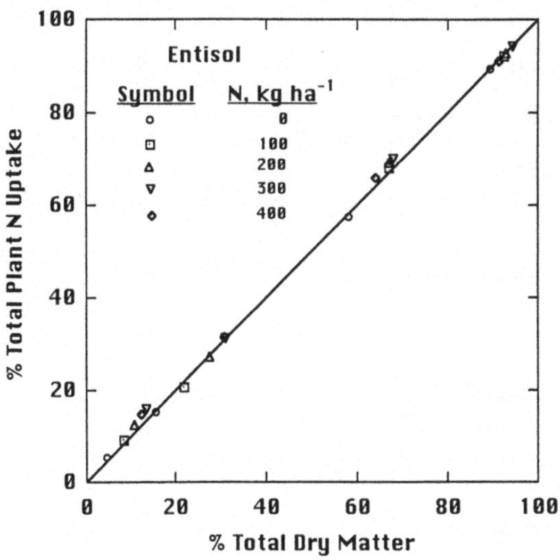

Fig. 6. Correlation of plant N uptake with dry matter for bahiagrass on an Entisol in Florida

Table 1. Summary of parameters for the extended probability model for bahiagrass grown with N = 400 kg ha^{-1} on two soils in Florida

Soil	Parameter		
	A (Mg ha^{-1})	μ (week)	σ (week)
Entisol	11.6	28.9	7.3
Spodosol	19.3	25.6	8.6

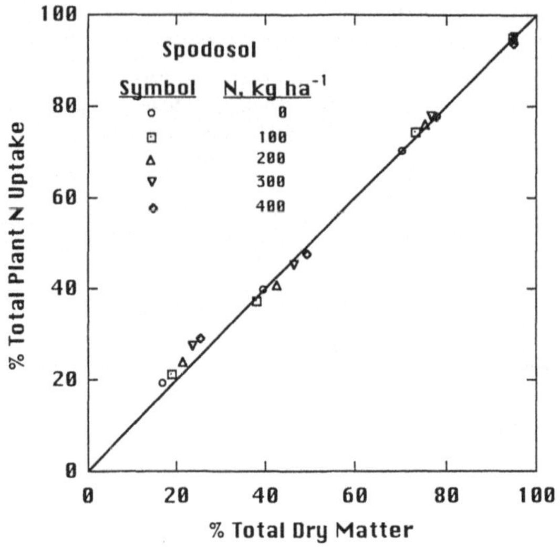

Fig. 7. Correlation of plant N uptake with dry matter for bahiagrass on a Spodosol in Florida

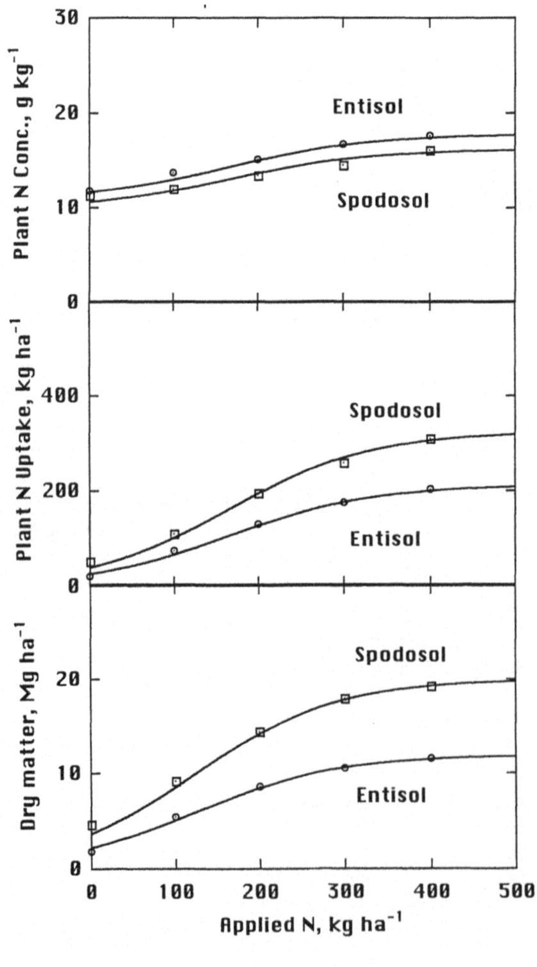

Fig. 8. Response of dry matter, plant N uptake, and plant N concentration to applied N for bahiagrass on two soils in Florida

and

$$\ln(A_n/N_u - 1) = b_n - cN = 1.96 - 0.0120N \tag{45}$$

in Figs. 9 and 10, with correlation coefficients of −0.9995 and −0.9972, respectively. The relationship of dry matter and plant N concentration to plant N uptake is shown in Fig. 11. Curves and lines are drawn from Eqs. (20) and (23) with parameters from Table 2, and clearly support the predictions of the model.

4.2 Warm-Season Annual

Overman and Rhoads (1991) analyzed data from a study conducted by Rhoads and Stanley (1979) with silage maize grown in Florida on Dothan loamy fine sand under irrigation. Applied N levels were 0, 90, 180, 270, and 450 kg ha^{-1}, with plant populations of 60 000 and 90 000 plants ha^{-1}. Results for the higher plant population are discussed here. Accumulation of dry matter, plant N

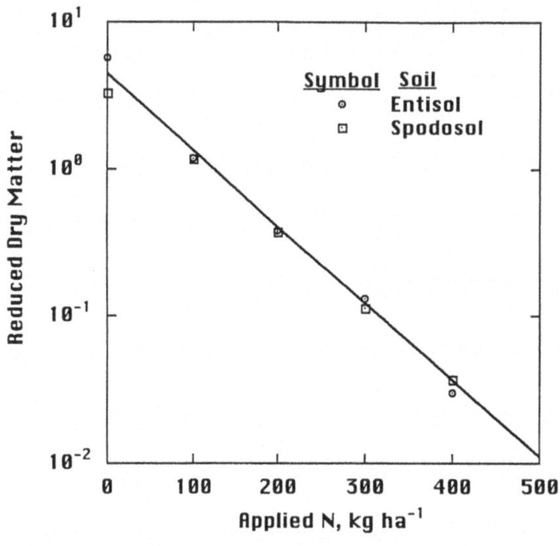

Fig. 9. Dependence of reduced dry matter on applied N for bahiagrass on two soils in Florida

Table 2. Summary of parameters for the extended logistic model for bahiagrass grown with N = 400 kg ha^{-1} on two soils in Florida

Soil	Parameter					
	A (Mg ha^{-1})	A_n (kg ha^{-1})	b	b_n	c (ha kg^{-1})	N_{cm} (g kg^{-1})
Entisol	11.9	212	1.45	1.96	0.0120	17.8
Spodosol	20.0	325	1.45	1.96	0.0120	16.2

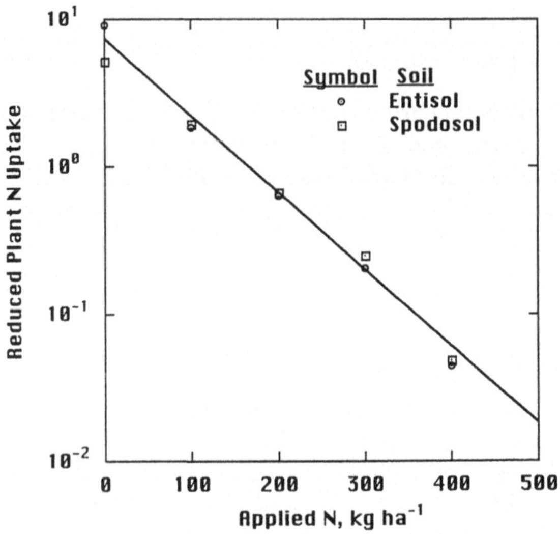

Fig. 10. Dependence of reduced plant N uptake on applied N for bahiagrass on two soils in Florida

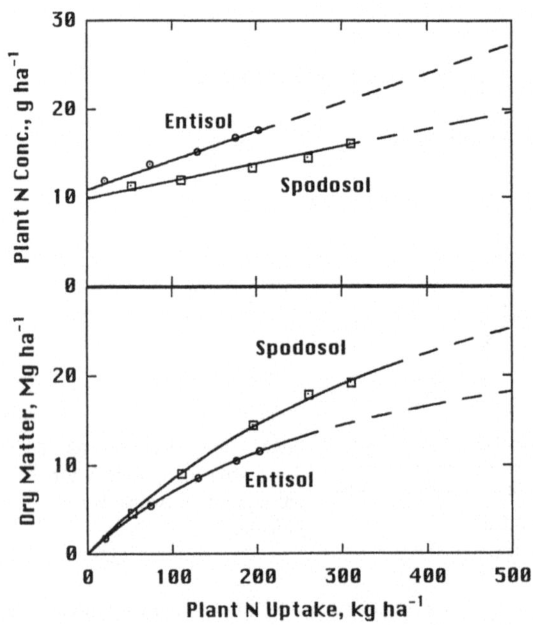

Fig. 11. Dependence of dry matter and plant N concentration on plant N uptake for bahiagrass on two soils in Florida

uptake, and plant N concentration with time are shown in Fig. 12 for N = 450 kg ha^{-1}, where the curves are drawn from Eqs. (1) and (4) with parameters listed in Table 3. Note that plant N concentration is defined as $N_c = N_u/Y$. The probability plots shown in Fig. 13 lend strong support to the extended probability model.

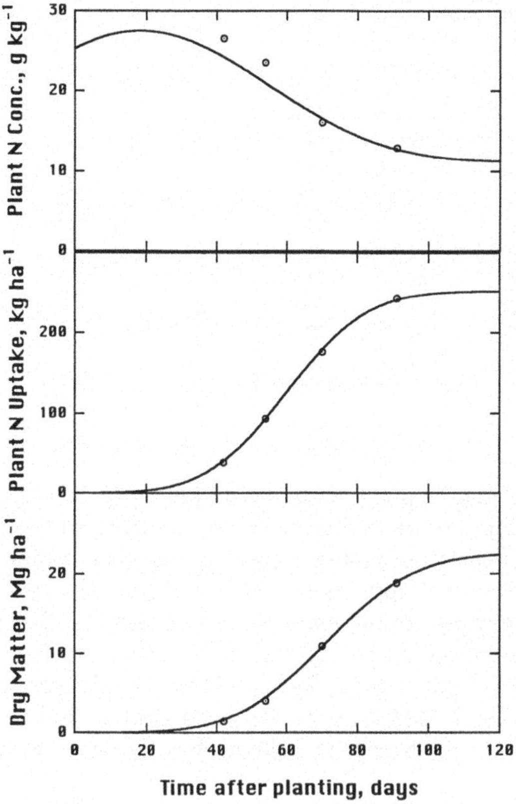

Fig. 12. Accumulation of dry matter and plant N with time for maize in Florida

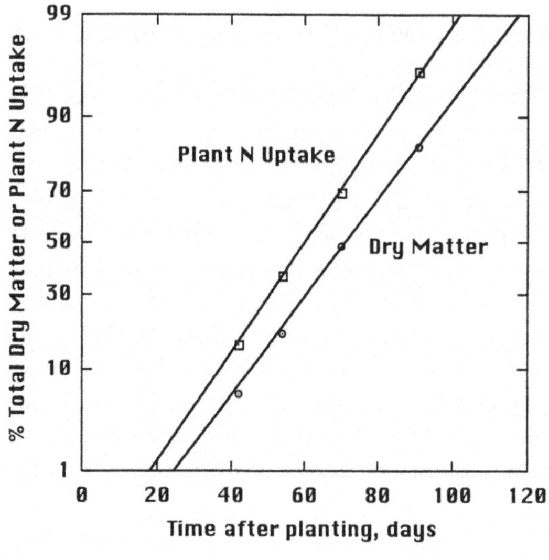

Fig. 13. Probability distribution of dry matter and plant N uptake for maize in Florida

Table 3. Summary of parameters for the extended probability model for maize grown in Florida

Parameter	Applied N ($kg\,ha^{-1}$)				
	0	90	180	270	450
A, $Mg\,ha^{-1}$	9.5	11.5	14.5	15.5	22.5
A_n, $kg\,ha^{-1}$	65	100	153	173	253
N_{cm}, $g\,kg^{-1}$	6.8	8.7	10.6	11.2	11.2
μ, d	69	67	67	64	71
σ, d	19	20	21	17	20
μ_n, d	58	55	55	52	60
σ_n, d	18	20	17	12	18

Response of seasonal total dry matter, plant N uptake, and plant N concentration to applied N is shown in Fig. 14, where the curves are drawn from Eqs. (18), (19), and (24) with parameters listed in Table 4. While there is some scatter in the data, the extended logistic model appears to describe the results reasonably well. The relationship of dry matter and plant N concentration to plant N uptake is shown in Fig. 15, where the curve and line are drawn from Eqs. (20) and (23) with $Y_m = 44.7\,Mg\,ha^{-1}$ and $K' = 271\,kg\,ha^{-1}$. This is in fair agreement with values of $45.4\,Mg\,ha^{-1}$ and $224\,kg\,ha^{-1}$ on irrigated Dothan soil in North Carolina (Overman et al. 1994b) based on field studies by Kamprath (1986).

4.3 Cool-Season Annual

The extended multiple logistic model is now applied to analysis of data from the literature. Walker and Morey (1962) conducted a 2-year study with winter rye (*Secale cereale* cv. Gator) at Tifton, Georgia on Norfolk loamy sand. The experimental design was an incomplete factorial with N = 0, 45, 90, 135, 180, and $225\,kg\,ha^{-1}$; P = 0, 20, 40, 60, 80, and $100\,kg\,ha^{-1}$; and K = 0, 37, 74, 111, 148, and $185\,kg\,ha^{-1}$. Measurements included dry matter and plant N, P, and K. Due to the structure of the experiment, it is possible to analyze response to each nutrient separately. Analysis of variance leads to the response shown in Figs. 16–18, where the curves are drawn from Eqs. (28)–(32) with model parameters listed in Table 5. A change of b_n' from 1.93 to 2.10 shifts results to the dotted curve. This illustrates the sensitivity of the intercept to the b parameter, which is caused by the absence of a check value (N = 0) for N_u. With this adjustment, the curves in Figs. 16–18 are seen to describe the data rather well. As already noted for bermudagrass, response is greater for applied N than for applied P and K.

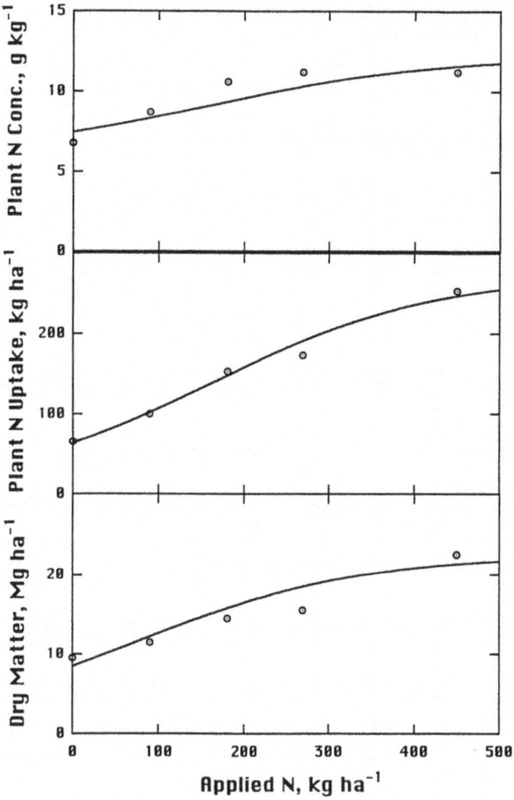

Fig. 14. Response of dry matter, plant N uptake, and plant N concentration to applied N for maize in Florida

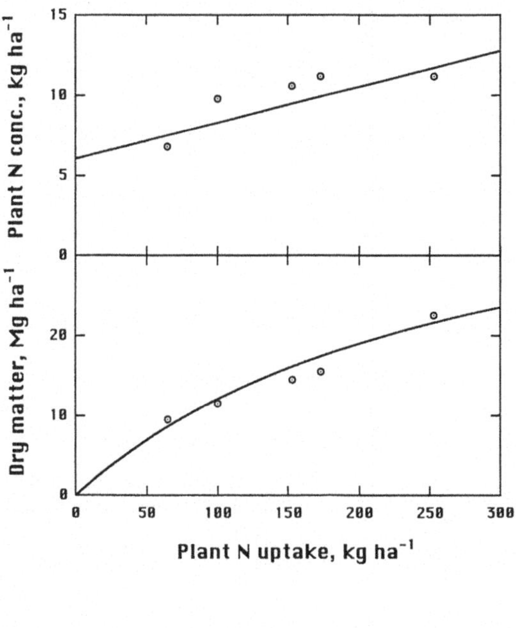

Fig. 15. Dependence of dry matter and plant N concentration on plant N uptake for maize in Florida

Table 4. Summary of parameters for the extended logistic model for maize grown in Florida

Parameter	A ($Mg\,ha^{-1}$)	A_n ($kg\,ha^{-1}$)	b	b_n	c ($ha\,kg^{-1}$)	N_{cm} ($g\,kg^{-1}$)
Value	22.5	275	0.50	1.20	0.0075	12.2

Table 5. Summary of parameters for the extended multiple logistic model for winter rye grown at Tifton, Georgia

Parameter	Dry matter	N uptake	P uptake	K uptake
A or A_n, $kg\,ha^{-1}$	5500	240	32	230
b_n	1.58	1.93	1.58	1.58
b_p	−0.14	−0.14	−0.14	−0.14
b_k	−0.91	−0.91	−0.91	0.46
c_n, $ha\,kg^{-1}$	0.0225	0.0225	0.0225	0.0225
c_p, $ha\,kg^{-1}$	0.0464	0.0464	0.0464	0.0464
c_k, $ha\,kg^{-1}$	0.0201	0.0201	0.0201	0.0201

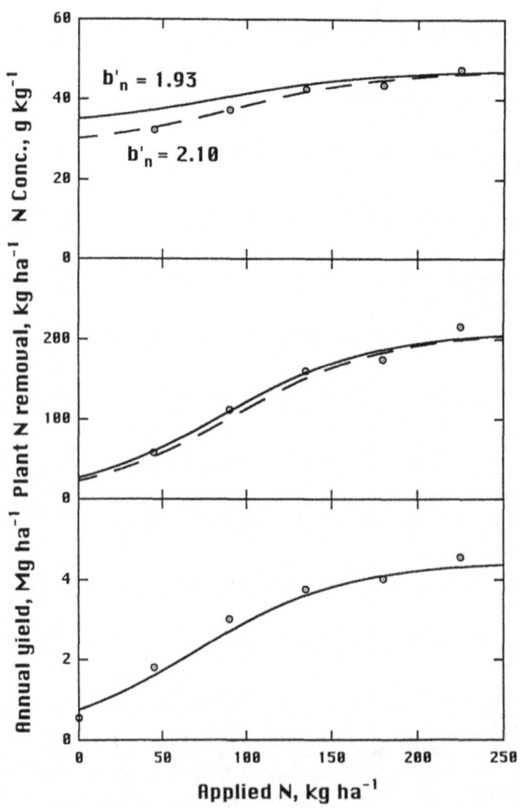

Fig. 16. Response of dry matter, plant N uptake, and plant N concentration to applied N for rye in Georgia

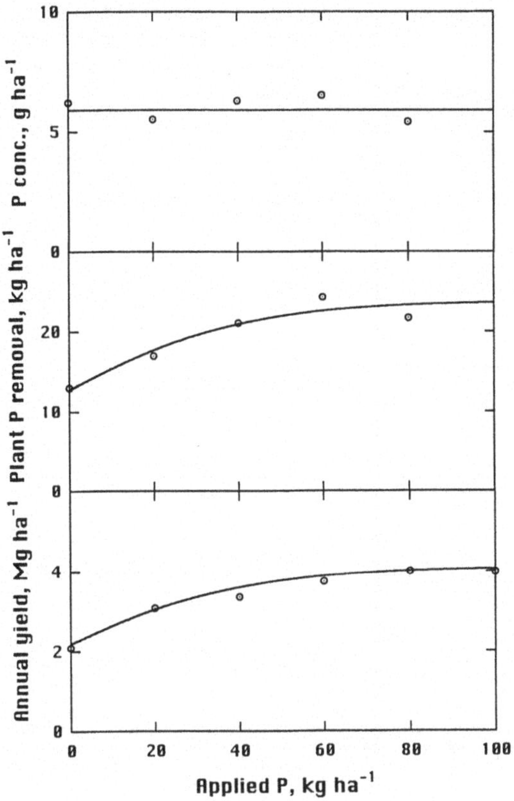

Fig. 17. Response of dry matter, plant P uptake, and plant P concentration to applied P for rye in Georgia

4.4 Harvest Interval

In this section we examine response of warm-season bermudagrass to harvest interval and of cool-season perennial ryegrass to applied N and harvest interval.

Burton et al. (1963) conducted a field study with Coastal bermudagrass at Tifton, Georgia, on Tifton loamy sand for the period of 1959–61 at an applied N rate of $672\,kg\,ha^{-1}$. Harvest intervals included 3, 4, 5, 6, 8, 12, and 24 weeks. Response of dry matter, plant N uptake, and plant N concentration are shown in Fig. 19. Construction of the curves is discussed below. Dry matter production appears to peak at a harvest interval of about 10 weeks. The first step is to calculate Y' and N_u' for Eqs. (41) and (42) by guessing a value of γ. After several tries a value of $\gamma = 0.077$ week^{-1} is chosen to plot results shown in Fig. 20. This produces the straight lines given by

$$Y' = 8.91 + 3.49\Delta t \tag{46}$$

and

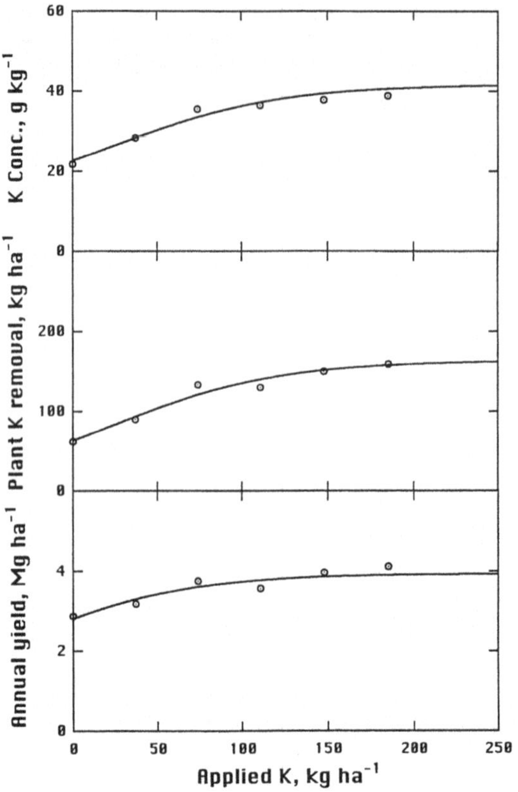

Fig. 18. Response of dry matter, plant K uptake, and plant K concentration to applied K for rye in Georgia

$$N_u' = 422 + 33.0\Delta t \tag{47}$$

with correlation coefficients of 0.9995 and 0.9780, respectively. These values of the parameters are substituted into Eqs. (38) through (40) to construct the curves shown in Fig. 19. Plant N concentration decreases with age and is bounded by $47.4 \geq N_c \geq 9.5\,\mathrm{g\,kg^{-1}}$. It can be shown that peak harvest interval, Δt_p, for dry matter yield is given by

$$\Delta t_p = 1/\gamma - \alpha/\beta, \tag{48}$$

which becomes $\Delta t_p = 10.4$ weeks. The corresponding estimated dry matter, plant N uptake, and plant N concentration are $Y = 20.3\,\mathrm{Mg\,ha^{-1}}$, $N_u = 344\,\mathrm{kg\,ha^{-1}}$, and $N_c = 16.9\,\mathrm{g\,kg^{-1}}$. Under these conditions the grass removes 344/672 = 51% of the applied N. Crude protein is only 10.5%, which may be low for cattle feed. Production can be adjusted by selecting the harvest interval as well as the level of applied N.

Bartholomew and Chestnutt (1977) conducted a filed study at Belfast, Northern Ireland, with perennial ryegrass (*Lolium perenne* L.) over the period 1972–74 at applied N rates of 0, 300, 600, 900, and 1200 kg ha^{-1} and harvest inter-

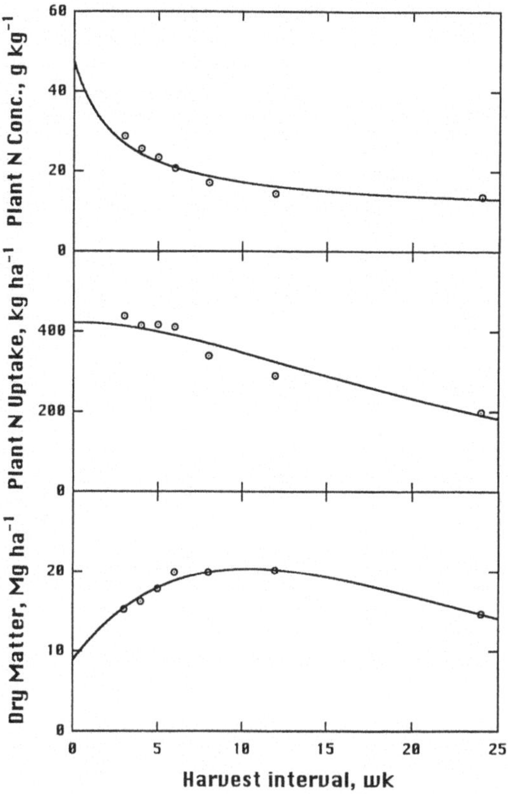

Fig. 19. Dependence of dry matter, plant N uptake, and plant N concentration on harvest interval for bermudagrass in Georgia

vals of 3.1, 4.0, 6.4, 10.7, and 16.0 weeks. Response of dry matter yield, plant N uptake, and plant N concentration to applied N for a harvest interval of 6.4 wk is shown in Fig. 21, where the curves are drawn from Eqs. (18), (19), and (24) with $b = 0.55$, $b_n = 1.20$, and $c = 0.0050\,\text{ha}\,\text{kg}^{-1}$ and other parameters from Table 6. The next step is to plot parameter values from Table 6 vs. harvest interval, as shown in Fig. 22. Following the same procedure as with bermudagrass, these results are linearized by choosing $\gamma = 0.075\ \text{week}^{-1}$, which leads to Fig. 23 for standardized maximum yield, A', and standardized maximum plant N uptake, A'_n, where the lines are given by

$$A' = 9.16 + 2.67\Delta t \tag{49}$$

and

$$A'_n = 743 + 26.9\Delta t \tag{50}$$

with correlation coefficients of 0.9993 and 0.9913, respectively. Curves in Fig. 22 are constructed with these parameters and provide excellent fit of the data. In this case peak dry matter yield occurs at a harvest interval of $\Delta t_p = 9.9$ weeks. The corresponding parameter values are $A = 17.0\,\text{Mg}\,\text{ha}^{-1}$, A_n

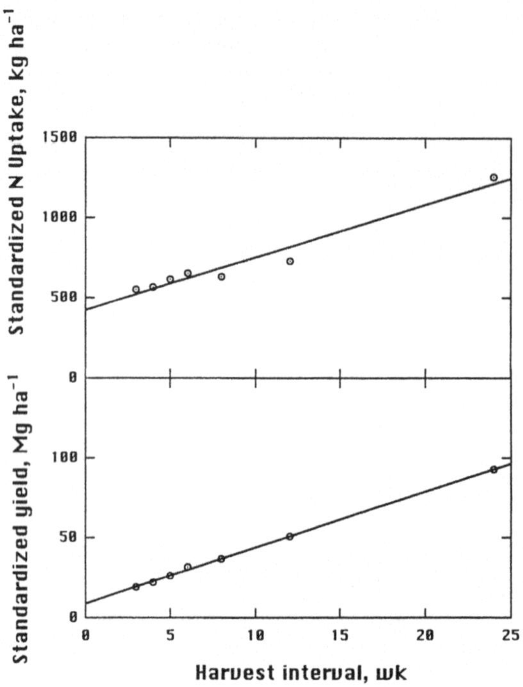

Fig. 20. Dependence of standardized dry matter and standardized plant N uptake on harvest interval for bermudagrass in Georgia

$= 480\,kg\,ha^{-1}$, and $N_{cm} = 28.3\,g\,kg^{-1}$. It may be noted that this level of plant N removal represents 480/743 = 65% of potential removal.

5 Summary

In this chapter we have focused on particular models which have proven useful in describing forage systems and which are relatively simple to use in practice. The models include probability, logistic, and hyperbolic equations. All of these are self-limiting, are bounded by zero, and asymptotically approach a maximum at long times or high input levels. Procedures have been discussed for parameter evaluation and for graphical presentation of results. The functional concept has been used rather than a cause and effect approach (Walker 1963), agreement with data serving as a guide to utility rather than detailed mechanism (Hunt 1982). A part of this process involves the search for patterns and relationships in form and numbers discussed by several authors (Judson 1987; Devlin 1994; Speyer 1994). In our view a truly mechanistic model of plant growth would be extremely complicated due to the complex geometry of plants (roots and tops) as well as detailed biochemical processes, which would necessarily involve quantum and statistical mechanics (Johnson et al. 1974).

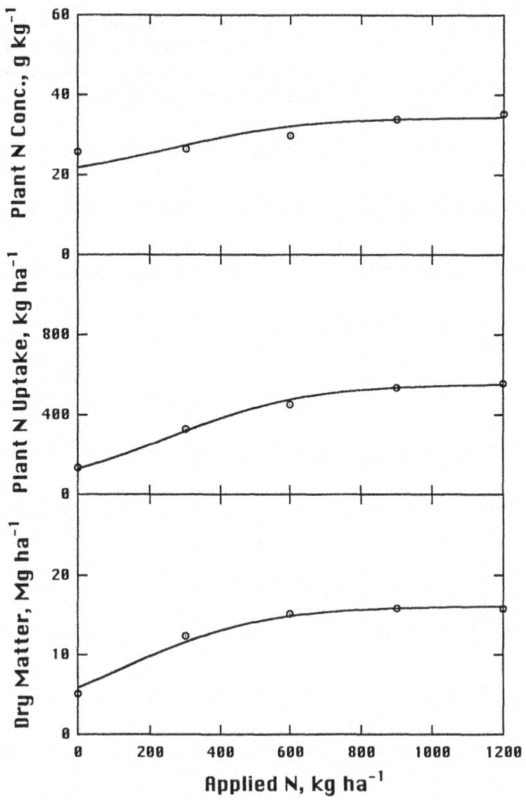

Fig. 21. Response of dry matter, plant N uptake, and plant N concentration to applied N at a harvest interval of 6.4 weeks for ryegrass in Belfast, Northern Ireland

Table 6. Summary of parameters for the extended logistic model for perennial ryegrass grown at Belfast, Northern Ireland

Harvest interval weeks	A ($Mg\,ha^{-1}$)	A_n ($kg\,ha^{-1}$)	N_{cm} ($g\,kg^{-1}$)
3.1	13.5	660	48.9
4.0	15.0	625	41.7
6.4	16.1	555	34.5
10.7	17.2	475	27.6
16.0	15.5	350	22.6

Models discussed in this chapter are useful for engineering application and in relating various components of the system. Allhands et al. (1995) have incorporated the logistic model into analysis of a municipal water reuse system which includes crop production. Willcutts (1994) has used the logistic model for the crop management component of a model of dairy waste management by land application. Overman (1995b) has shown how to couple various com-

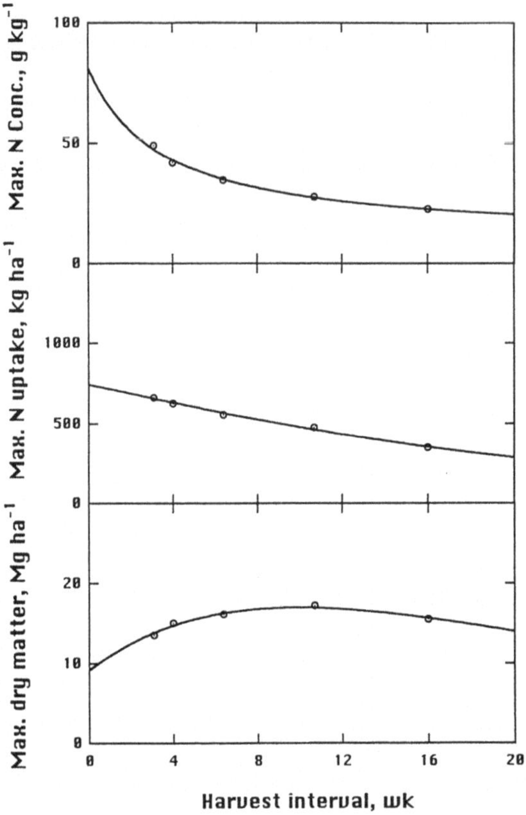

Fig. 22. Dependence of model parameters A, A_n, and N_{cm} on harvest interval for perennial ryegrass in Belfast, Northern Ireland

ponents of the soil-plant-atmosphere system by use of logistic and hyperbolic models, including response of soybeans [*Glycine max* (L.) Merr.] to levels of atmospheric CO_2, observed by Allen et al. (1990). This analysis showed that one half of maximum leaf area was reached at a CO_2 concentration of 255 μmol mol^{-1}. If this holds true for grasses, then it might explain why maximum observed yield (A) is approximately one half of maximum potential yield (Y_m). An unexpected consequence of the extended logistic model was the allometric relationship (France and Thornley 1984) given by Eq. (20) which relates dry matter and plant N uptake. In dynamical systems this is called a phase equation (Edwards and Hamson 1989; Stewart 1995). This followed from the symmetry between response of dry matter and plant nutrient accumulation to applied nutrient. The relationship appears to hold for a number of grasses. Since A has been shown to depend upon factors such as water availability and harvest interval, it follows that Y_m also depends upon these same factors. It should also be noted that the multiple logistic model replaces the semiquantitative limiting law of the minimum of Liebig (Sinclair and Park 1993).

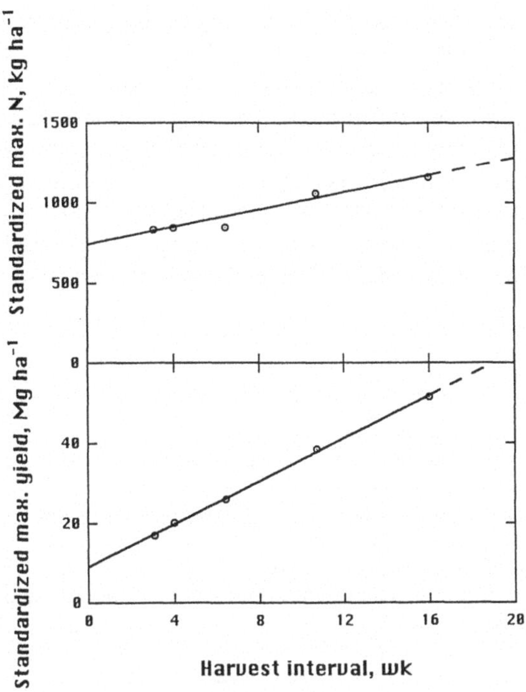

Fig. 23. Dependence of standardized model parameters A and A_n on harvest interval for perennial ryegrass in Belfast, Northern Ireland

Symmetry has figured prominently in the logistic models. The differential form [Eq. (27)] of the simple logistic equation exhibits bilateral (reflection) symmetry and leads to the sigmoid shape of Eq. (26). Another type of symmetry occurs between Eqs. (18) and (19) in that the form of dependence upon applied N is the same. Finally, Eqs. (28) and (29) for the multiple logistic model include permutation symmetry, i.e., exchange of N, P, and K terms leads to the same form of equation, and leads to sigmoid response to applied nutrient for each element. The symmetry property adds beauty to the models, as suggested by Dirac (Judson 1987) and Hardy (1967), as well as leading to well-behaved functions for descriptive purposes. The general subject of symmetry has been discussed by Hargittai (1986).

A major challenge is to compile model parameters for various crops and sites to search for commonality of values and any further connections among parameters. Work along these lines is presently underway in our laboratory. We believe that the utility of the simplified models described here fully justifies the effort.

Finally, a few words about modeling. Richard Feynman identified two approaches to physics (Mehra 1994), viz. the Greek method and the Babylonian method. The Greek way flows from first principles, such as axioms. The Babylonian way simply relates two things, such as a function between two variables. In our view there is a place for both of these in modeling complex systems such as plant growth. Progress is sometimes made by *reductionism* (reduce system to parts) and sometimes by *holism* (view system as whole), as noted by Kaku (1994). It has been common in applied science to guess relationships on intuitive grounds, some of which were later derived on rigorous grounds (Longair 1984). Implementation of ideas can require time and intense effort by a number of investigators (Hunt 1991). We should also note that in continuous form the logistic model is a well-behaved nonlinear differential equation, whereas in discrete form it can exhibit extremely chaotic behavior (Stewart 1989; Peitgen et al. 1992).

References

Abramowitz M, Stegun IA (1965) Handbook of mathematical functions. Dover, New York
Allen LH Jr, Bisbal EC, Campbell WJ, Boote KJ (1990) Carbon dioxide effects on soybean developmental stages and expansive growth. Soil Crop Sci Soc Fla Proc 49:124–131
Allhands MN, Allick SA, Overman AR, Leseman WG, Vidak W (1995) Municipal water reuse at Tallahassee, Florida. Trans Am Soc Agric Engin 38:411–418
Aris R (1994) Mathematical modeling techniques. Dover, New York
Ball DM, Hoveland CS, Lacefield GD (1991) Southern forages. Potash & Phosphate Institute and the Foundation for Agronomic Research, Atlanta
Bartholomew PW, Chestnutt DMB (1977) The effect of a wide range of fertilizer nitrogen application rates and defoliation intervals on the dry-matter production, seasonal response to nitrogen, persistence and aspects of chemical composition of perennial ryegrass (*Lolium perenne* cv. S.24). J Agric Sci 88:711–721
Blue WG (1987) Response of Pensacola bahiagrass (*Paspalum notatum* Flügge) to fertilizer nitrogen on an entisol and a spodosol in north Florida. Soil Crop Sci Soc Fla Proc 47:135–139
Buckner RC, Bush LP (eds) (1979) Tall fescue. American Society of Agronomy Madison, Wisconsin
Bunch B (1989) Reality's mirror: exploring the mathematics of symmetry. John Wiley, New York
Burton GW, Jackson JE, Hart RH (1963) Effects of cutting frequency and nitrogen on yield, in vitro digestibility, and protein, fiber, and carotene content of Coastal bermudagrass. Agron J 55:500–502
Carreker JR, Wilkinson SR, Barnett AP, Box JE (1977) Soil and water systems for sloping land. ARS-S-160. US Government Printing Office, Washington, DC
Caswell H (1989) Matrix population models. Sinauer Assoc, Sunderland, Massachusetts
Cerrato ME, Blackmer AM (1990) Comparison of models for describing corn yield response to nitrogen fertilizer. Agron J 82:138–143
Charles-Edwards DA, Doley D, Rimmington GM (1986) Modelling plant growth and development. Academic Press, Orlando, Florida
Davies P (1992) The mind of God: the scientific basis of a rational world. Simon & Schuster, New York
Davis PJ, Hersh R (1981) The mathematical experience. Houghton Mifflin, Boston
Devlin K (1994) Mathematics: The science of patterns. WH Freeman, New York
Dovrat A (1993) Irrigated forage production. Elsevier, Amsterdam
Edwards D, Hamson M (1989) Guide to mathematical modeling. CRC Press, Boca Raton, Florida
Eigen M, Winkler R (1981) Laws of the game: how the principles of nature govern chance. Princeton University Press, Princeton, New Jersey
Einstein A, Infeld L (1938) The evolution of physics. Simon & Schuster, New York
Fageria NK, Baligar VC, Jones CA (1991) Growth and mineral nutrition of field crops. Marcel Dekker, New York
Ferris T (1988) Coming of age in the Milky Way. Doubleday, New York

Feynman R (1986) The character of physical law. MIT Press Cambridge, Massachusetts
France J, Thornley JHM (1984) Mathematical models in agriculture. Butterworths, London
Freund RJ, Littell RC (1991) SAS System for regression. SAS Institute Cary, North Carolina
Ginzburg LR, Golenberg EM (1985) Lectures in theoretical population biology. Prentice-Hall, Englewood Cliffs, New Jersey
Glass ADM (1989) Plant nutrition. Jones and Bartlett, Boston
Gleick J (1987) Chaos: Making a new science. Viking Penguin, New York
Hanks J, Ritchie JT (1991) Modeling plant and soil systems. American Society of Agronomy Madison, Wisconsin
Hardy GH (1967) A mathematician's apology. Cambridge University Press, New York
Hargittai I (ed) (1986) Symmetry: unifying human understanding. Pergamon Press, New York
Heath ME, Barnes RF, Metcalfe DS (eds) (1985) Forages: the science of grassland agriculture. 4th edn. Iowa State University Press, Ames
Holt EC, Conrad BE (1986) Influence of harvest frequency and season on bermudagrass cultivar yield and forage quality. Agron J 78:433–436
Hosmer DW, Lemeshow S (1989) Applied logistic regression. John Wiley, New York
Hunt BJ (1991) The Maxwellians. Cornell University Press, Ithaca, New York
Hunt R (1982) Plant growth curves: the functional approach to plant growth analysis. Edward Arnold, London
Huxley JS (1993) Problems of relative growth. Johns Hopkins University Press, Baltimore
Johnson FH, Eyring H, Stover BJ (1974) The theory of rate processes in biology and medicine. John Wiley, New York
Jones JB Jr (1994) Plant nutrition manual. Micro-Macro Publishing, Athens, Georgia
Judson HF (1987) The search for solutions. Johns Hopkins University Press, Baltimore
Kaku M (1994) Hyperspace: a scientific Odyssey through parallel universes, time warps, and the 10th dimension. Oxford University Press, New York
Kamprath EJ (1986) Nitrogen studies with corn on Coastal plain soils. North Carolina Agric Res Ser Tech Bull 282. North Carolina State University, Raleigh, North Carolina
Keen RE, Spain JD (1992) Computer simulation in biology: a basic introduction. John Wiley, New York
Kingsland SE (1985) Modeling nature: episodes in the history of population ecology. University of Chicago Press, Chicago
Kleinbaum DG (1994) Logistic regression: a self-learning text. Springer, Berlin Heidelberg New York
Kramer EE (1981) The nature and growth of modern mathematics. Princeton University Press, Princeton
Krauss LM (1993) Fear of physics: a guide for the perplexed. Basic Books, New York
Lanczos C (1988) Applied analysis. Dover, New York
Lindley D (1993) The end of physics: the myth of a unified theory. Basic Books, New York
Longair MS (1984) Theoretical concepts in physics. Cambridge University Press, New York
Lorenz EN (1993) The essence of chaos. University of Washington Press, Seattle
Lotka AJ (1925) Elements of physical biology. Williams and Wilkins, Baltimore
Marschner H (1986) Mineral nutrition of higher plants. Academic Press, Orlando, Florida
Mays DA (ed) (1974) Forage fertilization. American Society of Agronomy, Madison, Wisconsin
Mays DA, Wilkinson SR, Cole CV (1980) Phosphorus nutrition of forages. In: Khasawneh FE, Sample EC, Kamprath EJ (eds) The role of phosphorus in agriculture. American Society of Agronomy, Madison, Wisconsin
Mehra J (1994) The beat of a different drummer: the life and science of Richard Feynman. Clarendon Press, Oxford
Mengel K, Kirkby EA (1987) Principles of plant nutrition. International Potash Institute, Bern, Switzerland
Murray JD (1993) Mathematical biology. Springer, Berlin Heidelbey, New York
Overman AR (1984) Estimating crop growth rate with land treatment. J Environ Engin Div Am Soc Civil Engin 110:1009–1012

Overman AR (1995a) Rational basis for the logistic model for forage grasses. J Plant Nutr 18:995–1012
Overman AR (1995b) Coupling among applied, soil, root, and top components for forage crop production. Commun Soil Sci Plant Anal 26:1179–1202
Overman AR, Blue WG (1991) Estimation of dry matter production and nitrogen uptake by Pensacola bahiagrass in Florida. Fla Agric Exp Stn Bull 880. University of Florida, Gainesville
Overman AR, Rhoads FM (1991) Estimation of dry matter production and nitrogen uptake by corn silage in Florida. Fla Agric Exp Stn Bull 882. University of Florida, Gainesville
Overman AR, Wilkinson SR (1989) Partitioning of dry matter between leaf and stem in Coastal bermudagrass. Agric Syst 30:35–47
Overman AR, Wilkinson SR (1995) Extended logistic model of forage grass response to applied nitrogen, phosphorus, and potassium. Trans Am Soc Agric Engin 38:103–108
Overman AR, Angley EA, Wilkinson SR (1988a) Empirical model of Coastal bermudagrass production. Trans Am Soc Agric Engin 31:466–470
Overman AR, Angley EA, Wilkinson SR (1988b) Evaluation of an empirical model of Coastal bermudagrass production. Agric Syst 28:57–66
Overman AR, Angley EA, Wilkinson SR (1989) A phenomenological model of Coastal bermudagrass production. Agric Syst 29:137–148
Overman AR, Angley EA, Wilkinson SR (1990a) Evaluation of a phenomenological model of Coastal bermudagrass production. Trans Am Soc Agric Engin 33:443–450
Overman AR, Martin FG, Wilkinson SR (1990b) A logistic equation for yield response of forage grass to nitrogen. Commun Soil Sci Plant Anal 21:595–609
Overman AR, Neff CR, Wilkinson SR, Martin FG (1990c) Water, harvest interval, and applied nitrogen effects on forage yield of bermudagrass. Agron J 82:1011–1016
Overman AR, Dagan A, Martin FG, Wilkinson SR (1991) A nitrogen-phosphorus-potassium model for forage yield of bermudagrass. Agron J 83:254–258
Overman AR, Evers GW, Wilkinson SR (1992) Yield response of bermudagrass and bahiagrass to applied nitrogen and overseeded clover. Agron J 84:998–1001
Overman AR, Wilkinson SR, Wilson DM (1994a) An extended model of forage grass response to applied nitrogen. Agron J 86:617–620
Overman AR, Wilson DM, Kamprath EJ (1994b) Estimation of yield and nitrogen removal by corn. Agron J 86:1012–1016
Overman AR, Evers GW, Wilkinson SR (1995a) Coupling of dry matter and nutrient accumulation in forage grass. J Plant Nutr 18:2629–2642
Overman AR, Robinson DL, Wilkinson SR (1995b) Coupling of dry matter and nitrogen accumulation in ryegrass. Fert Res 40:105–108
Pagels H (1985) Perfect symmetry. Simon & Schuster, New York
Pagels H (1988) The dreams of reason: the computer and the rise of the science of complexity. Bantam Books, New York
Pearl R (1925) The biology of population growth. Alfred A Knopf, New York
Peitgen HO, Jurgens H, Saupe D (1992) Chaos and fractals: new frontiers of science. Springer, Berlin Heidelberg New York
Prine GM, Burton GW (1956) The effect of nitrogen rate and clipping frequency upon the yield, protein content and certain morphological characteristics of Coastal bermudagrass [*Cynodon dactylon* (L.) Pers.]. Agron J 48:296–301
Ratkowsky DA (1983) Nonlinear regression modeling. Marcel Dekker, New York
Reck WR, Overman AR (1996) Estimation of corn response to water and applied nitrogen. J Plant Nutr 19:201–214
Rhoads FM, Stanley RL Jr (1979) Effect of population and fertility on nutrient uptake and yield components of irrigated corn. Soil Crop Sci Soc Fla Proc 38:78–81
Robinson DL, Wheat KG, Hubbert NL, Henderson MS, Savoy HJ Jr (1988) Dallisgrass yield, quality and nitrogen recovery response to nitrogen and phosphorus fertilizers. Commun Soil Sci Plant Anal 19:529–542

Rothman MA (1985) Conservation laws and symmetry. In: Besancon RM (ed) The encyclopedia of physics. Van Nostrand Reinhold, New York, pp 222–226

Rucker R (1987) Mind tools: the five levels of mathematical reality. Houghton Mifflin, Boston

Schwab GO, Fangmeier DD, Elliot WJ, Frevert RK (1993) Soil and water engineering, 4th edn. John Wiley, New York

Sinclair TR, Park WI (1993) Inadequacy of the Liebig limiting-factor paradigm for explaining varying crop yields. Agron J 85:742–746

Speyer E (1994) Six roads from Newton: great discoveries in physics. John Wiley, New York

Stewart I (1989) Does God play dice? the mathematics of chaos. Basil Blackwell, Cambridge, Massuchusetts

Stewart I (1995) Concepts of modern mathematics. Dover, New York

Thompson DW (1992) On growth and form: the complete revised edition. Dover, New York

Tilman D (1988) Plant strategies and the dynamics and structure of plant communities. Princeton University Press, Princeton, New Jersey

Tollenaar M (1986) Seeking the upper limit of corn productivity. Better Crops Plant Food 70:6–8

Verhulst PF (1838) Notice sur la loi que la population suit dans son accroissement. Corr Math Phys 10:113–121

Volterra V (1926) Variazionic fluttuazioni del numero d'individui in specie animali conviventi. Mem Acad Lindei 2:31–113. (Variations and fluctuations of a number of individuals in animal species living together. Translation. In: Chapman RN, Animal ecology. McGraw Hill, New York, 1931, pp 409–448)

Walker M (1963) The nature of scientific thought. Prentice-Hall, Englewood Cliffs, New Jersey

Walker ME, Morey DD (1962) Influence of rates of N, P, and K on forage and grain production of Gator rye in South Georgia. Ga Agric Exp Stn Circ NS 27. University of Georgia, Athens

Weinberg S (1993) Dreams of a final theory: the scientist's search for the ultimate laws of nature. Vintage Books, New York

Will CM (1986) Was Einstein right? Putting general relativity to the test. Basic Books, New York

Willcutts JF (1994) Nutrient cycling in land application dairy waste management systems. Master of Engineering Thesis. University of Florida, Gainesville, Florida

Subject Index

SPRINGER NATURE

GPSR Compliance

The European Union's (EU) General Product Safety Regulation (GPSR) is a set of rules that requires consumer products to be safe and our obligations to ensure this.

If you have any concerns about our products, you can contact us on ProductSafety@springernature.com

In case Publisher is established outside the EU, the EU authorized representative is:

Springer Nature Customer Service Center GmbH
Europaplatz 3
69115 Heidelberg, Germany

The manufacturer's authorised representative in the EU is Springer Nature Customer Service Centre GmbH, Europaplatz 3, 69115 Heidelberg, Germany. If you have any concerns regarding our products, please contact ProductSafety@springernature.com

Printed and bound by CPI Group (UK) Ltd, Croydon, CR0 4YY
22/09/2026
02220988-0008